Contents

CHAPTER 1 Careers in Automotive Technology ...1

CHAPTER 2 Introduction to Automotive Technology... 7

CHAPTER 3 Introduction to Automotive Safety ...15

CHAPTER 4 Personal Safety... 23

CHAPTER 5 Vehicle, Customer, and Service Information... 31

CHAPTER 6 Tools and Equipment... 40

CHAPTER 7 Vehicle Protection and Jack and Lift Safety... 53

CHAPTER 8 Vehicle Maintenance Inspection... 66

CHAPTER 9 Communication... 77

CHAPTER 10 Engine Mechanical Testing... 82

CHAPTER 11 Engine Removal and Replacement... 93

CHAPTER 12 Cylinder Head Components... 99

CHAPTER 13 Engine Block Components... 113

CHAPTER 14 Engine Machining... 127

CHAPTER 15 Engine Assembly... 135

CHAPTER 16 Automatic Transmission Fundamentals... 148

CHAPTER 17 Hydraulic Fundamentals... 157

CHAPTER 18 Hydraulically Controlled Transmission... 163

CHAPTER 19 Electronically Controlled Transmission... 171

CHAPTER 20 Servicing the Automatic Transmission/Transaxle... 180

CHAPTER 21 Hybrid and Continuously Variable Transmissions... 194

CHAPTER 22 Manual Transmission/Transaxle Principles... 201

CHAPTER 23 The Clutch System... 210

CHAPTER 24 Manual Transmissions/Transaxles Basic Diagnosis and Maintenance... 227

CHAPTER 25 Drive Train Components... 242

CHAPTER 26 Basic Drive Layouts... 254

CHAPTER 27 Servicing Wheels... 266

CHAPTER 28 Servicing the Steering System... 281

CHAPTER 29 Servicing the Suspension System... 297

CHAPTER 30 Principles of Braking... 315

CHAPTER 31 Hydraulic and Power Brakes... 325

CHAPTER 32 Disc Brake System... 339

CHAPTER 33 Drum Brake System... 353

CHAPTER 34 Wheel Bearings... 369

CHAPTER 35 Electronic Brake Control... 381

CHAPTER 36 Principles of Electrical Systems... 393

CHAPTER 37 Meter Usage and Circuit Diagnosis... 407

CHAPTER 38 Batteries, Starting, and Charging Systems... 417

CHAPTER 39 Lighting Systems... 435

CHAPTER 40 Body Electrical System... 446

CHAPTER 41 Principles of Heating and Air-Conditioning Systems... 458

CHAPTER 42 Heating and Air-Conditioning Systems and Service... 471

CHAPTER 43 Electronic Climate Control... 487

CHAPTER 44 Motive Power Types—Spark-Ignition (SI) Engines... 502

CHAPTER 45 Engine Lubrication... 515

CHAPTER 46 Engine Cooling... 531

CHAPTER 47 Ignition Systems Overview.......545

CHAPTER 48 Gasoline Fuel Systems 563

CHAPTER 49 On-Board Diagnostics...........583

CHAPTER 50 Induction and Exhaust..........591

CHAPTER 51 Emission Control...............604

CHAPTER 52 Alternative Fuel Systems.........619

CHAPTER 53 Compression-Ignition Engines....629

CHAPTER 54 Diesel Fuel Systems.............639

Careers in Automotive Technology

CHAPTER 1

Tire Tread: © AbleStock

Chapter Review

The following activities have been designed to help you refresh your knowledge of this chapter. Your instructor may require you to complete some or all of these activities as a regular part of your training program. You are encouraged to complete any activity that your instructor does not assign as a way to enhance your learning.

Matching

Match the following terms with the correct description or example.

A. Automotive Service Excellence (ASE)
B. Brake technician
C. Drivability technician
D. Heavy line technician
E. Lube technician
F. Service consultant/advisor
G. Shop foreman

_D_ 1. Specializes in major engine, transmission, and differential overhaul and repair.
_E_ 2. Carries out scheduled maintenance activities.
_G_ 3. Supervisor who oversees the work of technicians and staff.
_A_ 4. An independent, nonprofit organization dedicated to the improvement of vehicle repair through the testing and certification of automotive professionals.
_B_ 5. Specializes in working on vehicle brake systems.
_C_ 6. Diagnoses and identifies mechanical and electrical faults that affect vehicle performance and emissions.
_F_ 7. A service worker who works with both customers and technicians.

Multiple Choice

Read each item carefully, and then select the best response.

_B_ 1. Who is generally acknowledged to have invented the modern automobile around 1885?
   A. Henry Ford
   B. Karl Benz
   C. Armand Peugeot
   D. Charles Rolls

_B_ 2. Who is a repair shop's first point of contact for customers seeking vehicle repairs?
   A. Shop foreman
   B. Service consultant/advisor
   C. Drivability technician
   D. Service manager

Fundamentals of Automotive Technology Student Workbook

__D__ 3. What type of technician might specialize in particular vehicle systems, such as engines, transmissions, or final drives?
   A. Shop foreman
   B. Light line technician
   C. Drivability technician
   D. Heavy line technician

__C__ 4. What type of technician diagnoses and repairs faults, replaces or overhauls brake systems, and tests the components of disc, drum, and power brake systems used on all types of vehicles?
   A. Light line technician
   B. Heavy line technician
   C. Brake technician
   D. Drivability technician

__A__ 5. What type of technician works with computer-controlled engine management systems to service, identify, and repair faults on electronically controlled vehicle systems such as fuel injection, ignition, antilock braking, cruise control, and automatic transmissions?
   A. Electrical technician
   B. Heavy line technician
   C. Shop foreman
   D. Service manager

__C__ 6. What type of technician regularly uses meters, oscilloscopes, circuit wiring diagrams, and solder equipment?
   A. Light line technicians
   B. Transmission specialist
   C. Electrical technicians
   D. Heavy line technicians

__D__ 7. Which of the following is a key skill of a service manager?
   A. Communicating
   B. Motivating
   C. Creating positive work environments
   D. All of the above

__B__ 8. What types of shops are usually independent and focus on one type of service, such as transmission service, electrical system repair, or emission system diagnosis?
   A. Dealerships
   B. Specialty shops
   C. Franchises
   D. Fleet shops

__C__ 9. Which of the following programs receives access to new vehicle technology as well as manufacturer service information to help prepare students for working on today's vehicles and technology?
   A. National Automotive Technicians Education Foundation
   B. Automotive Service Excellence
   C. Automotive Youth Educational Systems
   D. Advanced Engine Performance certification

__A__ 10. Technicians who handle refrigerants or work on AC systems are required to have what?
   A. Environmental Protection Agency Section 609 certification
   B. ASE Advanced Engine Performance certification
   C. NATEF refrigerant certification
   D. AYES R134a certification

JONES & BARTLETT LEARNING
CDX Automotive

We support ASE program certification through

FUNDAMENTALS OF
Automotive Technology
Principles and Practice

STUDENT WORKBOOK

First Edition Revised

JONES & BARTLETT
LEARNING

World Headquarters
Jones & Bartlett Learning
5 Wall Street
Burlington, MA 01803
978-443-5000
info@jblearning.com
www.jblearning.com

Jones & Bartlett Learning books and products are available through most bookstores and online booksellers. To contact Jones & Bartlett Learning directly, call 800-832-0034, fax 978-443-8000, or visit our website, www.jblearning.com.

Substantial discounts on bulk quantities of Jones & Bartlett Learning publications are available to corporations, professional associations, and other qualified organizations. For details and specific discount information, contact the special sales department at Jones & Bartlett Learning via the above contact information or send an email to specialsales@jblearning.com.

Copyright © 2015 by Jones & Bartlett Learning, LLC, an Ascend Learning Company

All rights reserved. No part of the material protected by this copyright may be reproduced or utilized in any form, electronic or mechanical, including photocopying, recording, or by any information storage and retrieval system, without written permission from the copyright owner.

The content, statements, views, and opinions herein are the sole expression of the respective authors and not that of Jones & Bartlett Learning, LLC. Reference herein to any specific commercial product, process, or service by trade name, trademark, manufacturer, or otherwise does not constitute or imply its endorsement or recommendation by Jones & Bartlett Learning, LLC and such reference shall not be used for advertising or product endorsement purposes. All trademarks displayed are the trademarks of the parties noted herein. *Fundamentals of Automotive Technology: Principles and Practice, Student Workbook* is an independent publication and has not been authorized, sponsored, or otherwise approved by the owners of the trademarks or service marks referenced in this product.

There may be images in this workbook that feature models; these models do not necessarily endorse, represent, or participate in the activities represented in the images. Any screenshots in this product are for educational and instructive purposes only.

Production Credits
Chief Executive Officer: Ty Field
President: James Homer
Chief Product Officer: Eduardo Moura
Executive Publisher: Kimberly Brophy
Acquisitions Editor—CDX: Ian Andrew
Managing Editor—CDX Automotive: Amanda J. Mitchell
Senior Editorial Assistant: Marisa Hines
Associate Production Editor: Nora Menzi
Senior Marketing Manager: Brian Rooney
VP, Manufacturing and Inventory Control: Therese Connell
Composition: Cenveo® Publisher Services
Cover Design: Kristin E. Parker
Director of Photo Research and Permissions: Amy Wrynn
Cover Image: © Mark Evans/the Agency Collection/Getty Images
Printing and Binding: Edwards Brothers Malloy
Cover Printing: Edwards Brothers Malloy

ISBN: 978-1-284-05942-7

6048

Printed in the United States of America
18 17 16 15   10 9 8 7 6 5 4 3

Editorial Credits

Authors
Kirk T. VanGelder
Christopher W. Benson

Reviewers
Mark Mitchell
Columbus State Community College

David L. Stidham
Columbus North High School

Kristofer Kowalski
Argo Community High School

Danny Camden

True/False

If you believe the statement to be more true than false, write the letter "T" in the space provided. If you believe the statement to be more false than true, write the letter "F".

__T__ 1. Today's vehicles are assembled on high-volume production lines, with robots used for many of the assembly processes, including welding seams.

__F__ 2. Heavy line technicians diagnose and replace the mechanical and electrical components of motor vehicles, such as gaskets, belts, hoses, timing belts, water pumps, radiators, alternators, and starters.

__F__ 3. Drivability technicians perform wheel alignments and wheel balancing, and they diagnose and replace faulty steering system components.

__T__ 4. In larger shops, roles may be assigned to separate electrical and drivability technicians, whereas in smaller shops, one technician could perform both roles.

__F__ 5. Electrical technicians test and replace faulty charging system components, starter motors, and related items such as batteries.

__F__ 6. Light line technicians use electronic test equipment, scan tools, pressure transducers, exhaust gas analyzers, lab scopes, meters, and circuit wiring diagrams to locate electrical, fuel, and emission systems faults.

__F__ 7. A service manager's job is to oversee technicians' work in order to ensure that customers receive quality repair work.

__T__ 8. Because dealership technicians are working on the latest vehicles, they are right at the cutting edge of technology.

__F__ 9. The automotive service industry in the United States is generally not subject to licensure requirements.

__T__ 10. The Automotive Youth Educational Systems (AYES) is an independent, nonprofit organization dedicated to the improvement of vehicle repair through the testing and certification of automotive professionals.

Fill in the Blank

Read each item carefully, and then complete the statement by filling in the missing word(s).

1. Henry Ford applied two concepts that helped make the Model T affordable for the masses, __interchangeability__ and the __assembly line__.
2. A(n) __Lube technician__ changes oil and filters and carries out lubrication, fluid inspection, fluid service, and tire rotations.
3. A(n) __Chassis and brake technician__ diagnoses, repairs, and services steering system components and suspension systems on all types of vehicles.
4. A(n) __Electrical technician__ diagnoses, replaces, maintains, identifies faults with, and repairs electrical wiring and computer-based equipment in vehicles.
5. Electrical technicians use meters, oscilloscopes, test instruments, and circuit wiring diagrams to diagnose __electrical faults__.
6. A(n) __Drivability technician__ works with computer-controlled engine management systems to service, identify, and repair faults on electronically controlled vehicle systems such as fuel injection, ignition, and automatic transmissions.
7. A(n) __transmission specialist__ may also work on the other components of the drivetrain, including the drive shafts and differentials.
8. A(n) __shop foreman__ oversees the work of all types of technicians and staff, communicates with customers and external suppliers, and handles the various administrative duties involved with operating a business.
9. __Dealerships__ are affiliated with a specific vehicle manufacturer.
10. To earn __ASE certification__, technicians are required to pass one or more ASE certification tests and have 2 years of qualifying work experience as a technician.

Crossword Puzzle

Use the clues to complete the puzzle.

		1		2							
		t		E							
		r		l							
		a		e							
		n		c							
3 C	h	a	s	s	i	s	t	e	c	h	n	i	c	i	a	n
		m		r							
		i		i							
		s		c							
		s		a							
		i		l							
	6	o		t							
4 l i g h t	D	e	t	e	c	h	n	i	c	i	a	n
		s		c							
		p		h							
5 S e r v i c e m	a	n	a	g	e	r	
		c		i							
		i		c							
		a		i							
		l		a							
		i		n							
		s									
		t									

Across

3. A technician who specializes in working on vehicle suspension and steering systems.
4. A technician who diagnoses and replaces the mechanical and electrical components of motor vehicles.
5. The shop supervisor who is responsible for the management of the service department.

Down

1. A technician who diagnoses, overhauls, and repairs transmissions.
2. A technician who diagnoses, replaces, maintains, identifies fault with, and repairs electrical wiring and computer-based equipment in vehicles.

Chapter 1   Careers in Automotive Technology

ASE-Type Questions

Read each item carefully, and then select the best response.

__C__ 1. Tech A says that newer vehicles require less maintenance compared to older vehicles. Tech B says that service intervals for an older vehicle can be extended if new oils are used. Who is correct?
   A. Tech A
   B. Tech B
   C. Both A and B
   D. Neither A nor B

__B__ 2. Tech A says that Henry Ford is credited with the invention of the automobile. Tech B says that Carl Benz is credited with the invention of the automobile. Who is correct?
   A. Tech A
   B. Tech B
   C. Both A and B
   D. Neither A nor B

__C__ 3. Tech A says that the production of vehicles today requires a mix of robotic and human assembly to be profitable. Tech B says that most parts on a car are preassembled before they reach the assembly line for higher assembly numbers per day. Who is correct?
   A. Tech A
   B. Tech B
   C. Both A and B
   D. Neither A nor B

__C__ 4. Tech A says that the automotive industry is highly technical and only a certain few people will find jobs. Tech B says the automotive industry is wide open with job opportunities for almost every level of skill. Who is correct?
   A. Tech A
   B. Tech B
   C. Both A and B
   D. Neither A nor B

__A__ 5. Tech A says that a technician can specialize in different areas based on his or her interest and ability. Tech B says that when a technician specializes in a certain area, he or she will only work on certain vehicle models. Who is correct?
   A. Tech A
   B. Tech B
   C. Both A and B
   D. Neither A nor B

__B__ 6. Tech A says that the foreman is the frontline contact for customer relations. Tech B says that the service consultant is the frontline contact for customer relations. Who is correct?
   A. Tech A
   B. Tech B
   C. Both A and B
   D. Neither A nor B

__C__ 7. Tech A says that dealership technicians generally have access to manufacturers' training to help prepare them as technicians. Tech B says that an independent shop works on a wide variety of equipment that requires a broad skill level in technicians. Who is correct?
   A. Tech A
   B. Tech B
   C. Both A and B
   D. Neither A nor B

Fundamentals of Automotive Technology Student Workbook

___D___ 8. Tech A says that AYES certifies technicians. Tech B says that ASE certifications can help get you a job. Who is correct?
   A. Tech A
   B. Tech B
   C. Both A and B
   D. Neither A nor B

___B___ 9. Tech A says that the maintenance requirements of a vehicle have not changed since the creation of the automobile. Tech B says that manufacturers are predicting 25,000-mile (40,000-km) intervals. Who is correct?
   A. Tech A
   B. Tech B
   C. Both A and B
   D. Neither A nor B

___C___ 10. Tech A says that a technician can progress to different jobs within the industry. Tech B says that carriers in the automotive industry include new car assembly lines. Who is correct?
   A. Tech A
   B. Tech B
   C. Both A and B
   D. Neither A nor B

Introduction to Automotive Technology

CHAPTER 2

Chapter Review

The following activities have been designed to help you refresh your knowledge of this chapter. Your instructor may require you to complete some or all of these activities as a regular part of your training program. You are encouraged to complete any activity that your instructor does not assign as a way to enhance your learning.

Matching

Match the following terms with the correct description or example.

A. Differential gear set
B. Four-wheel drive
C. Horizontally opposed engine
D. In-line engine
E. Longitudinal
F. Piston engine
G. Rotary engine
H. Torque converter
I. Unibody design
J. V engine

__D__ 1. An engine in which the cylinders are arranged side by side in a single row.
__F__ 2. An internal combustion engine that uses cylindrical pistons moving back and forth in a cylinder to extract mechanical energy from chemical energy.
__J__ 3. A term used to describe an engine configuration that uses a single bank of cylinders staggered at a shallow 15-degree V.
__G__ 4. An engine that uses a triangular rotor turning in a housing instead of conventional pistons.
__E__ 5. A term used to describe the front-to-back engine orientation when mounted in the engine compartment.
__A__ 6. The arrangement of gears between two axles that allows each axle to spin at its own speed when the vehicle is going around a corner.
__B__ 7. A drive train layout in which the engine drive has either two wheels or four wheels depending on which mode is selected by the driver.
__I__ 8. A vehicle design that does not use a rigid frame to support the body. The body panels are designed to provide the strength for the vehicle.
__H__ 9. A device that is turned by the crankshaft and transmits torque to the input shaft of an automatic transmission.
__C__ 10. An engine with two banks of cylinders, 180 degrees apart, on opposite sides of the crankshaft. It is also called a flat engine or a boxer engine.

Multiple Choice

Read each item carefully, and then select the best response.

__B__ 1. Which vehicle design has an enclosed body with a maximum of four doors, and a trunk located in the rear of the vehicle accessible from a trunk lid?
   A. Coupe
   B. Sedan
   C. Hatchback
   D. Station wagon

Fundamentals of Automotive Technology Student Workbook

__C__ 2. Which type of vehicle acts like both a full-size van and a pickup in that it has a heavier-duty chassis so it can carry heavier loads?
   A. Sedan
   B. Hatchback
   C. Sport utility vehicle
   D. Minivan

__A__ 3. What type of chassis design was first used in aircraft and then spread to automobiles?
   A. Unibody
   B. Body on frame
   C. Dual shell
   D. Steel ladder

__C__ 4. In which type of drivetrain layout are all four wheels driven by the engine all of the time?
   A. Front-wheel drive
   B. Rear-wheel drive
   C. All-wheel drive
   D. Four-wheel drive

__B__ 5. All of the following criteria are used to define the drive train layout, *except*:
   A. Engine position
   B. Transmission type
   C. Engine orientation
   D. Type of drive

__C__ 6. Which of the following engine designs is the most powerful compared to its overall dimensions, but more complicated and expensive than the other engines?
   A. V8
   B. Flat 6
   C. W12
   D. In-line 4

__D__ 7. Which type of engine uses a single bank of cylinders, staggered at a shallow 15-degree V within the bank?
   A. Horizontally opposed
   B. W
   C. V
   D. VR

__A__ 8. Which type of axle uses the engine's torque to turn the wheels (drive the vehicle) and at the same time support the weight of the vehicle?
   A. Live axle
   B. Dead axle
   C. Transaxle
   D. Solid axle

__B__ 9. The twisting force applied to a shaft is known as what?
   A. Play
   B. Torque
   C. Collar
   D. Give

__D__ 10. What designation is used when measuring torque?
   A. Foot-pound
   B. Inch-pound
   C. Newton meter
   D. Any of the above

Chapter 2   Introduction to Automotive Technology

True/False

If you believe the statement to be more true than false, write the letter "T" in the space provided. If you believe the statement to be more false than true, write the letter "F".

__T__ 1. Reducing the number of doors to the passenger compartment makes the vehicle structure more rigid.
__F__ 2. In some vehicles, known as roadsters, the roof can be a series of folding steel or fiberglass panels.
__T__ 3. A station wagon has an extended roof that goes all the way to the rear of the vehicle. It is similar to a van but not as tall.
__T__ 4. Body-on-frame is the term used when a vehicle body is mounted on a rigid frame or chassis.
__F__ 5. Some high-performance racing cars today have no chassis at all.
__F__ 6. Mechanical energy can be converted into chemical energy in two primary ways: through the operation of an internal combustion engine or through the operation of an electric motor.
__T__ 7. The suspension system makes the connection between the steering wheel and the road wheels so the driver can point the vehicle in the intended direction of travel.
__F__ 8. A drive train is classified by type, cylinder arrangement, number of cylinders/rotors, and total engine displacement in cubic inches or liters.
__T__ 9. Multi-cylinder internal combustion automotive engines are produced in four common configurations.
__T__ 10. V engines have two banks of cylinders sitting side by side in a V arrangement sharing a common crankshaft.
__F__ 11. Horizontally opposed engines are very powerful for their size, but they do not use conventional pistons that slide back and forth inside a straight cylinder.
__T__ 12. The automatic transmission uses a torque converter instead of a clutch.
__T__ 13. Part-time 4WD means the vehicle is usually driven in two-wheel drive and switched to full-time when needed by engaging the transfer case.
__T__ 14. A transfer case locks the drive shafts together and directs torque through them to both axles.
__F__ 15. All transfer cases use a viscous coupling to split the drive between the front and rear wheels.

Fill in the Blank

Read each item carefully, and then complete the statement by filling in the missing word(s).

1. A(n) __coupe__ has only two doors.
2. A(n) __hatchback__ is available in three-door and five-door designs.
3. A(n) __convertible__ is an automobile that can convert from having an enclosed top to having an open top by means of a roof that can be removed, retracted, or folded away.
4. A(n) pickup, or __truck__, carries and tows cargo.
5. A(n) __sport__ __utility__ __vehicle__ can easily be used to carry out functions that would otherwise require several different vehicles.
6. A(n) __chassis__ is an underlying supporting structure for vehicles—similar to the skeleton of a human—on which additional components are mounted.
7. The __unibody__ design is constructed of a large number of steel sheet metal panels that are precisely formed in presses and spot-welded together into a structural unit.
8. Stored __converting__ energy is converted to mechanical energy to propel a vehicle down the road.
9. As the pistons move up and down, they rotate the crankshaft, turning the __crank shaft__ or flex plate, which is bolted to the engine crankshaft.
10. The __suspension__ system evens out the road shocks caused by irregular road surfaces.
11. Manufacturers mount engines in one of two orientations, __piston__ and __rotary__, depending on which design best fits the vehicle and the rest of the drive train.
12. In a piston engine, the way engine cylinders are arranged is called the engine __configuration__.
13. __Horizontally opposed__ engines are sometimes referred to as "flat" engines and are commonly found in 4- and 6-cylinder configurations.
14. Axles come in two configurations: __live__ axle and __dead__ axle.
15. A vehicle with a manual transmission uses a __clutch__ to engage and disengage the engine from the transmission.

Labeling

Label the following images with the correct terms.

1. Chassis, engines, and axles:

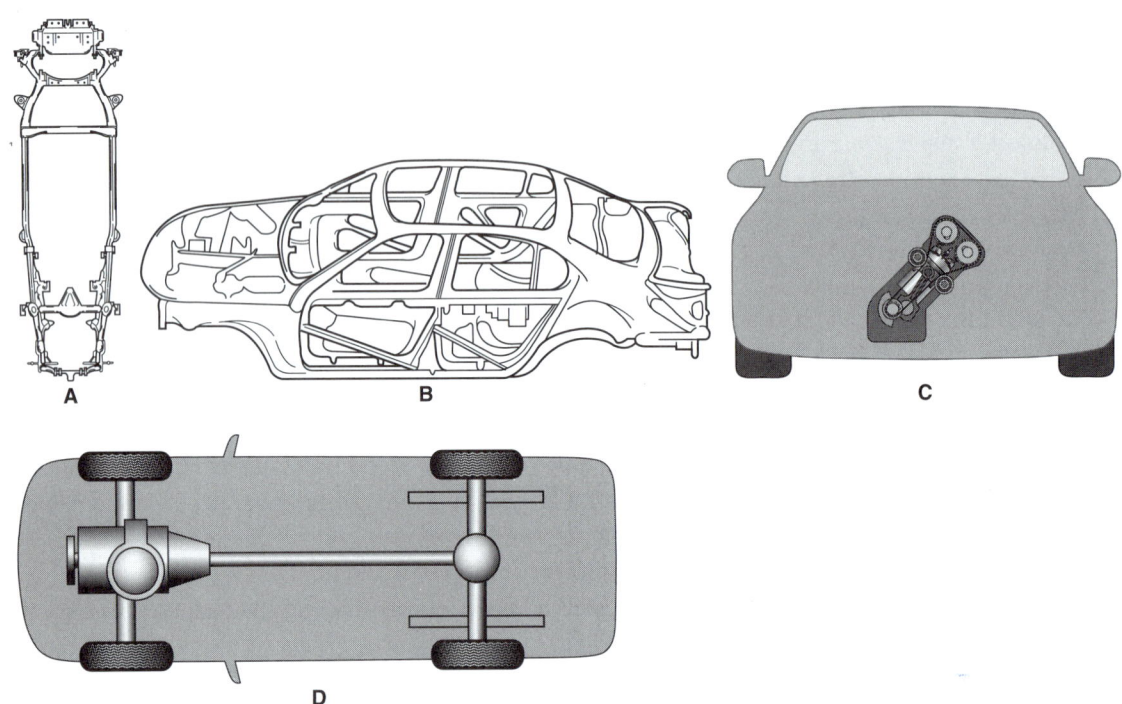

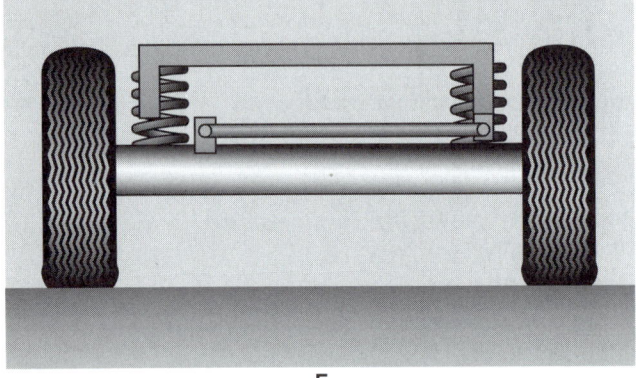

A. Steel ladder fram chassi
B. The unibody design
C. tilting cylinder banks
D. live axle
E. dead axle

Chapter 2 Introduction to Automotive Technology

2. Manual transmission clutch:

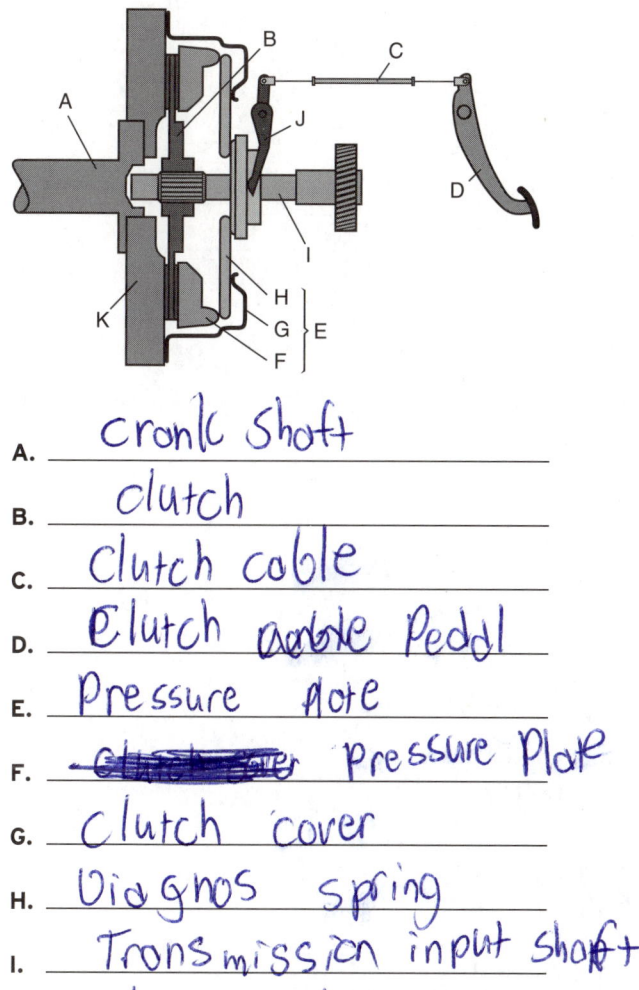

- A. Crank shaft
- B. clutch
- C. Clutch cable
- D. Clutch cable Pedal
- E. Pressure Plate
- F. ~~Clutch cover~~ pressure Plate
- G. Clutch cover
- H. Diagnos spring
- I. Transmission input shaft
- J. Clutch fork
- K. Fly wheel

3. Final drive assembly:

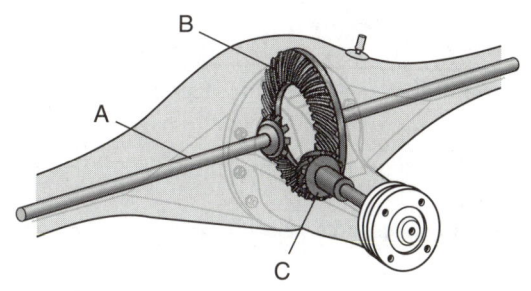

- A. Rear axle
- B. Ring Gear
- C. Pinion Gear

Crossword Puzzle

Use the clues to complete the puzzle.

Across

3. Twisting force.
5. A vehicle configuration that has an enclosed body, with a maximum of four doors to allow access to the passenger compartment.
6. A vehicle that converts from having an enclosed top to having an open top by a roof that can be removed, retracted, or folded away.
7. A term used to identify the engine, transmission/transaxle, differential, axles, and wheels.
10. A device that acts like a limited slip clutch.
13. The heavy disc bolted to the rear of the crankshaft that smooths out the power pulses and stores energy from the power stroke for use in keeping the crankshaft rotating through the other three strokes.
14. The main support frame in a vehicle. It includes the running gear, such as suspension, the engine, and the drive train.

Down

1. A term used to describe the side-to-side engine orientation when mounted in the engine compartment.
2. A component that provides a final gear reduction and allows for the difference in speed of each wheel when cornering.
4. A vehicle that has a shared passenger and cargo area; it typically is available in three- and five-door arrangements.
5. A passenger vehicle built on a light truck chassis; it is usually equipped with four-wheel drive and capable of hauling heavier loads than typical passenger vehicles.
8. A lighter-duty van used for carrying six to eight occupants or light cargo.
9. A shaft connected to wheels that transmits the driving torque to the wheels.
11. A two-door vehicle that has seating for two people and may have a small rear seat.
12. A vehicle that carries cargo; it has stronger chassis components and suspension than a sedan.

ASE-Type Questions

Read each item carefully, and then select the best response.

__B__ 1. Tech A says that most vehicles today are built with a ladder frame. Tech B says that most vehicles today do not have a frame. Who is correct?
   A. Tech A
   B. Tech B
   C. Both A and B
   D. Neither A nor B

__C__ 2. Tech A says that a roadster is a coupe. Tech B says that a station wagon and a hatchback are the same. Who is correct?
   A. Tech A
   B. Tech B
   C. Both A and B
   D. Neither A nor B

__A__ 3. Tech A says that gasoline is energy in chemical form. Tech B says that chemical energy is converted to mechanical energy in the combustion process of an engine. Who is correct?
   A. Tech A
   B. Tech B
   C. Both A and B
   D. Neither A nor B

__B__ 4. Tech A says that a live axle is an axle that can be steered. Tech B says that a dead axle supports the weight of the vehicle, but doesn't power it. Who is correct?
   A. Tech A
   B. Tech B
   C. Both A and B
   D. Neither A nor B

__C__ 5. Tech A says that the gear ratio in the transmission helps the vehicle gain speed. Tech B says that the gear ratio of the final drive can be selected with a gearshift lever. Who is correct?
   A. Tech A
   B. Tech B
   C. Both A and B
   D. Neither A nor B

__A__ 6. Tech A says that pushing on the brake pedal stops the vehicle by converting chemical energy into thermal energy. Tech B says that brake pedal pressure is transmitted to the brakes hydraulically. Who is correct?
   A. Tech A
   B. Tech B
   C. Both A and B
   D. Neither A nor B

14 Fundamentals of Automotive Technology Student Workbook

__B__ 7. Tech A says that one advantage of a mid-engine design is better weight distribution. Tech B says that the rear-engine design is commonly used in 4-wheel drive vehicles. Who is correct?
  A. Tech A
  B. Tech B
  C. Both A and B
  D. Neither A nor B

__C__ 8. Tech A says that the purpose of the battery is to charge the vehicle. Tech B says that in-line engines are generally easier to work on than V engines. Who is correct?
  A. Tech A
  B. Tech B
  C. Both A and B
  D. Neither A nor B

__D__ 9. Tech A says that the natural angle of a V6 is 90 degrees. Tech B says that the natural angle of a V10 is 60 degrees. Who is correct?
  A. Tech A
  B. Tech B
  C. Both A and B
  D. Neither A nor B

__A__ 10. Tech A says that the definition of torque is how far the crankshaft twists. Tech B says that torque can relate to tightness of bolts. Who is correct?
  A. Tech A
  B. Tech B
  C. Both A and B
  D. Neither A nor B

Introduction to Automotive Safety

CHAPTER 3

Chapter Review

The following activities have been designed to help you refresh your knowledge of this chapter. Your instructor may require you to complete some or all of these activities as a regular part of your training program. You are encouraged to complete any activity that your instructor does not assign as a way to enhance your learning.

Matching

Match the following terms with the correct description or example.

A. Environmental Protection Agency
B. Hazardous environment
C. Material safety data sheet
D. Occupational Safety and Health Administration
E. Personal protective equipment
F. Threshold limit value

_**F**_ 1. Equipment designed to protect the technician, such as safety boots, gloves, clothing, protective eyewear, and hearing protection.

_**C**_ 2. A sheet that provides information about handling, use, and storage of a material that may be hazardous.

_**F**_ 3. The maximum allowable concentration of a given material in the surrounding air.

_**D**_ 4. Government agency created to provide national leadership in occupational safety and health.

_**B**_ 5. A place where hazards exist.

_**A**_ 6. Federal government agency that deals with issues related to environmental safety.

Multiple Choice

Read each item carefully, and then select the best response.

_**D**_ 1. A safe work environment includes which of the following?
   A. A well-organized shop layout
   B. Good supervision
   C. Safety training
   D. All of the above

_**B**_ 2. Which federal government agency deals with issues related to environmental safety?
   A. MSDS
   B. EPA
   C. OSHA
   D. NIOSH

_**C**_ 3. A document that describes the steps required to safely use the vehicle hoist is an example of?
   A. Regulation
   B. Policy
   C. Procedure
   D. Compliance

A 4. A _____ can be used to identify hazards and risks within the work environment.
   A. Risk analysis
   B. Procedure
   C. Vulnerability analysis
   D. Threat assessment

D 5. All of the following are components of safety signs, *except*:
   A. Signal word
   B. Background color
   C. Warning light
   D. Pictorial message

D 6. Which signal word indicates a potentially hazardous situation, which, if not avoided, may result in minor or moderate injury?
   A. Danger
   B. Hazard
   C. Warning
   D. Caution

C 7. A droplight that is designed in such a way that the electrical parts can never come into contact with the outer casing of the device is called?
   A. Grounded
   B. Ground fault
   C. Double-insulated
   D. Cordless

B 8. What class of fires involves flammable liquids or gaseous fuels?
   A. Class A
   B. Class B
   C. Class C
   D. Class D

A 9. To operate a fire extinguisher, you should follow which of the following acronyms?
   A. PULL
   B. PASS
   C. PAGE
   D. APES

B 10. The concentration of hazardous material in the air you breathe in your shop must not exceed _____.
   A. The threshold limit value
   B. Ten percent
   C. Shop standards
   D. Fifteen percent

True/False

If you believe the statement to be more true than false, write the letter "T" in the space provided. If you believe the statement to be more false than true, write the letter "F."

F 1. MSDS refers to items of safety equipment like safety footwear, gloves, clothing, protective eyewear, and hearing protection.

T 2. There is the possibility of an accident occurring whenever work is undertaken.

T 3. Shop policies and procedures ensure that the shop operates according to OSHA and EPA laws and regulations.

F 4. An OSHA document for the shop that describes how the shop complies with legislation is an example of a procedure.

Chapter 3 Introduction to Automotive Safety   17

__T__ 5. There are three standard signal words—danger, warning, and caution.
__T__ 6. Signal words on a warning sign allow the safety message to be conveyed to people who are illiterate or who do not speak the local language.
__T__ 7. Whenever a vehicle's engine is running, toxic gases are emitted from its exhaust.
__F__ 8. An engine fitted with a catalytic converter can be run safely indoors.
__F__ 9. Always keep circuit breaker and electrical panel covers open to provide easy access.
__T__ 10. The danger of a gasoline fire is always present in an automotive shop.
__T__ 11. Fire extinguishers are marked with pictograms depicting the types of fires that the extinguisher is approved to fight.
__F__ 12. Eye wash stations are used to flush the eye with potassium chloride in the event that you get foreign liquid or particles in your eye.
__T__ 13. In the United States it is required that workplaces have an MSDS for every chemical that is on site.
__F__ 14. Always use compressed air to blow brake and clutch dust from components and parts before brake service.
__T__ 15. Used oil and fluids will often contain dangerous chemicals and impurities and need to be safely recycled or disposed of in an environmentally friendly way.

Fill in the Blank

Read each item carefully, and then complete the statement by filling in the missing word(s).

1. __Evacuation__ __Routes__ are a safe way of escaping danger and gathering in a safe place where everyone can be accounted for in the event of an emergency.
2. OSHA stands for the __occuption__ __saftey__ and __health__ __administration__.
3. EPA stands for the __Environmen__ __protection__ __agency__.
4. It is the __Person__'s responsibility to know and follow the rules.
5. The signal word __danger__ indicates a potentially hazardous situation, which, if not avoided, could result in death or serious injury.
6. __handrails__ are used to separate walkways and pedestrian traffic from work areas.
7. __Carbon__ __monoxide__ is extremely dangerous, as it is odorless and colorless and can build up to toxic levels very quickly in confined spaces.
8. All circuit breakers and fuses should be clearly __labeled__ so you know which circuits and functions they control.
9. Three elements must be present at the same time for a fire to occur: __fuel__, __oxygen__, and __heat__.
10. Class B fire extinguishers are designed for use on flammable liquids or gaseous fuels and are marked with a __red__ __square__.
11. __fire__ __blanket__ are designed to smother a small fire and are very useful in putting out a fire on a person.
12. MSDS stands for __material__ __Safety__ __data__ __sheets__.
13. __fire__ filters can trap very small particles and prevent them from being redistributed into the surrounding air.
14. Coming into frequent or prolonged contact with used __engine__ __oil__ can cause dermatitis and other skin disorders, including some forms of cancer.
15. Shop __Safety__ __Inspection__ are valuable ways of identifying unsafe equipment, materials, or activities so they can be corrected to prevent accidents or injuries.

Labeling

Label the following images with the correct terms.

To operate a fire extinguisher, follow PASS:

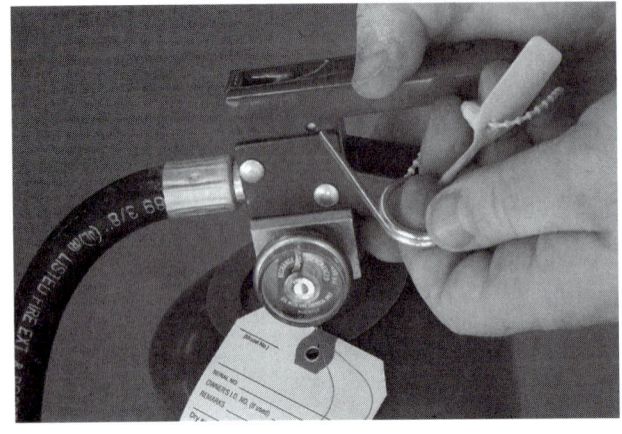

A

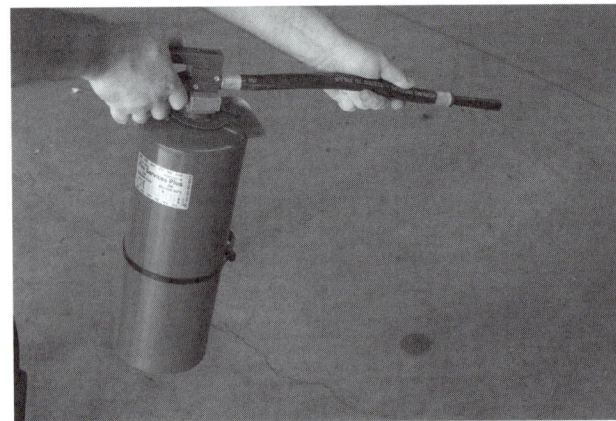

B

C

D

A. ~~aim~~ pull

B. ~~pull~~ aim

C. squeeze

D. sweep

Skill Drills

Place the skill drill steps in the correct order.

1. Identifying Hazardous Environments:

___2___ **A.** Check for air quality. Locate the extractor fans or ventilation outlets and make sure they are not obstructed in any way. Locate and observe the operation of the exhaust extraction hose, pump, and outlet used on the vehicle's exhaust pipes.

___1___ **B.** Familiarize yourself with the shop layout. Study and understand the various warning signs around your shop. Identify exits and plan your escape route.

___4___ **C.** Find out where flammable materials are kept, and make sure they are stored properly.

___6___ **D.** Identify caustic chemicals and acids associated with activities in your shop. Ask your supervisor for information on any special hazards in your particular shop and any special avoidance procedures, which may apply to you and your working environment.

___3___ **E.** Check the location, type, and operation of fire extinguishers in your shop.

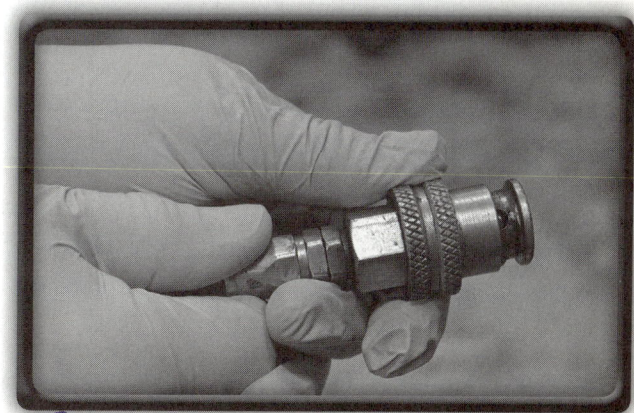

___5___ **F.** Check the hoses and fittings on the air compressor and air guns for any damage or excessive wear.

2. Safely Cleaning Brake Dust:

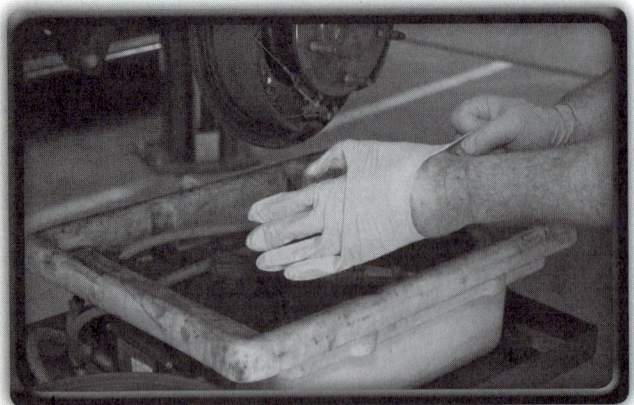

___1___ **A.** When performing any cleaning tasks on brake or clutch components, always wear a face mask, gloves, and eye protection.

___3___ **B.** Turn on the wash station pump and paint the solution over the components to wet and clean the components and remove the dust.

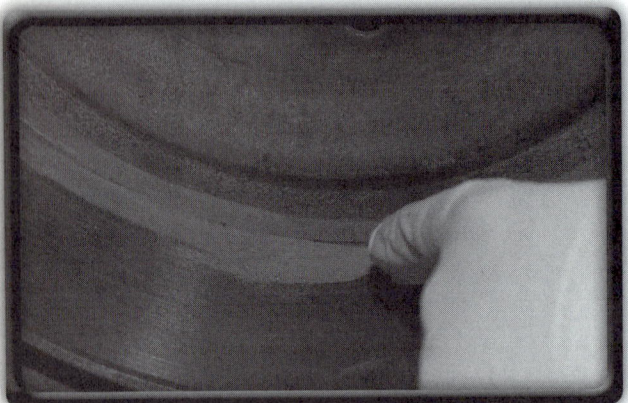

___2___ **C.** Position the brake wash station under the bottom of the backing plate. When cleaning brakes, remove the brake drum and check for the presence of dust and brake fluid. When cleaning a clutch, position the wash station underneath the bell housing.

___4___ **D.** Periodically dispose of the residue in an approved manner.

Crossword Puzzle

Use the clues to complete the puzzle.

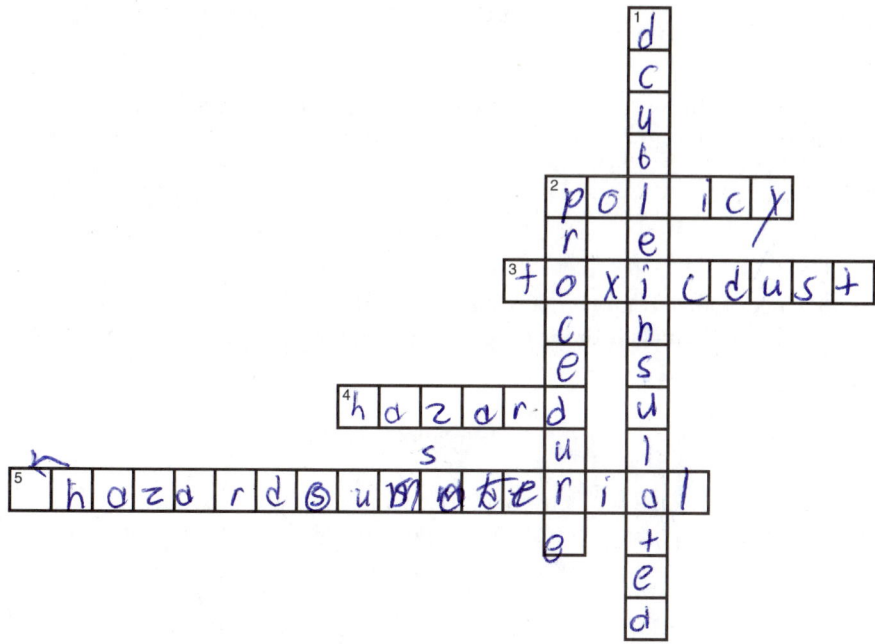

Across

2. A guiding principle that sets the shop direction.
3. Any dust that may contain fine particles that could be harmful to humans or the environment.
4. Anything that could hurt you or someone else.
5. Any material that poses an unreasonable risk of damage or injury to persons, property, or the environment if it is not properly controlled during handling, storage, manufacture, processing, packaging, use and disposal, or transportation.

Down

1. Tools or appliances that are designed in such a way that no single failure can result in a dangerous voltage coming into contact with the outer casing of the device.
2. A list of the steps required to get the same result each time a task or activity is performed.

ASE-Type Questions

Read each item carefully, and then select the best response.

____A____ 1. Tech A says that exposure to solvents may have long-term effects. Tech B says that accidents are almost always avoidable. Who is correct?
   A. Tech A
   B. Tech B
   C. Both A and B
   D. Neither A nor B

____A____ 2. Tech A says that after an accident you should take measures to avoid it in the future. Tech B says that it is OK to block an exit for a shop. Who is correct?
   A. Tech A
   B. Tech B
   C. Both A and B
   D. Neither A nor B

Fundamentals of Automotive Technology Student Workbook

___A___ 3. Tech A says that both OSHA and the EPA can inspect facilities for violations. Tech B says that a shop safety rule does not have to be reviewed once put in place. Who is correct?
   A. Tech A
   B. Tech B
   C. Both A and B
   D. Neither A nor B

___B___ 4. Tech A says that all hazards can be removed from a shop. Tech B says that it is a good practice to disconnect an air gun while inspecting it. Who is correct?
   A. Tech A
   B. Tech B
   C. Both A and B
   D. Neither A nor B

___A___ 5. Tech A says that both caution and danger indicate a potentially hazardous situation. Tech B says that an exhaust extraction hose is not needed if the vehicle is only going to run for a few minutes. Who is correct?
   A. Tech A
   B. Tech B
   C. Both A and B
   D. Neither A nor B

___D___ 6. Tech A says that if you are unsure of what personal protective equipment (PPE) to use to perform a job, you should just use what is nearby. Tech B says that air tools are less likely to shock you than electrically powered tools. Who is correct?
   A. Tech A
   B. Tech B
   C. Both A and B
   D. Neither A nor B

___A___ 7. Tech A says that firefighting equipment includes safety glasses. Tech B says that a class A fire extinguisher can be used to fight an electrical fire only. Who is correct?
   A. Tech A
   B. Tech B
   C. Both A and B
   D. Neither A nor B

___D___ 8. Tech A says that a material safety data sheet (MSDS) contains information on procedures to repair a vehicle. Tech B says that you only need an MSDS if your safety may be in danger. Who is correct?
   A. Tech A
   B. Tech B
   C. Both A and B
   D. Neither A nor B

___B___ 9. Tech A says that a good way to clean dust off brakes is with compressed air. Tech B says that asbestos may be in current auto parts. Who is correct?
   A. Tech A
   B. Tech B
   C. Both A and B
   D. Neither A nor B

___B___ 10. Tech A says that when cleaning brake and clutch components, the wash station should be placed directly under the component. Tech B says that you should follow state and local regulations when disposing of used oil. Who is correct?
   A. Tech A
   B. Tech B
   C. Both A and B
   D. Neither A nor B

Personal Safety

CHAPTER 4

Tire Tread:
© AbleStock

Chapter Review

The following activities have been designed to help you refresh your knowledge of this chapter. Your instructor may require you to complete some or all of these activities as a regular part of your training program. You are encouraged to complete any activity that your instructor does not assign as a way to enhance your learning.

Matching

Match the following terms with the correct description or example.

A. Barrier cream
B. Complicated fracture
C. Ear protection
D. External bleeding
E. First-degree burns
F. Gas welding goggles
G. Heat buildup
H. Open fracture
I. Second-degree burns
J. Third-degree burns

__C__ 1. Protective gear worn when the sound levels exceed 85 decibels, when working around operating machinery for any period of time, or when the equipment you are using produces loud noise.

__E__ 2. Burns that show reddening of the skin and damage to the outer layer of skin only.

__F__ 3. Protective gear designed for gas welding; they provide protection against foreign particles entering the eye and are tinted to reduce the glare of the welding flame.

__G__ 4. A dangerous condition that occurs when a glove can no longer absorb or reflect heat and heat is transferred to the inside of the glove.

__D__ 5. The loss of blood from an external wound; blood can be seen escaping.

__I__ 6. Burns that involve blistering and damage to the outer layer of skin.

__A__ 7. A cream that looks and feels like a moisturizing cream but has a specific formula to provide extra protection from chemicals and oils.

__H__ 8. A fracture in which the bone is protruding through the skin or there is severe bleeding.

__J__ 9. Burns that involve white or blackened areas and damage to all skin layers and underlying structures and tissues.

__B__ 10. A fracture in which the bone has penetrated a vital organ.

Multiple Choice

Read each item carefully, and then select the best response.

__D__ 1. Proper footwear provides protection against which of the following?
   A. Chemicals
   B. Cuts
   C. Slips
   D. All of the above

__B__ 2. What type of gloves will protect your hands from burns when welding or handling hot components?
   A. Light-duty
   B. Leather
   C. Chemical
   D. General-purpose cloth

__C__ 3. What type of protectant prevents chemicals from being absorbed into your skin and should be applied to your hands before you begin work?
   A. Sunscreen
   B. Moisturizer
   C. Barrier cream
   D. Sanitizer

__B__ 4. What type of gloves should be used to protect your hands from exposure to greases and oils?
   A. Chemical
   B. Light-duty
   C. Leather
   D. General-purpose cloth

__C__ 5. When using a _____ always make sure the cartridge is the correct type for the contaminant in the atmosphere.
   A. dust mask
   B. welding helmet
   C. respirator
   D. all of the above

__A__ 6. The most common type of eye protection is a pair of safety glasses, which must be marked with _____ on the lens and frame.
   A. Z87
   B. PPE
   C. UV
   D. Polarized

__D__ 7. What type of protection can be worn instead of a welding mask when using or assisting a person using an oxyacetylene welder?
   A. Respirator
   B. Full face shield
   C. Safety glasses
   D. Gas welding goggles

__B__ 8. Which of the following is an example of a mechanical means of ventilation?
   A. Open door
   B. Fume hood
   C. Open window
   D. Both A and C

__A__ 9. Before lifting anything, you can reduce the risk of injury by:
   A. Breaking down the load into smaller quantities
   B. Seeking assistance
   C. Using a mechanical device
   D. All of the above

__C__ 10. In the event of an accident the first step is to:
   A. Remove the injured person
   B. Perform first aid
   C. Survey the scene
   D. Call 9-1-1

Chapter 4 Personal Safety   25

__C__ 11. In the event of a medical emergency, always have a bystander call _____, unless you are alone.
   A. the fire department
   B. the shop owner
   C. 9-1-1
   D. The shop manager

__B__ 12. What type of fracture involves bleeding or the protrusion of bone through the skin?
   A. Simple
   B. Open
   C. Complicated
   D. All of the above

__C__ 13. What type of injury symptoms include pain or tenderness around the area, inability to move the joint, deformity of the joint, and swelling and discoloration over the joint?
   A. Sprain
   B. Strain
   C. Dislocation
   D. Open fracture

__A__ 14. What type of burns show reddening of the skin and damage to the outer layer of skin only?
   A. First degree
   B. Second degree
   C. Third degree
   D. Fourth degree

__D__ 15. Burns are caused by:
   A. Electricity
   B. Excessive heat
   C. Chemicals
   D. All of the above

True/False

If you believe the statement to be more true than false, write the letter "T" in the space provided. If you believe the statement to be more false than true, write the letter "F".

__T__ 1. PPE includes clothing, shoes, safety glasses, hearing protection, masks, and respirators.
__F__ 2. Leather gloves should always be worn when using solvents and cleaners.
__F__ 3. Avoid picking up very hot metal with leather gloves because it causes the leather to harden, making it less flexible during use.
__F__ 4. Light-duty gloves are designed for use in cold temperatures, particularly during winter, so that cold tools do not stick to your skin.
__F__ 5. If a proper barrier cream is not available, use a standard moisturizer as a suitable replacement.
__F__ 6. Ear protection that covers the entire outer ear usually has higher noise-reduction ratings than in-the-ear type.
__T__ 7. A disposable dust mask should not be used if chemicals, such as paint solvents, are present in the atmosphere.
__T__ 8. When grinding, you should wear a pair of safety glasses underneath your face shield for added protection.
__F__ 9. Tinted safety glasses are not designed to be worn outside in bright sunlight conditions.
__T__ 10. Ultraviolet radiation can burn your skin like a sunburn.
__T__ 11. Safety goggles provide much the same eye protection as safety glasses but with added protection against harmful chemicals that may splash up behind the lenses of glasses.
__T__ 12. Always remove watches, rings, and jewelry before starting work.
__F__ 13. When attending to an injured victim, never send for assistance until after administering CPR.
__T__ 14. If an object punctures the victim's skin and becomes embedded in the victim's body, remove the object and cover with a sterile dressing.
__T__ 15. Burns are classified as either superficial, partial thickness, or full thickness.

Fill in the Blank

Read each item carefully, and then complete the statement by filling in the missing word(s).

1. It is a good idea to keep a spare set of __protective__ __clothing__ in the workshop in case a toxic or corrosive fluid is spilled on the ones you are wearing.
2. Your __hard__ __hat__ can protect you from bumping your head on a vehicle when the vehicle is raised on a hoist.
3. Never use __solvent__ such as gasoline or kerosene to clean your hands.
4. There are two types of breathing devices: disposable __dust__ __masks__ and __respirator__.
5. A(n) __respirator__ has removable cartridges that can be changed according to the type of contaminant being filtered.
6. The light from a welding arc is very bright and contains high levels of __ultraviolet__ radiation.
7. It is necessary to use a full __face__ __shield__ when using solvents and cleaners, epoxies, and resins or when working on a battery.
8. You should always think __safety__ __first__ and then act safely.
9. In high-exposure situations, such as vehicles running in the shop, a mechanical means of __ventilation__ is required.
10. Good __housekeeping__ is about always making sure the shop and your work surroundings are neat and kept in good order.
11. Bleeding is divided into two categories: internal and __external__.
12. Symptoms of __internal__ __bleeding__ include bruising, a painful or tender area, coughing frothy blood, and vomiting blood.
13. If a chemical splashes into the eyes, you may be able to flush it out using a(n) __eye__ __wash__ __station__.
14. Full-thickness burns, or __third__-__degree__ burns, involve white or blackened areas and include damage to all skin layers and underlying structures and tissues.
15. If clothing is burning, have the victim roll on the ground using the __stop__, __drop__, and __roll__ method.

Labeling

Label the following images with the correct terms.

Level of burns:

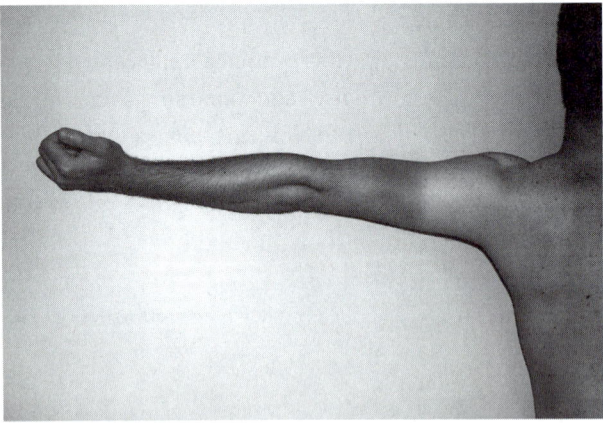

A. __first degree__

Chapter 4 Personal Safety

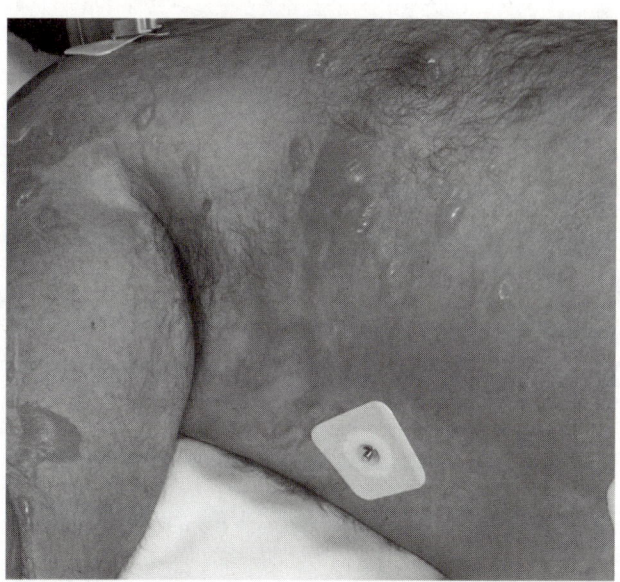
© E. M. Singletary, MD. Used with permission.

B. _Second degree_

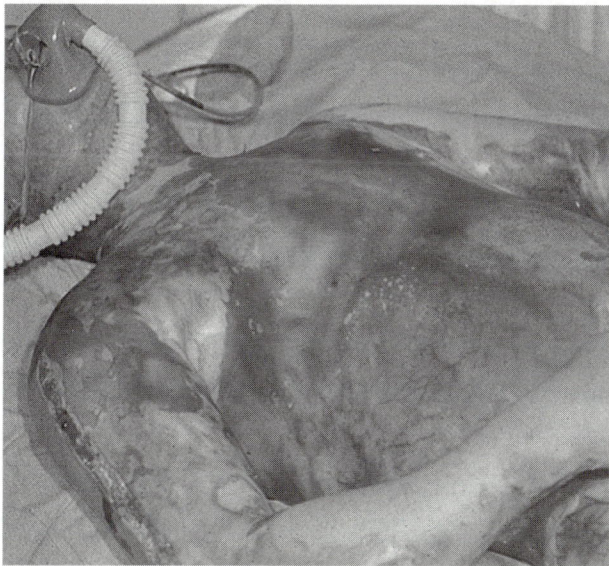
Courtsty of AAOS

C. _Third degree_

Crossword Puzzle

Use the clues to complete the puzzle.

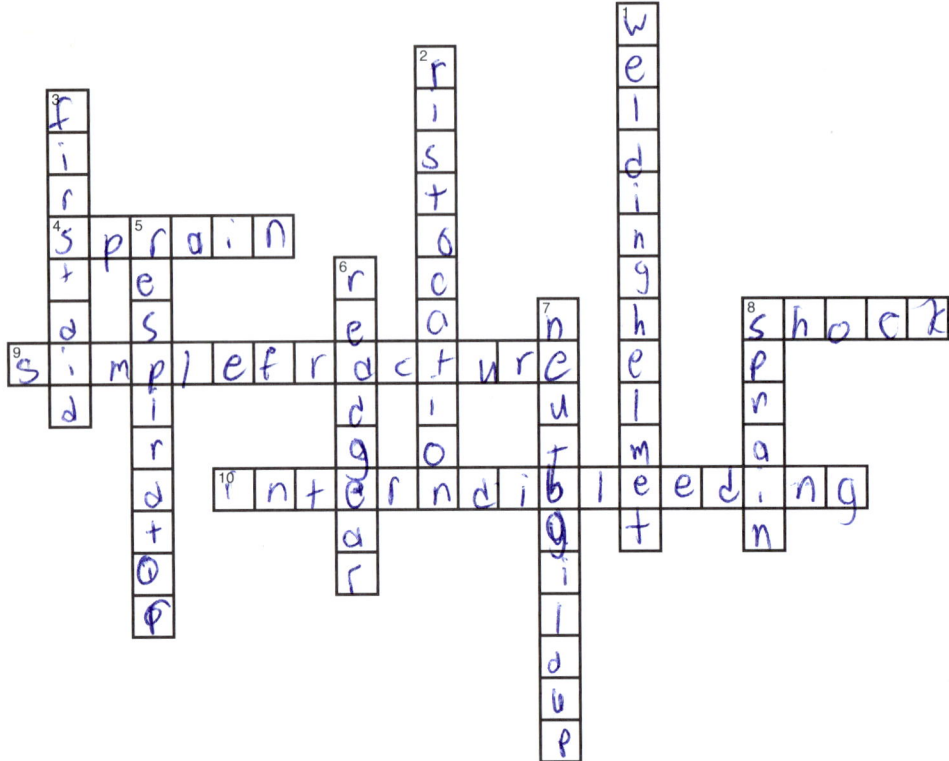

Across

4. An injury in which a joint is forced beyond its natural movement limit.
8. Inadequate tissue oxygenation resulting from serious injury or illness.
9. A fracture that involves no open wound or internal or external bleeding.
10. The loss of blood into the body cavity from a wound; there is no obvious sign of blood.

Down

1. Protective gear designed for arc welding; it provides protection against foreign particles entering the eye, and the lens is tinted to reduce the glare of the welding arc.
2. The displacement of a joint from its normal position; it is caused by an external force stretching the ligaments beyond their elastic limit.
3. The immediate care given to an injured or suddenly ill person.
5. Protective gear used to protect the wearer from inhaling harmful dusts or gases.
6. Protective gear that includes items like hairnets, caps, or hard hats.
7. A dangerous condition that occurs when the glove can no longer absorb or reflect heat and heat is transferred to the inside of the glove.
8. An injury caused by the overstretching of muscles and tendons.

ASE-Type Questions

Read each item carefully, and then select the best response.

__B__ 1. Tech A says that personal protective equipment (PPE) does not include clothing. Tech B says that the PPE used should be based on the task you are performing. Who is correct?
  A. Tech A
  B. Tech B
  C. Both A and B
  D. Neither A nor B

__C__ 2. Tech A says that protective clothing that is not in good condition should be replaced. Tech B says that safety glasses are adequate to protect your eyes regardless of the activity. Who is correct?
  A. Tech A
  B. Tech B
  C. Both A and B
  D. Neither A nor B

__B__ 3. Tech A says that appropriate work clothes include loose-fitting clothing. Tech B says that you should always wear cuffed pants when working in a shop. Who is correct?
  A. Tech A
  B. Tech B
  C. Both A and B
  D. Neither A nor B

__C__ 4. Tech A says that proper footwear may include both leather- and steel-toed shoes. Tech B says that leather-soled shoes provide slip resistance. Who is correct?
  A. Tech A
  B. Tech B
  C. Both A and B
  D. Neither A nor B

__C__ 5. Tech A says that a hat can help keep your hair clean when working on a vehicle. Tech B says that chemical gloves may be used when working with solvent. Who is correct?
  A. Tech A
  B. Tech B
  C. Both A and B
  D. Neither A nor B

__B__ 6. Tech A says that you should only wear gloves when it is absolutely necessary. Tech B says that leather gloves are used to pick up very hot pieces of metal. Who is correct?
  A. Tech A
  B. Tech B
  C. Both A and B
  D. Neither A nor B

__C__ 7. Tech A says that barrier creams are used to make cleaning your hands easier. Tech B says that hearing protection only needs to be worn by people operating loud equipment. Who is correct?
  A. Tech A
  B. Tech B
  C. Both A and B
  D. Neither A nor B

__A__ 8. Tech A says that dust masks should be used when painting. Tech B says that a respirator should be used when the TLV for a chemical is exceeded. Who is correct?
  A. Tech A
  B. Tech B
  C. Both A and B
  D. Neither A nor B

D 9. Tech A says that tinted safety glasses can be worn when working outside. Tech B says that welding can cause a sunburn. Who is correct?
   A. Tech A
   B. Tech B
   C. Both A and B
   D. Neither A nor B

A 10. Tech A says that you should put tools away when done using them. Tech B says that a bystander can perform first aid. Who is correct?
   A. Tech A
   B. Tech B
   C. Both A and B
   D. Neither A nor B

Vehicle, Customer, and Service Information

CHAPTER 5

Chapter Review

The following activities have been designed to help you refresh your knowledge of this chapter. Your instructor may require you to complete some or all of these activities as a regular part of your training program. You are encouraged to complete any activity that your instructor does not assign as a way to enhance your learning.

Matching

Match the following terms with the correct description or example.

- A. Belt routing label
- B. Lemon law buyback
- C. Refrigerant label
- D. Service campaign and recall
- E. Shop or service manual
- F. Technical Service Bulletin (TSB)
- G. Vehicle Emission Control Information (VECI) label
- H. Title history
- I. Vehicle Identification Number (VIN)
- J. Vehicle Safety Certification (VSC) label

_____ 1. A detailed account of a vehicle's past.

_____ 2. A label that lists the type and total capacity of refrigerant that is installed in the air conditioning system.

_____ 3. Information issued by manufacturers to alert technicians of unexpected problems or changes to repair procedures.

_____ 4. A label that lists a diagram of the serpentine belt routing for the engine accessories.

_____ 5. A consumer protection law used in some states to identify a new vehicle that has undergone several unsuccessful attempts to repair the same fault.

_____ 6. A unique serial number that is assigned to each vehicle produced.

_____ 7. A label certifying that the vehicle meets the Federal Motor Vehicle Safety, Bumper, and Theft Prevention Standards in effect at the time of manufacture.

_____ 8. A corrective measure conducted by manufacturers when a safety issue is discovered with a particular vehicle.

_____ 9. Manufacturer's or after-market information on the repair and service of vehicles.

_____ 10. A label used by technicians to identify engine and emission control information for the vehicle.

Multiple Choice

Read each item carefully, and then select the best response.

_____ 1. What type of information publication comes in two types—factory and aftermarket?
- A. Owner's manual
- B. Service manual
- C. Technical service bulletin
- D. Labor guide

_____ 2. Who pays for the costs associated with a mandatory recall?
- A. Consumer
- B. Repair shop
- C. Manufacturer
- D. Insurance company

_____ 3. If a customer wants to know how much it will cost to replace a leaking intake manifold gasket on a particular vehicle, a technician can look up this procedure in the _____ to help estimate the cost.
   A. owner's manual
   B. labor guide
   C. service manual
   D. service information program

_____ 4. When determining labor costs, every tenth of an hour equal _____ minutes.
   A. 4
   B. 5
   C. 6
   D. 7

_____ 5. Initial information on a(n) _____ includes customer and vehicle details, along with a brief description of the customer's complaint(s).
   A. repair order
   B. service bulletin
   C. insurance claim
   D. recall notice

_____ 6. Which of the following is *not* required to determine the total cost of service?
   A. Labor costs
   B. Tax amounts
   C. Cost of gas and consumables used to service the vehicle
   D. Odometer reading

_____ 7. The _____ can provide potential new owners of used vehicles an indication of how well the vehicle was maintained.
   A. service history
   B. owner's manual
   C. repair order
   D. service manual

_____ 8. The vehicle identification number is designed for what type of motor vehicle?
   A. Trucks
   B. Motorcycles
   C. Cars
   D. All of the above

_____ 9. The first digit of a North American vehicle identification number indicates?
   A. Manufacturer
   B. Country of origin
   C. Model
   D. Body type

_____ 10. Which of the three Cs stands for understanding the reason that there is a fault?
   A. Connect
   B. Check
   C. Cause
   D. Control

True/False

If you believe the statement to be more true than false, write the letter "T" in the space provided. If you believe the statement to be more false than true, write the letter "F".

_____ 1. Usually a factory service manual is specific to one year and make of vehicle.

_____ 2. The information found in shop manuals provides a systematic procedure and identifies special tools, safety precautions, and specifications relevant to the task.

_____ 3. To use a technical service bulletin, you need to have a basic understanding of how to start and use a computer.

_____ 4. A typical owner's manual contains step-by-step procedures and diagrams on how to identify if there is a fault and perform an effective repair.

_____ 5. Online versions of labor guides can be updated as new models of vehicles are released or updates are made by the manufacturer.
_____ 6. If you are replacing the brake pads on a vehicle with wheel locks, the customer should be charged for the extra time it takes to find the lock key and to remove and install the wheel locks.
_____ 7. A labor guide is a computer application that is used to identify part numbers for vehicle components.
_____ 8. Most manufacturers store all service history performed in their dealerships.
_____ 9. The VIN is a 12-character identification code composed of letters and digits.
_____ 10. The Vehicle Safety Certification label is used by technicians to identify engine and emission control information for the vehicle.

Fill in the Blank

Read each item carefully, and then complete the statement by filling in the missing word(s).

1. A(n) _____ _____ includes an overview of the controls and features of the vehicle; the proper operation, care, and maintenance of the vehicle; owner service procedures; and specifications or technical data.
2. _____ _____ programs are computer applications used to provide technical information for the repair and maintenance of vehicles.
3. _____ can be mandatory and enforced by law or voluntary in order to ensure the safe operation of the vehicle or damage to their business and product image.
4. When working on vehicles that have not been recalled, check to see if a _____ _____ _____ has been issued for that vehicle and type of repair.
5. _____ _____ are essentially electronic versions of a parts manual, available via a CD/DVD, a computer network, or the Internet.
6. A(n) _____ _____ is used by the technician to guide him or her to the problem, and by the customer service staff to create the invoice when the work is completed.
7. The _____ section contains information about the methods of payment, which can be cash, credit card, or account.
8. A(n) _____ _____ tells you that the vehicle has been wrecked and suffered irreparable damage.
9. The _____ _____ _____ label certifies that the vehicle meets the Federal Motor Vehicle Safety, Bumper, and Theft Prevention Standards in effect at the time of manufacture.
10. The 3 Cs are an easy way to learn the fundamental steps in conducting repairs. They stand for _____, _____, and _____.

Labeling

Label the following images with the correct terms.

1. VIN standards:

    A.

1	2	3	4	5	6	7	8	9	10	11	12	13	14	15	16	17
Manufacturer Identifier			Vehicle Attributes						Check Character	Model Year	Plant Code	Sequential Number					

    B.

1	2	3	4	5	6	7	8	9	10	11	12	13	14	15	16	17
World Manufacturer Identifier (WMI)			Vehicle Descriptor System (VDS)						Vehicle Identifier System (VIS)							

    A. _____

    B. _____

2. Vehicle information labels:

A. _____

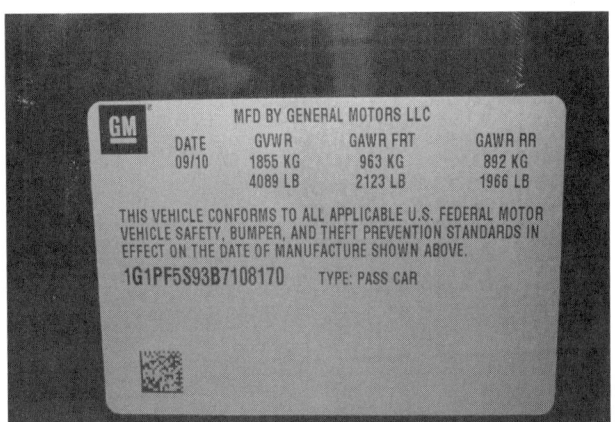

B. _____

C. _____

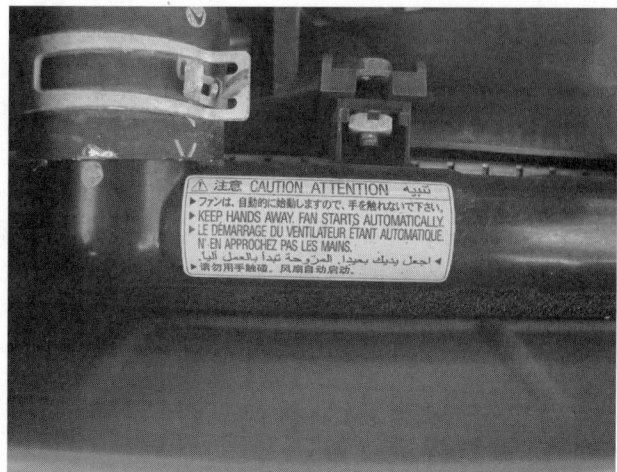

D. _____

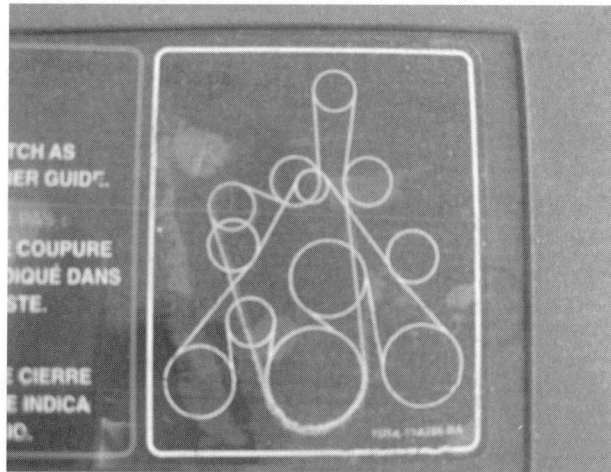

E. _____

Skill Drills

Read each item carefully, and then place each procedure in the correct sequence.

1. Identifying Correct Service Procedures Using a Shop Manual:
   _____ **A.** Locate the correct section that will contain the information you need. The first page of the shop manual is usually a table of contents.
   _____ **B.** Decide what information you need to know about the job and about the vehicle. Make sure you know the make, model, and year of manufacture of the vehicle, and the type and size of the engine. You should also have the vehicle identification number handy.
   _____ **C.** Locate the vehicle specifications by consulting the specifications page in the proper section.
   _____ **D.** Find the appropriate shop manual for the make, model, and year of the vehicle you are working on.
   _____ **E.** Locate the correct section that will contain the information you need.

2. Using a Service Information Program:
   _____ **A.** The search engine will provide a list of possible matches for you to select from. If the initial search does not produce what you are looking for, try changing the search criteria. Keep searching until you find the information.
   _____ **B.** Enter the vehicle identification information into the system in the appropriate places: year, make, model, engine, and possibly VIN.

_____ **C.** Log in to the application using the appropriate username and password.
_____ **D.** Finally, once the general details for the item are displayed, gather the specific information on the specifications or repairs. You may need more than one piece of information.
_____ **E.** Search for the information you require to perform the service or repair.
_____ **F.** If necessary, start the computer and select the service information program.
_____ **G.** Print out or write down the information needed. Put this on a clipboard and take it with you to perform the service or repair.

3. Using a Labor Guide:
_____ **A.** Check for any "combination" time that would need to be added to the base job when a related job is also being completed.
_____ **B.** Find the labor operation either by working your way through the menu tree or by typing a keyword into the search bar.
_____ **C.** Calculate the total time and multiply it by the shop's hourly labor rate. You now have the correct figure to estimate the charge for the particular service.
_____ **D.** Once you locate the labor operation, there are usually two columns that list the time. The first one is "warranty time." The second time listed is the "customer pay" time.
_____ **E.** Decide what specific labor operations you need to locate. Make sure you know the year, make, model, engine, and any other pertinent details of the vehicle.
_____ **F.** Enter the vehicle information into the system.
_____ **G.** Check for any "additional" time. This is extra time needed to deal with situations that occur on a relatively common basis, such as vehicle-installed options that are not common to all vehicles, like wheel locks.
_____ **H.** Log in to the labor estimating system.

4. Locating Parts Information on the Computer:
_____ **A.** Search for the parts you require to conduct the service or repair.
_____ **B.** Gather information on the identified parts, including part numbers, location, availability, and cost.
_____ **C.** Enter the year, make, model, and engine and VIN number information into the system in the appropriate places.
_____ **D.** Log in to the application using the appropriate username and password.
_____ **E.** The search engine will provide a list of possible matches for you to select from. If the initial search does not produce what you are looking for, try changing the search criteria. Keep searching until you find the information.

5. Applying the 3 Cs:
_____ **A.** Retest the vehicle to be sure the fault has been corrected.
_____ **B.** Review the information you collect from the tests. To review effectively, you need to understand how the systems work and interact.
_____ **C.** Identify and document the correction required, including work activities and parts required or used. Make repairs or replace parts to complete the repair.
_____ **D.** Identify and document the root cause of the concern. Research shop manuals and conduct tests to identify the cause of the problem. This may require a number of tests across multiple systems.
_____ **E.** Using the 3 Cs, document the repair process required for a repair order.
_____ **F.** Fill in the required repair order with details of the work conducted.
_____ **G.** Identify and document the concern. This should be on the repair order. Obtain as much information as possible, as this will help you to understand the problem. Identify what the problem is and which vehicle systems are involved.
_____ **H.** Always work safely and use the proper tools and correct personal protective equipment (PPE).

Crossword Puzzle

Use the clues to complete the puzzle.

Across

2. An information guide supplied by the manufacturer; it contains basic vehicle operating information.
4. A form used by shops to collect information regarding a vehicle coming in for repair, also referred to as a work order.
5. A label that lists the type of coolant installed in the cooling system.
7. Also called a branded title; a record that a vehicle has been severely damaged or deemed a total loss by an insurance company.
8. The person who serves customers at the parts counters.

Down

1. A computer software program for identifying and ordering replacement vehicle parts.
3. A complete list of all the servicing and repairs that have been performed on a vehicle.
6. A guide that provides information to make estimates for repairs.

ASE-Type Questions

Read each item carefully, and then select the best response.

_____ 1. Tech A says that the owner's manual will have the oil pan drain plug torque information. Tech B says that oil pan capacity information for that specific vehicle will be in the owner's manual. Who is correct?
   A. Tech A
   B. Tech B
   C. Both A and B
   D. Neither A nor B

_____ 2. Tech A says that vehicle security PIN codes can be found in the service manual. Tech B says that vehicle security PIN codes can be found in the owner's manual. Who is correct?
   A. Tech A
   B. Tech B
   C. Both A and B
   D. Neither A nor B

_____ 3. Tech A says that paper service manuals are gone and electronic versions of the service manual are now available. Tech B says that online manuals are more current, as they can be updated periodically. Who is correct?
   A. Tech A
   B. Tech B
   C. Both A and B
   D. Neither A nor B

_____ 4. Tech A says that service information programs are extremely helpful, as the technician can use a laptop at the repair for quick access to the procedure to perform a repair. Tech B says that service information programs allow a technician to know the labor guide and, with the support of the program, perform the task in the time allowed. Who is correct?
   A. Tech A
   B. Tech B
   C. Both A and B
   D. Neither A nor B

_____ 5. Tech A says that TSBs are updates to the owner's manual. Tech B says that TSBs are generally updated information on model changes that do not affect the technician. Who is correct?
   A. Tech A
   B. Tech B
   C. Both A and B
   D. Neither A nor B

_____ 6. Tech A says that labor guides are necessary for the service writer to quote prices for a customer on the repair bill. Tech B says that labor guides are what the customer pays and that the warranty pays more for labor using a different labor guide. Who is correct?
   A. Tech A
   B. Tech B
   C. Both A and B
   D. Neither A nor B

_____ 7. Tech A says that locating the part number needed by the technician requires computer knowledge and mechanical knowledge. Tech B says that anybody can be a parts person. Who is correct?
   A. Tech A
   B. Tech B
   C. Both A and B
   D. Neither A nor B

_____ 8. Tech A says that the repair order is just a piece of paper telling the technician what to do. Tech B says that the repair order is a legal and binding contract between the customer and the repair facility. Who is correct?
   A. Tech A
   B. Tech B
   C. Both A and B
   D. Neither A nor B

_____ 9. Tech A says that customer authorization for a change in the repair order after initial authorization is the responsibility of the service writer or foreman. Tech B says that changes to the repair order can be dealt with after the repair is complete, as the customer will be satisfied if more repairs are completed. Who is correct?
   A. Tech A
   B. Tech B
   C. Both A and B
   D. Neither A nor B

_____ 10. Tech A says that the VIN number on a vehicle can help identify which engine is installed in the chassis. Tech B says that most digits in the VIN number can be an identifier for information pertinent to that vehicle. Who is correct?
   A. Tech A
   B. Tech B
   C. Both A and B
   D. Neither A nor B

CHAPTER 6
Tools and Equipment

Tire Tread:
© AbleStock

Chapter Review

The following activities have been designed to help you refresh your knowledge of this chapter. Your instructor may require you to complete some or all of these activities as a regular part of your training program. You are encouraged to complete any activity that your instructor does not assign as a way to enhance your learning.

Matching

Match the following terms with the correct description or example.

- A. Bottoming tap
- B. Chassis dynamometer
- C. Cross-arm
- D. Dead blow hammer
- E. Dial bore gauge
- F. Die stock
- G. Flashback arrestor
- H. Intermediate tap
- I. Micrometer
- J. Parallax error
- K. Peening
- L. Pullers
- M. Solder
- N. Split ball gauge
- O. Telescoping gauge
- P. Tensile strength
- Q. Vernier calipers
- R. Wad punch

_____ 1. A spring-loaded valve installed on oxyacetylene torches as a safety device to prevent flame from entering the torch hoses.

_____ 2. A mixture of lead and tin with a low melting point for connecting wires.

_____ 3. An accurate measuring device for inside bores, usually made with a dial indicator attached to it.

_____ 4. An accurate measuring device for internal, external, and depth measurements that incorporates fixed and adjustable jaws.

_____ 5. A thread-cutting tap designed to cut threads to the bottom of a blind hole.

_____ 6. A gauge that expands and locks to the internal diameter of bores; a caliper or outside micrometer is used to measure its size.

_____ 7. A type of hammer that has a cushioned head to reduce the amount of head bounce.

_____ 8. An accurate measuring device for internal and external dimensions.

_____ 9. In reference to fasteners, the amount of force it takes before a fastener breaks.

_____ 10. A description for an arm that is set at right angles or 90 degrees to another component.

_____ 11. A type of punch that is hollow for cutting circular shapes in soft materials such as gaskets.

_____ 12. One of a series of taps designed to cut an internal thread. Also called a plug tap.

_____ 13. A term used to describe the action of flattening a rivet through a hammering action.

_____ 14. A machine with rollers that allows a vehicle to attain road speed and load while sitting still in the shop.

_____ 15. A generic term to describe hand tools that mechanically assist the removal of bearings, gears, pulleys, and other parts.

_____ 16. A handle for securely holding dies to cut threads.

_____ 17. A measuring device used to accurately measure small holes.

_____ 18. A visual error caused by viewing measurement markers at an incorrect angle.

Multiple Choice

Read each item carefully, and then select the best response.

_____ 1. What device is fitted to compressed air systems to remove the moisture or water from the compressed air that is condensed as a result of compressing air from the atmosphere?
   A. Spit filter
   B. Air drier
   C. Relief valve
   D. Strainer

_____ 2. Which of the following is a type of fastener?
   A. Bolt
   B. Nut
   C. Stud
   D. All of the above

_____ 3. In the metric system, which of the following is measured by the distance between the peaks of threads in millimeters?
   A. Thread pitch
   B. Tensile strength
   C. Torque
   D. Thread count

_____ 4. What type of wrench is also known as a tension wrench?
   A. Open-end wrench
   B. Ratcheting box-end wrench
   C. Torque wrench
   D. Pipe wrench

_____ 5. As long as a bolt is not tightened too much it will return to its original length when loosened; this is called?
   A. Torque
   B. Elasticity
   C. Yield
   D. Play

_____ 6. A(n) _____ is the fastest way to spin a fastener on or off a thread by hand, but it cannot apply much torque to the fastener.
   A. breaker bar
   B. sliding T-handle
   C. speed brace
   D. lug wrench

_____ 7. What type of pliers are used for cutting wire and cotter pins?
   A. Diagonal
   B. Snap ring
   C. Flat-nosed
   D. Needle-nosed

_____ 8. A screw or bolt with a cross-shaped recess requires a(n)_____?
   A. flat blade screwdriver
   B. offset screwdriver
   C. Phillips head screwdriver
   D. Allen wrench

_____ 9. What type of screwdriver fits into spaces where a straight screwdriver cannot and is useful where there is not much room to turn it?
   A. Ratcheting screwdriver
   B. Impact driver
   C. Phillips head screwdriver
   D. Offset screwdriver

Fundamentals of Automotive Technology Student Workbook

_____ 10. What kind of tools are composed of a strong metal and used as a lever to move, adjust, or pry?
  A. Cold chisels
  B. Pry bars
  C. Drift punches
  D. Speed brace

_____ 11. What type of file is thinner than other files, comes to a point, and is used for working in narrow slots?
  A. Warding file
  B. Triangular file
  C. Thread file
  D. Square file

_____ 12. What type of file cleans clogged or distorted threads on bolts and studs?
  A. Triangular file
  B. Warding file
  C. Thread file
  D. Square file

_____ 13. The name for this type of clamp comes from its shape, it can hold parts together while they are being assembled, drilled, or welded.
  A. J-clamp
  B. C-clamp
  C. D-clamp
  D. K-clamp

_____ 14. What type of tap narrows at the tip to give it a good start in the hole where the thread is to be cut?
  A. Intermediate tap
  B. Taper tap
  C. Plug tap
  D. Bottoming tap

_____ 15. What type of tool consists of three main parts: jaws, a cross-arm, and a forcing screw?
  A. Tap and die set
  B. Bench vice
  C. Flaring tool
  D. Gear puller

_____ 16. What tool is used to measure the gap between a straight edge and the surface being checked for flatness?
  A. Steel rule
  B. Caliper
  C. Feeler gauge
  D. Split ball gauge

_____ 17. What type of grinder uses discs rather than wheels?
  A. Bench grinder
  B. Angle grinder
  C. Straight grinder
  D. Pedestal grinder

_____ 18. What tool uses high-pressure to blast small abrasive particles to clean the surface of parts?
  A. Power washer
  B. Angle grinder
  C. Sand blaster
  D. Solvent tank

_____ 19. What type of torch is occasionally used by technicians to heat, braze, weld, and cut metal?
  A. Ultrasonic
  B. Magnesium
  C. Plasma
  D. Oxyacetylene

_____ 20. What type of welder has a filler rod that automatically feeds into the welding joint via the hand piece?
  A. Plasma welder
  B. Wire feed welders
  C. Arc welder
  D. Oxyacetylene welder

Chapter 6 Tools and Equipment

True/False

If you believe the statement to be more true than false, write the letter "T" in the space provided. If you believe the statement to be more false than true, write the letter "F".

_____ 1. Serious, sometimes fatal, injuries can be caused by compressed air being injected into the body through the skin or into a body opening, such as your mouth or ear.

_____ 2. Thread pitch is a way of defining how much a fastener should be tightened.

_____ 3. The higher the grade number of a fastener, the higher the tensile strength.

_____ 4. Torque wrenches come in various types: beam style, clicker, dial, and electronic.

_____ 5. The open-end wrench fits fully around the head of the bolt or nut and grips each of the six points at the corners just like a socket.

_____ 6. Six- and 12-point sockets fit the heads of hexagonal shaped fasteners.

_____ 7. Arc joint pliers, also called vice grips, are general-purpose pliers used to clamp and hold one or more objects.

_____ 8. Allen wrenches are sometimes called hex keys.

_____ 9. When a large chisel needs a really strong blow, it is time to use a dead blow hammer.

_____ 10. When marks need to be drawn on an object like a steel plate to help locate a hole to be drilled, a drift punch can be used to mark the points so they will not rub off.

_____ 11. A cold chisel gets its name from the fact it is used to cut cold metals, rather than heated metals.

_____ 12. A bottoming tap is used to tap a thread into a hole that does not come out the other side of the material.

_____ 13. Always use a single flare if the tubing is to be used for higher pressures such as in a brake system.

_____ 14. A tubing cutter is more convenient and neater than a saw when cutting pipes and metal tubing.

_____ 15. When soldering, always apply flux to the joint if cored solder is used.

_____ 16. Depth micrometers are used to measure inside dimensions.

_____ 17. Morse taper is a system for securing drill bits to drills.

_____ 18. A solvent tank is a cleaning tank that is filled with a suitable solvent to clean parts by removing oil, grease, dirt, and grime.

_____ 19. Thread repair is used in situations where it is not possible to replace a damaged component.

_____ 20. Batteries are filled with sulphuric acid, so if the hydrogen explodes, the battery case can then rupture and spray everything and everyone nearby with this dangerous and corrosive liquid.

Fill in the Blank

Read each item carefully, and then complete the statement by filling in the missing word(s).

1. _____ / _____ is an umbrella term that describes a set of safety practices and procedures that are intended to reduce the risk of technicians inadvertently using tools, equipment, or materials that have been determined to be unsafe.

2. A(n) _____ _____ is designed to regularly oil an air tool or air equipment so it does not have to be done manually, before or during its use.

3. A(n) _____ is a cylindrical piece of metal with a hexagonal head on one end and a thread cut into the shaft at the other end.

4. Each bolt diameter in the standard system can have one of two thread pitches, _____ or _____.

5. _____-to-_____ means that a fastener is torqued to, or just beyond, its yield point.

6. A(n) _____ wrench has an open-end head on one end and a box-end head on the other end.

7. _____ is a hand tool designed to hold, cut, or compress materials.

8. End cutting pliers, also called _____, have a cutting edge at right angles to their length.

9. A(n) _____ _____ is used when a screw or a bolt is rusted/corroded in place or overtightened and needs a tool that can apply more force.

10. A(n) _____ _____ hammer is designed not to bounce back when it hits something.

11. _____ are used when the head of the hammer is too large to strike the object being hit without causing damage to adjacent parts.

12. A(n) _____ _____ has a hardened, sharpened blade and is designed to remove a gasket without damaging the sealing face of the component.
13. A screw _____ is a device designed to remove screws, studs, or bolts that have broken off in threaded holes.
14. To cut a brand new thread on a blank rod or shaft, a die held in a _____ _____ is used.
15. A measuring _____ is a flexible type of ruler and a common measuring tool.
16. A(n) _____ _____ can measure how round something is.
17. A(n) _____ _____ is used to measure the width of gaps, such as the clearance between valves and rocker arms.
18. The most common air tool in an automotive shop is the air _____ _____.
19. A(n) _____ _____ works by applying a pressurized gas such as air, oxygen, nitrogen, or argon through a nozzle located in the center of a hand piece with an electrode.
20. A(n) _____ _____ incorporates microprocessors to monitor and control the charge rate so a battery receives the correct amount of charge depending on its state of charge.

Labeling

Label the following diagrams with the correct terms.

1. Anatomy of a socket:

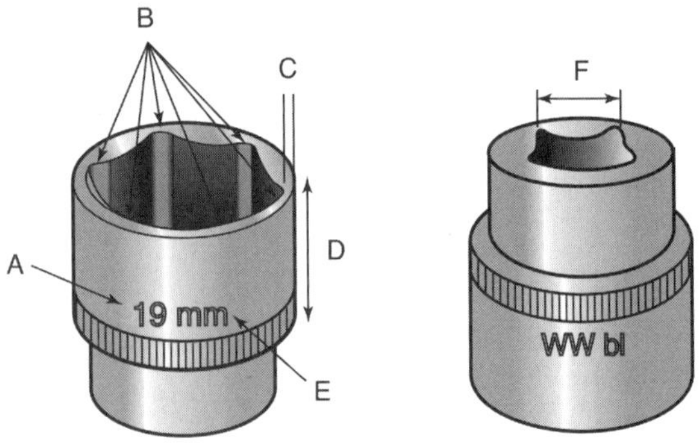

   A. _____
   B. _____
   C. _____
   D. _____
   E. _____
   F. _____

2. Components of a flare tool:

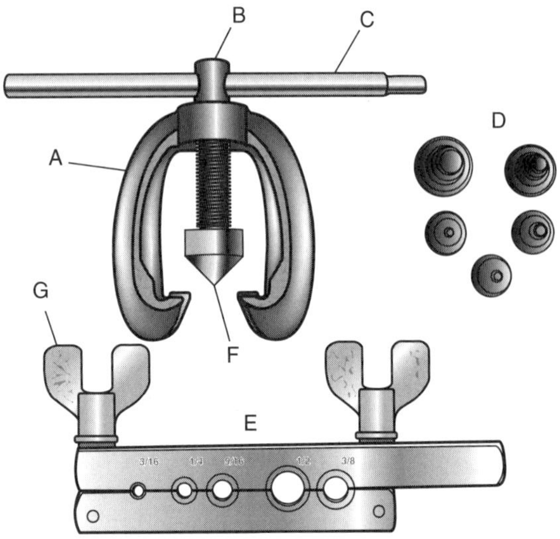

   A. _____
   B. _____
   C. _____
   D. _____
   E. _____
   F. _____
   G. _____

Chapter 6 Tools and Equipment

3. Anatomy of a rivet:

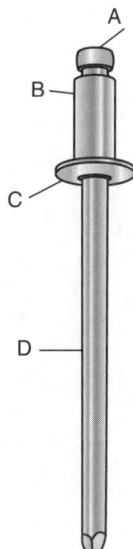

A. _____

B. _____

C. _____

D. _____

4. Air tools:

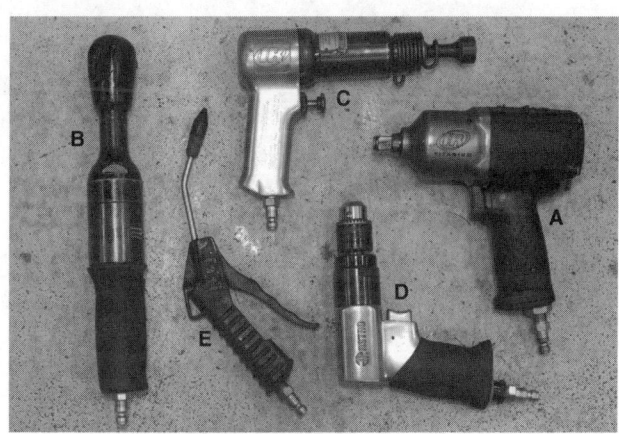

A. _____

B. _____

C. _____

D. _____

E. _____

5. Oxyacetylene tips:

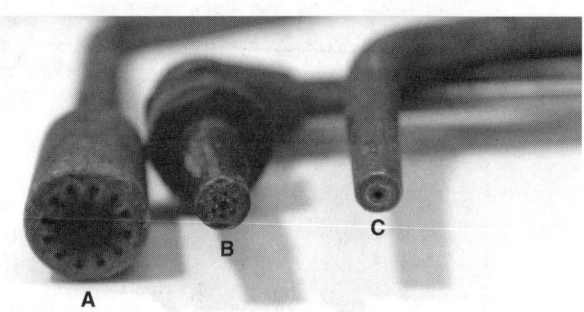

A. _____

B. _____

C. _____

Skill Drills

Place the skill drill steps in the correct order.

1. Safe Handling and Use of Tools:

_____ **A.** Return tools to correct storage locations.

_____ **B.** Clean tools prior to use if necessary.

_____ **C.** Select the correct tool(s) to undertake tasks. Inspect tools prior to use to ensure they are in good working order. If tools are faulty, remove them from service according to shop procedures.

_____ **D.** Use tools to complete the task while ensuring manufacturer and shop procedures are followed. Always use tools safely to prevent injury and damage. Report, tag, and remove damaged tools.

2. Using a Torque Wrench:

_____ **A.** Check the specifications for the bolt or fastener you are using. Tighten the bolt to the specified torque. If the component requires multiple bolts or fasteners, tighten them all to the same torque value in the sequence and steps that are specified by the manufacturer.

_____ **B.** Turn the torque wrench the specified number of degrees as indicated on the angle gauge. If the component requires multiple bolts or fasteners, make sure to tighten them all to the same torque angle in the sequence that is specified by the manufacturer. Some torqueing procedures could call for four or more steps to complete the torqueing process properly.

_____ **C.** Install the torque angle gauge over the head of the bolt, and then put the torque wrench on top of the gauge and zero it, if necessary.

3. Using Gear Pullers:

_____ **A.** Position the forcing screw. Use the appropriate wrench to run the forcing screw down to touch the shaft. Check that the point of the forcing screw is centered on the shaft. If not, adjust the jaws and cross-arms until the point is in the center of the shaft. If it is not, adjust the jaws and cross-arms until the point is in the center of the shaft.

_____ **B.** Examine the gear puller and ensure the jaws will fit the part you want to remove. Select the right size wrench to fit the nut on the end of the forcing screw. Adjust and fit the puller so that it fits tightly around the part to be removed. The arms of the jaws should be pulling against the component at close to right angles.

_____ **C.** Tighten the forcing screw slowly and carefully onto the shaft. Check that the puller is not going to slip off center or off the pulley. If the forcing screw and puller jaws remain in the correct position, tighten the forcing screw and pull the part off the shaft.

4. Using Soldering Tools:

_____ **A.** Clean any excess flux from the joint.

_____ **B.** Apply flux to the wires or metal to be soldered. This may not be necessary if you are using cored solder.

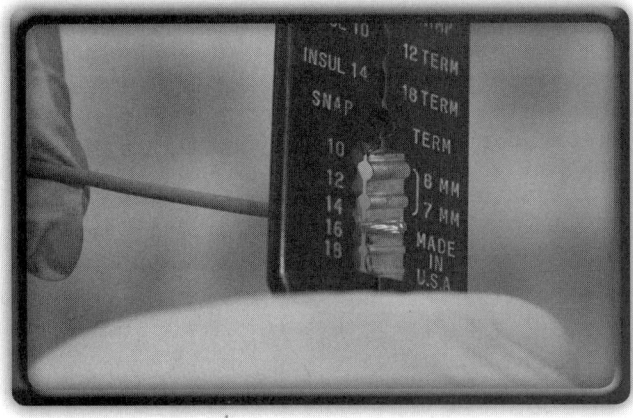

_____ **C.** Prepare the materials to be soldered. Strip wires or clean metal parts before soldering.

_____ **D.** Apply the hot solder iron tip to heat the joint, then apply solder to the joint (not the iron). If the solder does not melt within a few seconds, remove it and allow the joint to heat further before reapplying.

_____ **E.** Prepare the soldering iron by ensuring the correctly sized tip is fitted and is clean. Tin the soldering iron tip by melting some solder to it and wiping any excess from the tip.

_____ **F.** Once the solder has been applied, ensure the joint does not move until the solder has cooled sufficiently to set. Once cooled, inspect the joint; it should be shiny and firm.

5. Using Dial Indicators:

_____ **A.** Continue the rotation and make sure the needle does not go below zero. If it does, rezero the indicator and remeasure the point of maximum variation. Check your readings against the manufacturer's specifications. If the deviation is greater than the specifications allow, consult your supervisor.

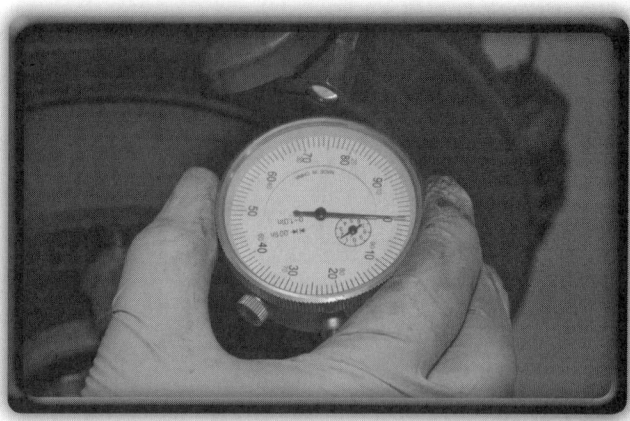

_____ **B.** Rotate the part one complete turn and locate the low spot. Zero the indicator.

_____ **C.** Find the point of maximum height and note the reading. This will indicate the runout value.

_____ **D.** Select the gauge type, size, attachment, and bracket that fit the part you are measuring. Mount the dial indicator firmly to keep it stationary.

_____ **E.** Adjust the indicator so that the plunger is at 90 degrees to the part you are measuring and lock it in place.

Crossword Puzzle

Use the clues to complete the puzzle.

Across

4. The amount of stretch or give a material has.
7. A seal that is made at the end of metal tubing or pipe.
9. A fastener with a hexagonal head and internal threads for screwing on bolts.
10. The shaft of a pop rivet.
11. A type of threaded fastener with a thread on one end and a hexagonal head on the other.
13. A type of punch in various sizes with a straight or parallel shaft.
14. A tool designed to remove wheel lug nuts and commonly shaped like a cross.

Down

1. A device that converts and stores electrical energy through chemical reactions.
2. An enclosed metal tube commonly with 6 or 12 points to remove and install bolts and nuts.
3. A compressed air-powered drill.
5. A type of hexagonal drive mechanism for fasteners.
6. Less sharp than a prick punch, this makes a bigger indentation that centers a drill bit at the point where a hole is required to be drilled.
8. A device fitted to compressed air lines to remove moisture.
11. A device that securely holds material in jaws while it is being worked on.
12. A liquid or paste that protects a soldering or welding joint from oxidization.

ASE-Type Questions

Read each item carefully, and then select the best response.

_____ 1. Tech A says that knowing how to use tools correctly creates a safe working environment. Tech B says that a flare nut wrench is used to loosen very tight bolts and nuts. Who is correct?
   A. Tech A
   B. Tech B
   C. Both A and B
   D. Neither A nor B

_____ 2. Tech A says that lockout/tagout is a safety procedure for a safe working environment. Tech B says that a soldering iron is used to weld, cut, and braze steel components. Who is correct?
   A. Tech A
   B. Tech B
   C. Both A and B
   D. Neither A nor B

_____ 3. Tech A says that torque wrenches need to be calibrated periodically to ensure proper torque values. Tech B says that when a bolt is torqued the bolt stretches beyond its yield point and the bolt has to be replaced when removed. Who is correct?
   A. Tech A
   B. Tech B
   C. Both A and B
   D. Neither A nor B

_____ 4. Tech A says that head bolts are tightened past their yield point. Tech B says that head bolts are torqued then tightened to their yield point. Who is correct?
   A. Tech A
   B. Tech B
   C. Both A and B
   D. Neither A nor B

_____ 5. Tech A says that a box-end wrench is more likely to round the head of a bolt than an open-end wrench. Tech B says that 6-point sockets and wrenches have more surface area on the bolt and will hold more firmly when removing and tightening. Who is correct?
   A. Tech A
   B. Tech B
   C. Both A and B
   D. Neither A nor B

_____ 6. Tech A says that it is usually better to pull a wrench to tighten or loosen a bolt. Tech B says that pushing a wrench will protect your knuckles if the wrench slips. Who is correct?
   A. Tech A
   B. Tech B
   C. Both A and B
   D. Neither A nor B

_____ 7. Tech A says that when jump starting a vehicle, a spark typically occurs when making the last jumper cable connection. Tech B says that all four of the jumper cable connections should be made at the battery terminals. Who is correct?
   A. Tech A
   B. Tech B
   C. Both A and B
   D. Neither A nor B

_____ 8. Tech A says that a dead blow hammer reduces rebound of the hammer. Tech B says that a dead blow hammer should be used to cut the head of a bolt off with a chisel. Who is correct?
   A. Tech A
   B. Tech B
   C. Both A and B
   D. Neither A nor B

_____ 9. Tech A says that gaskets can be removed quickly and safely with a portable grinder as long as the grinding wheel isn't too coarse. Tech B says that extreme care must be used when removing a gasket on an aluminum surface. Who is correct?
   A. Tech A
   B. Tech B
   C. Both A and B
   D. Neither A nor B

_____ 10. Tech A says that when using a file, apply pressure to the file in the direction of the cut and no pressure when pulling the file back. Tech B says that file cards are used to file uneven surfaces. Who is correct?
   A. Tech A
   B. Tech B
   C. Both A and B
   D. Neither A nor B

Vehicle Protection and Jack and Lift Safety

CHAPTER 7

Chapter Review

The following activities have been designed to help you refresh your knowledge of this chapter. Your instructor may require you to complete some or all of these activities as a regular part of your training program. You are encouraged to complete any activity that your instructor does not assign as a way to enhance your learning.

Matching

Match the following terms with the correct description or example.

- A. Four-post hoist
- B. Hydraulic jack
- C. Pneumatic jack
- D. Safe working load
- E. Single-post hoist
- F. Two-post hoist
- G. Vehicle inspection pit

_____ 1. A type of vehicle jack that uses compressed gas or air to lift a vehicle.

_____ 2. A type of vehicle jack that uses oil under pressure to lift vehicles.

_____ 3. A trench permanently fitted into the floor of the shop to allow easy work access to the vehicle's underside.

_____ 4. The maximum safe lifting load for lifting equipment.

_____ 5. A type of vehicle hoist that uses two parts (one on each side of vehicle) and four arms to lift the vehicle.

_____ 6. A type of vehicle hoist that uses a single central platform to lift a vehicle.

_____ 7. A type of hoist onto which a vehicle is driven, that uses two long, narrow platforms to lift the vehicle.

Multiple Choice

Read each item carefully, and then select the best response.

_____ 1. It is good practice for the service advisor to perform a(n) _____ with the customer to point out any existing damage or missing components on the vehicle.
  - A. detailed inspection
  - B. vehicle walk-around
  - C. audit
  - D. performance analysis

_____ 2. Battery electrolyte contains acid, which can be cleaned up by neutralizing it with which of the following materials?
  - A. An alkaline
  - B. Engine oil
  - C. Gasoline
  - D. Water

_____ 3. There are three main types of mechanisms that provide the lifting action for vehicle jacks including all of the following, *except*:
   A. Hydraulic
   B. Pneumatic
   C. Electric
   D. Mechanical

_____ 4. What type of jack is mounted on four wheels, two of which swivel to provide a steering mechanism?
   A. Bottle jack
   B. Sliding bridge jack
   C. Floor jack
   D. Scissor jack

_____ 5. Often used on farms, what type of jack is designed to lift, winch, clamp, pull, and push?
   A. High-lift jack
   B. Scissor jack
   C. Sliding bridge jack
   D. Bottle jack

_____ 6. What type of jack uses compressed air to either operate a large ram or inflate an expandable air bag to lift the vehicle?
   A. Bottle jack
   B. Air jack
   C. High-lift jack
   D. Floor jack

_____ 7. Which lifting device is useful for raising a vehicle to a height that removes the need for the technician to bend down?
   A. Engine hoist
   B. Jack stand
   C. Farm jack
   D. Vehicle hoist

_____ 8. What type of hoist comes in two configurations—symmetrical and asymmetrical?
   A. Four-post hoist
   B. Two-post hoists
   C. Farm hoist
   D. Single-post hoists

_____ 9. The engine or component to be lifted is attached to the lifting arm of an engine hoist by which of the following?
   A. Sling
   B. Rope
   C. Lifting chain
   D. Either A or C

_____ 10. The size of the vehicle jack you use will be determined by:
   A. The length of the axles
   B. The length of the vehicle
   C. The weight of the vehicle
   D. The type of vehicle

Chapter 7 Vehicle Protection and Jack and Lift Safety

True/False

If you believe the statement to be more true than false, write the letter "T" in the space provided. If you believe the statement to be more false than true, write the letter "F".

_____ 1. Customers expect their vehicles to be treated with care and respect while in your shop.
_____ 2. Fender, carpet, seat, and steering wheel covers should be the first thing on and the last thing removed when working on vehicles.
_____ 3. Never place tools in your back pocket.
_____ 4. Only the most skilled and experienced drivers available should be allowed to test-drive higher performance vehicles.
_____ 5. When multiple pieces of lifting equipment are used, the safe working load is determined by the highest rated piece of equipment.
_____ 6. Vehicle jacks may be used to support the weight of the vehicle during any task that requires you to get underneath any part of the vehicle.
_____ 7. High-lift jacks are usually fitted in pairs to four-post hoists as an accessory to allow the vehicle to be lifted off the drive-on hoist runways.
_____ 8. Tall jack stands are used to stabilize a vehicle up on a hoist that is having a heavy component, such as a transaxle, removed or installed.
_____ 9. Since the vehicle rests on its wheels on the four-post hoist, the wheels cannot be removed, unless the hoist is fitted with sliding bridge jacks.
_____ 10. Every vehicle hoist in the shop must have a built-in mechanical locking device so that the vehicle hoist can be secured at the chosen height after the vehicle is raised.

Fill in the Blank

Read each item carefully, and then complete the statement by filling in the missing word(s).

1. _____ _____ are a protective layer used to cover the fenders when work is conducted around the engine bay.
2. If spills do occur, be sure to clean them up thoroughly using appropriate methods, which can usually be found in the _____ _____ _____ sheets for each material.
3. Special _____ materials in granular form can be used on some liquid spills such as engine oil.
4. Often _____ _____ will be used by someone directing you to maneuver the vehicle.
5. When using lifting equipment always maintain some _____ _____ as an extra safety margin.
6. Lifting equipment should be periodically _____ and _____ to make sure it is safe.
7. A(n) _____ jack is a portable jack that usually has either a mechanical screw or a hydraulic ram mechanism that rises vertically from the center of the jack as the handle is operated.
8. _____ two-post hoists have arms that are of approximately equal length so that the vehicle is roughly centered lengthwise between the posts.
9. The lifting arm of the _____ _____ is moved by a hydraulic cylinder and is adjustable for length.
10. A(n) _____ _____ _____ allows the technician to access the underside of the vehicle without the need for a hoist or jacks to raise the vehicle.

Labeling

Label the following images with the correct terms.

1. Jacks:

A. _____   B. _____

C. _____

2. Hoists:

A. _____   B. _____

Skill Drills

Place the skill drill steps in the correct order.

1. Applying Fender Covers and Floor Mats:

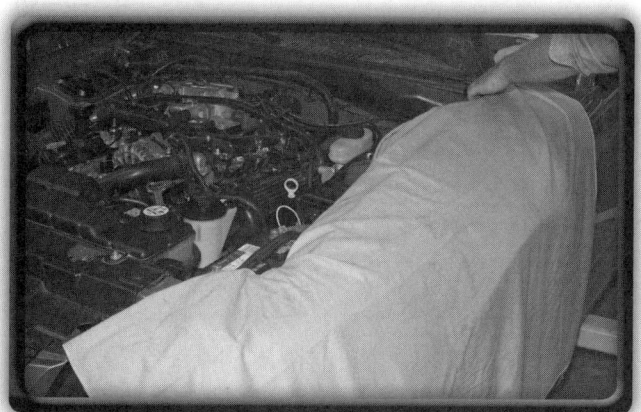

_____ **A.** Position the fender covers and floor mats so they provide adequate protection. Ensure that fender covers and floor mats stay in position, providing protection while the vehicle is in the shop.

_____ **B.** Remove fender covers and floor mats prior to customer pickup.

_____ **C.** Prepare the fender covers and floor mats by checking that they are clean and in good condition. Select appropriate fender covers and floor mats for the vehicle and type of repair. Inspect the fender cover backs for rocks, metals, or fluids that would damage the vehicle.

2. Lifting and Securing a Vehicle with a Vehicle Jack and Jack Stands:

_____ **A.** When the repairs are complete, use the jack to raise the vehicle off the jack stands. Slide the jack stands from under the vehicle. Make sure no one goes under the vehicle or puts any body parts under the vehicle since the jack could fail or slip.

_____ **B.** Position the vehicle on a flat, solid surface. Put the vehicle into neutral or park and set the parking brake. Place wheel chocks in front of and behind the wheels that are not going to be raised off the ground.

_____ **C.** Roll the vehicle jack under the vehicle, and position the lifting pad correctly under the frame or cross member. Turn the jack handle clockwise, and begin pumping the handle up and down until the lifting pad touches and begins to lift the vehicle.

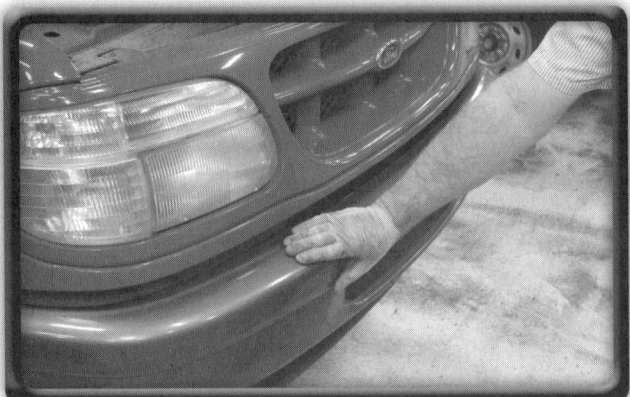

_____ **D.** Slide the two jack stands underneath the vehicle and position them to support the vehicle's weight. Slowly turn the jack handle counterclockwise to open the release valve and gently lower the vehicle onto the jack stands. When the vehicle has settled onto the jack stands, lower the vehicle jack completely and remove it from under the vehicle. Gently push the vehicle sideways to make sure it is secure. Repeat this process to lift the other end of the vehicle.

_____ **E.** Once the wheels lift off the floor, stop and check the placement of the lifting pad under the vehicle to make sure there is no danger of slipping. Double-check the position of the wheel chocks to make sure they have not moved. If the vehicle is stable, continue lifting it until it is at the height at which you can safely work under it.

_____ **F.** Slowly turn the jack handle counterclockwise to gently lower the vehicle to the ground. Return the jack, jack stands, and wheel chocks to their storage area before you continue working on the vehicle.

_____ **G.** Select two jack stands of the same type, suitable for the weight of the vehicle. Place one jack stand on each side of the vehicle at the same point, and adjust them so that they are both the same height.

_____ **H.** Before you try to use the jack, check for leaks in the hydraulic system. Check the pad, or saddle, and the wheels of the jack. They should rotate freely and show no signs of damage. **(There is no image associated with this step.)**

_____ **I.** Check the manufacturer's label on the vehicle jack. Refer to the owner's manual to find out where you can safely place the vehicle jack. **(There is no image associated with this step.)**

3. Lifting a Vehicle Using a Two-Post Hoist:

_____ **A.** Prepare to use the two-post hoist. Check the hoist and check the vehicle clearance. Carefully drive the vehicle so that it is centered between the two posts, left and right. Also ensure that it is positioned properly, front to back, for the type of hoist and vehicle you are using. Leave the vehicle in neutral and apply the emergency brake.

_____ **B.** Before the two-post hoist is lowered, remove all tools and equipment from the area and wipe up any spilled fluids. Raise the hoist to unlock the lift before lowering it. Make sure no one is near the vehicle before lowering it. Once the vehicle is on the ground, remove the lifting arms and drive it away.

_____ **C.** Make sure no one is near the vehicle and then raise the vehicle until the wheels are a couple of inches off the floor. Check the position of the lifting pads, and shake the vehicle gently to confirm that it is stable. Lift the vehicle to slightly above working height, and then lower it onto the locks or safety device.

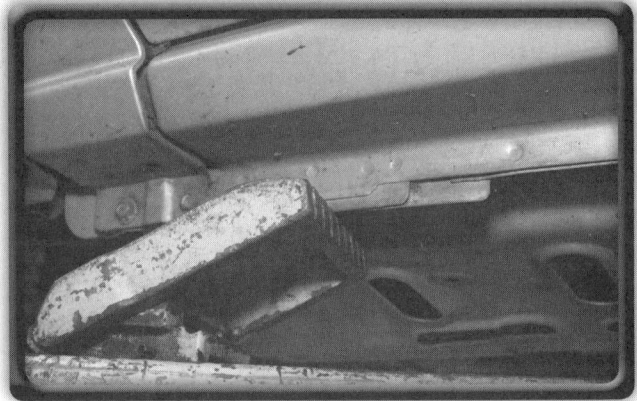

_____ **D.** Position the lifting pads under the vehicle lifting points. Make sure the lifting pads are adjusted to the same height for both sides of the vehicle. Move to the operating controls and raise the two-post hoist just far enough to come into contact with the vehicle.

4. Using Four-Post Hoists to Lift a Vehicle:

_____ A. Get out of the vehicle and check that it is correctly positioned on the platform. If it is, apply the emergency brake and select first gear for a manual transmission or park for an automatic.

_____ B. Make sure the four-post hoist area is clear. Move to the controls and lift the vehicle until it reaches the appropriate work height. If the four-post hoist has a manual safety mechanism, lock it in place to engage whatever safety device is used.

_____ C. Before the four-post hoist is lowered, remove all tools and equipment from the area and wipe up any spilled fluids. Remove the safety device or unlock the lift before lowering it. Make sure no one is near the area. Once the four-post hoist is fully lowered, with the help of a guide, you can carefully back the vehicle off the hoist.

_____ D. Prepare to safely use the vehicle hoist. With the aid of an assistant guiding the driver, or a large mirror in front of the hoist, drive the vehicle slowly and carefully onto the four-post hoist and position it centrally. If the vehicle has front wheel restraints, drive the vehicle forward until the wheels lock onto the brackets.

5. Using Vehicle Inspection Pits:

_____ A. Get out of the vehicle and check that it is correctly positioned on the platform. If it is, apply the emergency brake and select first gear on a manual transmission or park with an automatic transmission.

_____ B. Ensure that the vehicle inspection pit and surrounding area are clear of obstructions and that you have a clear pathway to drive the vehicle over the vehicle inspection pit.

_____ C. Turn the lights on in the vehicle inspection pit. Get someone to help guide you as you drive the vehicle over the vehicle inspection pit. Drive the vehicle slowly and carefully and position it centrally.

_____ D. Before driving the vehicle off the vehicle inspection pit, remove all tools and equipment from the vehicle inspection pit area and clean up any spilled fluids. Make sure no one is near the vehicle or in the vehicle inspection pit. Have someone guide you off the pit area. **(There is no image associated with this step.)**

Crossword Puzzle

Use the clues to complete the puzzle.

Across

2. A type of vehicle lifting tool designed to lift the entire vehicle.
4. A type of vehicle jack that uses mechanical leverage to lift a vehicle.
5. Metal stands with adjustable height to hold a vehicle once it has been jacked up.
6. A small crane used to lift engines.

Down

1. A tool for lifting a vehicle.
3. A certificate issued when lifting equipment has been checked and deemed safe.

ASE-Type Questions

Read each item carefully, and then select the best response.

_____ 1. Tech A says that you should always deal with the customer's valuables according to company policy. Tech B says that you should protect a customer's vehicle by washing it when you are finished with it. Who is correct?
   A. Tech A
   B. Tech B
   C. Both A and B
   D. Neither A nor B

_____ 2. Tech A says that it is a good practice to perform a walk-around inspection of the vehicle with the customer. Tech B says that fender covers will protect the fenders from dents when working on the engine. Who is correct?
   A. Tech A
   B. Tech B
   C. Both A and B
   D. Neither A nor B

_____ 3. Tech A says that a vehicle jack can be used to support the vehicle while working under it. Tech B says that a jack stand automatically adjusts to the vehicle's height. Who is correct?
   A. Tech A
   B. Tech B
   C. Both A and B
   D. Neither A nor B

_____ 4. Tech A says that hoists should be inspected and certified periodically. Tech B says that safety locks do not need to be applied before working under the vehicle, unless you will be working for more than 10 minutes. Who is correct?
   A. Tech A
   B. Tech B
   C. Both A and B
   D. Neither A nor B

_____ 5. Tech A says that an engine hoist can lift more weight when the legs and arms are extended. Tech B says that the bolts used to mount an engine to a stand should complete at least six turns. Who is correct?
   A. Tech A
   B. Tech B
   C. Both A and B
   D. Neither A nor B

_____ 6. Tech A says that gasoline vapors are lighter than air, so inspection pits do not have a fire hazard like above ground hoists. Tech B says that you should never lift a fully loaded vehicle off the ground. Who is correct?
   A. Tech A
   B. Tech B
   C. Both A and B
   D. Neither A nor B

_____ 7. Tech A says that you should always ensure that the vehicle has enough ground clearance before driving on a lift. Tech B says that you should center the vehicle on the lift before raising it. Who is correct?
   A. Tech A
   B. Tech B
   C. Both A and B
   D. Neither A nor B

_____ 8. Tech A says that you should always inspect a lifting device for leaks and operation before using it. Tech B says that all mechanical safety locks on a hoist should be in place before getting under the vehicle. Who is correct?
   A. Tech A
   B. Tech B
   C. Both A and B
   D. Neither A nor B

Chapter 7　Vehicle Protection and Jack and Lift Safety

_____ **9.** Tech A says that an engine sling should have an angle greater than 90 degrees. Tech B says that all slings and lifting chains should be inspected for damage prior to use. Who is correct?
   A. Tech A
   B. Tech B
   C. Both A and B
   D. Neither A nor B

_____ **10.** Tech A says that you should have a coworker help guide you onto an inspection pit. Tech B says that all lights should be on in the pit before driving over it. Who is correct?
   A. Tech A
   B. Tech B
   C. Both A and B
   D. Neither A nor B

CHAPTER 8
Vehicle Maintenance Inspection

Tire Tread:
© AbleStock

Chapter Review
The following activities have been designed to help you refresh your knowledge of this chapter. Your instructor may require you to complete some or all of these activities as a regular part of your training program. You are encouraged to complete any activity that your instructor does not assign as a way to enhance your learning.

Matching
Match the following terms with the correct description or example.

- **A.** Body Control Module (BCM)
- **B.** Coolant
- **C.** CV joint
- **D.** Diesel Exhaust Fluid (DEF)
- **E.** Hydrometer
- **F.** Hygroscopic
- **G.** Pinion shaft

_____ **1.** A property of a substance or liquid that causes it to absorb moisture (water), as a sponge absorbs water.

_____ **2.** A type of universal joint used on the drive axles or half-shafts of a vehicle. Usually refers to front-wheel drive vehicles.

_____ **3.** On a drive axle using a ring-and-pinion gear assembly, the input component that drives the ring gear.

_____ **4.** An onboard computer that controls many vehicle functions including the vehicle interior and exterior lighting, horn, door locks, power seats, and windows.

_____ **5.** The resulting mixture when anti-freeze concentrate is mixed with water.

_____ **6.** A tool that measures the specific gravity of a liquid.

_____ **7.** A mixture of urea and water that is injected into the exhaust system of a late-model diesel-powered vehicle to reduce exhaust nitrogen oxide emissions.

Multiple Choice
Read each item carefully, and then select the best response.

_____ **1.** If the oil in an engine's lubrication system is too high, the oil will be struck by the crankshaft and will:
- **A.** Burn
- **B.** Smoke
- **C.** Coagulate
- **D.** Foam

_____ **2.** Do not remove the _____ when the engine is warm or hot.
- **A.** oil dipstick
- **B.** radiator cap
- **C.** master cylinder cover
- **D.** transmission dipstick

Chapter 8 Vehicle Maintenance Inspection

_____ 3. The anti-freeze protection level can be checked with which of the following?
   A. Hydrometer
   B. Hydroscope
   C. Refractometer
   D. Either A or C

_____ 4. If an automatic transmission does not have a dipstick, the fluid level is checked using a _____ on the side of the transmission.
   A. master cylinder
   B. reservoir
   C. fill plug
   D. sight glass

_____ 5. Some late-model diesel-powered vehicles use a fluid called _____, which is injected into the exhaust stream to reduce oxides of Nitrogen during certain driving conditions.
   A. diesel exhaust fluid
   B. oxide reducer
   C. nitrous oxide
   D. octane booster

_____ 6. What type of drive belt has a flat profile with a number of grooves running lengthwise along the belt?
   A. V-type
   B. Serpentine type
   C. Variable diameter type
   D. Toothed

_____ 7. The _____ is the most important component driven by the drive belt, and the engine will quickly overheat if the belt breaks or comes off.
   A. alternator
   B. air conditioning compressor
   C. water pump
   D. power steering pump

_____ 8. What is the shininess on the surface of a belt where it comes in contact with the pulley called?
   A. Shine
   B. Gloss
   C. Soaking
   D. Glazing

_____ 9. What device on a multiport fuel-injected vehicle is typically located in a rectangular box within the air induction system?
   A. Fuel filter
   B. Air cleaner
   C. Serpentine belt
   D. Carburetor

_____ 10. What type of fluid is normally green, orange, or yellow in color?
   A. DOT 4
   B. Power steering fluid
   C. Coolant
   D. Automatic transmission fluid

True/False

If you believe the statement to be more true than false, write the letter "T" in the space provided. If you believe the statement to be more false than true, write the letter "F".

_____ 1. Component damage or failure is often caused by a lack of service or low fluid level in the related system.
_____ 2. Always check engine oil with the engine on.
_____ 3. If the brake fluid level gets too low, air can be pulled into the hydraulic system.
_____ 4. Most brake fluid is dark or black.
_____ 5. Some automatic transmissions do not have a dipstick.
_____ 6. Transmission fluid is added through the transmission reservoir.
_____ 7. Many newer vehicles have transmissions that are considered sealed and lubricated for the life of the vehicle.
_____ 8. There are two types of drive belts: the V-type and the serpentine type.
_____ 9. The engine should be hot when inspecting radiator hoses.
_____ 10. The warning lamps perform a self-check each time the ignition is switched on or the engine is cranked.

Fill in the Blank

Read each item carefully, and then complete the statement by filling in the missing word(s).

1. Nearly all of the _____ in a vehicle, except some used in automatic transmissions/transaxles get old and wear out, requiring replacement.
2. The level of _____ in an engine's lubrication system is critical to the engine's operation.
3. Engine _____ must be controlled to prevent overheating and to maintain proper exhaust emission levels.
4. A hydraulic braking system depends on a special fluid called _____ _____.
5. When checking power steering fluid levels the engine should be _____ and the fluid hot.
6. Driving in dusty or wet conditions may require that the _____ _____ _____ be checked more often.
7. Engine _____ _____ are used to operate the various accessories on the engine, such as the water pump, power steering pump, air conditioner compressor, and alternator.
8. _____ that exceed a certain number per inch in a belt indicate that the belt may soon fail and should be replaced.
9. The _____ _____ should flex at the hinge and be held firmly against the windshield by the wiper arm spring.
10. The _____ _____ acts as a lever to increase the force applied to the brake assemblies by the driver.

Labeling

Label the following diagrams with the correct terms.

Power steering fluid reservoir:

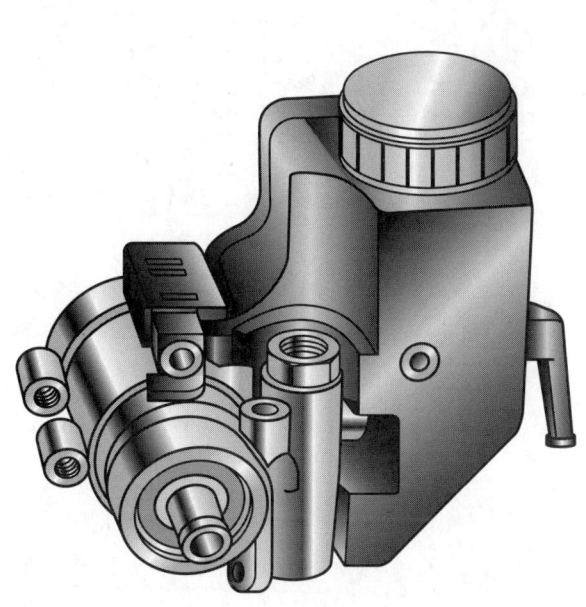
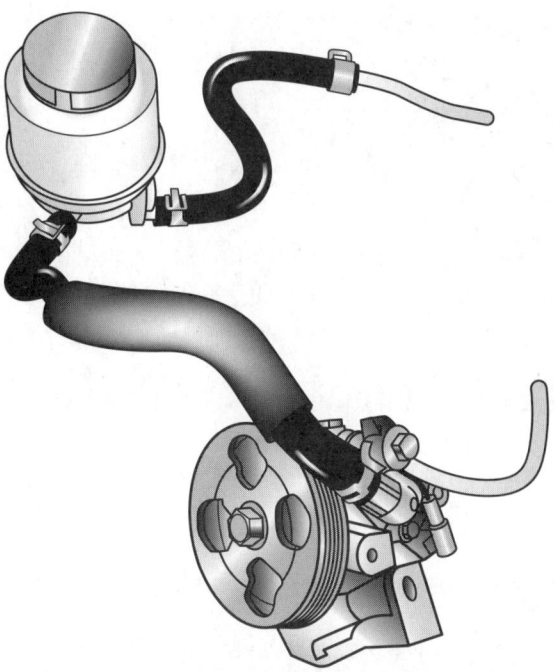

A. _____   B. _____

Skill Drills

Place the skill drill steps in the correct order.

1. Changing the Air Filter:

_____ **A.** On carbureted or throttle body injected engines, remove the top of the air filter by unscrewing the wing nut, and remove the air filter.

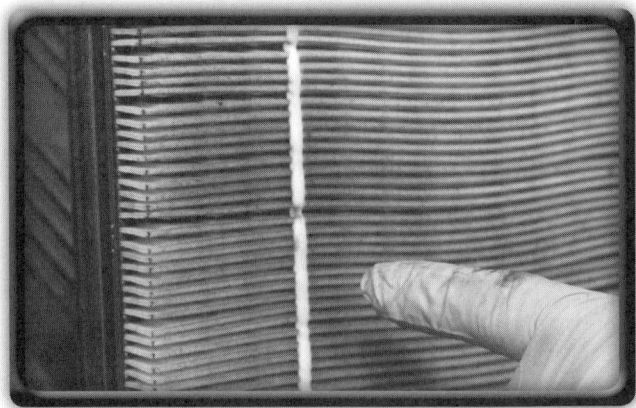

_____ **B.** Inspect the air cleaner element by holding the filter element up to light and looking through it. If it is bright with no tears or cracks, it can be reused. If it is dark or damaged in any way, it will need to be replaced.

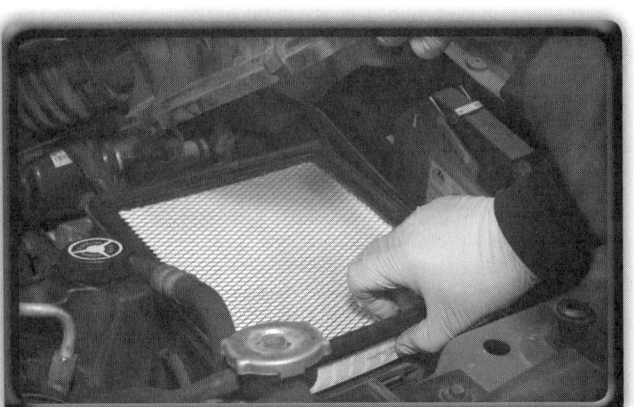

_____ **C.** Place the new air filter inside the filter housing. Make sure it is aligned properly on both sides. Replace the cover of the air filter housing and tighten the latches, screws, or wing nut until completely closed. Reinstall any induction tubing or clamps.

_____ **D.** Clean the inside of the air filter housing to make sure the new air filter will seal properly when fitted. Also inspect the housing and any ducts for cracks. If the air filter is being replaced, obtain a new air filter and compare it with the old one to ensure that they are exactly the same.

_____ **E.** On fuel-injected engines, unlatch or unscrew the filter housing fasteners to remove the air filter. It may be necessary to loosen the clamps and hoses on the induction tubing to remove the filter housing cover.

2. Performing a Visual Inspection:

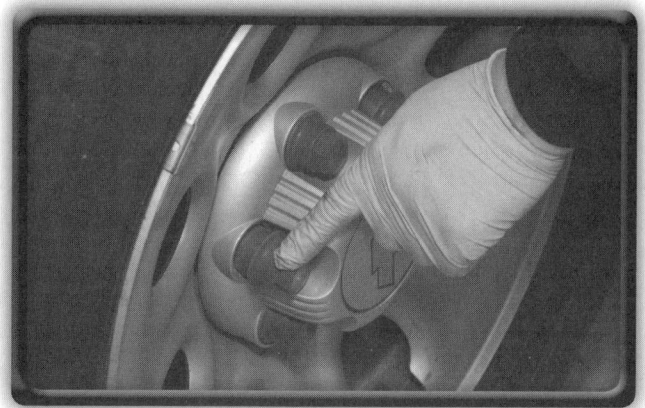

_____ **A.** Check exterior component and system operation. Check the body condition to make sure all the body components are secure. Look for loose plastic trim.

_____ **B.** Prepare the vehicle. Park the vehicle in a well-lit area. Turn the engine off and unlock the doors and trunk or rear hatch.

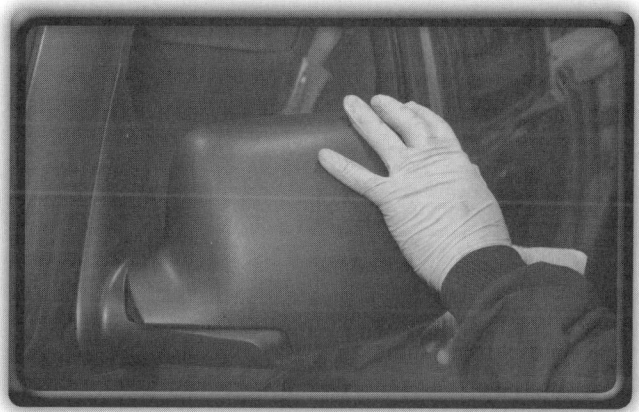

_____ **C.** Inspect the external mirrors to ensure that they are secure and not broken.

_____ **D.** Push and pull on the bumpers or fenders to ensure that they are secure.

_____ **E.** Open and close doors to check that they are operating correctly.

_____ **F.** Walk around the vehicle, observing any obvious items that need attention.

3. Checking and Replacing the Windshield Wiper Blades:

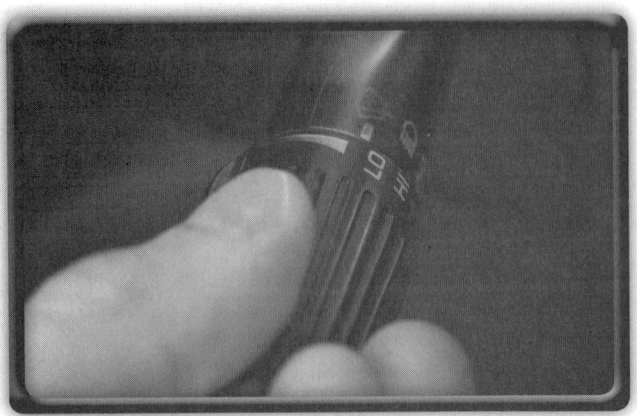

_____ **A.** Wet the windshield with a hose or with the washers and switch the wipers on. If the windshield is being wiped cleanly, do not replace the wiper blades. If the wiper blades are not wiping the glass evenly or are smearing, replace the blades.

_____ **B.** Obtain and install the appropriate replacement blades. Test the wiper blades.

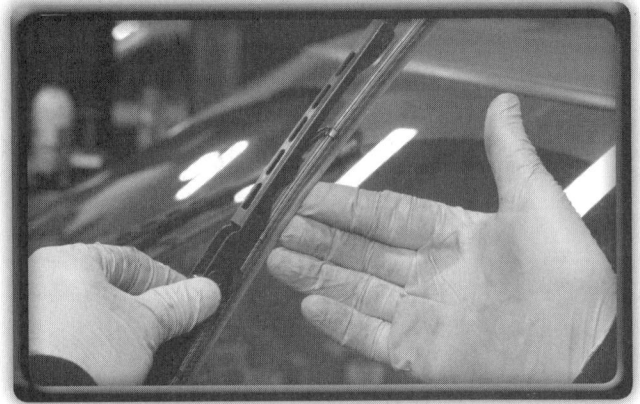

_____ **C.** Check the windshield wiper blades. Lift the wiper arm away from the windshield and inspect the condition of the blades. Look for damage or loss of resilience in the material.

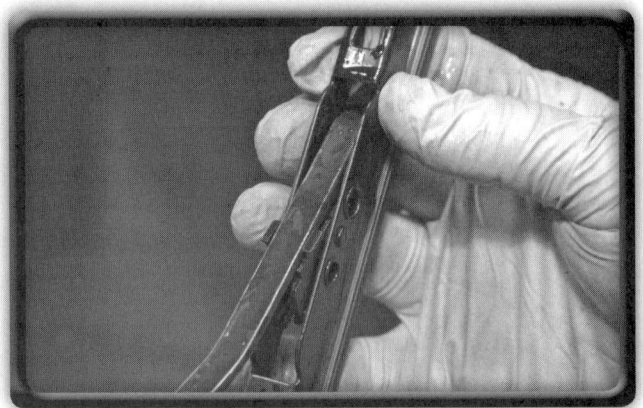

_____ **D.** Remove the blade assembly.

4. Inspecting the Interior Lights:

_____ **A.** Operate the courtesy lights from any other switches, which may be found on the lights themselves or on the dash.

_____ **B.** Park the vehicle inside or in a shaded area. Using the remote key fob or door key, unlock the doors. On most vehicles, unlocking the doors will cause the interior lights to come one. Check that each light works as intended. On some vehicles, the lights will not come on until the door is actually opened. Check each door on the vehicle.

_____ **C.** Enter the vehicle and close the door. Wait to see that the lights go off after a time. Repeat this check with each door on the vehicle.

5. Inspecting the Exterior Lights:

_____ **A.** Depress the brake pedal to make sure the brake lights work. Check that the third (center) brake light works.

_____ **B.** Park the vehicle inside or in a shaded area. Have someone stand behind the vehicle to report any problems while you turn the ignition on. Switch the lighting switch to the park light position. Check that the taillights and any side markers come on and are equal in brightness.

_____ **C.** Check the rear license plate lights to be sure they are operating. Put the turn signal switch in the left and then right turn position, and check that the signals flash equally on each side.

_____ **D.** Make sure the high and low headlight beams, park lights, side markers, and turn indicators are all working properly.

Chapter 8 Vehicle Maintenance Inspection

ASE-Type Questions

Read each item carefully, and then select the best response.

_____ 1. Tech A says that during an exterior inspection of a vehicle you should open and close doors to check that they are operating correctly. Tech B says you should never pull on bumpers or fenders. Who is correct?
   A. Tech A
   B. Tech B
   C. Both A and B
   D. Neither A nor B

_____ 2. Tech A says wiper blade condition cannot cause a vehicle to fail a safety inspection in some states. Tech B says check wiper blades for wear and tear. Who is correct?
   A. Tech A
   B. Tech B
   C. Both A and B
   D. Neither A nor B

_____ 3. Tech A says that a low oil level is bad for the engine but that it is OK for the level to be too high. Tech B says that overfilling the engine oil is bad for the engine. Who is correct?
   A. Tech A
   B. Tech B
   C. Both A and B
   D. Neither A nor B

_____ 4. Tech A says that improper handling of a windshield wiper can lead to a broken windshield. Tech B says you should place a fender cover on the windshield to prevent damage while working on windshield wipers. Who is correct?
   A. Tech A
   B. Tech B
   C. Both A and B
   D. Neither A nor B

_____ 5. While servicing an automatic transaxle–equipped vehicle, the transaxle dipstick cannot be found. Tech A says that some of these vehicles do not have a transaxle dipstick. Tech B says that the dipstick may have fallen off, since all vehicles have a transaxle dipstick. Who is correct?
   A. Tech A
   B. Tech B
   C. Both A and B
   D. Neither A nor B

_____ 6. Tech A says to test new wiper blades against a dry windshield to ensure they seat properly. Tech B says to wet the windshield and operate the wipers to check their performance. Who is correct?
   A. Tech A
   B. Tech B
   C. Both Tech A and Tech B
   D. Neither Tech A nor Tech B

_____ 7. An engine serpentine belt has broken and come off of the pulleys. Tech A says that it is OK to drive the vehicle without the belt. Tech B says that the belt only runs the charging system, so it is OK to drive a short distance without the belt. Who is correct?
   A. Tech A
   B. Tech B
   C. Both A and B
   D. Neither A nor B

_____ 8. Tech A says to clean a windshield first, and then inspect the windshield for chips, scratches, or etching. Tech B says to always clean a windshield first and then inspect the windshield for cracks or signs of delamination. Who is correct?
   A. Tech A
   B. Tech B
   C. Both A and B
   D. Neither A nor B

_____ 9. Tech A says that unlocking doors on most vehicles turns on the interior lights. Tech B says on some vehicles the interior light will not turn on until a door is actually opened. Who is correct?
   A. Tech A
   B. Tech B
   C. Both A and B
   D. Neither A nor B

_____ 10. Tech A says you should test the rear lights with the help of an assistant. Tech B says some shops have a mirror mounted in the service bay so the technician can check rear lights. Who is correct?
   A. Tech A
   B. Tech B
   C. Both A and B
   D. Neither A nor B

Communication

CHAPTER 9

Tire Tread:
© AbleStock

Chapter Review

The following activities have been designed to help you refresh your knowledge of this chapter. Your instructor may require you to complete some or all of these activities as a regular part of your training program. You are encouraged to complete any activity that your instructor does not assign as a way to enhance your learning.

Multiple Choice

Read each item carefully, and then select the best response.

_____ 1. The first step in good communication is:
   A. Nonverbal communication
   B. Asking good questions
   C. Providing feedback
   D. Active listening

_____ 2. Attempting to see a situation from someone else's point of view is known as which of the following?
   A. Sympathy
   B. Empathy
   C. Nonverbal communication
   D. Feedback

_____ 3. Phrases such as "I see" or "Tell me more" that indicate you are paying attention are examples of a(n)_____?
   A. validating statement
   B. supporting statement
   C. closed question
   D. open question

_____ 4. Statements like "Go on" and "Give me an example" that let the speaker know you would like more detail because you are genuinely interested in finding a solution are examples of a(n)_____?
   A. validating statement
   B. supporting statement
   C. closed question
   D. open question

_____ 5. What type of question usually begins with the words when or where?
   A. Open questions
   B. Yes/no questions
   C. Closed questions
   D. Good questions

_____ 6. What aspect of our physical appearance creates the first impression made in any encounter?
   A. Clothes
   B. Hairstyle
   C. Demeanor
   D. All of the above

_____ 7. Looking through the table of contents, introduction, conclusion, headings, and index until we find what we are looking for is the quickest way to use what type of reading?
   A. Absorbing
   B. Comprehending
   C. Selective
   D. Quick

_____ 8. Which of the following is like conducting an investigation?
   A. Reading
   B. Researching
   C. Comprehending
   D. Skimming

_____ 9. What online reference organization consists of over 75,000 active members, with more than 1.7 million years of combined experience, who share their knowledge with each other on over a dozen forums?
   A. International Automotive Technicians Network
   B. National Association for Stock Car Auto Racing
   C. Occupational Safety and Health Administration
   D. National Automotive Parts Association

_____ 10. Which of the following should you do when you come across any defective equipment?
   A. Tag the defective item
   B. Complete a defective equipment report
   C. Notify your supervisor
   D. All of the above

True/False

If you believe the statement to be more true than false, write the letter "T" in the space provided. If you believe the statement to be more false than true, write the letter "F".

_____ 1. Learning and applying good communication skills will save you time and help you avoid or get through tricky situations.
_____ 2. Most people tend to believe the actual words expressed over the nonverbal message you are sending.
_____ 3. Your body position, eye contact, and facial expression can all set the direction of a conversation.
_____ 4. Speaking is often referred to as a science.
_____ 5. Closed questions allow individuals to answer with a simple yes or no.
_____ 6. Nonverbal communication is not possible during a phone conversation.
_____ 7. Being part of a team allows us to complement each other's strengths and weaknesses.
_____ 8. If you have to take customers into the shop, always escort them, and be sure to keep them safe.
_____ 9. Routinely texting your friends or taking personal calls during work hours is stealing from your employer.
_____ 10. The CSI rating is reported each month and used to evaluate individual technicians and the entire service facility.
_____ 11. The person who wants the job done right will probably be very keen to have the repairs finished on schedule.
_____ 12. The faster you try to read, the more you are likely to be able to concentrate on the meaning.
_____ 13. A repair order can become a legal document that will be used by the court to determine if the shop has any liability in a lawsuit situation.
_____ 14. Safety inspection forms are a way of keeping up on routine maintenance tasks such as changing the oil or replacing the belt of an air compressor.
_____ 15. If a problem is noticed with a piece of equipment, the lockout/tagout procedure should be followed.

Fill in the Blank

Read each item carefully, and then complete the statement by filling in the missing word(s).

1. Mental _____ are thoughts and feelings that interfere with our listening, such as our own assumptions, emotions, and prejudices.
2. As a person is speaking to you, you need to provide _____ feedback, which indicates to the speaker that you are engaged in what he or she is saying.
3. _____ feedback includes very simple spoken signals that can enhance the conversation and let the person know you comprehend.
4. Before speaking, take a moment to consider that your _____ of _____ reveals a lot about your feelings and adds significant meaning to your message.
5. A(n) _____ question encourages people to speak freely so we can gather facts, insights, and opinions from them.
6. When taking a _____ _____ for someone else, make sure you have the caller's name and organization, contact details, the date and time of the call, and a summary of the caller's message.
7. _____ should contain information about who, what, when, where, and why, and direction on how a task should be completed.
8. Good _____ skills involve setting a clear vision of the goals, empowering each team member to contribute his or her best efforts, and recognizing each team member's strengths and weaknesses.
9. Our _____ is the image we present of ourselves to the public.
10. _____ means showing up on time (typically 5–10 minutes early to get ready to start work at the appointed time).
11. _____ reading is reading only the parts we need to know. This method is useful when looking for a particular piece of information.
12. Before you spend too much time researching information, it is important that you _____ the problem.
13. When completing a repair order these elements constitute what are called the three Cs: _____, _____, and _____.
14. If someone is injured in the workplace a(n) _____ _____ should be completed by those involved, both the victim and witnesses, if possible.
15. When performing a(n) _____, you need to check that all major components and systems are operational, secured, and safe in accordance with the vehicle manufacturer's recommendations.

Skill Drills

Read each item carefully, and then place the procedure in the correct sequence.

Properly identifying faulty equipment:

_____ **A.** Power tools that have been identified as faulty, due to failure of parts, should also be tagged and set aside. The tool can only be used again after an authorized agent has made the repair.

_____ **B.** Basic workshop tools that are broken or worn should be replaced. Make sure you tag the tool as faulty or broken and do not use it until you buy a replacement. Then discard the tool.

_____ **C.** Remove the keys and lock the vehicle, if appropriate. Attach a tag to the keys that identifies the vehicle they belong to. Store the keys in the key organizer, and notify your supervisor.

_____ **D.** Isolation tags are also used on disabled vehicles or vehicles undergoing a repair. In this case, you will have to locate and complete the "Disabled Vehicle" warning notice. Write the license number of the vehicle and the nature of the defect. Write your name and then the date and time you completed the notice. Attach the notice to the steering wheel or driver's window.

ASE-Type Questions

Read each item carefully, and then select the best response.

_____ 1. Tech A says that repair orders are legal documents so they need to be filled out accurately and carefully. Tech B says that ensuring the repair order is well written, clear, and concise promotes a professional reputation. Who is correct?
   A. Tech A
   B. Tech B
   C. Both A and B
   D. Neither A nor B

_____ 2. Tech A says that you shouldn't waste time inspecting a vehicle in for a repair unless the customer requests it. Tech B says that performing an inspection in addition to a repair can lead to discovery of additional concerns. Who is correct?
   A. Tech A
   B. Tech B
   C. Both A and B
   D. Neither A nor B

_____ 3. Tech A says that the proper way to listen to a customer is to maintain eye contact with the customer in between taking notes. Tech B says that eye contact distracts you from taking good notes. Who is correct?
   A. Tech A
   B. Tech B
   C. Both A and B
   D. Neither A nor B

_____ 4. Tech A says it is best to quietly listen to the customer and refrain from asking questions as much as possible. Tech B says it is constructive to ask questions throughout the conversation to obtain more details. Who is correct?
   A. Tech A
   B. Tech B
   C. Both A and B
   D. Neither A nor B

_____ 5. Tech A says that maintaining an appearance of neatness is important as it conveys to the customer the idea of careful, professional technicians. Tech B says that a dirty and cluttered shop indicates that the shop gets a lot of quality work done. Who is correct?
   A. Tech A
   B. Tech B
   C. Both A and B
   D. Neither A nor B

_____ 6. Tech A says researching the service information is a waste of time. Tech B says that researching the service information saves time. Who is correct?
   A. Tech A
   B. Tech B
   C. Both A and B
   D. Neither A nor B

_____ 7. Tech A says that an example of an open question is: "What are the conditions like when your A/C is not working?" Tech B says that an example of an open question is: "Does your A/C work at all?" Who is correct?
   A. Tech A
   B. Tech B
   C. Both A and B
   D. Neither A nor B

_____ 8. Tech A says that one customer who has a bad experience stemming from miscommunication and a misdiagnosed repair due to a technician's failure to listen has more of an effect on the repair shop than several good and happy customers. Tech B says that it is only one customer, and since the happy ones paid their bills, all is well. Who is correct?
   A. Tech A
   B. Tech B
   C. Both A and B
   D. Neither A nor B

_____ 9. Tech A says that it is primarily the job of the service writer to be customer-oriented and learn listening skills. Tech B says that customer service is the responsibility of all service employees to achieve the goal of customer satisfaction. Who is correct?
   A. Tech A
   B. Tech B
   C. Both A and B
   D. Neither A nor B

_____ 10. Tech A says that the 3 Cs are the "customer, complaint, and concern." Tech B says that the 3 Cs are "concern, cause, and correction." Who is correct?
   A. Tech A
   B. Tech B
   C. Both A and B
   D. Neither A nor B

CHAPTER 10
Engine Mechanical Testing

Tire Tread:
© AbleStock

Chapter Review

The following activities have been designed to help you refresh your knowledge of this chapter. Your instructor may require you to complete some or all of these activities as a regular part of your training program. You are encouraged to complete any activity that your instructor does not assign as a way to enhance your learning.

Matching

Match the following terms with the correct description or example.

A. Compression tester
B. Cylinder leakage tester
C. Data link connector (DLC)
D. Pressure transducer
E. Vacuum gauge

_____ 1. The connector through which the scan tool communicates to the vehicle's computers; it will display the readings from the various sensors and can retrieve trouble codes, freeze-frame data, and system monitor data.

_____ 2. A device used to measure the amount of pressure a cylinder can generate.

_____ 3. A device used to measure the amount of vacuum an engine can generate during various operating conditions.

_____ 4. A device that pumps air into the cylinder and measures the percentage of air that is leaking out of the cylinder.

_____ 5. A device used to measure engine vacuum and display it graphically on a lab scope.

Multiple Choice

Read each item carefully, and then select the best response.

_____ 1. What tool, available in both standard and electronic versions is used by technicians to listen for unusual noises in a vehicle?
A. Microphone
B. Stethoscope
C. Sound transducer
D. Amplifier

_____ 2. When using fluorescent dye to help locate the leak, the dye can only be seen under what kind of light?
A. Fluorescent
B. Infrared
C. Ultraviolet
D. Incandescent

_____ 3. Many newer vehicles are programmed with a _____ capability that effectively shuts off the fuel injectors as long as the throttle is held to the floor before the ignition key is turned to the run or crank position.
   A. time delay
   B. cold start
   C. crank mode
   D. clear flood

_____ 4. A low and steady reading gauge reading during a vacuum test indicates?
   A. Everything is good
   B. Possible late valve or ignition timing
   C. Possible worn rings
   D. Burned or leaking valve

_____ 5. The typical vacuum reading for a properly running engine at idle is a steady _____ of vacuum.
   A. 10" to 12"
   B. 17" to 21"
   C. 4" to 8"
   D. 20" to 23"

_____ 6. Using a pressure transducer and _____ is a similar process to using a vacuum gauge, except it is much more accurate and allows you to look at the vacuum graphically.
   A. lab scope
   B. stethoscope
   C. amp probe
   D. monitor

_____ 7. Which test identifies which cylinder(s) are not operating properly and gives a general indication of each cylinder's overall health?
   A. Vacuum gauge test
   B. Cylinder leakage test
   C. Pressure transducer test
   D. Power balance test

_____ 8. In which test is a high-pressure hose hand threaded into the spark plug hole of the cylinder to be tested and then connected to the compression gauge?
   A. Power balance test
   B. Vacuum gauge test
   C. Cranking or running compression test
   D. Cylinder leakage test

_____ 9. What test checks an engine's ability to move air into and out of the cylinder?
   A. Running compression test
   B. Vacuum gauge test
   C. Power balance test
   D. Running compression test

_____ 10. Black exhaust is an indication of?
   A. Coolant leak
   B. Excessive rich fuel mixture
   C. Burning engine oil
   D. Blown head gasket

True/False

If you believe the statement to be more true than false, write the letter "T" in the space provided. If you believe the statement to be more false than true, write the letter "F".

_____ 1. Mechanical testing starts with a good visual inspection.
_____ 2. A mechanic could diagnose a burned valve with just a piece of bubble gum and its wrapper.
_____ 3. The smell of a leaking fluid can give you a clue as to its source.
_____ 4. Seals leak only when the engine is running.
_____ 5. You can disable the ignition system so that it will crank but not start.
_____ 6. A sharp oscillation back and forth in the needle or a dip in the vacuum gauge reading could relate to a problem such as a restricted exhaust system.
_____ 7. Using a vacuum gauge is much more accurate than using a pressure transducer and lab scope.
_____ 8. Disabling a cylinder that is not operating correctly will not produce much, if any rpm drop.
_____ 9. A cranking compression test is performed when indications show a misfiring or dead cylinder that is not caused by an ignition or fuel problem.
_____ 10. A rod bearing makes an evenly spaced single knock while a main bearing generally makes a double knock.

Fill in the Blank

Read each item carefully, and then complete the statement by filling in the missing word(s).

1. Engine mechanical testing uses a series of tests to assess the mechanical _____ of the engine.
2. Sometimes adding a special fluorescent _____ to the fluid will help locate a leak.
3. If one or more cylinders have _____ _____, the engine will run rough, in many cases misleading the customer to request a "tune-up."
4. A(n) _____ _____ can be used so that the vacuum gauge can be connected and, at the same time, vacuum can still be supplied to the existing component.
5. A vacuum _____ reading shows the difference between outside atmospheric pressure and the amount of manifold pressure in the engine.
6. During an engine vacuum test, all vacuum gauges are calibrated to, and all instructions and readings are referenced to _____ _____.
7. The purpose of the _____ _____ test is to see whether the cylinders are creating equal amounts of power and, if not, to isolate the problem to a particular cylinder or cylinders.
8. When a low-compression cylinder is found, you should put a couple squirts of oil into the spark plug hole, crank the engine a few turns, and recheck the compression. if the compression rises significantly, the problem is typically worn _____ _____.
9. The _____ _____ test is performed by leaving all of the spark plugs in the engine except for the one in the cylinder that you are testing.
10. The _____ _____ test is performed on a cylinder with low compression to determine the severity of the compression leak and where the leak is located.

Chapter 10   Engine Mechanical Testing

Labeling
Label the following images with the correct terms.

Testing tools:

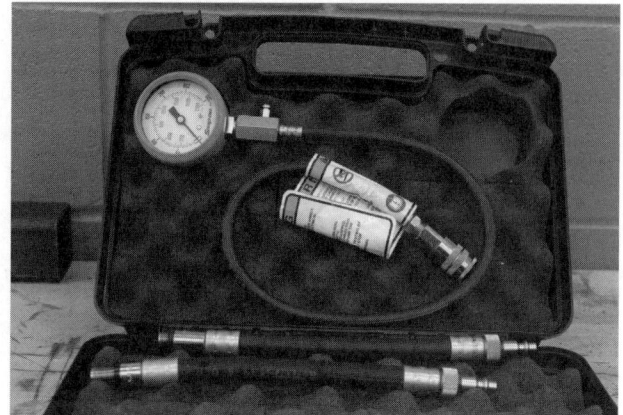

A. _____

B. _____

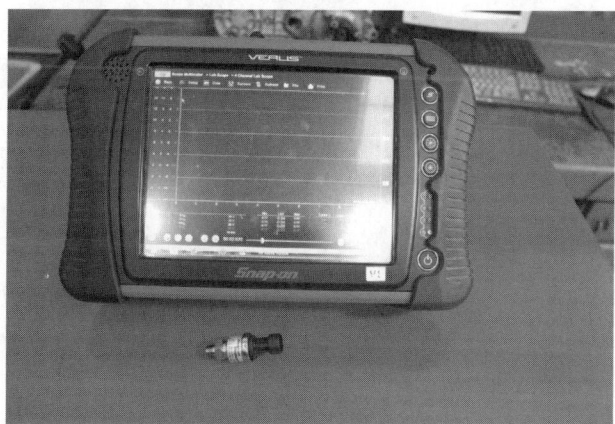

C. _____

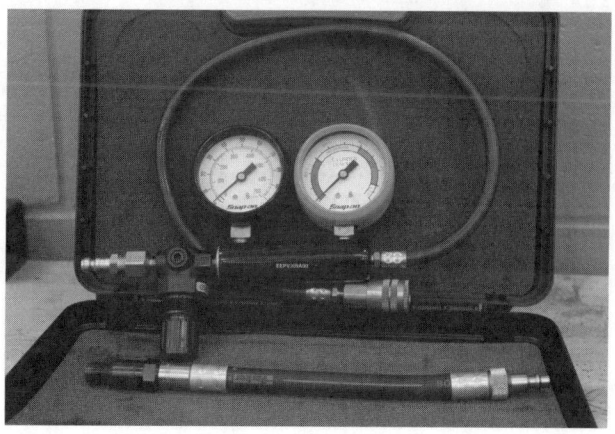

D. _____

Skill Drills

Place the skill drill steps in the correct order.

1. Performing a Fluid Leak Inspection by Looking under the Hood:

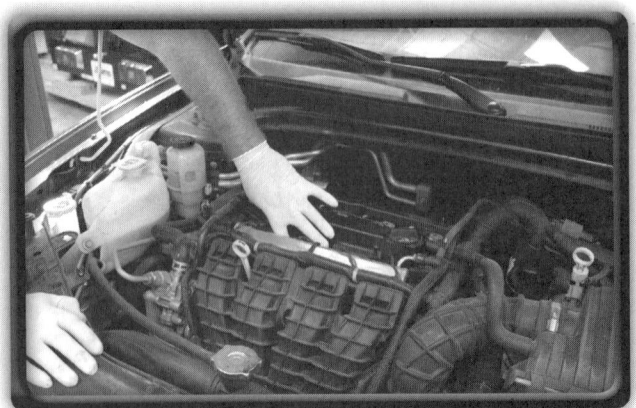

_____ **A.** Check for engine oil leaks or seepage at the valve covers, intake manifold, cam seal, and so on.

_____ **B.** Check for power steering leaks at the power steering pump, lines, and steering box or rack and pinion.

_____ **C.** Raise the hood and make sure it is secure. Check for any coolant leaks. Check the radiator, radiator hoses, heater hoses, water pump, heater control valve, and any coolant lines.

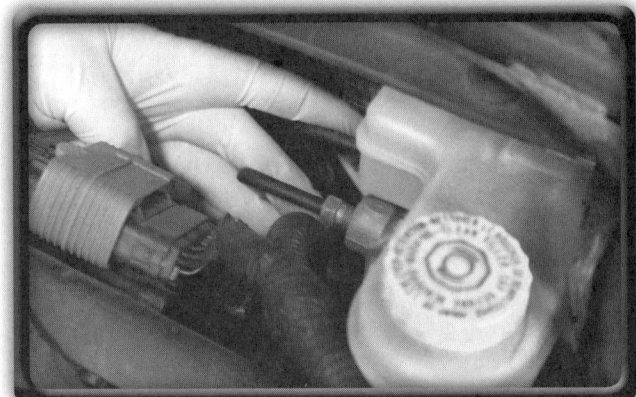

_____ **D.** Check the master cylinder brake lines. Check the rear seal of the brake master cylinder, looking for signs of seepage between the master cylinder and the vacuum booster.

2. Performing a Fluid Leak Inspection by Looking under the Vehicle:

_____ **A.** Inspect the front and rear differentials if equipped.

_____ **B.** Inspect for transmission and transaxle leaks.

_____ **C.** Inspect for engine oil leaks around the front main seal, oil pan, oil filter, oil pressure switch, and rear main seal.

_____ **D.** Raise the vehicle using a hoist, making sure the vehicle is being lifted on the proper lift points. Check for any coolant seepage or leakage on each side of the block around the soft plugs.

3. Testing Engine Vacuum Using a Pressure Transducer:

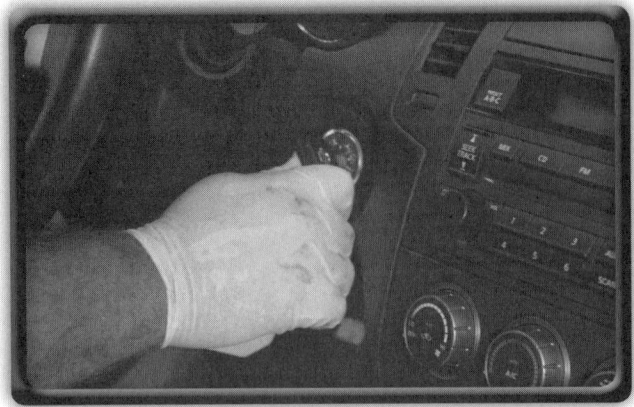

_____ **A.** Start the engine and let it idle. Adjust the lab scope so that the screen captures the vacuum pulses for all of the cylinders. Observe the vacuum trace and compare it to known good readings.

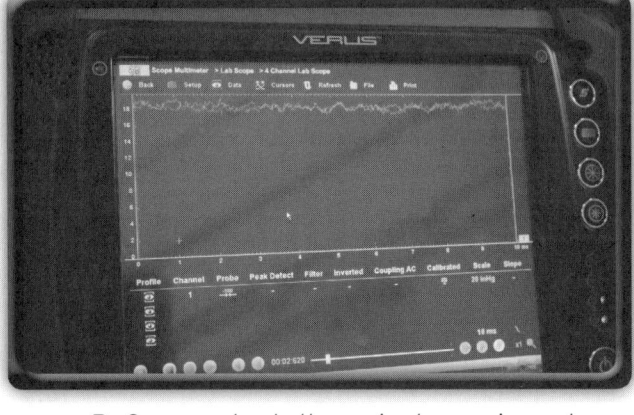

_____ **B.** Snap accelerate the engine by opening and closing the throttle, and compare the trace to known good readings.

_____ **C.** Connect the second channel of the lab scope to the ignition system so it can identify cylinder 1.

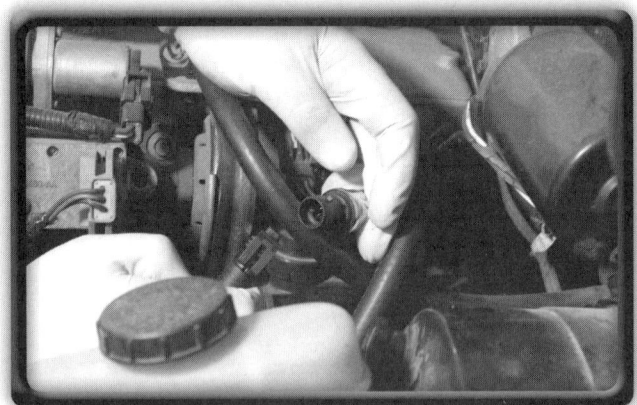

_____ **D.** Connect the pressure transducer to the intake manifold and the lab scope.

_____ **E.** Hold the throttle steady at 1200–1500 rpm and compare the trace to known good readings. Then hold the throttle steady at 2500 rpm and compare the trace to known good readings.

4. Performing a Cylinder Power Balance Test:

_____ **A.** Using the method chosen to disable cylinders, disable the first cylinder and record the rpm. (Do not leave the cylinder disabled for more than a few seconds.)

_____ **B.** Visually inspect the engine to determine the best method to disable the cylinders. If necessary, disable the idle control system. Start the engine and allow it to idle. Record the idle rpm.

_____ **C.** Reactivate the cylinder and allow the engine to run for approximately 10 seconds to stabilize. Repeat the steps on each of the cylinders and record your readings. Determine any necessary action.

5. Performing a Running Compression Test:

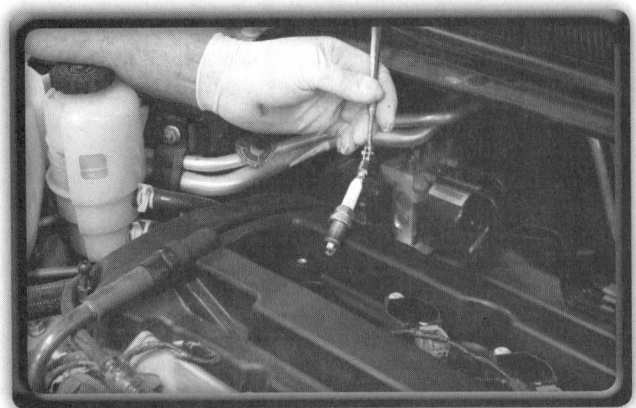

_____ **A.** Remove the spark plug on the cylinder that you are testing and ground the spark plug wire.

_____ **B.** Have your partner quickly snap the throttle open for about 1 second and then quickly close it. (Make sure the key can be turned off quickly if the throttle sticks.) Record the reading. Repeat the process on the other cylinders. Determine any necessary action

_____ **C.** Install the proper hose and compression tester into the spark plug hole. Start the engine, allow it to idle, press and release the bleed valve, and record the reading.

ASE-Type Questions

Read each item carefully, and then select the best response.

_____ 1. Tech A says that a cranking sound diagnosis can be used to diagnose problems in the ignition system. Tech B says that a cranking sound diagnosis can indicate differences in compression. Who is correct?
   A. Tech A
   B. Tech B
   C. Both A and B
   D. Neither A nor B

_____ 2. Tech A says that a power balance test is a good way to narrow a misfire down to a particular cylinder or cylinders. Tech B says that a cylinder power balance test measures the volumetric efficiency of the cylinder being tested. Who is correct?
   A. Tech A
   B. Tech B
   C. Both A and B
   D. Neither A nor B

_____ 3. Tech A says that a cranking compression wet test can indicate if the cylinder has worn piston rings. Tech B says that the throttle should be held wide open during a cranking compression check. Who is correct?
   A. Tech A
   B. Tech B
   C. Both A and B
   D. Neither A nor B

_____ 4. Tech A says that low compression on a single cylinder will cause an engine not to start. Tech B says that low compression on a single cylinder will affect the engine's cranking sound. Who is correct?
   A. Tech A
   B. Tech B
   C. Both A and B
   D. Neither A nor B

_____ 5. Tech A says that a cylinder leakage test is performed on a cylinder with low compression to determine the severity of the leak and where it is located. Tech B says that manufacturers will consider up to 50% cylinder leakage past the piston rings acceptable. Who is correct?
   A. Tech A
   B. Tech B
   C. Both A and B
   D. Neither A nor B

_____ 6. Tech A says that a scan tool connected to the data link connector (DLC) will perform a cylinder power balance test and report cylinder pressures. Tech B says that a scan tool will perform a cylinder power balance test and report whether the rings or valves have failed. Who is correct?
   A. Tech A
   B. Tech B
   C. Both A and B
   D. Neither A nor B

_____ 7. Tech A says that researching related service information for a vehicle repair will assist the technician. Tech B says that research should be done only when the technician needs direction. Who is correct?
   A. Tech A
   B. Tech B
   C. Both A and B
   D. Neither A nor B

_____ 8. Tech A says that a vacuum gauge needle that dips 4–8 inches rhythmically can indicate a burned valve. Tech B says that a stethoscope can be used to determine the source of unusual engine noises. Who is correct?
   A. Tech A
   B. Tech B
   C. Both A and B
   D. Neither A nor B

_____ **9.** Tech A says that a vacuum test can determine exhaust restriction. Tech B says that when performing a cylinder power balance test, results should be 5% or less. Who is correct?
   A. Tech A
   B. Tech B
   C. Both A and B
   D. Neither A nor B

_____ **10.** Tech A says that a bad cam lobe or broken valve spring will show up during a running compression test. Tech B says that when performing a cylinder leakage test the engine must be running to get proper results. Who is correct?
   A. Tech A
   B. Tech B
   C. Both A and B
   D. Neither A nor B

CHAPTER 11

Engine Removal and Replacement

Tire Tread:
© AbleStock

Chapter Review

The following activities have been designed to help you refresh your knowledge of this chapter. Your instructor may require you to complete some or all of these activities as a regular part of your training program. You are encouraged to complete any activity that your instructor does not assign as a way to enhance your learning.

Matching

Match the following terms with the correct description or example.

- **A.** Date link connector (DLC)
- **B.** Diagnostic trouble code (DTC)
- **C.** Malfunction indicator light (MIL)
- **D.** Subframe

_____ **1.** A mount attached to the vehicle that is used to support the engine and transaxle assembly.

_____ **2.** An underdash connector for connecting a diagnostic scanner.

_____ **3.** A dash light usually indicating the presence of a diagnostic trouble code or malfunction.

_____ **4.** A code set by the computer indicating system or component malfunction.

Multiple Choice

Read each item carefully, and then select the best response.

_____ **1.** If the engine will need to be pulled from the top of the engine compartment you will need to use a(n) _____?
- **A.** scissor jack
- **B.** sling bridge
- **C.** engine hoist
- **D.** single-post hoist

_____ **2.** Which of the following tools is required if an engine is being removed by lowering it from underneath the vehicle?
- **A.** Scissor jack
- **B.** Sling bridge
- **C.** Two-post hoist
- **D.** Both A and C

_____ **3.** The first step in the engine removal process is to remove the _____?
- **A.** tires
- **B.** hood
- **C.** transmission
- **D.** muffler

_____ **4.** What is the typical boiling point for automotive anti-freeze or coolant?
- **A.** 212°F
- **B.** 100°F
- **C.** 223°F
- **D.** 225°F

_____ 5. For each pound of pressure maintained by the radiator cap, the boiling point of the coolant increases about _____.
   A. 3°F
   B. 39°F
   C. 45°F
   D. 1.67°F

_____ 6. To begin the exhaust removal process, identify the _____ that need to be removed or disconnected.
   A. brackets
   B. oxygen sensors
   C. pipes
   D. bolts

_____ 7. The point where the engine and the transmission/transaxle are mounted is called the _____, and the starter is usually bolted to it.
   A. Torque converter
   B. Subframe
   C. Bell housing
   D. Starter mount

_____ 8. When using an engine hoist you must use at least _____ bolts, and make sure they screw in at least five full turns.
   A. seven
   B. grade 5
   C. six inch
   D. grade 3

_____ 9. When reinstalling the torque converter make sure to pour enough _____ into the torque converter to fill it about halfway.
   A. power steering fluid
   B. coolant
   C. motor oil
   D. transmission fluid

_____ 10. What should you do during start-up if the gauges indicate overheating or low oil pressure?
   A. Add oil
   B. Check the wiring
   C. Shut it down
   D. Tap the gauges

Chapter 11 Engine Removal and Replacement

True/False

If you believe the statement to be more true than false, write the letter "T" in the space provided. If you believe the statement to be more false than true, write the letter "F".

_____ 1. Some engines have pulling brackets that allow the chains to be attached by hooks when pulling the engine.

_____ 2. Most engines in rear-wheel drive vehicles are best removed from the bottom of the engine compartment.

_____ 3. You can block off and seal the fuel supply line from the fuel tank by wedging a tapered punch inside the line, or taping over the end of the line with duct tape.

_____ 4. An automobile may have as many as 50 separate microprocessors onboard.

_____ 5. When disconnecting the battery, remove the positive terminal of the battery first.

_____ 6. In late-model vehicles, the electrical plug ends are interchangeable so you need to label them before disconnecting.

_____ 7. Always make sure the engine is cool before removing the radiator cap.

_____ 8. You must separate the transmission/transaxle from the engine before the transmission is supported and the engine is on the hoist.

_____ 9. Before starting an engine after installation you should verify that the brake pedal is not spongy.

_____ 10. There are significant differences in timing methods from one decade to another.

Fill in the Blank

Read each item carefully, and then complete the statement by filling in the missing word(s).

1. A(n) _____ _____ light is a dash light that indicates the presence of a diagnostic trouble code or a malfunction such as dead engine cylinders.

2. If the engine is removed from the _____ of the engine compartment, the engine and transaxle are usually removed as a unit attached to the engine cradle.

3. Always put _____ first in each step of the engine removal and installation process.

4. You will need to relieve the pressure on the _____ system before disconnecting the lines.

5. If power is removed from some electronic components such as radios and navigation devices, they will not work until the proper _____ _____ is keyed into it.

6. Before disconnecting electrical wire harness plug ends, it is a good idea to _____ _____.

7. The _____ _____ is normally located where the engine and the transmission/transaxle are mounted together.

8. A _____ is a mount used to support the engine and transaxle assembly and is attached to the vehicle.

9. An important task to complete after you have started a replaced engine is connecting a diagnostic scan tool to the _____ _____ connector.

10. If you find that a _____ _____ _____ was set during the engine break-in period, you should diagnose and correct the condition.

Labeling

Label the following images with the correct descriptions.

1. Engine removal tools:

A. _____

B. _____

2. Front accessories disconnection: radiator

A. _____

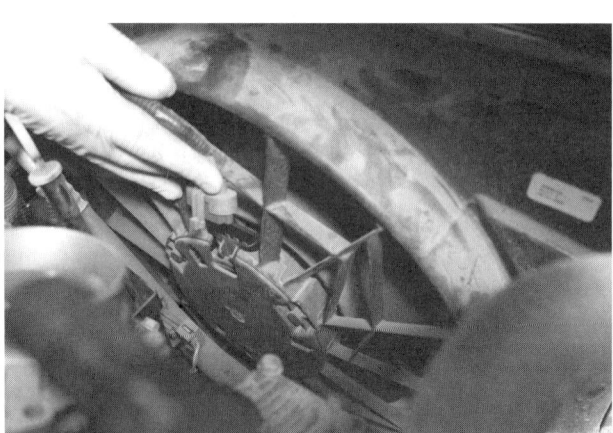

B. _____

C. _____

Skill Drills

Place the skill drill steps in the correct order.

Prestart engine check:
_____ **A.** Verify that the brake pedal is not spongy.
_____ **B.** Check all fluid levels, especially the engine oil and coolant levels.
_____ **C.** Check the serpentine belt and/or other belts and pulleys for proper installation.
_____ **D.** Make sure the fuel is supplied to the injection system by turning the key to the run position, and check that there are no fuel leaks at the connections and fittings.
_____ **E.** Check that the battery terminals are connected firmly and the battery is charged.
_____ **F.** Determine that electrical connectors and vacuum hoses are properly connected.
_____ **G.** If the ignition timing is adjustable, make sure it is set close enough to start and then hook up a timing light to make an adjustment when the engine starts.
_____ **H.** Check all of the auxiliary components, such as the air conditioner and the power steering pump, for proper installation and hose routing.
_____ **I.** Connect the exhaust removal hose to the exhaust pipe.

ASE-Type Questions

Read each item carefully, and then select the best response.

_____ 1. Tech A says that the first step in performing an engine removal and installation procedure is to research the appropriate service information. Tech B says that looking up service information is a waste of time for an experienced technician. Who is correct?
  A. Tech A
  B. Tech B
  C. Both A and B
  D. Neither A nor B

_____ 2. Tech A says that front-wheel drive, rear-wheel drive, and all-wheel drive are examples of drive train configurations. Tech B says that before an engine fails, there are usually warning signs such as dash warning lights indicating low oil pressure, high operating temperature, or abnormal engine noises. Who is correct?
  A. Tech A
  B. Tech B
  C. Both A and B
  D. Neither A nor B

_____ 3. Tech A says that when disconnecting the battery, remove the positive terminal from the battery first. Tech B says that you should disconnect the negative terminal first when preparing to disconnect the battery. Who is correct?
  A. Tech A
  B. Tech B
  C. Both A and B
  D. Neither A nor B

_____ 4. Tech A says that a fuel injection system automatically relieves pressure when the engine is turned off. Tech B says that you will need to relieve fuel pressure from most fuel-injected vehicles before disconnecting the lines. Who is correct?
  A. Tech A
  B. Tech B
  C. Both A and B
  D. Neither A nor B

_____ 5. Tech A says that if refrigerant removal is necessary, air-conditioning refrigerant must be removed with an approved recycling machine. Tech B says that if refrigerant removal is needed, it is much better to release it into the atmosphere because modern refrigerants are safe. Who is correct?
  A. Tech A
  B. Tech B
  C. Both A and B
  D. Neither A nor B

_____ 6. Tech A says that the boiling point of engine coolant is increased by the coolant properties, and the cooling system pressure is maintained by the pressure cap. Tech B says that if a technician were to remove a cooling system pressure cap from a hot engine, the sudden reduction in the coolant boiling point could boil the water violently and scald the technician. Who is correct?
   A. Tech A
   B. Tech B
   C. Both A and B
   D. Neither A nor B

_____ 7. Tech A says that after starting the engine, it should be revved higher than 4,000 rpm for a few minutes to break in the rings. Tech B says that after starting the engine, connect a diagnostic scan tool to the DLC and check for DTCs. Who is correct?
   A. Tech A
   B. Tech B
   C. Both A and B
   D. Neither A nor B

_____ 8. Tech A says that a technician needs to make sure an engine hoist is rated for the lifting capacity needed for the engine being removed. Tech B says that when installing a manual transmission, a clutch alignment tool is needed. Who is correct?
   A. Tech A
   B. Tech B
   C. Both A and B
   D. Neither A nor B

_____ 9. Tech A says to check all fluid levels before starting a newly installed engine, especially the engine oil and coolant levels. Tech B says that to ensure that the piston rings are seated properly and that the camshaft and other internal components are not damaged, it is important to follow appropriate engine break-in procedures. Who is correct?
   A. Tech A
   B. Tech B
   C. Both A and B
   D. Neither A nor B

_____ 10. Tech A says that a torque converter should engage at three different levels as it is being installed in a transmission. Tech B says that the torque converter requires an alignment tool when installing it in the transmission. Who is correct?
   A. Tech A
   B. Tech B
   C. Both A and B
   D. Neither A nor B

Cylinder Head Components

CHAPTER 12

Tire Tread:
© AbleStock

Chapter Review

The following activities have been designed to help you refresh your knowledge of this chapter. Your instructor may require you to complete some or all of these activities as a regular part of your training program. You are encouraged to complete any activity that your instructor does not assign as a way to enhance your learning.

Matching

Match the following terms with the correct description or example.

- A. Anisotropic
- B. Bedding-in
- C. Camshaft follower
- D. Canted valves
- E. Concentricity
- F. Flame front
- G. Flame propagation
- H. Flat tappet
- I. Head gasket
- J. Keeper groove
- K. L-head
- L. Poppet valve
- M. Pushrod
- N. Quench or squish area
- O. Valve face
- P. Valve guide
- Q. Valve keeper
- R. Valve margin
- S. Valve stem
- T. Valve train

_____ 1. A device used to keep the valve spring retainers attached to the valve while in the cylinder head.

_____ 2. A cam-operated, spring-loaded mushroom-type valve used to control intake into, and exhaust out of, the combustion chamber.

_____ 3. The process of a valve wearing into the valve seat and creating a positive seal around the whole diameter.

_____ 4. The portion of the valve between the valve face and the valve head.

_____ 5. The portion of the valve that does the actual sealing to the valve seat.

_____ 6. An object that has unequal physical properties along its various axes. Used in head gaskets to pull heat laterally from the edge surrounding the combustion chamber to the water jacket.

_____ 7. A camshaft specifically designed to push on flat bottom lifters (nonroller types). It typically has a higher rolling resistance than the roller-style camshafts.

_____ 8. A system encompassing all of the parts used in the opening and closing of the valves.

_____ 9. The movement of the flame through the combustion chamber during the combustion process.

_____ 10. A valve arrangement in the cylinder head where the valves are at an angle to the cylinder bore, this can make for a straighter path for air to flow into and out of the intake and exhaust ports.

_____ 11. The front edge of the burning air/fuel mixture in the combustion chamber.

_____ 12. The long shaftlike portion of the valve.

_____ 13. A long, often hollow, metal tube that transfers the force from the valve lifter to the rocker arm assembly.

_____ 14. A term used to describe a valve seat where the valve seat and valve stem share a common center. In this design, the valve can seal against the seat no matter how it is rotated.

_____ 15. A thin piece of material, often a multilayered, bimetallic sheet used to seal the cylinder head assembly to the engine block.

_____ 16. The narrow area between the top of the piston at top dead center and the cylinder head.

_____ 17. A groove machined into the top of the valve stem near the tip of the valve that is used to "lock in" the valve keepers to help retain the valve spring onto the valve assembly.

_____ 18. A slider or roller placed in direct contact with the lobes of the OHC camshaft that pushes on the tip of the valve to open it.

_____ 19. A hole machined into the cylinder head or a hollow metal tube pressed into the cylinder head in which the valve stem rides, holding the stem in the proper position within the cylinder head.

_____ 20. A type of four-stroke internal combustion engine having both intake and exhaust valves located in one side of the engine block, which are operated by lifters actuated by a single camshaft.

Multiple Choice

Read each item carefully, and then select the best response.

_____ 1. The cylinder head forms the top of the _____.
   A. carburetor
   B. head gasket
   C. combustion chamber
   D. intake port

_____ 2. Which type of engine has valves that are positioned in the cylinder head assembly directly over the top of the piston, operated by a camshaft that is located in the cylinder block?
   A. Overhead valve engine
   B. Overhead cam engine
   C. Dual overhead cam engine
   D. None of the above

_____ 3. In the cylinder head, swirl is initially created by the shape and angle of the _____.
   A. intake valves
   B. combustion chamber
   C. quench area
   D. intake port

_____ 4. What type of combustion chamber is shaped like an inverted antique bathtub?
   A. Hemispherical
   B. Oval
   C. Wedge
   D. Elliptical

_____ 5. The portion of the valve stem that comes in direct contact with the rocker arm, cam follower, or bucket lifter is called the _____.
   A. valve stem
   B. valve tip
   C. keeper
   D. valve margin

_____ 6. What type of valve stem seals are used in today's engines?
   A. O-ring
   B. Umbrella-style
   C. Positive
   D. Integral

_____ 7. When the valve and seat are machined about 1 degree different than each other for sealing purposes the difference is called _____.
   A. bedding-in
   B. concentricity
   C. positive seal
   D. interference angle

_____ 8. Which term refers to how far the intake lobe is offset from the exhaust lobe?
   A. Lobe separation angle
   B. Lobe interference angle
   C. Lobe profile
   D. Lobe nose

_____ 9. The _____ closes the valve and holds the valve firmly against the valve seat.
   A. rocker arm
   B. valve spring
   C. push rod
   D. camshaft lobe

_____ 10. The metal spring wrapped circularly around the inside of a seal that applies a small, constant pressure to keep the lip in contact with the rotating part it is sealing is called the _____.
   A. keeper
   B. seal spring
   C. garter spring
   D. retainer spring

True/False

If you believe the statement to be more true than false, write the letter "T" in the space provided. If you believe the statement to be more false than true, write the letter "F".

_____ 1. The dual overhead valve engine contains two camshafts per cylinder head.
_____ 2. The valve keeper actuates a valve by pivoting near the center and pushing on the tip of the valve stem to open it.
_____ 3. GDI refers to gasoline injected directly into the combustion chamber just above the piston.
_____ 4. The use of smaller intake and exhaust ports allows the engine to develop more torque at low engine speeds.
_____ 5. The valve spring is compressed when the valve is closed.
_____ 6. The intake valve does not get as hot as the exhaust valve does.
_____ 7. Integral valve guides can be removed and replaced as needed for repairs.
_____ 8. In some cases intake valve angles are different than exhaust valve angles.
_____ 9. Solid valve lifters are constructed as a centrally located plunger inside a hollow cylindrical body.
_____ 10. A camshaft follower performs the same basic function as a rocker arm.
_____ 11. Valve float is a condition that occurs when the valves cannot be closed by the valve springs as fast as is needed before the next stroke.
_____ 12. Valve keepers lock the valve stem to the retainer.
_____ 13. Gaskets are used to seal the rotating parts of an engine.
_____ 14. When RTV is applied, it has the consistency of a gel, and as it sets, it becomes rubbery.
_____ 15. The presence of a crack in a cylinder head automatically condemns it to the scrap bin.

Fill in the Blank

Read each item carefully, and then complete the statement by filling in the missing word(s).

1. Cylinder heads can be made of cast _____ or _____ alloy.
2. A(n) _____ or pent-roof combustion chamber has the intake valve on one side of the chamber and the exhaust valve on the other.
3. The _____ _____ is responsible for moving air and fuel through the engine.
4. The portion of the valve between the valve face and the valve head is called the valve _____.
5. A poppet valve must be able to stand up to both the _____ pressure trying to pull the valve apart and the _____ pressure trying to force it into the valve seat.
6. The non-replaceable _____ valve guide is cast into the cylinder head when the cylinder head is formed and is machined into the casting.
7. The _____ _____ is the part of the cylinder head that mates with the face of the valve when it is fully closed.
8. The _____ is the highest point of the cam lobe and determines how far the valve opens.
9. _____ refers to how long the valve is held open.
10. If the camshaft is located on top of and centered over the valves, the engine will use _____-_____ lifters.

Fundamentals of Automotive Technology Student Workbook

11. A _____ _____ is a lever that actuates a valve by pivoting inward near the center, and pushing on the tip of the valve to open it.
12. The valve spring _____ is used along with valve keepers to hold the valve spring in place, centered around the valve stem.
13. _____ form a seal by being compressed between stationary parts where liquid or gases could pass.
14. The most widely used seal for rotating parts is the _____ - _____ dynamic oil seal.
15. Aerosol _____ _____ is capable of adhering to many different surfaces, including felt, cork, metal, paper, rubber, and asbestos gaskets.

Labeling

Label the following diagrams with the correct terms.

1. Dual overhead cam (DOHC) engine:

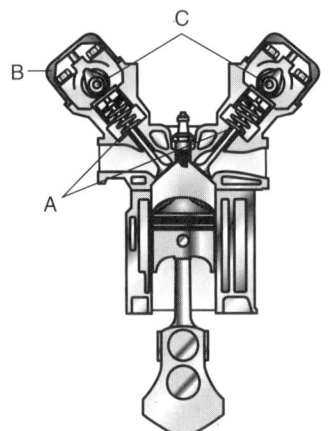

A. _____

B. _____

C. _____

2. Valve train:

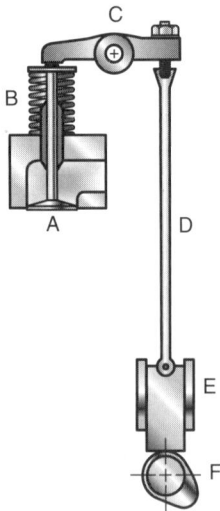

A. _____

B. _____

C. _____

D. _____

E. _____

F. _____

3. Parts of the valve:

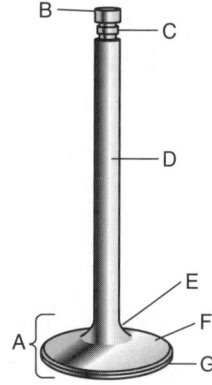

A. _____

B. _____

C. _____

D. _____

E. _____

F. _____

G. _____

4. Valve-to-valve seat interference angle:

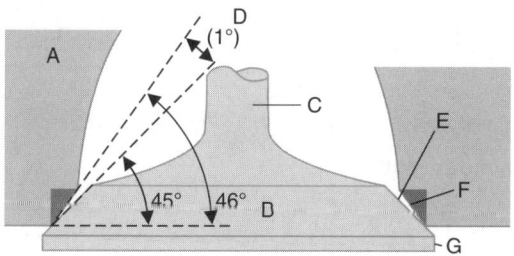

A. _____

B. _____

C. _____

D. _____

E. _____

F. _____

G. _____

104 Fundamentals of Automotive TECHnology Student Workbook

5. Bucket-style lifters:

A. _____
B. _____
C. _____
D. _____
E. _____
F. _____
G. _____
H. _____
I. _____
J. _____

Skill Drills

Test your knowledge of skill drills by filling in the correct words in the photo captions.

1. Removing Cylinder Heads:

Step 1: Research the procedure for _____ the cylinder head(s) in the appropriate service information. Determine whether the head bolts are torque-to-yield (TTY) bolts. If so, you will need to discard them and replace them with new TTY bolts upon reassembly. Determine if the head bolts need to be _____ in a specified sequence. **(There is no image associated with this step.)**

Step 2: Before the _____ come off, put an identifying mark on at least one of the heads. Follow the specified procedure to remove all of the head _____, and set them aside.

Step 3: If dealing with a _____ _____ that is stuck to the head and block surface, reinstall two corner head bolts _____ or _____ turns in their respective holes.

Chapter 12 Cylinder Head Components

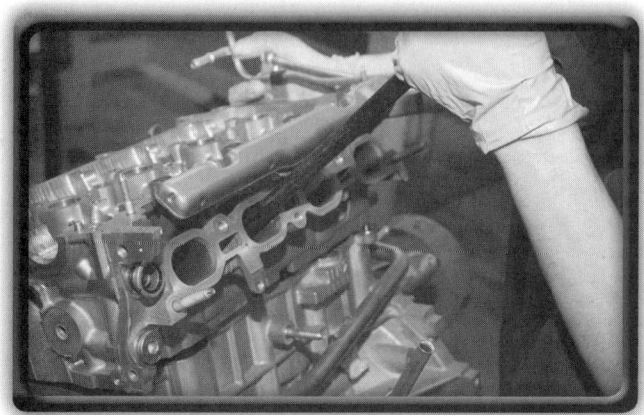

Step 4: If the engine is mounted on an engine stand, check that the _____ _____ is in your engine stand. Insert a _____ _____ or a long breaker bar handle into one of the intake port openings of the head, and give it a firm push.

Step 5: Remove the safety bolts from the _____ and carefully lift it. Inspect the _____ _____ visually for any unusual conditions.

2. Inspecting Valves and Valve Seats:

Step 1: Inspect each valve for signs of _____, _____, and excessive face, stem, and tip wear. If found, replace the valves with new ones.

Step 2: Inspect each valve seat for signs of burning, leakage, or excessive wear. If found, the seat must be _____ or _____.

Step 3: Using a micrometer, measure each valve stem diameter in _____ places where the valve rides in the valve guide, and _____ your answers.

Step 4: Measure the valve _____ with a machinist's rule and record your readings.

Step 5: Measure the _____ of the valve seats and record your readings. Compare readings to the _____, and determine needed actions.

3. Reassembling a Cylinder Head:

Step 1: Ensure that all _____ and _____ have been cleaned, checked for defects, and replaced if necessary. If any _____ are found, correct them now.

Step 2: Dip the _____ stem that is about to be installed in _____ engine oil or use the appropriate assembly lubrication.

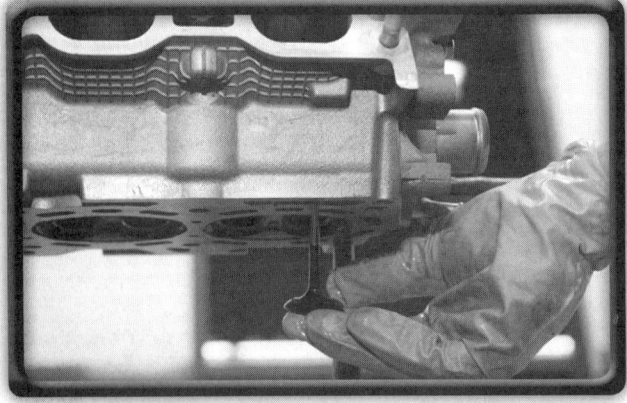

Step 3: Insert the corresponding valve into the valve guide, and place the _____ sleeve over the grooves near the end of the valve stem.

Step 4: Dip the valve seal in clean engine oil and install it down over the _____ top of the valve guide assembly, using a(n) _____ if necessary.

Chapter 12   Cylinder Head Components

Step 5: Remove the protective sleeve and place the valve _____ and _____ over the installed valve.

Step 6: Compress the valve spring and retainer with the valve spring _____, and install the valve keepers. Be sure the _____ are locked into their groove before releasing _____ on the valve spring compressor. Repeat this process for the remaining valves.

4. Inspecting and/or Measuring the Camshaft:

Step 1: Inspect the camshaft journals and _____ for obvious damage or wear. If found, replace the camshaft. Install the _____ on a set of V-blocks so that two specified cam journals are resting in the Vs.

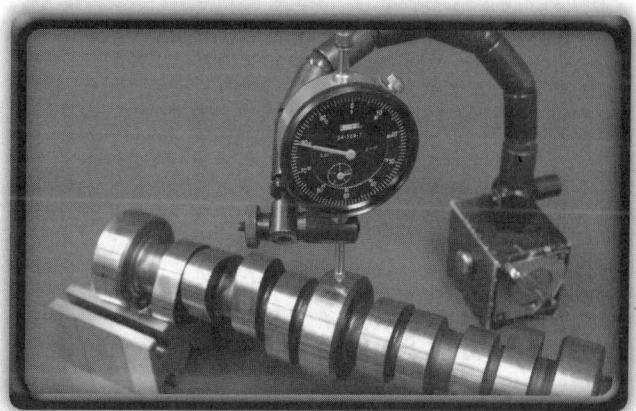

Step 2: Set the _____ indicator so it is engaged on one of the other cam journals _____.

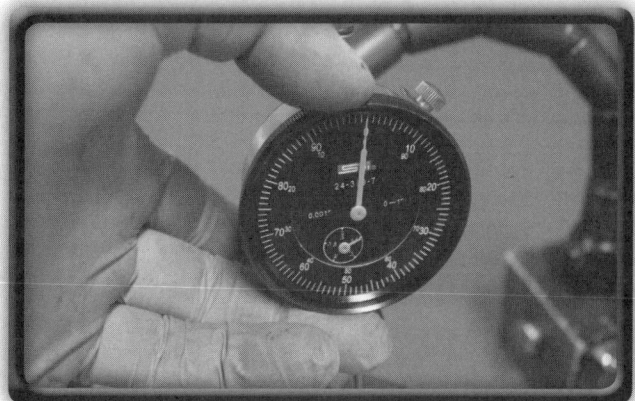

Step 3: Rotate the _____ until the dial indicator reads the _____ point, and zero the dial.

Step 4: Rotate the camshaft until the dial indicator reads the _____ point, and record the reading. Continue rotating the camshaft to verify that the needle does not go _____ zero. If it does, zero it again, and _____ the runout. Perform this process on all of the remaining journals, and compare to the runout specifications.

Step 5: Using a _____, measure each journal in two places _____ degrees apart to determine any out-of-round condition. Record your readings, and compare to the specifications.

Step 6: Reset the dial indicator so it is engaged on one of the cam _____ perpendicularly. _____ the camshaft until the dial indicator reads the lowest point, and zero the dial.

Step 7: Rotate the camshaft until the dial indicator reads the _____ point, and record the reading. Continue to rotate the camshaft to verify that the needle does not go _____ zero. If it does, zero it again, and remeasure the lobe. Measure each of the remaining cam lobes, and compare your readings to specifications.

5. Assembling Cam Follower/Rocker Arm–Style Heads:

Step 1: Temporarily install the _____ into the cam bearings and install the _____ _____. **(There is no image associated with this step.)**

Step 2: Make sure the camshaft turns freely in the bearings. If not, you may need to lightly _____ the cam bearings or replace the camshaft if it is _____. **(There is no image associated with this step.)**

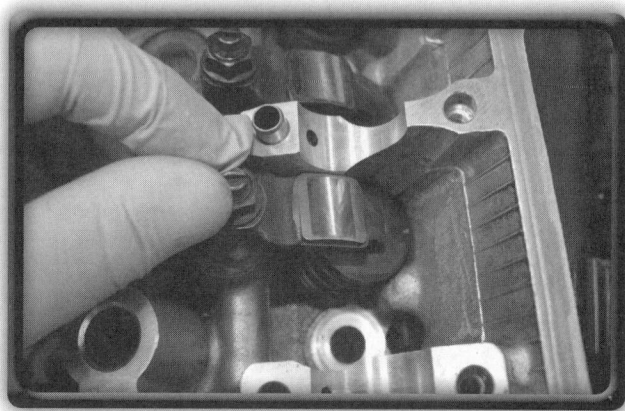

Step 3: _____ the rocker arm pivot points and the valve tips using engine assembly _____ or clean engine _____.

Step 4: Install the rocker arms into the _____ points and over the tip of the valve.

Step 5: Lubricate the _____ arm where the camshaft lobe will ride using engine _____ lube or clean engine oil. Lubricate the camshaft bearings, journals, and cam lobes with engine _____ lube or clean engine oil. Install the camshaft into the camshaft bearings.

Step 6: Install the camshaft _____ _____. Tighten the support cap bolts finger tight, and torque the bolts to the manufacturer's specification.

Step 7: Use a _____ _____ to measure the valve clearance following the manufacturer's recommended procedure.

Step 8: Adjust the valve clearance to the manufacturer's specifications. Be sure to _____ the adjustment jam nut to prevent the clearance from _____.

Crossword Puzzle

Use the clues to complete the puzzle.

Across

2. Any seal used to seal oil in and dirt, moisture, and debris out.
5. The valve through which air and fuel enter the combustion chamber.
7. A device used to control the flow of air and fuel into the combustion chamber and exhaust gases out of the combustion chamber.
8. The portion of the valve that does the actual sealing to a valve seat.
9. A lever that actuates a valve by pivoting near the center and pushing on the top of the valve to open it.
11. The port through which the air or air/fuel mixture travels from the throttle body area to the combustion chamber.
12. Any material used on an engine to seal a fluid or gas into a particular area.

Down

1. The valve through with exhaust gases are forced out of the combustion chamber.
3. Pressure exerted by a metal coil usually measured in pounds.

4. Steel rings integrated into the cylinder head gasket nearest the combustion chambers that provide extra sealing to seal in the high combustion pressures.
6. A metal coil spring that returns valves to their fully closed positions after being opened.
10. The eccentric "egg-shaped" portion of the camshaft that pushes on the valve lifter or camshaft follower.

ASE-Type Questions

Read each item carefully, and then select the best response.

1. Tech A says that variable valve timing allows better control. Tech B says that in-block cams usually include variable valve timing. Who is correct?
   A. Tech A
   B. Tech B
   C. Both A and B
   D. Neither A nor B

2. Tech A says that the cylinder head design determines the output power of the engine. Tech B says that some engines use three valves: one intake and two exhaust. Who is correct?
   A. Tech A
   B. Tech B
   C. Both A and B
   D. Neither A nor B

3. Tech A says that the head gasket is designed to seal between the head and the block. Tech B says that most head gaskets require that you coat them with special sealers before installation. Who is correct?
   A. Tech A
   B. Tech B
   C. Both A and B
   D. Neither A nor B

4. Tech A says that some head bolts can be reused. Tech B says that some head bolts are torqued to their yield point. Who is correct?
   A. Tech A
   B. Tech B
   C. Both A and B
   D. Neither A nor B

5. Tech A says that an interference angle is the difference in angle between the valve face and valve seat. Tech B says an interference angle is when a piston hits a valve. Who is correct?
   A. Tech A
   B. Tech B
   C. Both A and B
   D. Neither A nor B

6. Tech A says that soft plugs allow the block to expand and contract without cracking. Tech B says that soft plugs will prevent damage to the head or block if coolant freezes. Who is correct?
   A. Tech A
   B. Tech B
   C. Both A and B
   D. Neither A nor B

7. Tech A says that valve seals are designed to stop combustion gases from entering the valve train area. Tech B says that valve retainers hold the valve in the spring. Who is correct?
   A. Tech A
   B. Tech B
   C. Both A and B
   D. Neither A nor B

_____ 8. Tech A says that gas-direct injection engines have injectors mounted in the cylinder heads. Tech B says that turbulence in the combustion area creates a better burn of the air/fuel mixture. Who is correct?
   A. Tech A
   B. Tech B
   C. Both A and B
   D. Neither A nor B

_____ 9. Tech A says that when the lifter is on the base circle of the cam lobe, the valve is closed. Tech B says that the valve has a small amount of clearance when the valve is fully open. Who is correct?
   A. Tech A
   B. Tech B
   C. Both A and B
   D. Neither A nor B

_____ 10. Tech A says that smaller ports and valves in a cylinder head will create more power and torque at high rpm. Tech B says that smaller ports and valves in a cylinder head will create more power and torque at low rpm. Who is correct?
   A. Tech A
   B. Tech B
   C. Both A and B
   D. Neither A nor B

Engine Block Components

CHAPTER 13

Tire Tread:
© AbleStock

Chapter Review

The following activities have been designed to help you refresh your knowledge of this chapter. Your instructor may require you to complete some or all of these activities as a regular part of your training program. You are encouraged to complete any activity that your instructor does not assign as a way to enhance your learning.

Matching

Match the following terms with the correct description or example.

- A. Back clearance
- B. Bearing crush
- C. Billet
- D. Core plug
- E. Crank core
- F. Deck
- G. Expander
- H. Fillet
- I. Flex hone
- J. Gravity pouring
- K. Interference fit
- L. Magnafluxing
- M. Mallory metal
- N. Piston slap
- O. Reciprocation
- P. Rod beam
- Q. Sintering process
- R. Tang
- S. Water jacket
- T. Welch plug

_____ 1. The area of connecting rod between the big and small ends.
_____ 2. A machine tool with abrasive stones that is used to refinish the inside of the cylinder walls.
_____ 3. The force created to seat the bearing by the extra bearing material when the ends of the bearing inserts touch each other and are forced against each other.
_____ 4. An engine noise caused by excessive clearance between the piston skirt area and the cylinder wall.
_____ 5. A part of the bearing insert that helps to lock the bearing insert into the bearing saddles and caps.
_____ 6. The part of an oil control ring that holds the ring against the cylinder wall.
_____ 7. A soft, round, metal disc pressed into the engine block to plug a water jacket or oil passageway.
_____ 8. A metal hardening process in which the metal is fused together without melting.
_____ 9. The roughly shaped steel piece that has been cast or forged but has not undergone its final machining.
_____ 10. A metal cap for the holes that are used to remove core sand used during the casting process.
_____ 11. A tungsten alloy of copper and nickel that is used as a metal substitute added to the crankshaft counterweight for balancing purposes.
_____ 12. The radius portion of an inside corner that reduces stress at the corner.
_____ 13. A passageway for coolant to flow inside the engine block that is formed when the block is cast.
_____ 14. The casting process used for creating metal parts.
_____ 15. The area or space behind the piston rings when the rings are in the piston ring grooves.
_____ 16. The back-and-forth movement of the piston assembly inside the cylinder.
_____ 17. An electromagnetic process used to locate cracks in ferrous engine blocks and cylinder heads and other ferrous metal parts.
_____ 18. The rough, unfinished crankshaft assembly that has just left the foundry or forging area.

_____ 19. The surface area at the top of the engine block against which the cylinder head seals.

_____ 20. A fastening between two parts that is activated by friction after the two parts are pushed together.

Multiple Choice

Read each item carefully, and then select the best response.

_____ 1. The largest part of the engine and the main supporting structure for all engine parts is the _____.
   A. chassis
   B. crankshaft
   C. engine block
   D. main bearing cradle

_____ 2. Sometimes called a boxer engine, these engines have cylinders that are positioned side to side.
   A. Two-stroke engine
   B. Horizontally opposed engine
   C. Rotary engine
   D. Diesel engine

_____ 3. Holes in the sides and ends of the engine block casting can be sealed with _____.
   A. core plugs
   B. welch plugs
   C. soft plugs
   D. all of the above

_____ 4. When casting an engine block all of the following structures are used to provide rigidity to the casting while reducing overall engine weight, *except*:
   A. Pillars
   B. Ribs
   C. Fillets
   D. Webs

_____ 5. The job of the _____ is to extract the thermal energy from the burning fuel and convert it into mechanical energy, which can be used to power the vehicle.
   A. camshaft
   B. valve train
   C. piston
   D. connecting rod

_____ 6. The tubular structure that connects the piston to the top of the connecting rod is called the _____.
   A. piston pin
   B. connecting pin
   C. wrist pin
   D. Either A or C

_____ 7. The difference between the volume in the combustion chamber above the piston when the piston is at bottom dead center and the volume of the combustion chamber above the piston when the piston is at top dead center is called the _____.
   A. power stroke
   B. compression ratio
   C. range
   D. combustion volume

_____ 8. What type of piston ring prevents excessive oil from working up into the combustion chambers?
   A. Oil control rings
   B. Fire rings
   C. Compression rings
   D. Expander rings

_____ 9. Which type of connecting rod is the most common?
   A. Cast rod
   B. Powdered metal
   C. Aluminum rod
   D. Forged rod

_____ 10. The main role of the _____ is to transfer the up-and-down strokes of the pistons into rotary motion.
   A. connecting rod
   B. crankshaft
   C. camshaft
   D. flywheel

_____ 11. The heavy metal disc that bolts to the rear of the crankshaft of all engines equipped with a manual transmission/transaxle is called the _____.
   A. harmonic balancer
   B. flex plate
   C. camshaft
   D. flywheel

_____ 12. Also known as the crankshaft damper, this part is located on the front of the crankshaft assembly; its function is to reduce vibration from the crankshaft and piston assembly.
   A. Flywheel
   B. Harmonic balancer
   C. Flex plate
   D. None of the above

_____ 13. To address the problem of second-order vibration, engine manufacturers have developed a _____ with counterweights spaced so as to cancel out the inherent vibrations.
   A. balance shaft
   B. flywheel
   C. harmonic balancer
   D. fluid damper

_____ 14. A cutting tool that will cut into the metal ridge and carbon buildup at the top of the cylinder and remove the lip is called a _____.
   A. honing stone
   B. wedge
   C. chamfer bore
   D. ridge reamer

_____ 15. Cracks in cast iron parts can be detected by using an electromagnetic process called _____.
   A. electroplating
   B. Zyglo™
   C. magnafluxing
   D. ferropolarization

True/False

If you believe the statement to be more true than false, write the letter "T" in the space provided. If you believe the statement to be more false than true, write the letter "F".

_____ 1. Combustion is carefully timed to occur when the piston is near the top of its travel on the compression stroke.

_____ 2. The most common methods used for engine block casting are the green sand, shell, and lost foam processes.

_____ 3. Main bearing caps can be fastened using a two-, four-, or eight-bolt design.

_____ 4. In a dry sleeve only the outer surface is in direct contact with the coolant in the water jacket that surrounds the cylinder.

_____ 5. The wrist pin design can be a floating pin or a press-fit pin.

_____ 6. With the stronger hypereutectic piston, the higher compression ratios that create hotter exhaust gases and higher pressures can be used without piston damage.

_____ 7. Piston rings are not a complete circle; they are split and must have a specified piston ring end gap.

_____ 8. The main journals are ground to manufacturer specifications from the crank core.

_____ 9. Crankshafts can be machined out of a billet, which is a solid piece of metal.

_____ 10. Most crankshafts use Mallory metal journal surfaces to give good wear qualities.

_____ 11. All main bearings and rod bearings are held in place by bearing crush.

_____ 12. In automatic transmission engines, a flex plate is used instead of a flywheel.
_____ 13. In engines that use a timing chain or timing gears, the front main seal is located in a housing bolted to the front of the engine block.
_____ 14. The harmonic balancer is attached to the crankshaft using a Woodruff key or a square key and center bolt.
_____ 15. The harmonic balancer is used to cancel out second-order vibrations in the engine.

Fill in the Blank

Read each item carefully, and then complete the statement by filling in the missing word(s).

1. _____ _____ create passageways around the cylinders and throughout the engine block that hold coolant to keep engine temperature in functional ranges.
2. Some engines use an integrated _____ _____ _____ that has all the main caps cast in a single supporting structure.
3. The _____ _____ sleeve is not held in place by an interference fit; instead, it has a flange at the top of the cylinder that is used to lock the sleeve into the engine block.
4. The most common piston arrangement has two thin _____ ring grooves and one larger _____ ring groove.
5. Piston slap occurs as the piston pivots on the pin near _____ _____ _____.
6. The most popular piston coating is a black _____.
7. A compression ring must have _____ _____ inside the piston groove to accommodate heat expansion of the piston.
8. The _____ _____ connects the piston to the crankshaft.
9. As the crankshaft rotates, the crankshaft _____ rotates on bearing inserts inside the connecting rod's big end.
10. Down the center of the connecting rod mold, there is an imperfection separating one half of the rod from the other referred to as a _____ _____.
11. Connecting rods are attached to offset journals called _____.
12. Main bearing caps mate to _____ _____, which are smooth, curved, machined areas located at the bottom of the engine block.
13. Rod throws, also called rod journals, are located in the _____ _____ of the crankshaft assembly, as opposed to the main journals that sit in a straight line with the centerline of the crankshaft assembly.
14. The crankshaft requires a _____ _____, which limits the end play movement of the crankshaft.
15. The two most popular _____ _____ are the trimetal and the bimetal types.

Chapter 13   Engine Block Components

Labeling

Label the following diagrams with the correct terms.

1. The engine block of an overhead cam engine:

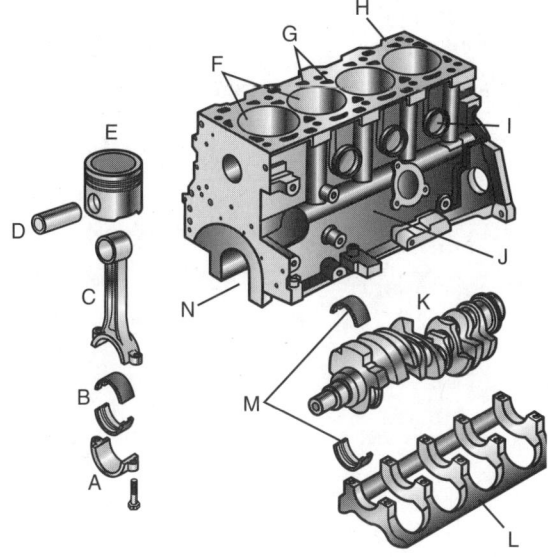

A. _____
B. _____
C. _____
D. _____
E. _____
F. _____
G. _____
H. _____
I. _____
J. _____
K. _____
L. _____
M. _____
N. _____

2. A horizontally opposed engine:

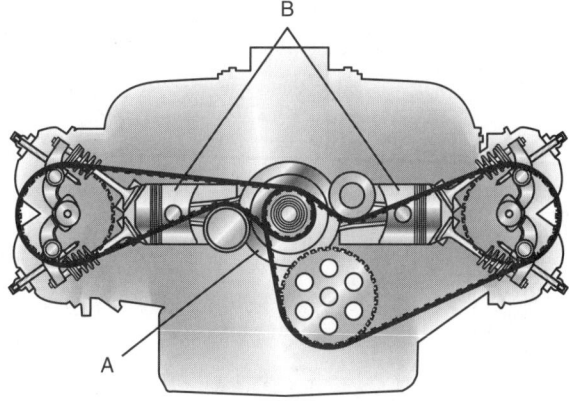

A. _____
B. _____

3. Cylinder sleeves:

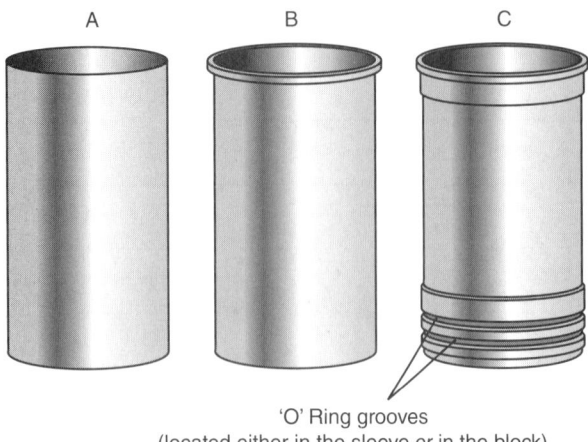

'O' Ring grooves
(located either in the sleeve *or* in the block)

A. _____
B. _____
C. _____

4. Oil passageway drilled in the crankshaft:

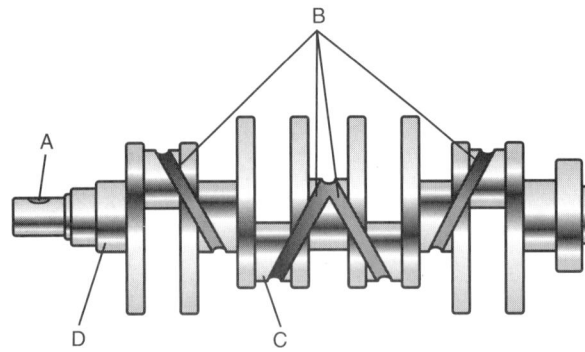

A. _____
B. _____
C. _____
D. _____

5. Bearing wear patterns:

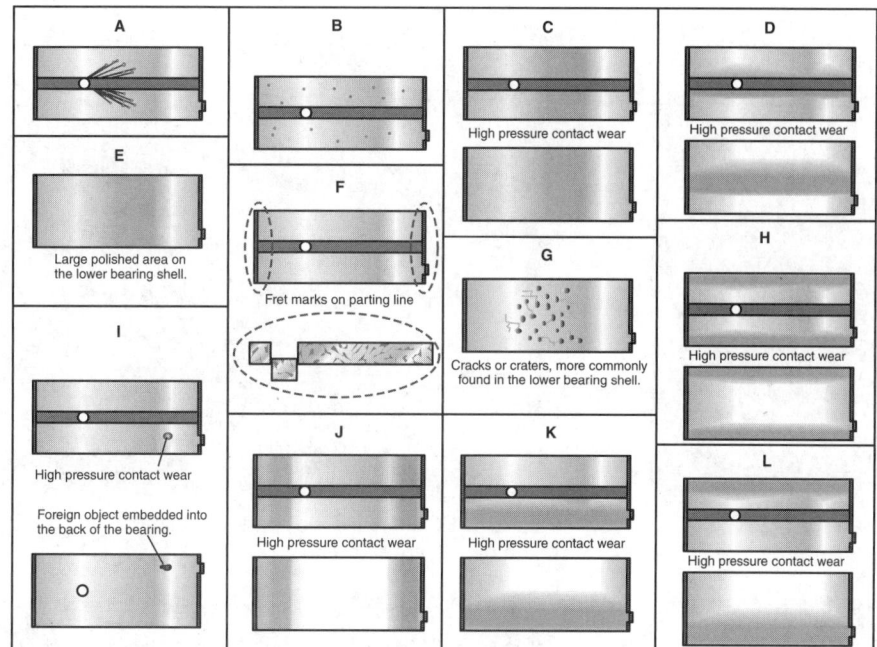

A. _____

B. _____

C. _____

D. _____

E. _____

F. _____

G. _____

H. _____

I. _____

J. _____

K. _____

L. _____

Skill Drills

Place the skill drill steps in the correct order.

1. Inspecting the Crankshaft:

_____ **A.** Visually inspect the crankshaft for cracks; magnaflux, if necessary.

_____ **B.** Measure the rod and journals using a micrometer, and record your findings.

_____ **C.** Make sure the oil passages are clear. Check the crankshaft position sensor reluctor ring to make sure there is no damage.

_____ **D.** Visually inspect the square-cut or Woodruff key and keyway for damage.

_____ **E.** Visually inspect the engine block for any damage, excessive wear, or cracks. If none are visible, it should be checked for hard-to-see hairline fractures or cracks by magnafluxing or a dye check method. **(There is no image associated with this step.)**

_____ **F.** Check the crankshaft for straightness with a dial indicator.

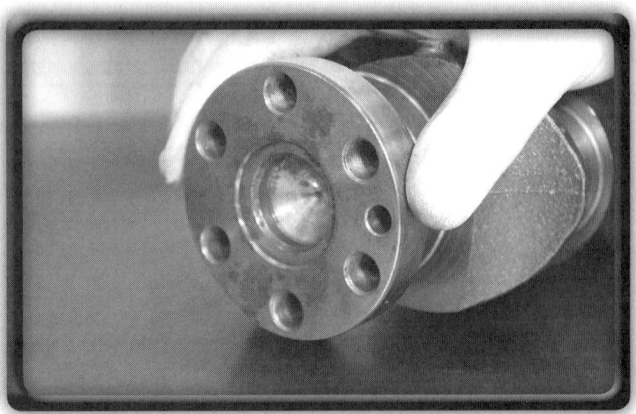

_____ **G.** Visually inspect the thrust flange, bolt holes, and rear seal surface for nicks or scratches.

2. Removing and Replacing the Piston Pin:

_____ **A.** Position the piston so it is in the proper relationship to the connecting rod (see service information).

_____ **B.** Install the second snap ring, and move the piston and connecting rod, rocking the rod back and forth to ensure that it moves freely and smoothly.

_____ **C.** Push on the wrist pin to separate the wrist pin from the piston and connecting rod.

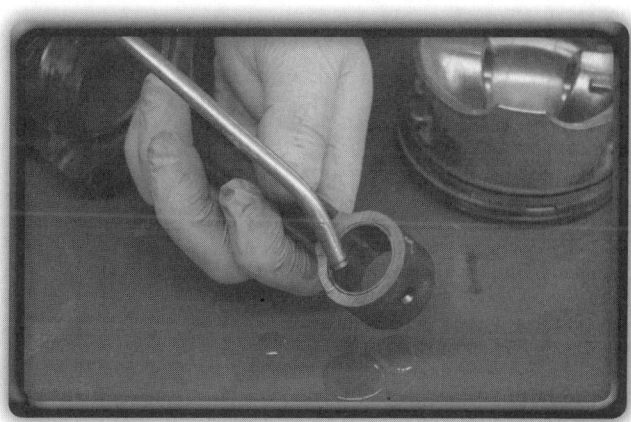

_____ **D.** Install the floating pin design. Lubricate the wrist pin hole on the piston and the small end of the connecting rod along with the wrist pin. Using your snap ring pliers, install one snap ring on the piston.

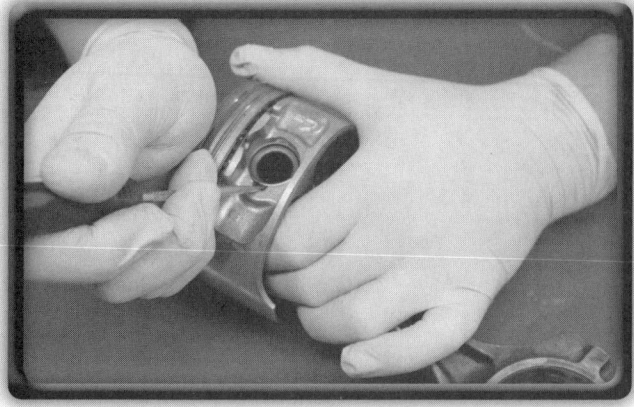

_____ **E.** Using the proper snap ring pliers that fit the holes in the snap ring, remove both snap rings.

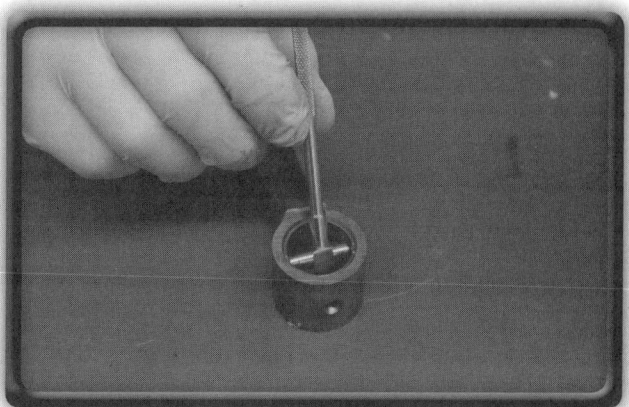

_____ **F.** Using a bore gauge or inside micrometer, measure the inside of the small end of the connecting rod. Record the clearance.

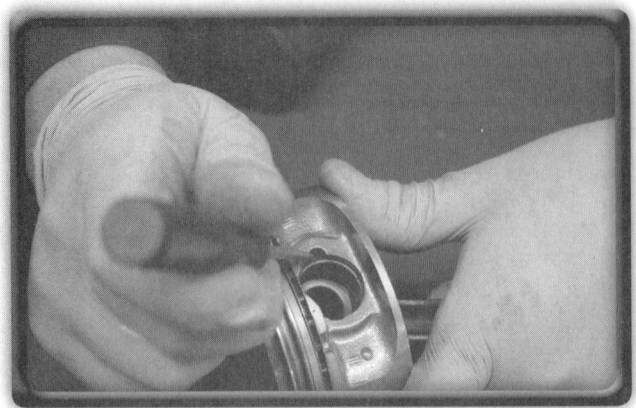

_____ **G.** Place the small end of the connecting rod between the piston pin bosses. Using your fingers or a wrist pin driver, push the wrist pin through the piston and small end of the rod until it stops against the snap ring.

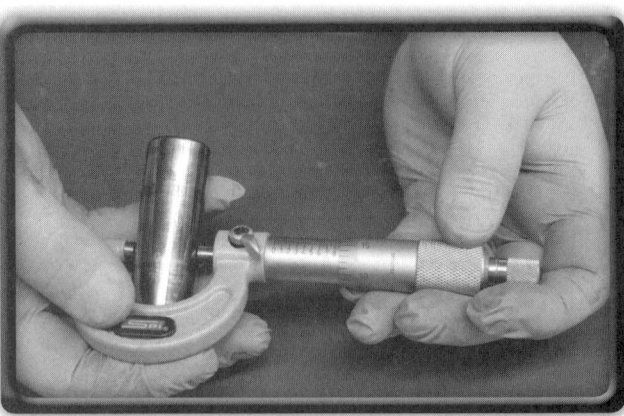

_____ **H.** Using an outside micrometer, measure the piston pin outside diameter and record your findings.

3. Inspecting and Measuring the Cylinder Walls/Sleeves:

_____ **A.** Measure each cylinder at the top, center, and bottom, 90 degrees across the thrust side and pin side at each level. Compare the readings to specifications.

_____ **B.** If there is obvious scoring or damage, then the block will need to be machined. If there is no obvious damage, set a dial bore gauge or inside micrometer to measure the cylinders.

4. Inspecting and Measuring Piston Skirts:

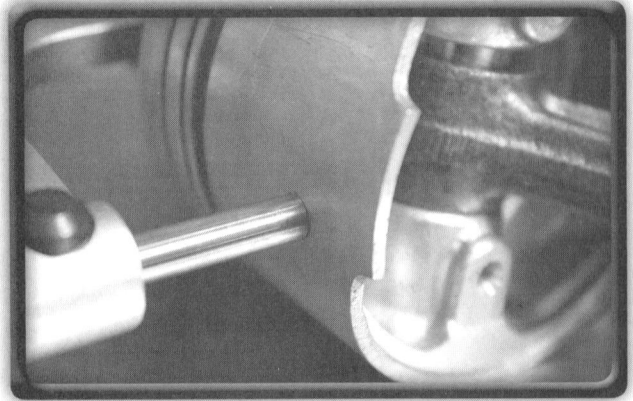

_____ **A.** With the piston inverted, measure the diameter of the bottom of the piston skirt. This reading should never be smaller than the previous reading.

_____ **B.** With the piston inverted, measure the diameter of the skirt just above the pin bore with a micrometer.

Chapter 13   Engine Block Components

_____ **C.** With the piston inverted, measure the diameter of the piston skirt just below the oil ring groove and compare to specifications. Replace the pistons if necessary.

5. Deglazing and Cleaning a Cylinder Wall:

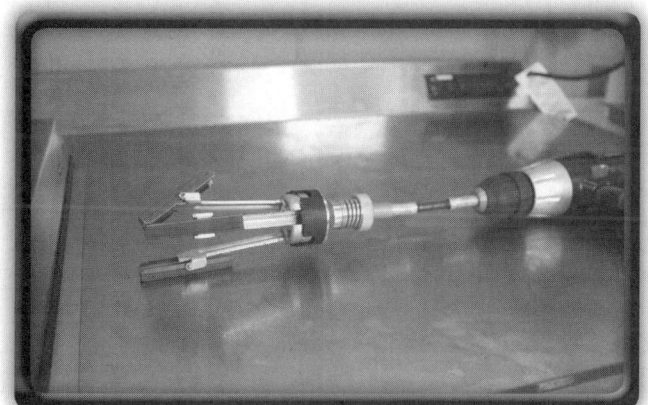

_____ **A.** Using a drill, attach the proper-sized flex hone. Apply honing oil to the cylinder walls.

_____ **B.** For final cleaning, use a pressure wash cabinet for the block. Clean the cylinders with soap and water using a cylinder block brush; rinse and blow-dry. Wipe down the cylinders with an antirusting spray.

_____ **C.** Using the drill with the attached flex hone, stroke up and down in the cylinder that you are honing for approximately 10 to 15 seconds, at a speed at which you get a good crosshatch. Pull the flex hone (ball hone) out of the bore before stopping the drill. Use a rag to wipe the cylinder out so you can inspect the cylinder walls. Repeat until a good crosshatch pattern is achieved.

Crossword Puzzle

Use the clues to complete the puzzle.

Across

2. A specially designed metal piece that supports circular moving parts.
5. The roughly shaped steel piece that has been cast or forged but has not undergone its final matching.
8. A metal surfacing process that uses nitrogen ammonia gas to create a very thin, highly hardened surface.
10. A metal casting process that uses a foaming process to create the pattern of the part being cast in metal.
11. The bottom area of the engine cylinder block where the crankshaft is located.
12. An area on the crankshaft that is offset from the crankshaft centerline and that becomes a counterweight.
14. An electromagnetic process used to locate cracks in ferrous engine blocks and cylinder heads and other ferrous metal parts.
15. The use of three different types of metals to build up a bearing insert to give it long-lasting wear characteristics.

Down

1. Aluminum, tin, and silicon alloy metals placed together with steel to form one piece of material and used for bearing materials.
3. A fastening between two parts that is activated by friction after the two parts are pushed together.

4. The back-and-forth movement of the piston assembly inside the cylinder.
6. A metal device connected to the bottom of the engine cylinder block to create strength for the main bearing caps.
7. A metal ring that is placed in a square groove around a piston for sealing purposes.
9. The surface area at the top of the engine block against which the cylinder head seals.
13. The radius portion of an inside corner that reduces stress at the corner.

ASE-Type Questions

Read each item carefully, and then select the best response.

_____ 1. Tech A says that soft plugs or freeze plugs are designed to prevent the coolant from freezing. Tech B says that soft plugs or freeze plugs were designed to plug holes or galleys. Who is correct?
   A. Tech A
   B. Tech B
   C. Both A and B
   D. Neither A nor B

_____ 2. Tech A says that most pistons use one compression ring and one oil control ring. Tech B says that most pistons use two compression rings and one oil control ring. Who is correct?
   A. Tech A
   B. Tech B
   C. Both A and B
   D. Neither A nor B

_____ 3. Tech A says that the thrust bearing should be aligned by tapping the crankshaft forward and backward before tightening the main bearing cap. Tech B says that the thrust bearing is located on the flywheel. Who is correct?
   A. Tech A
   B. Tech B
   C. Both A and B
   D. Neither A nor B

_____ 4. Tech A says that using a thicker head gasket would lower the compression ratio. Tech B says that the effective compression ratio is affected by valve timing. Who is correct?
   A. Tech A
   B. Tech B
   C. Both A and B
   D. Neither A nor B

_____ 5. Tech A says that when removing the pistons, you must first remove the piston rings. Tech B says that cracks in a cylinder can be measured with a dial indicator, and are OK as long as they are less than 0.003" wide. Who is correct?
   A. Tech A
   B. Tech B
   C. Both A and B
   D. Neither A nor B

_____ 6. Tech A says that a piston is manufactured slightly oval in shape. Tech B says that the piston pin is slightly oval in shape to allow for heat expansion and lubrication. Who is correct?
   A. Tech A
   B. Tech B
   C. Both A and B
   D. Neither A nor B

_____ 7. Tech A says that cylinder walls are manufactured with a slight taper so that the piston rings will seal tighter at TDC. Tech B says that piston coatings assist in the control of piston temperature, which allows for tighter bore clearances and greater engine efficiency. Who is correct?
   A. Tech A
   B. Tech B
   C. Both A and B
   D. Neither A nor B

_____ 8. Tech A says that fractured rods have a unique mating surface and that only the cap that was fractured is the one that will fit that rod. Technician B says that a powdered metal rod is less expensive, lighter, and as strong as a forged rod. Who is correct?
   A. Tech A
   B. Tech B
   C. Both A and B
   D. Neither A nor B

_____ 9. Tech A says that when grinding main journals, an undersized main bearing will be required. Technician B says that the main bearing journals should be measured for minimum diameter, out-of-round, and taper. Who is correct?
   A. Tech A
   B. Tech B
   C. Both A and B
   D. Neither A nor B

_____ 10. Tech A says that a failed harmonic balancer can cause a crankshaft to break. Tech B says that the crankshaft rod journals are typically measured using a dial indicator. Who is correct?
   A. Tech A
   B. Tech B
   C. Both A and B
   D. Neither A nor B

_____ 11. Tech A says that main and rod bearings are glued in place with special bearing cement. Tech B says that main and rod bearings are held in place by a precise amount of "bearing crush." Who is correct?
   A. Tech A
   B. Tech B
   C. Both A and B
   D. Neither A nor B

Engine Machining

CHAPTER 14

Tire Tread:
© AbleStock

Chapter Review

The following activities have been designed to help you refresh your knowledge of this chapter. Your instructor may require you to complete some or all of these activities as a regular part of your training program. You are encouraged to complete any activity that your instructor does not assign as a way to enhance your learning.

Matching

Match the following terms with the correct description or example.

- **A.** Ball hone
- **B.** Bob weights
- **C.** Crosshatch
- **D.** Cubic boron nitride cutter
- **E.** Engine block deck
- **F.** Induction hardening
- **G.** Interference angle
- **H.** Nitriding

_____ 1. A metal surface-hardening process using nitrogen and high temperatures.

_____ 2. An engine block resurfacing tool used on cast iron parts.

_____ 3. Weights used during the balancing process to mimic the exact weight of the piston and connecting rod weights for that particular rod bearing journal.

_____ 4. The portion of the engine cylinder block the head gasket lies on and the cylinder head is bolted to.

_____ 5. A pattern of lines placed at angles to each other, appearing as a series of Xs across a surface.

_____ 6. A process for hardening metal by touching it to a device that raises the temperature of the part with heat or magnets and then suddenly cooling it in water, oil, or another chemical.

_____ 7. An assembly of metal rods, with abrasive stones attached, that is inserted into the cylinder and spun to break the glazing off the cylinder walls and make a new crosshatched pattern.

_____ 8. The angle formed between the valve face and the valve seat; usually it is ½ to 1 degree.

Multiple Choice

Read each item carefully, and then select the best response.

_____ 1. What type of engine block cutting bit would you choose for an aluminum engine block?
  - **A.** Tungsten carbide cutter
  - **B.** Polycrystalline diamond cutter
  - **C.** Ball hone
  - **D.** Cubic boron nitride cutter

_____ 2. What type of engine block cutting bit would you choose for a cast iron engine block?
  - **A.** Ball hone
  - **B.** Polycrystalline diamond cutter
  - **C.** Tungsten carbide cutter
  - **D.** Cubic boron nitride cutter

____ 3. Dry cylinder sleeves require the cylinder to be bored out an additional 0.002" to 0.0025", this extra machining is called ____.
   A. overbore
   B. sleeve clearance
   C. interference fit
   D. taper

____ 4. A long bar used to position and align a single-point tool, such as an engine block boring machine is called a(n) ____.
   A. boring bar
   B. reference bar
   C. engine block dowel
   D. mainline bar

____ 5. What term refers to the finish produced when the crosshatch scratches have been worn away, leaving the cylinder smooth and shiny and unable to retain oil?
   A. Glazing
   B. Enameled
   C. Hone
   D. Varnished

____ 6. To correct a bore alignment issue, the engine block must be ____.
   A. line bored
   B. ball honed
   C. line honed
   D. either A or C

____ 7. What process adds materials to the journal by melting and spraying molten metal onto the crankshaft journal surface so there is material that can be ground?
   A. Submerged arc welding
   B. Plasma welding
   C. Spray welding
   D. Chrome plating

____ 8. This process takes that sharp edge at the top of the oil hole and makes it more like a 45-degree angle.
   A. Chamfering
   B. Honing
   C. Polishing
   D. Reaming

____ 9. A good way to check the valve seat contact on the valve is to use ____.
   A. dry-erase marker
   B. plastigauge
   C. prussian blue
   D. either A or C

____ 10. Weights used during the balancing process that are used to mimic the exact weight of the piston and connecting rod weights for that particular rod bearing journal are called ____.
   A. rotating weights
   B. lead shot
   C. bob weights
   D. reciprocating weights

True/False

If you believe the statement to be more true than false, write the letter "T" in the space provided. If you believe the statement to be more false than true, write the letter "F".

____ 1. It is best to leave machining to the Automotive Service Excellence (ASE) Master Certified Machinist professionals.

____ 2. A 60-Ra finish is smoother than a 30-Ra finish.

____ 3. With some of the newer CNC-style machines, engine block measurements are taken from the top of the engine block dowel pinhole.

_____ 4. Machining a cylinder bore restores it to its factory dimensions allowing the original size pistons to be reused.
_____ 5. A portable boring bar should not be used on modern thinner-walled engine blocks.
_____ 6. A good crosshatch helps to trap the oil and retain it in the cylinder bores where it is needed.
_____ 7. The wider the Plastigauge, the larger the bearing clearance.
_____ 8. Nitriding produces a harder, more durable surface than induction hardening.
_____ 9. The press-fit wrist pin is typically held in place with a snap ring on each side of the piston pin boss.
_____ 10. Pressure testing is normally used on cast iron cylinder heads since the magnaflux process cannot be used.
_____ 11. Overhead cam engines have camshafts that typically ride in removable bearings in the cylinder head.
_____ 12. Knurling is when a bit with a spiral groove is threaded through or run through the valve guide.
_____ 13. When using grinding stones, a three-angle grind is typically used to properly form the valve seat.
_____ 14. Some engines are internally balanced, while others are externally balanced.
_____ 15. Reciprocating weight is the amount of weight that is moving in a circular motion.

Fill in the Blank

Read each item carefully, and then complete the statement by filling in the missing word(s).

1. _____ _____ average is a measure of how rough a surface is at the microscopic level.
2. The desired procedure in surfacing the engine block's deck is to surface the engine block parallel to the _____ where the crankshaft is installed.
3. The computer _____ _____ machine is a tool that uses computer programs to automatically execute a series of machining operations.
4. Using _____ allows the original size of pistons to be used, restoring the factory bore dimensions.
5. Hone stones are used to sand down and file away the cylinder walls, leaving a unidirectional finish also known as a _____ pattern.
6. Bore alignment can be checked for straightness by using a straightedge with a _____ _____.
7. _____ is a thin calibrated piece of soft plastic string that is laid on the main bearing journal of the crankshaft and squished between the journal and the bearing.
8. Powdered metal connecting rods are _____ _____ to ensure perfect alignment.
9. Before any machining can begin after the disassembly and cleaning of the cylinder head, the cylinder head should be _____ or dye checked to locate any external cracks.
10. A cylinder head can be surfaced using a surface grinder using a stone wheel or a _____ - _____ surfacer using a CBN or a PCD cutting bit.

Labeling

Label the following diagrams with the correct terms.

1. Measuring the deck height from the setting fixture:

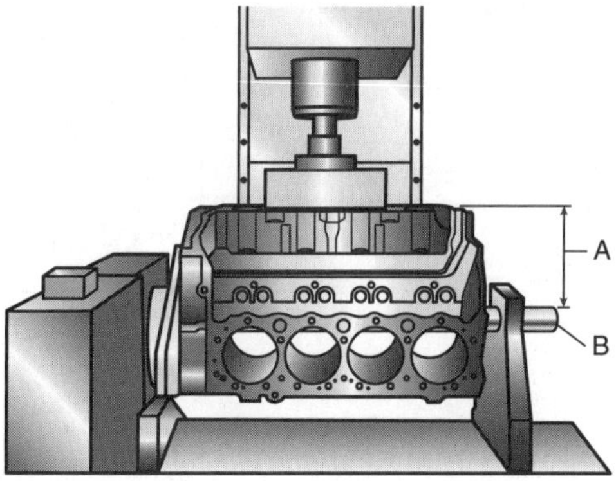

   A. _____
   B. _____

2. Cylinder sleeves:

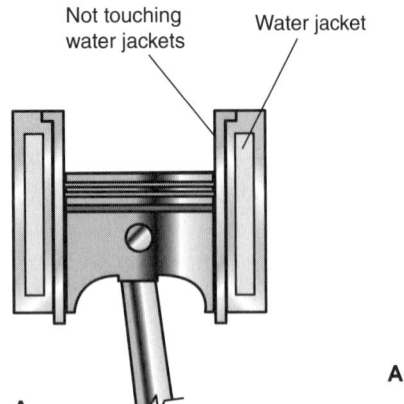

A. _____

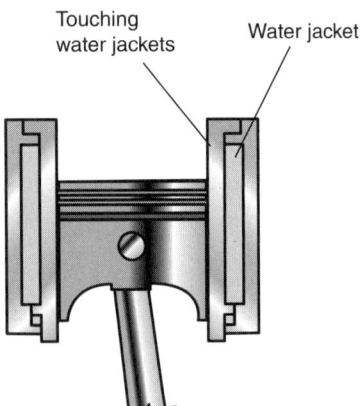

B. _____

3. Cam bores using removable bearing caps:

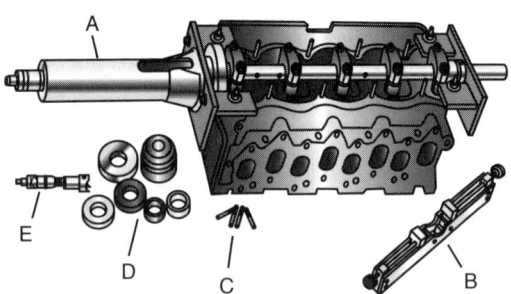

A. _____

B. _____

C. _____

D. _____

E. _____

4. Valve seat with proper angles:

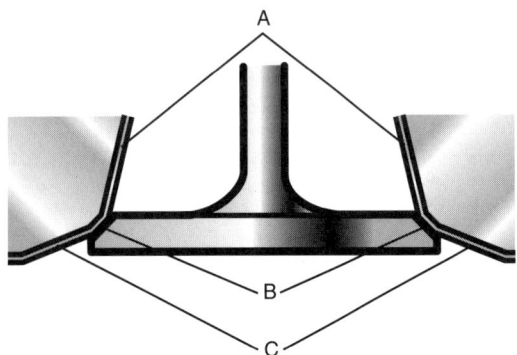

A. _____

B. _____

C. _____

Chapter 14 Engine Machining

5. Engine balancing:

A. _____

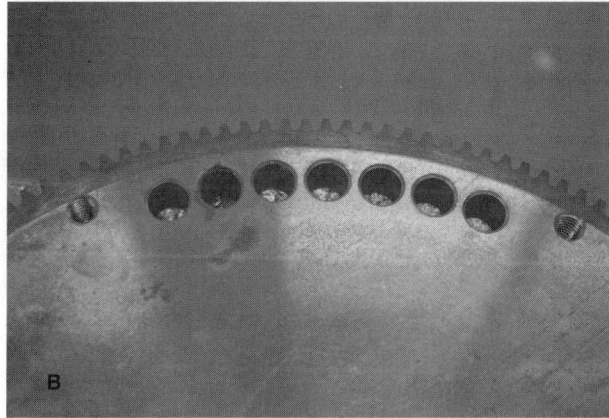

B. _____

Crossword Puzzle

Use the clues to complete the puzzle.

Across

5. An expansive bench unit complete with computer controls, a lathe, and drill press for machining metal for the balancing procedure, and all necessary adapters for most applications of crankshaft assemblies.

7. A process in which a drill bit with a spiral groove is threaded through or run through the valve guide to create raised ridges on either side of the groove, effectively shrinking the size of the valve guide.

8. The movement of the piston up and down.

9. A 2" (51 mm) thick plate that is bolted where the cylinder head is fastened on the engine block.

10. Metals that have an iron compound in their makeup or that are magnetic.

13. The surface area at the top of the engine block against which the cylinder head seals.

15. An object that is smaller in diameter at one end.

Down

1. The smoothing-out process of the cylinder walls performed after the boring process. This is the final step in refinishing the cylinder; it creates the proper crosshatched pattern necessary for the piston rings to seal and seat into the combustion chamber to give the proper seal.

2. The surface against which a valve rests during the portion of the engine operating cycle when that valve is closed.

3. The area of the crankshaft that meets up with the rod bearing journal or main bearing journal.

4. A process for checking the aluminum cylinder heads for cracks using air pressure.

Chapter 14   Engine Machining

6. A tool used to make a hole in or through an object.
11. A straight line extending from the center of a circle to its edge or from the center of a sphere to its surface.
12. The imaginary centerline down the engine cylinder block.
14. A cone-shaped tool that encloses and grips a rod or shaft when inserted into the sleeve of a lathe or other machine.

ASE-Type Questions

Read each item carefully, and then select the best response.

_____ 1. Tech A says that failure analysis will assist the technician in making an estimate of repairs. Tech B says that failure analysis will help prevent reoccurence of the problem. Who is correct?
  A. Tech A
  B. Tech B
  C. Both A and B
  D. Neither A nor B

_____ 2. Tech A says that Ra is critical in the machining process to ensure gasket sealing. Tech B says that an Ra rating of 0-1 is the best sealing surface. Who is correct?
  A. Tech A
  B. Tech B
  C. Both A and B
  D. Neither A nor B

_____ 3. Tech A says that full floating pistons use keepers to retain the piston pin. Tech B says that some piston pins are press-fit in the rod. Who is correct?
  A. Tech A
  B. Tech B
  C. Both A and B
  D. Neither A nor B

_____ 4. Tech A says that when measurement of the cylinder bore is out of specifications, just install an oversized piston. Tech B says that cylinders can typically be bored out to only a few specific sizes. Who is correct?
  A. Tech A
  B. Tech B
  C. Both A and B
  D. Neither A nor B

_____ 5. Tech A says that after using a portable boring bar, the cylinder bore will need to be honed by 0.0015" to 0.002". Tech B says that the portable boring bar is the best way to go as it can be used on all current engines. Who is correct?
  A. Tech A
  B. Tech B
  C. Both A and B
  D. Neither A nor B

_____ 6. Tech A says that when boring a cylinder, it should be bored to the final cylinder size. Tech B says that it should be bored smaller than the final size to allow for honing of the cylinder. Who is correct?
  A. Tech A
  B. Tech B
  C. Both A and B
  D. Neither A nor B

_____ 7. Tech A says that a torque plate bolts to the bottom of the block to strengthen the block when boring. Tech B says that the use of a torque plate compensates for the cylinder head being torqued in place. Who is correct?
  A. Tech A
  B. Tech B
  C. Both A and B
  D. Neither A nor B

_____ 8. Tech A says the proper crosshatch pattern is critical for proper lubrication of the pistons and rings. Tech B says that the smoother the cylinder walls, the better the rings will seal. Who is correct?
   A. Tech A
   B. Tech B
   C. Both A and B
   D. Neither A nor B

_____ 9. Tech A says that main bearing clearance can be calculated by knowing the inner diameter of the main bearing and the outer diameter of the main journal. Tech B says that rod bearing and main bearing clearance can be measured with Plastigauge. Who is correct?
   A. Tech A
   B. Tech B
   C. Both A and B
   D. Neither A nor B

_____ 10. Tech A says that bearing crush is when the main caps are torqued and the bearing is crushed against the crankshaft. Tech B says that crankshaft journals can be machined undersize and undersized bearings can be used. Who is correct?
   A. Tech A
   B. Tech B
   C. Both A and B
   D. Neither A nor B

Engine Assembly

CHAPTER 15

Chapter Review

The following activities have been designed to help you refresh your knowledge of this chapter. Your instructor may require you to complete some or all of these activities as a regular part of your training program. You are encouraged to complete any activity that your instructor does not assign as a way to enhance your learning.

Matching

Match the following terms with the correct description or example.

- **A.** Degree wheel
- **B.** Outside micrometer
- **C.** Pip mark
- **D.** Torque angle
- **E.** Torque-to-yield

_____ **1.** A tightening procedure in which a bolt is designed to be slightly elastic when tightened; the elastic bolt retains an even pressure on the head gasket.

_____ **2.** A precision measuring instrument meant to measure the outside of components. It is usually accurate to 0.0001" (0.0025 mm).

_____ **3.** A disc with 360 one-degree markings near its outer edge; it bolts to the front of the crankshaft and is used to check valve and cam timing.

_____ **4.** A small indent or dimple on the piston ring that indicates which side of the ring is installed upward. It is also used on some timing sprockets.

_____ **5.** A tightening procedure in which a bolt is torqued to a set torque value and then tightened using a measured angle instead of a torque value. Example: A bolt is torqued to 40 ft-lb and then tightened another 90 degrees.

Multiple Choice

Read each item carefully, and then select the best response.

_____ **1.** Which tool would you use to measure piston size?
   - **A.** Depth gauge
   - **B.** Inside micrometer
   - **C.** Bore gauge
   - **D.** Outside micrometer

_____ **2.** Which tool would you use to measure cylinder size?
   - **A.** Feeler gauge
   - **B.** Outside micrometer
   - **C.** Bore gauge
   - **D.** Tape measure

____ 3. What tool is used to measure the width of the ring end gap?
   A. Steel rule
   B. Feeler gauge
   C. Inside micrometer
   D. Gap tool

____ 4. In which piston design is the pin retained by a pair of snap rings?
   A. Floating pin
   B. Semi-floating pin
   C. Semi-press-fit
   D. Press-fit

____ 5. If you need to check the bearing clearance with the crankshaft installed, then you should use a(n) _____.
   A. outside micrometer
   B. bore gauge
   C. feeler gauge
   D. plastic gauging material

____ 6. All of the following tools may be used to measure the inside diameter of the main bearing bore, *except*:
   A. Inside micrometer
   B. Snap gauge
   C. Feeler gauge
   D. Bore gauge

____ 7. An extruded plastic thread that comes in different sizes, used to check clearances between parts is called _____.
   A. prussian blue
   B. plastigauge
   C. teflon tape
   D. clearance tape

____ 8. Make sure that the _____ are facing up when installing the compression rings.
   A. tabs
   B. tangs
   C. pip marks
   D. arrows

____ 9. If any of the head bolts go into holes that extend into the _____, it is essential to put some nonhardening sealer on the threads of the bolts to keep the coolant from leaking past the threads.
   A. water jacket
   B. engine block
   C. water pump
   D. wet sleeve

____ 10. The _____ may be located in the oil pan, on the side of the block, or in the timing cover.
   A. timing chain
   B. oil pump
   C. crankshaft
   D. camshaft

True/False

If you believe the statement to be more true than false, write the letter "T" in the space provided. If you believe the statement to be more false than true, write the letter "F".

____ 1. Measuring the piston-to-cylinder wall clearance and the piston ring end gap ensures that the machining was performed properly.

____ 2. Some snap rings are directional, which means they must be installed with the correct side facing the wrist pin.

____ 3. The main bearing oil hole provides the supply of oil for the crankshaft.

____ 4. Before using a micrometer to measure the crankshaft, it is important to make sure the crankshaft is clean and free of oil and grease.

Chapter 15 Engine Assembly

_____ 5. The same method used to check main bearing clearances can be used to check the rod bearing clearance.
_____ 6. The elastic gauge method is used to check clearance by putting clay on top of the piston where the valves will open.
_____ 7. Any threaded plugs such as the rear oil gallery plugs should be coated with Teflon tape prior to installation.
_____ 8. Never reuse torque-to-yield bolts, as they will likely fail in use.
_____ 9. Most engines use a composition-type gasket that requires sealant on both sides.
_____ 10. The torque angle method is a more precise way than the torque-to-yield method of torquing a standard fastener to a predetermined tension.
_____ 11. On OHC engines, the timing chains or belts are installed after the cylinder head is installed.
_____ 12. Use a thin coat of RTV silicone when installing neoprene gaskets with multi-sealing edges.
_____ 13. One way to pre-oil the engine is to spin the oil pump with a drill attached to the pump drive.
_____ 14. The camshaft and drive belt/chain assembly link the valve train system to the crankshaft and determine the right time in the engine cycle for the valves to open and close.
_____ 15. It is sometimes necessary to replace the valve stem seals on an assembled engine.

Fill in the Blank

Read each item carefully, and then complete the statement by filling in the missing word(s).

1. Before installing piston rings on the piston, you need to check the _____ _____ _____.
2. The _____-_____ pin moves freely in the piston bore but is press-fit in the connecting rod bore.
3. It is very important to perform a _____ _____ _____ of the rotating assembly; it will determine if additional machining or exchange of any parts will be necessary.
4. To insert the main bearing, start the _____ of the bearing into the main journal slot.
5. If the thrust bearing is not a flange thrust bearing, _____ will need to be applied to the back side of the bearing to hold the bearing in place until the crankshaft is installed or the main cap is installed.
6. Use a(n) _____ _____ to measure each main bearing journal.
7. When you subtract the crankshaft main journal readings from the main bearing bore readings, the difference will give you the _____ _____.
8. Checking the thrust bearing end play can be done with a _____ _____ after the crankshaft is installed.
9. Most engines have a specification for piston height as compared to the cylinder block _____.
10. Once the piston rings are aligned and lubricated, they need to be compressed with a _____ _____ so they will fit into the cylinder.
11. OHC chain drives typically are of the _____ _____ style, which looks a lot like a bicycle chain.
12. The _____ _____ generally includes a strainer that prevents chunks of debris from entering the oil pump.
13. To help hold the gaskets in place, you can use a product made by 3M called _____ _____, which is contact cement in a liquid or spray form.
14. There is a sequence that must be followed during installation of oil pans and _____ _____ since the parts can overlap one another.
15. Most newer engines do not have an adjustable _____ _____ sensor, so indexing is usually needed only on sensors that can be manually adjusted.

Labeling

Match the following images with the correct descriptions below.

1. Piston checks:

A. _____

B. _____

C. _____

D. _____

1. Measuring the piston-to-cylinder clearance using a bore gauge
2. Checking compression ring end gap
3. Measuring a piston
4. Squaring up piston rings

2. Installing main bearings:

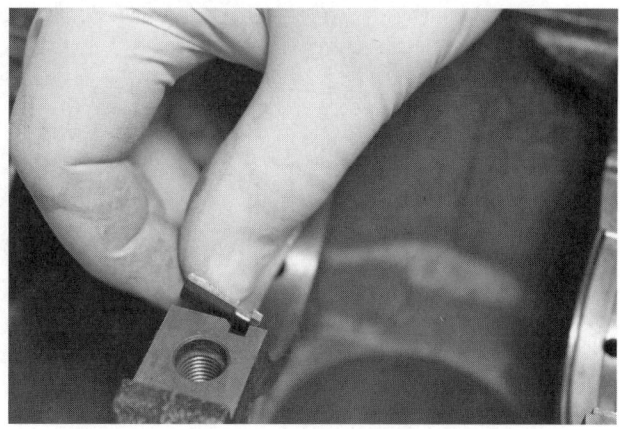

A. _____

B. _____

1. Oil hole lined up
2. Installing main bearing insert

Skill Drills

Test your knowledge of skill drills by filling in the correct words in the photo captions.

1. Checking Piston-to-Valve Clearance (Clay Method):

Step 1: _____ the valve train for the number _____ cylinder.

Step 2: Install _____ test springs. Put a little piece of wood, or something similar, under the valve to keep it in the _____ position while you install the lightweight springs.

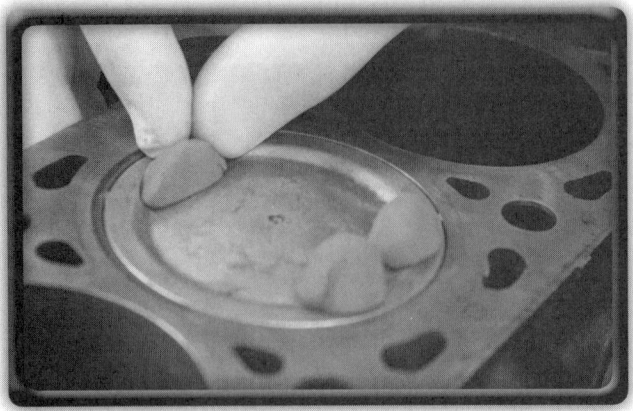

Step 3: Make balls using modeling clay, one for each valve. Stick them to the _____ in the valve reliefs that are cut into the _____ of the piston, and coat the clay with clean engine oil.

Step 4: Place the cylinder head on the _____. Tighten the _____ bolts by hand.

Step 5: Install the valve train and adjust the valves to the specified _____, or if using hydraulic lifters, zero _____.

Step 6: Carefully rotate the _____ until you see the exhaust and _____ valves open and close two times each.

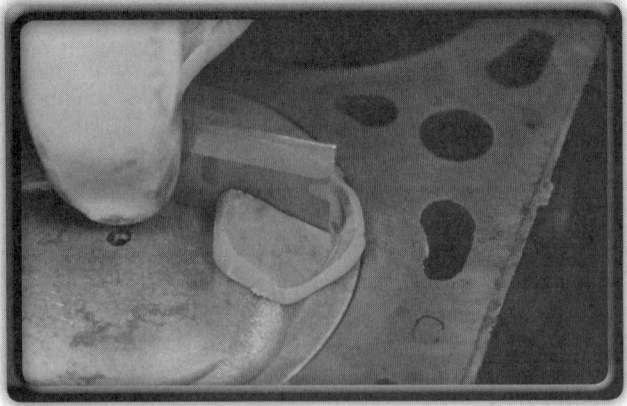

Step 7: Remove the valve train and head. Cut each piece of _____ and remove half. Measure the distance from the piston to the top of the valve imprint with a ruler or the _____ end of a dial caliper.

Chapter 15 Engine Assembly

2. Installing Engine Plugs:

Step 1: Start with the _____ block sitting on the floor with the rear area facing _____. Apply a small amount of _____ to the threaded oil gallery plugs and screw them in.

Step 2: Clean the edges of the _____ water jacket holes before you install the _____ plugs.

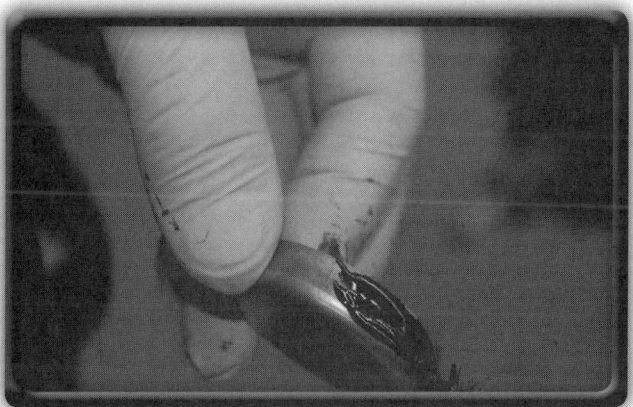

Step 3: Coat the soft plug with a nonhardening _____. Wipe it thin and even, being careful not to leave any bare spots.

Step 4: Carefully tap in each _____ _____ using an impact socket and hammer.

Step 5: Turn the block over so that the _____ is facing up, and select a punch that is smaller than the inner _____ of the front oil gallery plugs. Wipe sealer evenly around each plug, and tap it in until it is 1/16" (1.6 mm) below the rim.

Step 6: When the oil gallery plugs are installed, use a _____ to indent the surface of the block around each opening in at least two places. This will keep the _____ from backing out when the oil pressure builds. **(There is no image associated with this step.)**

Step 7: Turn the block on its side. Clean, apply _____, and install the soft plugs in the side of the block, making sure not to miss any.

Step 8: Check for miscellaneous plugs. Almost every engine has some hidden plug, like an _____ _____ plug, in an odd place. Carefully review your disassembly notes and the service information to make sure you do not miss any plugs. **(There is no image associated with this step.)**

3. Installing the Heads:

Step 1: Clean all the head bolt threads and _____. Use sealer on any bolts that extend into the _____ _____. Lubricate the threads and under the heads on other bolts, if specified.

Step 2: Clean the deck surface of the block and head, and install the head gaskets. Look for any _____ _____ _____ or _____ labels on the head gaskets.

Step 3: On OHC _____ _____, make sure the crankshaft is set to the _____ position before installing the head.

Step 4: On OHC engines, make sure the camshaft is set to the specified position before installing the _____. Failure to do so could lead to _____ or _____ valves.

Step 5: Gently lay the head over the alignment _____ of the block, and start threading in the bolts. If _____-_____ head bolts are used, tighten each bolt in the specified sequence to the specified torque. Torquing may need to be performed in several stages.

Step 6: Install the torque angle gauge, and tighten each bolt in the specified sequence to the specified angle. If torquing _____ head bolts, torque them in the specified sequence in three increments, resetting the torque wrench and increasing by _____-_____ each time.

4. Pre-oiling the Engine:

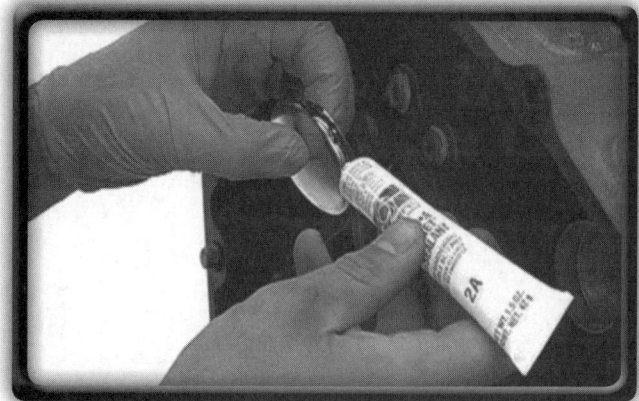

Step 1: Install the rear camshaft plug, if it has not already been done. Spread a _____ layer of _____ sealant on the edge of the plug itself, and hammer it in place.

Step 3: Install an oil pressure gauge into one of the _____ that lead to the _____ _____. **(There is no image associated with this step.)**

Step 2: Install the oil filter. _____ the rubber seal of the oil filter with a little engine oil, and _____ it on hand tight.

Step 4: If you have not done so already, install the _____-_____ rear main seal to finish closing off the _____ _____. **(There is no image associated with this step.)**

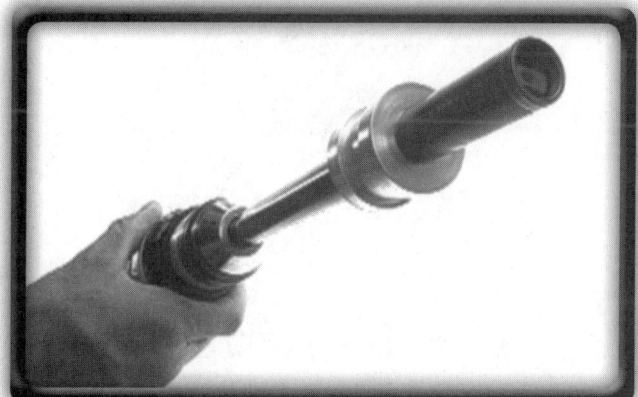

Step 5: Fill the crankcase with the type and _____ of engine oil recommended in the repair information. If the oil pump is driven by the distributor, use a drill to _____ the oil pump.

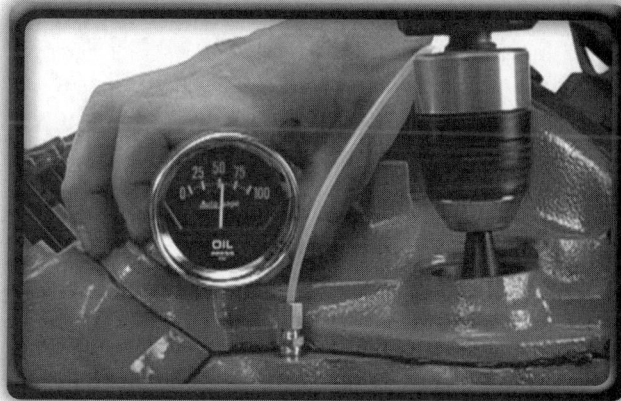

Step 6: Check that oil is reaching the _____ _____. You may need to rotate the engine two turns to get oil to flow to each valve.

Step 7: If the oil pump is driven by the crankshaft, use a pressurized _____-_____ to pressurize the _____ system. Once the engine has been pre-oiled, install the _____ _____.

5. Inspecting and Replacing the Camshaft and Drive Belt/Chain:

Step 1: Determine all specifications for timing chain or belt _____ assembly according to the manufacturer of the engine assembly being serviced. Remove all _____. Remove all components that cover the timing chain/belt assembly, such as the harmonic balancer, water pump, alternators, and power steering pump.

Step 2: Remove the camshaft timing chain/belt cover. Inspect the _____ for _____ _____, and replace if worn. Look for causes for the worn cover.

Step 3: With the timing chain/belt cover off and the timing gears and belts/chains in full view, turn the engine over _____ to line up the timing marks of the crankshaft and camshaft _____ with the appropriate marks on the block and head.

Step 4: Inspect the belt/chain and measure for _____, and replace if not within specifications. Measure the _____ between tensioners, if applicable, and replace if not within specifications.

Step 5: Remove the _____ or _____ following the specified procedure. Use the proper cam holding tool, if specified, to prevent damage to valves.

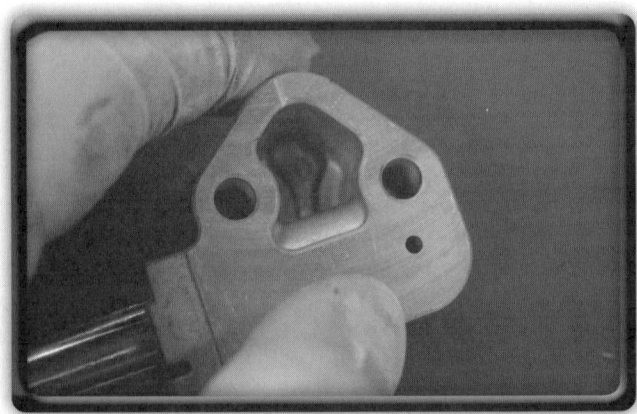

Step 6: If the tensioner is _____ operated, check the oil passages to the _____ for clogs or buildup of dirt sludge, and clean or replace according to recommendations.

Chapter 15   Engine Assembly

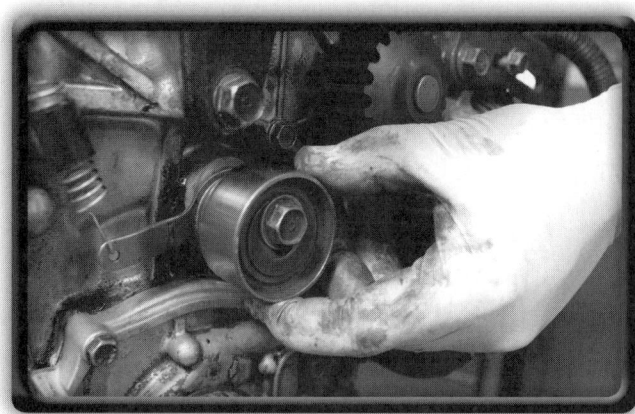

Step 7: Inspect any guide pulleys for smooth _____ on their bearings, and replace if damaged. With the chain/belt removed, inspect the cam sprockets visually for wear, cracking, and damage. Inspect _____ for backlash and end play, if applicable, on the vehicle.

Step 8: On engines with any type of variable valve timing, check components for worn and damaged parts on the gears, inspect any _____ _____ _____ for leaks, and perform other tests on components according to the manufacturer's recommendations.

Step 9: Reinstall new parts according to the manufacturer's recommendations. Reassemble the timing chain/belt assembly. Turn the crankshaft _____ complete revolutions by hand, and recheck the timing marks and belt/chain tension. _____ components following the specified procedure.

Crossword Puzzle

Use the clues to complete the puzzle.

Across

3. A raised or slightly bent corner on an engine bearing shell that fits into a matching recess in the bearing mounting surface to keep the bearing from rotating in the bore.
4. An engine assembly that includes the engine block, camshaft, timing set, pistons, rods, and crankshaft installed.
5. That area of the engine block casting that holds the engine coolant; cylinder head bolt holes are sometimes drilled through to the water jacket, requiring the bolt threads to be sealed.

Down

1. An engine assembly that includes the short block and adds the valve train cylinder heads, timing cover, oil pan, and, in some cases, manifolds.
2. The trademarked name for a plastic gauging material used to check the clearances between two surfaces.

ASE-Type Questions

Read each item carefully, and then select the best response.

_____ 1. Tech A says that measuring piston ring end gap in the cylinder bore will verify that the rings are properly sized for the bore. Tech B says that measuring piston-to-cylinder bore clearance determines the compression ratio of the cylinder. Who is correct?
   A. Tech A
   B. Tech B
   C. Both A and B
   D. Neither A nor B

_____ 2. Tech A says that semi-floating wrist pins can be pressed into the rods. Tech B says that a rod heater can be used to install semi-floating wrist pins. Who is correct?
   A. Tech A
   B. Tech B
   C. Both A and B
   D. Neither A nor B

_____ 3. Tech A says that the bearing tang's purpose is to be used as a balance pad that can be ground, if needed, to lighten the bearing. Tech B says that bearing tangs are designed to prevent the bearing from spinning in its bore. Who is correct?
   A. Tech A
   B. Tech B
   C. Both A and B
   D. Neither A nor B

_____ 4. Tech A says that the wider the Plastigauge is squished out, the smaller the bearing clearance. Tech B says that the narrower the Plastigauge is squished out, the smaller the bearing clearance. Who is correct?
   A. Tech A
   B. Tech B
   C. Both A and B
   D. Neither A nor B

_____ 5. Tech A says that when checking rod bearing clearance, it is best to install Plastigauge on the bearing in the rod. Tech B says that when checking rod bearing clearance, it is best to install Plastigauge on the cap bearing. Who is correct?
   A. Tech A
   B. Tech B
   C. Both A and B
   D. Neither A nor B

_____ 6. Tech A says that rod bearings are held in place by the torque of the rod bolts crushing the bearing into the bore. Tech B says that piston rings are installed so that all ring gaps are not lined up with each other. Who is correct?
   A. Tech A
   B. Tech B
   C. Both A and B
   D. Neither A nor B

_____ 7. Tech A says that when installing a performance camshaft, the cam-to-lifter clearance should be measured and recorded. Tech B says that when installing a performance camshaft, piston-to-valve clearance should be checked. Who is correct?
   A. Tech A
   B. Tech B
   C. Both A and B
   D. Neither A nor B

_____ 8. Tech A says that when installing rod caps on a rod, the bearing tangs should be installed on the same side of the rod. Tech B says that when installing rod caps on a rod, the bearing tangs should be installed on opposite sides of the rod. Who is correct?
   A. Tech A
   B. Tech B
   C. Both A and B
   D. Neither A nor B

_____ 9. Tech A says that torque-to-yield head bolts are special bolts that are reusable. Tech B says that head bolts should be torqued in only one step to provide the proper bolt stretch. Who is correct?
   A. Tech A
   B. Tech B
   C. Both A and B
   D. Neither A nor B

_____ 10. Tech A says that the use of some RTV sealants to seal components on an engine can damage the oxygen sensor. Tech B says that prelubing an engine is very important. Who is correct?
   A. Tech A
   B. Tech B
   C. Both A and B
   D. Neither A nor B

CHAPTER 16

Automatic Transmission Fundamentals

Tire Tread:
© AbleStock

Chapter Review

The following activities have been designed to help you refresh your knowledge of this chapter. Your instructor may require you to complete some or all of these activities as a regular part of your training program. You are encouraged to complete any activity that your instructor does not assign as a way to enhance your learning.

Matching

Match the following terms with the correct description or example.

- **A.** Extension housing
- **B.** Front hydraulic pump
- **C.** Gear ratio
- **D.** Helical-cut gear
- **E.** One-way clutch
- **F.** Planetary gears
- **G.** Ravigneaux gear set
- **H.** Simpson gear set
- **I.** Spur gear
- **J.** Thrust bearing

_____ 1. A type of gear in which the teeth of the gear are cut in a straight line down the axis of the gear.

_____ 2. A hydraulic pump used to supply lubrication oil and hydraulic pressure to the different components inside the automatic transmission.

_____ 3. A type of gear in which the teeth are cut in a spiral down the axis of the gear.

_____ 4. Also called a Torrington bearing, a small roller bearing assembly with the rollers laid flat axially around the centerline of the bearing. The bearing is used to control forward and backward movement of a part in an automatic transmission.

_____ 5. The small gears in a planetary gear set that revolve around the sun gear; also known as pinion gears.

_____ 6. A component of the automatic transmission housing that covers the output shaft of the transmission. The extension housing also supports the end of the driveshaft and may hold components such as the vehicle speed sensor, speedometer drive assembly, and governor assembly.

_____ 7. A type of compound planetary gear set that uses two different sun gears and two different-diameter planets while only using one ring gear.

_____ 8. The ratio of the size or teeth of one gear compared to the size or teeth of a mating gear.

_____ 9. A type of gear set with two planetary gear sets that share a common sun gear.

_____ 10. A type of holding device used by an automatic transmission to stop the movement of one component of a planetary gear set. It allows free spinning in one direction but will lock up when the part attempts to spin in the opposite direction.

Multiple Choice

Read each item carefully, and then select the best response.

_____ 1. Transmissions that use two pulleys that change diameter in response to vehicle load and speed are known as _____.
- **A.** dual-clutch transmissions
- **B.** continuously variable transmissions
- **C.** power-splitting transmissions
- **D.** automatic transmissions

Chapter 16 Automatic Transmission Fundamentals

_____ 2. With which type of hybrid drive train does the engine have two or more methods for the power to flow through the transmission?
   A. Series hybrid
   B. Parallel hybrid
   C. Dual-clutch
   D. Continuously variable

_____ 3. Early automatic transmissions used a _____ rather than a torque converter.
   A. fluid coupler
   B. flywheel
   C. transaxle
   D. countershaft

_____ 4. In its simplest form, a single-stage torque converter has three elements including all of the following, *except*:
   A. Turbine
   B. Impeller
   C. Fluid coupler
   D. Stator

_____ 5. The _____ has a small set of curved blades attached to a central hub and is positioned between the impeller and the turbine.
   A. fluid coupler
   B. pinion
   C. servo
   D. stator

_____ 6. An operating condition where the turbine is stationary and the engine throttle is wide open, making the rotational speed of the impeller as high as possible is called _____.
   A. stall
   B. torque
   C. choke
   D. turbo

_____ 7. Automatic transmission vehicles use a(n) _____, sometimes called a transmission cooler, in one tank of the radiator.
   A. evaporator
   B. heat exchanger
   C. blend box
   D. condenser

_____ 8. The difference in diameter between the driving gear and the driven gear is known as _____.
   A. leverage
   B. torque conversion
   C. gear ratio
   D. reduction

_____ 9. What type of gears are often used to change the direction of power flow by 90 degrees?
   A. Helical-cut gears
   B. Sun gears
   C. Planetary gears
   D. Hypoid gears

_____ 10. A friction-lined steel belt that wraps around the outside of a drum inside the automatic transmission is called a _____.
   A. band
   B. clutch
   C. pinion
   D. strap

_____ 11. Which of the following is a type of front hydraulic pump used in automatic transmissions?
   A. Gear pump
   B. Rotor pump
   C. Vane pump
   D. All of the above

Fundamentals of Automotive Technology Student Workbook

_____ **12.** Which of the following should never be used to lubricate an internal transmission seal?
   A. ATF fluid
   B. Grease
   C. Petroleum jelly
   D. Automatic transmission assembly lubricant

_____ **13.** What type of transmission seals come in several styles—continuous, butt cut, scarf cut, and step joint?
   A. Lip seals
   B. Square-cut seals
   C. Teflon seals
   D. O-rings

_____ **14.** Which of the following is an example of a compound planetary gear set?
   A. Ravigneaux gear set
   B. Continuously variable transmission
   C. Simpson gear set
   D. Both A and C

_____ **15.** What component is used in automatic transmissions to convert the hydraulic pressure acting on a piston to a mechanical force that is then applied to a brake band?
   A. Pinions
   B. Servos
   C. Simpson gears
   D. Fluid couplers

True/False

If you believe the statement to be more true than false, write the letter "T" in the space provided. If you believe the statement to be more false than true, write the letter "F".

_____ **1.** An automatic transmission can select and shift gears without input from the driver.
_____ **2.** In a manual transmission, engine torque is controlled by two, typically wet, clutches.
_____ **3.** Modern transmissions no longer use fluid couplers.
_____ **4.** During acceleration or hill climbing, the torque needed by the driveshaft can exceed the engine output torque.
_____ **5.** Planetary gears are held together in a planet carrier.
_____ **6.** Helical-cut gears tend to be much louder and do not offer as much strength as the other types of gears.
_____ **7.** Automatic transmission fluids are typically dyed purple for easy identification.
_____ **8.** On a manual transmission there is no flywheel; there is a thin, lightweight steel flex plate.
_____ **9.** If the tangs or flats are not engaged properly and the transmission is reinstalled into the vehicle, the front pump will be destroyed when the engine is first started.
_____ **10.** A basic planetary gear set has a sun gear, which meshes with planetary gears, also called planet pinions.
_____ **11.** A direct drive, or 1:1 ratio, is obtained by locking together any two members of the planetary gear set.
_____ **12.** Thrust bearings are a friction type of bearing similar to engine main bearings, they can be made of plastic or steel coated with a bronze bearing material.
_____ **13.** There are two types of clutches: multidisc and one-way.
_____ **14.** All bands inside an automatic transmission are externally contracting types.
_____ **15.** A servo-operated band can only hold a member stationary, but multidisc clutches can hold or drive individual members of the planetary set.

Fill in the Blank

Read each item carefully, and then complete the statement by filling in the missing word(s).

1. _____-_____ transmissions are a newer type of automatic transmission that uses two, typically wet, clutches.

2. In a _____ hybrid drive train, the electric motor is not typically able to propel the vehicle on its own. This electric motor is often placed between the engine and the transmission.

Chapter 16   Automatic Transmission Fundamentals

3. A series-parallel hybrid drive train uses what is called a _____-_____ transmission.
4. The _____ _____ is mounted between the engine and the transmission, in the same place as a manual transmission clutch.
5. A _____ _____ is basically two fans facing each other.
6. The _____ has a large number of vanes attached to the torque converter housing to form the driving member.
7. Combining rotary flow and vortex flow produces a progressive circular, or spiraling, motion known as _____ flow.
8. Planetary gears revolve inside a larger _____ gear that wraps around the outside of the whole planetary gear set.
9. _____ gears were used in early transmissions due to ease of manufacturing and lower cost.
10. _____ clutches are unique in that they not only can be used to hold a member of the planetary gear set, but they are the only type of component that can be used to drive a member of the planetary gear set.
11. One-way clutches can be either one-way _____ or _____.
12. A(n) _____ _____ vane pump reduces the pumping load on the engine when the transmission requires less fluid to be pumped.
13. When the driver shifts the transmission into park, a lever, called a _____ _____, is forced into notches cut into a hardened steel drum on the output shaft of the transmission.
14. The _____ _____ bushing is used to support the torque converter and sees the largest amount of wear in the transmission.
15. The number of plates installed determines the _____ _____ of the clutch.

Labeling

Label the following diagrams with the correct terms.

1. Dual-clutch transmission:

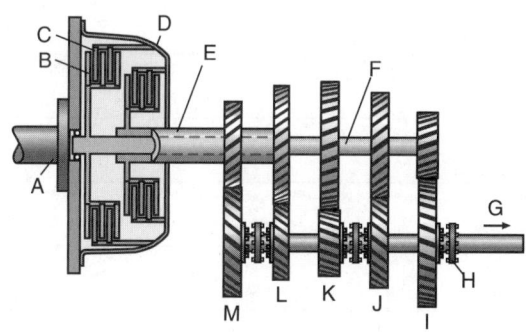

A. _____
B. _____
C. _____
D. _____
E. _____
F. _____
G. _____
H. _____
I. _____
J. _____
K. _____
L. _____
M. _____

2. Torque converter mounted between the engine and the transmission:

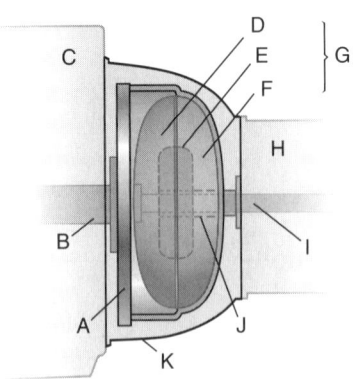

A. _____
B. _____
C. _____
D. _____
E. _____
F. _____
G. _____
H. _____
I. _____
J. _____
K. _____

3. Fluid flow through a torque converter:

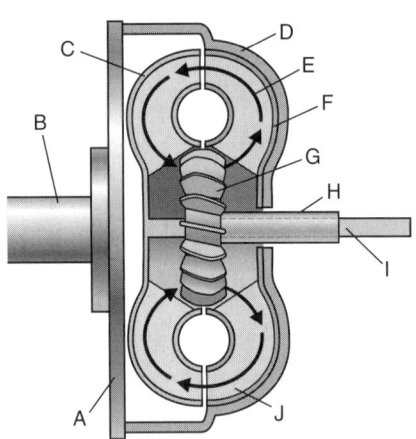

A. _____
B. _____
C. _____
D. _____
E. _____
F. _____
G. _____
H. _____
I. _____
J. _____

Chapter 16   Automatic Transmission Fundamentals

4. Simple planetary gear set:

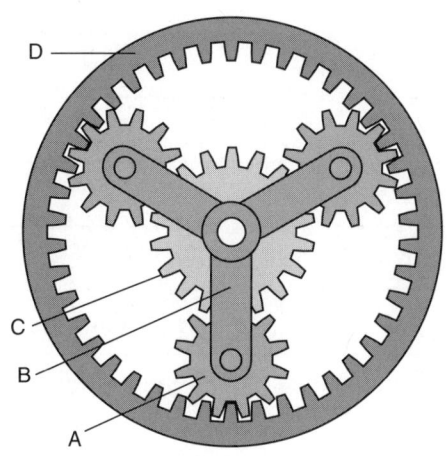

A. _____
B. _____
C. _____
D. _____

5. Gear set styles:

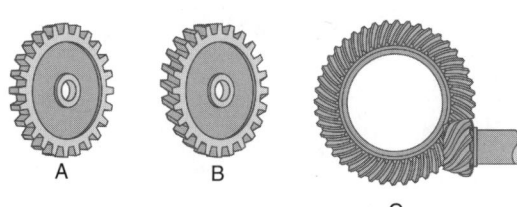

A. _____
B. _____
C. _____

6. A hypoid gear arrangement:

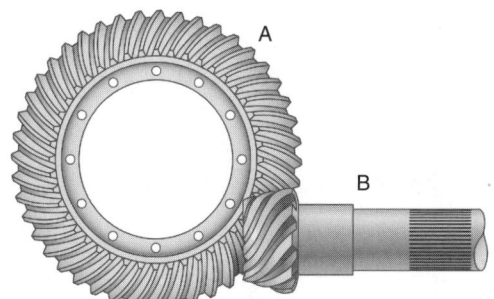

A. _____
B. _____

7. Typical transmission band:

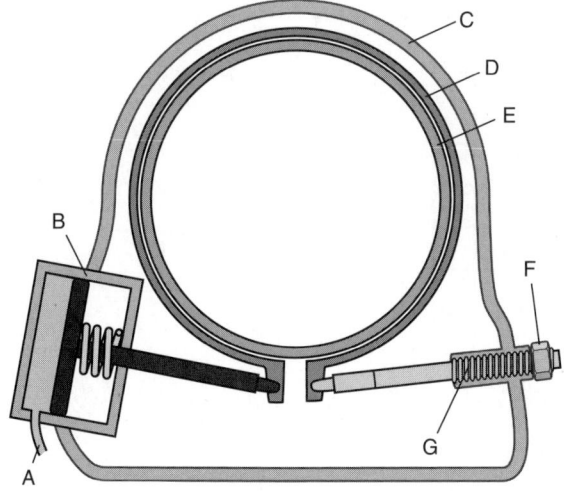

A. _____
B. _____
C. _____
D. _____
E. _____
F. _____
G. _____

8. Vane pump:

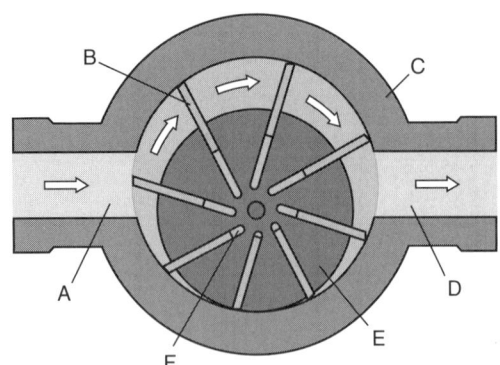

A. _____
B. _____
C. _____
D. _____
E. _____
F. _____

Crossword Puzzle

Use the clues to complete the puzzle.

Across

2. The small gears in a planetary gear set that revolve around the sun gear, also known as planetary gears.
4. A hydraulic device consisting of a piston inside a cylinder that is used to apply a band. The servo also has a spring that releases the servo when it is no longer needed.
6. A type of front pump used on some vehicles that uses two rotating gears to force fluid out of the pump.
9. The center gear of a planetary gear seat around which the other gears rotate.
12. A value assigned to materials to describe the amount of friction when two objects slide against each other.
13. A type of helical gear used to change the direction of motion 90 degrees. The axis of the input gear does not line up on the centerline of the output gear.
14. Often referred to as stall speed, the maximum rpm difference between the turbine and the impeller in the torque converter.
15. A rubber, cork, or paper spacer that goes between two parts to seal the gap between the parts.

Down

1. A type of hydraulic coupling used on vintage vehicles to connect and transfer power from the engine to the transmission.
3. Any gear ratio that results in a torque reduction with a speed increase. Overdrive is used on vehicles to reduce the engine speed when traveling at highway speed in order to save fuel.
5. State in which the fluid in the torque converter is traveling between the turbine, stator, and impeller.
7. A type of fluid flow in a torque converter in which fluid flows around the centerline of the torque converter in a circle.
8. A type of holding device used by an automatic transmission to stop the movement of one component of a planetary gear set. It uses several thin friction discs and thin steel plates that are squeezed together when hydraulic pressure is applied to a piston in the clutch.
10. A condition in which a gear is spinning but not moving.
11. A type of transmission typically used in front-wheel drive vehicles in which the transmission also includes the differential and final drive gear.

ASE-Type Questions

Read each item carefully, and then select the best response.

_____ 1. Tech A says that the higher the numerical gear ratio (4:1), the more torque that will be applied to the wheels. Tech B says that the lower the numerical gear ratio (2:1), the more torque that will be applied to the wheels. Who is correct?
  A. Tech A
  B. Tech B
  C. Both A and B
  D. Neither A nor B

_____ 2. Tech A says that helical cut gears are stronger than straight cut gears. Technician B says that helical cut gears are noisier than straight cut gears. Who is correct?
  A. Tech A
  B. Tech B
  C. Both A and B
  D. Neither A nor B

_____ 3. Tech A says that a transaxle must have a type of differential assembly to allow the front wheels to turn at different speeds while driving around a curve. Tech B says that a conventional rear-wheel drive vehicle will typically use a differential assembly that is located in the rear axle assembly. Who is correct?
  A. Tech A
  B. Tech B
  C. Both A and B
  D. Neither A nor B

____ 4. Tech A says that a parallel hybrid transmission has two or more different devices to propel the vehicle. Tech B says that hybrid power splitting transmissions use two adjustable pulleys and a heavy metal belt. Who is correct?
   A. Tech A
   B. Tech B
   C. Both A and B
   D. Neither A nor B

____ 5. Tech A says that planetary gear sets are used in all automatic transmissions. Tech B says that planetary gear sets are often combined together to create the needed gear ratios for a transmission. Who is correct?
   A. Tech A
   B. Tech B
   C. Both A and B
   D. Neither A nor B

____ 6. Tech A says that the torque converter hub drives the front pump in an automatic transmission. Tech B says that the turbine shaft drives the front pump. Who is correct?
   A. Tech A
   B. Tech B
   C. Both A and B
   D. Neither A nor B

____ 7. Tech A says that most automatic transmissions have a ring gear for the starter as part of the flex plate. Tech B says that the ring gear is typically replaceable (interference fit) on a flex plate. Who is correct?
   A. Tech A
   B. Tech B
   C. Both A and B
   D. Neither A nor B

____ 8. Tech A says that a knocking noise in the engine/transmission area is typically a bad front pump in the transmission. Tech B says that a knocking noise in the engine/transmission area is more likely to be loose flex plate bolts. Who is correct?
   A. Tech A
   B. Tech B
   C. Both A and B
   D. Neither A nor B

____ 9. Tech A says that painting a rebuilt transmission helps to identify it as being a rebuilt transmission. Tech B says that painting a rebuilt transmission prevents the housing from leaking due to case porosity. Who is correct?
   A. Tech A
   B. Tech B
   C. Both A and B
   D. Neither A nor B

____ 10. Tech A says that the torque converter multiplies the amount of torque transmitted from the engine to the transmission. Tech B says that the torque converter couples and uncouples the engine and transmission as the vehicle stops and starts in traffic. Who is correct?
   A. Tech A
   B. Tech B
   C. Both A and B
   D. Neither A nor B

Hydraulic Fundamentals

CHAPTER 17

Tire Tread:
© AbleStock

Chapter Review

The following activities have been designed to help you refresh your knowledge of this chapter. Your instructor may require you to complete some or all of these activities as a regular part of your training program. You are encouraged to complete any activity that your instructor does not assign as a way to enhance your learning.

Matching

Match the following terms with the correct description or example.

A. Electronic pressure control solenoid
B. Mechanical pressure control regulator valve
C. Pascal's law
D. Separator plate
E. Spool valve
F. Surface filter

_____ 1. Sometimes called a spacer plate, a thin sheet metal plate installed between the valve body and the transmission case.

_____ 2. A type of valve commonly used in automatic transmissions that resembles the spool that thread or fishing line comes on.

_____ 3. One of the fundamental physics principles behind the operation of an automatic transmission, this law states that if fluid is placed in a confined space, pressure will be transferred equally in all directions inside the space.

_____ 4. A filter that is a simple screen mechanism to catch dirt and other particles in the hydraulic oil as it passes through.

_____ 5. A solenoid used to regulate line pressure on a computerized transmission.

_____ 6. A spool valve and spring assembly that are used to control the amount of line pressure in a transmission.

Multiple Choice

Read each item carefully, and then select the best response.

_____ 1. One of the fundamental physics principles behind the operation of an automatic transmission is called _____.
   A. Ohm's law
   B. Pascal's law
   C. Faraday's principle
   D. Newton's law

_____ 2. If a master cylinder produces 120 pounds per square inch (psi) of hydraulic pressure, how much pressure will be produced in the calipers and wheel cylinders?
   A. 60 psi
   B. 30 psi
   C. 120 psi
   D. 240 psi

_____ 3. If 100 psi of pressure is applied to a piston that has an area of 5 in^2, what will the resulting force be?
   A. 500 psi
   B. 50 psi
   C. 100 lbs.
   D. 500 lbs.

____ 4. A piston that has a radius of 1.75" would have an area of _____.
   A. 5.495 in²
   B. 9.616 in²
   C. 4.808. in³
   D. 3.525 in²

____ 5. If 100 psi is applied to a piston with a radius of 2" the total force output of the piston will be _____.
   A. 2,512 lbs.
   B. 628 lbs²
   C. 100 psi
   D. 1,256 lbs.

____ 6. The hydraulic pump used in modern automatic transmissions is referred to as a _____.
   A. front pump
   B. rotor vane pump
   C. t-pump
   D. gear displacement pump

____ 7. The _____ operates back and forth inside of a bore in the valve body or the front pump.
   A. separator plate
   B. gear shift
   C. spool valve
   D. PEC valve

____ 8. Which of the following components are pulse-width modulated?
   A. EPC solenoid
   B. Spool valve
   C. Gear pump
   D. All of the above

____ 9. In an automatic transmission, which of the following devices can be used to hold a member of the planetary gear set?
   A. Multidisc clutch
   B. Band
   C. One-way roller or sprag
   D. All of the above

____ 10. Which of the following is an electromagnetic valve?
   A. Spool valve
   B. Solenoid valve
   C. Check valve
   D. All of the above

True/False

If you believe the statement to be more true than false, write the letter "T" in the space provided. If you believe the statement to be more false than true, write the letter "F".

____ 1. Fluids are noncompressible and therefore transmit force effectively.

____ 2. If 125 psi of pressure is applied to a piston that has an area of 4 in² the resulting force will be 250 lbs.

____ 3. A piston with a radius of .75" has a diameter of 2.5".

____ 4. Some vehicles have both an internal and an external transmission filter.

____ 5. Inside the transmission, there is often a magnet stuck to the transmission pan.

____ 6. Aluminum pump housings tend to resist wear very well and require replacement less often than cast iron pump housings.

____ 7. Light scratches and grooves can be removed on some pumps by rubbing a whet stone across the pump housing while running solvent over the pump in the solvent tank.

____ 8. A 50% pulse width modulation means that the valve is open 50% of the time and closed 50% of the time.

____ 9. A small piece of lint can cause a valve to stick in position and not operate, affecting the operation of the transmission.

____ 10. Accumulators are used to dampen the application of clutches and bands.

Chapter 17 Hydraulic Fundamentals

Fill in the Blank

Read each item carefully, and then complete the statement by filling in the missing word(s).

1. Pascal's law states that if we place fluid in a confined space, _____ will be transferred equally in all directions inside the space.
2. A(n) _____ type transmission filter is often a simple fine-mesh screen made out of metal or a plastic such as polyester.
3. A(n) _____ type transmission filter has a thick filter medium that the fluid must pass through.
4. Fluid pressure in an automatic transmission needs to be _____ to help control the quality of the shift.
5. When rebuilding a clutch pack, the _____ needs to be carefully inspected for cracks.
6. The majority of automatic transmission control valves are located inside the _____ _____.
7. _____ pressure is a type of hydraulic pressure that increases in a nearly linear manner as vehicle speed increases.
8. _____ valves can be one of two types in an automatic transmission: one-way or two-way.
9. _____ are often used in conjunction with spool valves to restrict fluid flow for timing purposes or to prevent valve fluctuations.
10. Manufacturer's clutch and band application charts are sometimes referred to as a _____ chart.

Labeling

Identify the correct component as shown in the following illustrations.

1. Transmission figures:

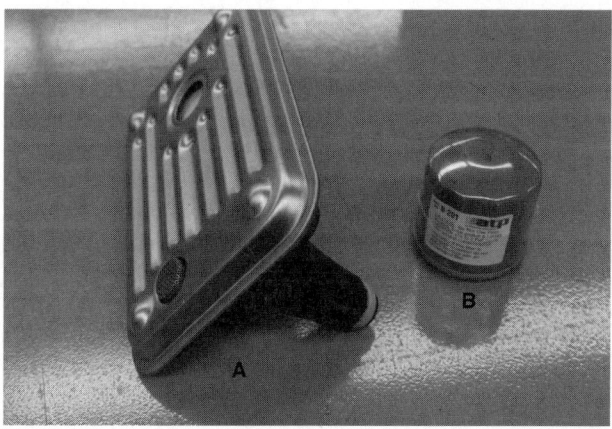

A. _____
B. _____

2. Mechanical pressure control regulator valve:

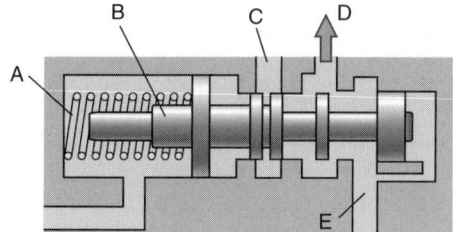

A. _____
B. _____
C. _____
D. _____
E. _____

160 Fundamentals of Automotive Technology Student Workbook

3. Pressure regulation with a variable displacement pump:

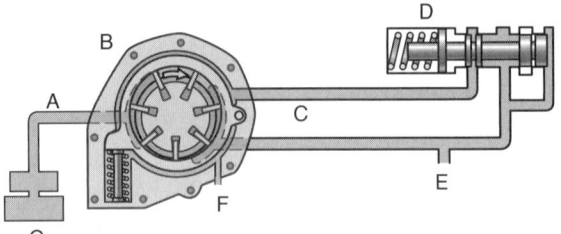

A. _____
B. _____
C. _____
D. _____
E. _____
F. _____
G. _____

4. EPC solenoid and fluid passages to and from the valve:

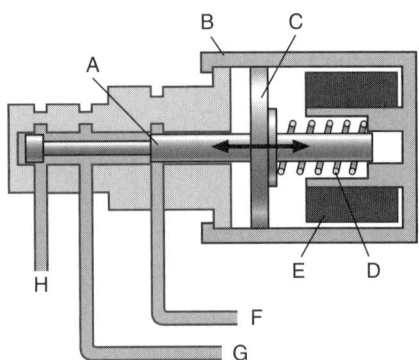

A. _____
B. _____
C. _____
D. _____
E. _____
F. _____
G. _____
H. _____

5. Accumulator operation in a servo circuit:

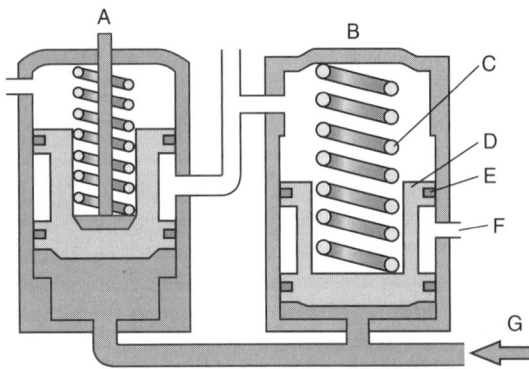

A. _____
B. _____
C. _____
D. _____
E. _____
F. _____
G. _____

Crossword Puzzle

Use the clues to complete the puzzle.

Across

2. A device used to reduce the speed of clutch or band application to help prevent harsh shifting.
5. A spool valve commonly used in automatic transmissions that resembles the spool that thread or fishing line comes on.
6. A hydraulic filter that has thick filter media to trap dirt and other particles of various sizes as they pass through the filter.

Down

1. An aluminum or cast iron housing inside the transmission that houses the majority of the valves that control transmission operation.
3. Also known as a shuttle valve, a valve that allows the flow of hydraulic oil in one direction only. It is typically used to allow oil to escape quickly from an application device when shifting gears.
4. A precisely sized hole used to reduce the speed and pressure of hydraulic oil flowing to a clutch or band.

ASE-Type Questions

Read each item carefully, and then select the best response.

_____ 1. Tech A says that band and clutch application charts can help to diagnose faults in transmissions. Tech B says that 100 psi of pressure on a 2-square-inch piston results in 50 lb of force. Who is correct?
   A. Tech A
   B. Tech B
   C. Both A and B
   D. Neither A nor B

_____ 2. Tech A says that hydraulic pressure is the same in all directions within a closed container. Tech B says that one benefit of using hydraulic pressure is that it is not affected by leaks in the system. Who is correct?
   A. Tech A
   B. Tech B
   C. Both A and B
   D. Neither A nor B

_____ 3. Tech A says that when performing a routine change of the transmission fluid and filter, the cooler should also be flushed. Tech B says that if the transmission pan is removed, the magnet must be replaced. Who is correct?
   A. Tech A
   B. Tech B
   C. Both A and B
   D. Neither A nor B

_____ 4. Tech A says that the greater the clearance between the gear pump housing and the gear, the more fluid can be pumped through the circuit. Tech B says that the front pump is turned by the torque converter hub. Who is correct?
   A. Tech A
   B. Tech B
   C. Both A and B
   D. Neither A nor B

_____ 5. Tech A says that spring tension against a spool valve regulates system pressure. Tech B says that a 50% pulse width modulation means the valve is closed 50% of the time. Who is correct?
   A. Tech A
   B. Tech B
   C. Both A and B
   D. Neither A nor B

_____ 6. Tech A says that transmission failures are often caused by failed lip seals preventing full pressure from being applied to the clutches. Tech B says that fluid leaks are the greatest cause of clutch slippage. Who is correct?
   A. Tech A
   B. Tech B
   C. Both A and B
   D. Neither A nor B

_____ 7. Tech A says that spool valves tend to be self-cleaning. Tech B says that park is obtained by mechanically locking one band and one clutch together. Who is correct?
   A. Tech A
   B. Tech B
   C. Both A and B
   D. Neither A nor B

_____ 8. Tech A says that multi-plate clutches must have a specified amount of clearance when they are installed. Tech B says that there should be no clearance when they are installed because the clutch would slip. Who is correct?
   A. Tech A
   B. Tech B
   C. Both A and B
   D. Neither A nor B

_____ 9. Tech A says that an orifice is a restriction. Tech B says that orifices are used to slow down clutch engagement. Who is correct?
   A. Tech A
   B. Tech B
   C. Both A and B
   D. Neither A nor B

_____ 10. Tech A says that accumulators are used to filter contaminates. Tech B says that accumulators are used to soften clutch engagement. Who is correct?
   A. Tech A
   B. Tech B
   C. Both A and B
   D. Neither A nor B

Hydraulically Controlled Transmission

CHAPTER 18

Chapter Review

The following activities have been designed to help you refresh your knowledge of this chapter. Your instructor may require you to complete some or all of these activities as a regular part of your training program. You are encouraged to complete any activity that your instructor does not assign as a way to enhance your learning.

Matching

Match the following terms with the correct description or example.

- **A.** Governor pressure
- **B.** Modulator pressure
- **C.** Reverse boost valve
- **D.** Throttle valve
- **E.** Throttle valve pressure
- **F.** Vacuum modulator

_____ 1. A component of the pressure regulator valve that increases line pressure when the vehicle is in reverse.

_____ 2. The pressure created by the throttle valve that is proportional to throttle opening.

_____ 3. The pressure created by the governor, which is used to make the shift valves upshift and is proportional to vehicle speed.

_____ 4. A device on a hydraulically controlled transmission that converts engine manifold vacuum into an engine load signal called modulator pressure.

_____ 5. A pressure used to delay transmission upshifting based upon engine load. It may also be used to raise line pressure to more firmly apply bands and clutches under higher engine loads.

_____ 6. A type of spool valve that is connected to the throttle that creates a pressure proportional to throttle opening and is used to delay upshifting based on throttle opening.

Multiple Choice

Read each item carefully, and then select the best response.

_____ 1. A hydro-mechanical valve that produces a variable pressure, based on vehicle speed is called a _____.
- **A.** flex plate
- **B.** servo
- **C.** spool valve
- **D.** governor

_____ 2. The variable pressure produced by a governor is called _____.
- **A.** line pressure
- **B.** governor pressure
- **C.** output pressure
- **D.** negative pressure

_____ 3. A type of governor in which the assembly uses two valves, a primary valve and a secondary valve is called _____.
   A. multi-valve governor
   B. output shaft-mounted governor
   C. staged governor
   D. in-line governor

_____ 4. The suction created in gasoline engines when the piston moves downward in the cylinder bore on the intake stroke is called _____.
   A. intake pressure
   B. manifold vacuum
   C. modulation
   D. throttle pressure

_____ 5. On many hydraulically controlled transmissions the engine vacuum signal is converted into hydraulic pressure using a _____.
   A. vacuum modulator
   B. governor
   C. pressure converter
   D. throttle valve

_____ 6. If the diaphragm on a vacuum modulator becomes ripped, what might the customer complain of?
   A. White exhaust smoke
   B. Dark exhaust smoke
   C. Delayed or harsh shifting
   D. Either A or C

_____ 7. Some hydraulically controlled transmissions use a(n) _____ in place of, or in addition to, a vacuum modulator.
   A. staged governor
   B. kickdown valve
   C. throttle valve
   D. orifice

_____ 8. On older hydraulically controlled transmissions, what device allows the transmission to downshift when you press the accelerator to the floor in order to accelerate faster?
   A. Modulator
   B. Kickdown valve
   C. Governor
   D. Servo

_____ 9. A device that cushions and absorbs application pressures within the transmission is called a(n) _____.
   A. detent valve
   B. orifice
   C. separator plate
   D. accumulator

_____ 10. In which gear are governor pressure and modulator pressure/throttle valve pressure not used?
   A. First gear
   B. Second gear
   C. Third gear
   D. Reverse

True/False

If you believe the statement to be more true than false, write the letter "T" in the space provided. If you believe the statement to be more false than true, write the letter "F".

_____ 1. Automobile manufacturers in the United States no longer produce fully hydraulically controlled transmissions.

_____ 2. The case-mounted governor is mounted in line with the output shaft.

_____ 3. Placing weaker springs in the governor will cause a delay in the transmission shift.

_____ 4. Governors have trouble accurately measuring ground speed when the vehicle is moving slowly.
_____ 5. A manual transmission can vary its shift points based on engine load.
_____ 6. When an engine is operating under a light load, the engine manifold vacuum is low or even zero.
_____ 7. The throttle valve has a rubber diaphragm that is compressed to the bottom of the housing by a large spring when there is no engine vacuum.
_____ 8. Replacement vacuum modulators are often adjustable to fine-tune the shift timing.
_____ 9. Throttle valve cables and linkages are a critical adjustment on a hydraulically controlled transmission.
_____ 10. Fluid leaving the transmission pan can be extremely hot, so it first goes to the transmission cooler.

Fill in the Blank

Read each item carefully, and then complete the statement by filling in the missing word(s).

1. Both hydraulic and electronic transmissions with multi-plate clutches and bands use _____ _____ to apply the bands and clutches.
2. The governor is supplied with line pressure and then uses _____ force to vary the line pressure in proportion to ground speed.
3. Springs behind the weights help to _____ the governor to the particular engine that is installed in the vehicle.
4. A(n) _____ governor enables more precise control of governor pressure and, therefore, the transmission shift.
5. When engine vacuum is _____, the modulator produces a low modulator pressure.
6. As the throttle opening increases, throttle _____ _____ increases.
7. In some transmissions the kickdown valve is called a _____ valve.
8. A(n) _____ provides a restriction in line with the device being applied.
9. A(n) _____ _____ will either block or allow the passage of line pressure to a particular clutch or band in order to apply the gear.
10. The _____ valve uses a simple spring-loaded valve that dumps fluid above the intended pressure back to the oil pan.

Labeling

Identify the correct component as shown in the following illustrations.

1. Hydraulic diagram of a governor valve:

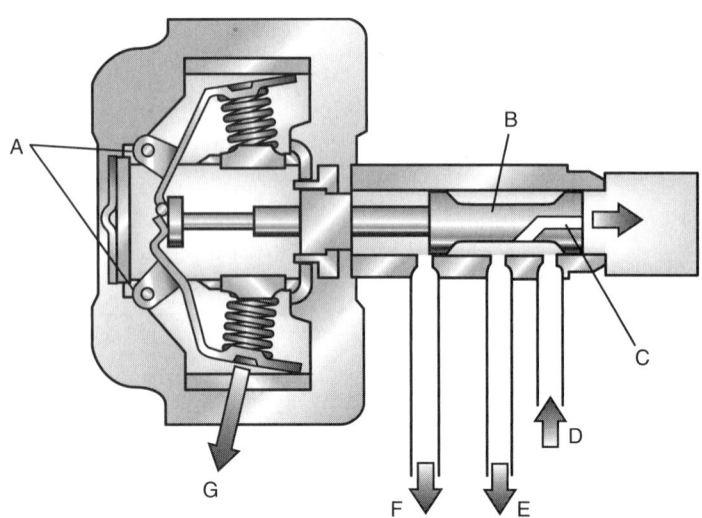

A. _____

B. _____

C. _____

D. _____

E. _____

F. _____

G. _____

2. Hydraulic throttle valve:

A. _____

B. _____

C. _____

D. _____

E. _____

F. _____

Chapter 18 Hydraulically Controlled Transmission

3. Accumulator action:

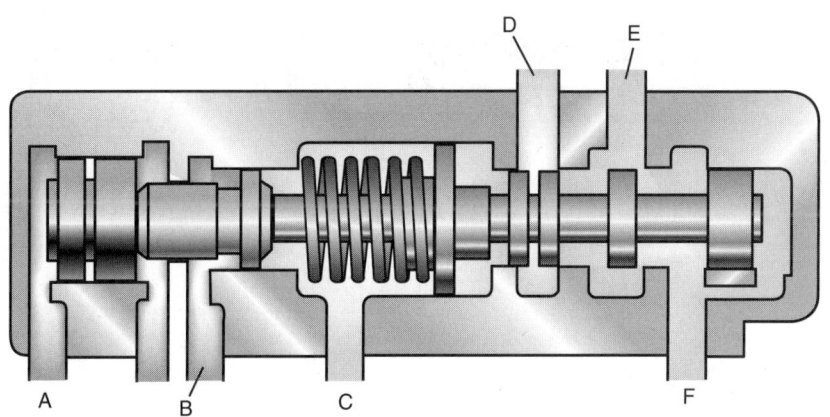

A. _____
B. _____
C. _____
D. _____
E. _____

4. Pressure regulator valve in a hydraulically controlled transmission:

A. _____
B. _____
C. _____
D. _____
E. _____
F. _____

5. Hydraulic diagram of a transmission in park:

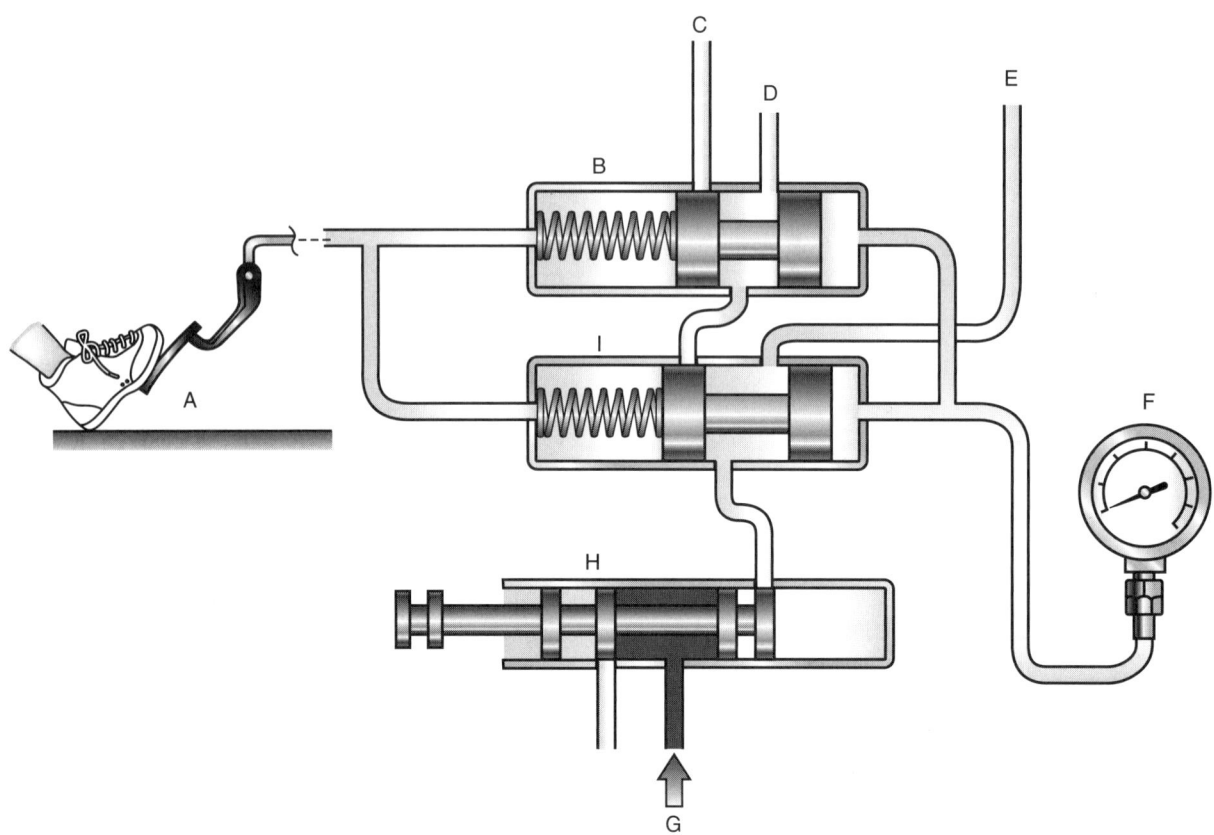

A. _____

B. _____

C. _____

D. _____

E. _____

F. _____

G. _____

H. _____

I. _____

Crossword Puzzle

Use the clues to complete the puzzle.

Across

3. A governor in which the assembly uses two valves: a primary valve and a secondary valve.
4. A hydraulic pressure that is used to apply bands and clutches, and is regulated by the pressure regulator valve.
6. Another name for kickdown valve.

Down

1. A type of spool valve that is connected to the throttle on the vehicle. The valve is used to force a downshift when the throttle is opened all the way, assuming the governor pressure is below a specified point.
2. A mechanical device that creates a pressure using centrifugal force. The pressure is proportional to vehicle speed.
5. A type of spool valve that has multiple fluid inputs and a spring; it is used to direct hydraulic pressure to a clutch or band needed for a shift.

ASE-Type Questions

Read each item carefully, and then select the best response.

_____ 1. Tech A says that measuring line pressure and comparing it to specifications is a good diagnostic procedure. Tech B says that the manual valve is moved by governor pressure to force a shift. Who is correct?
   A. Tech A
   B. Tech B
   C. Both A and B
   D. Neither A nor B

_____ 2. Tech A says that governor pressure is controlled by torque converter rpm. Tech B says that governor pressure is controlled by output shaft speed. Who is correct?
   A. Tech A
   B. Tech B
   C. Both A and B
   D. Neither A nor B

_____ 3. Tech A says that reciprocating force controls governor pressure. Tech B says that centrifugal force controls governor pressure. Who is correct?
   A. Tech A
   B. Tech B
   C. Both A and B
   D. Neither A nor B

_____ 4. Tech A says that modifying governor weights will alter transmission shifting. Tech B says that excessive modification of the governor weights may cause the transmission to not shift. Who is correct?
   A. Tech A
   B. Tech B
   C. Both A and B
   D. Neither A nor B

_____ 5. Tech A says that the converter and cooler operate at line pressure. Tech B says that line pressure causes the manual valve to move during upshifts and downshifts. Who is correct?
   A. Tech A
   B. Tech B
   C. Both A and B
   D. Neither A nor B

_____ 6. Tech A says that the throttle valve is controlled by a cable or linkage hooked to the engine throttle linkage. Tech B says that a misadjusted throttle valve cable can prevent wide open throttle (WOT) on an engine. Who is correct?
   A. Tech A
   B. Tech B
   C. Both A and B
   D. Neither A nor B

_____ 7. Tech A says that line pressure is reduced during light throttle conditions to improve fuel economy. Tech B says that as the engine speeds up in rpm, the transmission line pressure gradually reduces to maintain clutch pack pressure. Who is correct?
   A. Tech A
   B. Tech B
   C. Both A and B
   D. Neither A nor B

_____ 8. Tech A says that when governor pressure on one side of a shift valve overcomes spring pressure and throttle pressure on the other side, an upshift occurs. Tech B says when reverse boost pressure on one side of the reverse valve overcomes governor pressure on the other side, shift into reverse occurs. Who is correct?
   A. Tech A
   B. Tech B
   C. Both A and B
   D. Neither A nor B

_____ 9. Tech A says that fluid pressure in an automatic transmission is controlled by engine rpm. Tech B says that modulator pressure changes with engine vacuum. Who is correct?
   A. Tech A
   B. Tech B
   C. Both A and B
   D. Neither A nor B

_____ 10. Tech A says that in reverse, governor pressure regulates line pressure. Tech B says that in reverse, modulator pressure regulates line pressure. Who is correct?
   A. Tech A
   B. Tech B
   C. Both A and B
   D. Neither A nor B

Electronically Controlled Transmission

CHAPTER 19

Chapter Review

The following activities have been designed to help you refresh your knowledge of this chapter. Your instructor may require you to complete some or all of these activities as a regular part of your training program. You are encouraged to complete any activity that your instructor does not assign as a way to enhance your learning.

Matching

Match the following terms with the correct description or example.

- **A.** Crankshaft position (CKP) sensor
- **B.** Electronic pressure control (EPC) solenoid
- **C.** Engine coolant temperature (ECT) sensor
- **D.** Manifold absolute pressure (MAP) sensor
- **E.** Manual lever position (MLP) switch
- **F.** Mass airflow (MAF) sensor
- **G.** Power train control module (PCM)
- **H.** Throttle position sensor (TPS)
- **I.** Torque converter clutch (TCC)
- **J.** Transmission control module (TCM)
- **K.** Transmission oil temperature (TOT) sensor
- **L.** Vehicle speed sensor (VSS)

_____ 1. A sensor used by the PCM to measure vehicle speed. It is often located in the transmission extension housing.

_____ 2. A vacuum sensor that is attached to the intake manifold by a passageway or vacuum hose. The sensor measures engine intake manifold pressure to determine engine load and sends a corresponding signal to the PCM.

_____ 3. A pulse-width–modulated solenoid used to control transmission line pressure in an automatic transmission.

_____ 4. A sensor located in the air intake system that is used to measure the mass of the air flowing into the engine.

_____ 5. A type of variable resistor used inside the transmission to monitor oil temperature.

_____ 6. A sensor that changes resistance based upon coolant temperature, also known as a thermistor.

_____ 7. A hydraulically operated clutch located inside the torque converter that applies at predetermined conditions and stops torque converter slippage.

_____ 8. A type of potentiometer used by the PCM to measure throttle angle.

_____ 9. A computer module that controls the engine and possibly the transmission on a computer-controlled vehicle.

_____ 10. A sensor used by the PCM to monitor engine speed. It can be one of three types of sensors—Hall effect, magnetic pickup, or optical.

_____ 11. A switch that is used by the PCM to tell which gear range the driver has selected with the shift lever.

_____ 12. A computer module that controls the transmission operation. It may be integrated into the PCM.

Multiple Choice

Read each item carefully, and then select the best response.

_____ 1. In a fully electronically controlled transmission, the shift points are controlled by the _____.
   A. power train control module
   B. driver
   C. transmission control module
   D. either A or C

_____ 2. In a computerized transmission, the governor is replaced by a _____.
   A. manual lever position switch
   B. vehicle speed sensor
   C. shift solenoid
   D. crankshaft position sensor

_____ 3. Most vehicle speed sensors are a type of sensor called a _____.
   A. magnetic pickup
   B. potentiometer
   C. magnetic reluctance sensor
   D. either A or C

_____ 4. What type of electrical wave pattern is produced by a reed switch?
   A. DC square wave
   B. AC sine wave
   C. AC triangle wave
   D. DC sine wave

_____ 5. What type of sensor is a line pressure sensor?
   A. Thermistor
   B. Transducer
   C. Magnetic pickup
   D. Potentiometer

_____ 6. What tool is recommended to verify the operation of a throttle position sensor?
   A. Digital voltage meter
   B. Ohm meter
   C. Digital storage oscilloscope
   D. Scan tool

_____ 7. What type of sensor is used on drive-by-wire vehicles to determine the driver's intent as related to acceleration/deceleration?
   A. Throttle position sensor
   B. Vehicle speed sensor
   C. Input shaft speed sensor
   D. Accelerator position sensor

_____ 8. What type of sensors use a flexible silicon chip within a sealed chamber that changes resistance when pressure flexes the chip?
   A. Manifold absolute pressure sensor
   B. Line pressure sensor
   C. Throttle position sensor
   D. Mass airflow sensor

_____ 9. Crankshaft position sensors are what type of sensor?
   A. Pickup-style sensor
   B. Hall-effect sensor
   C. Optical-style sensor
   D. Any of the above

_____ 10. Many four-speed automatic transmissions use two _____ to control the transmission.
   A. transducers
   B. shift solenoids
   C. potentiometers
   D. electronic pressure control solenoids

Chapter 19   Electronically Controlled Transmission

True/False

If you believe the statement to be more true than false, write the letter "T" in the space provided. If you believe the statement to be more false than true, write the letter "F".

_____ 1. Fully computer-controlled transmissions allow manufacturers to install smaller transmissions into vehicles.
_____ 2. Most vehicle manufacturers have integrated the transmission control module into the construction of the power train control module.
_____ 3. As the teeth of the reluctor wheel get closer to the iron core, a negative voltage is produced. As the tooth moves away, the voltage becomes positive.
_____ 4. Magnetic pickup sensors do not need a separate wire to supply them with a reference voltage.
_____ 5. Some power train control modules compare the input shaft speed sensor to the output shaft speed sensor to determine gear ratios.
_____ 6. If the transmission oil is hotter, the PCM will allow higher engine rpm before shifting the transmission and allow more time for the transmission to complete the shift.
_____ 7. When there is no pressure in a hydraulic circuit, a transmission pressure switch is closed, and when pressure is applied, it opens.
_____ 8. An engine coolant temperature sensor cannot accurately read engine temperature when the coolant level is low.
_____ 9. Mass airflow sensors measure airflow in grams per square inch.
_____ 10. On some newer vehicles, the brake switch is used to put the transmission into neutral while idling at a stoplight.
_____ 11. Many manufacturers recommend using overdrive when pulling a trailer or heavy loads.
_____ 12. Shift solenoids are typically an electromagnetic type of valve that is either open or closed.
_____ 13. A solenoid with a 25% pulse-width modulation is off 25% of the time and on 75% of the time.
_____ 14. In time, the power train control module learns how a particular driver uses the vehicle.
_____ 15. A newer power train control module can apply logic to the operating conditions it sees.

Fill in the Blank

Read each item carefully, and then complete the statement by filling in the missing word(s).

1. The power train control module receives various _____ from the engine and transmission to determine the shift timing and the firmness of the shift.
2. A(n) _____ _____ _____ is sometimes called an output shaft speed sensor.
3. A(n) _____ changes its resistance based upon its temperature.
4. A(n) _____ varies its resistance based on its position.
5. The _____ _____ temperature sensor is installed in an engine coolant passage, often near the thermostat housing on an engine.
6. To meet more stringent emission standards, most vehicles come equipped with a _____ _____ sensor, rather than a MAP sensor, to measure engine load.
7. The _____ _____ _____ switch is often mounted directly on the transmission case where the shift linkage connects to the transmission.
8. _____-_____ _____ means that the solenoid is not on all of the time; instead it pulses rapidly on and off to control pressure.
9. In many transmissions the solenoid allows hydraulic oil to flow to a _____ _____ that moves, allowing line pressure to flow past and engage the clutch or band.
10. Early torque converters always had some slippage, so in the late 1970s and early 1980s manufacturers added a _____-_____ _____ to the torque converter.
11. The _____ _____ section of the power train control module is responsible for sending the proper reference voltage to many of the sensors.
12. The _____ _____ typically send either simple on/off signals or pulse-width–modulated signals to the actuators, depending on the actuator being controlled.

13. Identification of _____ means identifying unusual ambient conditions in which the vehicle is operating.
14. By having a completely _____-controlled transmission and engine, the PCM is able to reduce the amount of torque being produced by the engine right before the transmission shifts.
15. A fail-safe or _____-_____ mode allows a vehicle to be driven even if the computer for the transmission or the part of the PCM that controls the transmission fails.

Labeling

Label the following diagrams with the correct terms.

1. Input shaft speed sensor versus output shaft speed sensor:

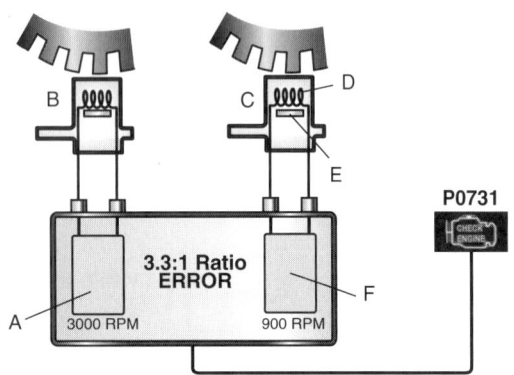

A. _____
B. _____
C. _____
D. _____
E. _____
F. _____

2. Two TPS sensor patterns from a DSO:

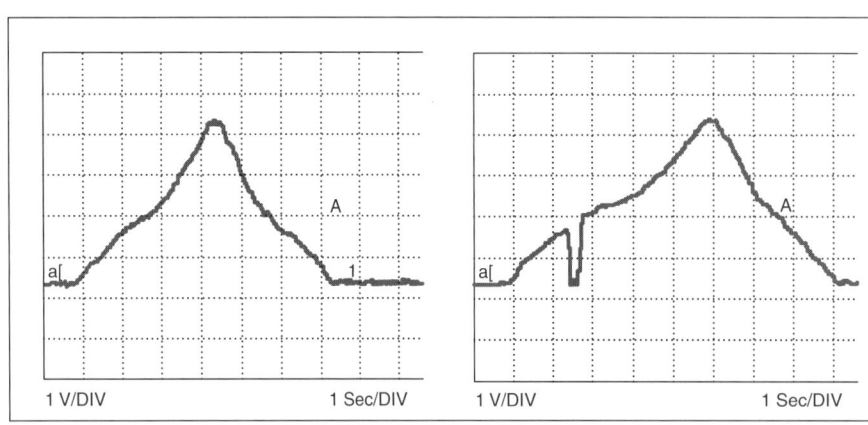

A. _____

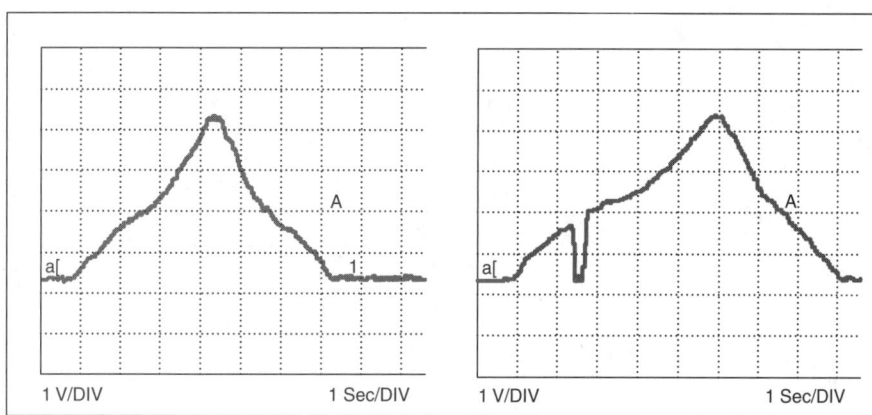

B. _____

3. A common brake on/off switch wiring diagram:

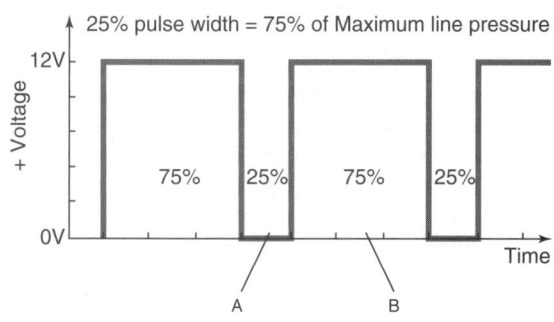

A. _____

B. _____

C. _____

4. A pulse-width–modulated pattern of an electronic pressure control (EPC) solenoid:

A. _____

B. _____

5. Simplified hydraulic diagram for a transmission in which the shift solenoids indirectly operate the clutches or bands in first gear:

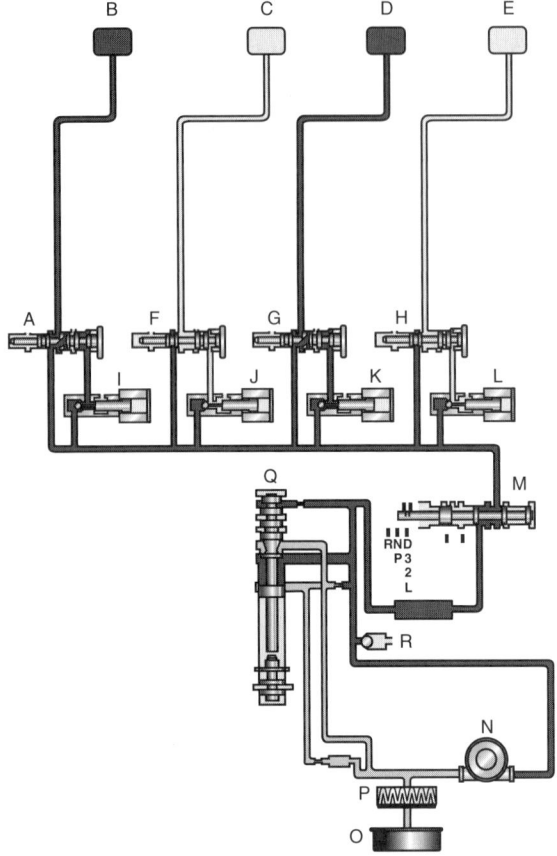

A. _____
B. _____
C. _____
D. _____
E. _____
F. _____
G. _____
H. _____
I. _____
J. _____
K. _____
L. _____
M. _____
N. _____
O. _____
P. _____
Q. _____
R. _____

Crossword Puzzle

Use the clues to complete the puzzle.

Across

2. A type of speed sensor that uses a magnetic field to open and close a movable set of contacts. It is used with a rotating magnet to measure rpm of a shaft and send the signal to the PCM.
5. A transmission operating mode in which limited computer controls are needed to operate for the purpose of getting the vehicle to a shop.
6. A sensor inside the transmission that measures the rpm of the input shaft. Also called a turbine shaft sensor.

Down

1. A type of variable resistor that increases or decreases its resistance as it is turned, which creates a varying voltage signal.
3. An electromechanical device used to control oil flow to bands and clutches in an automatic transmission to help shift the transmission.
4. A variable resistor sensor used to monitor line pressure. It sends a signal back to the PCM where it can be translated into a psi reading.

ASE-Type Questions

Read each item carefully, and then select the best response.

_____ 1. Tech A says that manufacturers developed the electronically controlled automatic transmission to improve fuel economy. Tech B says that manufacturers have added more gears to their electronically controlled transmission to improve fuel economy. Who is correct?
   A. Tech A
   B. Tech B
   C. Both A and B
   D. Neither A nor B

_____ 2. A vehicle comes in only operating in 3rd gear. Tech A says the transmission will need to be replaced. Tech B says that the transmission can be operating in limp-in mode. Who is correct?
   A. Tech A
   B. Tech B
   C. Both A and B
   D. Neither A nor B

_____ 3. Tech A says that reed switches produce an AC signal. Tech B says that a magnetic VSS can be diagnosed by observing the pattern on a lab scope and comparing it to a known good pattern. Who is correct?
   A. Tech A
   B. Tech B
   C. Both A and B
   D. Neither A nor B

_____ 4. Tech A says that transmissions with an input shaft speed sensor and a vehicle speed sensor (VSS) can use this information to determine shift solenoid failure. Tech B says the input shaft speed sensor is also called a turbine speed sensor. Who is correct?
   A. Tech A
   B. Tech B
   C. Both A and B
   D. Neither A nor B

_____ 5. Tech A says that delayed shifts could be caused by a malfunctioning TPS. Tech B says that this information can be used to determine clutch slippage. Who is correct?
   A. Tech A
   B. Tech B
   C. Both A and B
   D. Neither A nor B

_____ 6. Tech A says that the ECT and the TOT will delay transmission shifting. Tech B says that the ECT and the TOT could cause the transmission to shift sluggishly. Who is correct?
   A. Tech A
   B. Tech B
   C. Both A and B
   D. Neither A nor B

_____ 7. Tech A says that on most vehicles if the EPC solenoid fails the vehicle will have high line pressure. Tech B says that if an EPC fails it will cause a delayed shift in most vehicles. Who is correct?
   A. Tech A
   B. Tech B
   C. Both A and B
   D. Neither A nor B

_____ 8. A vehicle stalls when coming to a stop. After test-driving the vehicle, it is found that the TCC solenoid is remaining engaged as the vehicle slows down. Tech A says that the vehicle probably needs a new torque convertor. Tech B says that the TCC solenoid may be sticking and needs to be replaced. Who is correct?
   A. Tech A
   B. Tech B
   C. Both A and B
   D. Neither A nor B

_____ 9. A vehicle comes in with the check engine light on. The vehicle is scanned and has transmission ratio error codes for 1st, 2nd, 3rd, and 4th gears. Tech A says that the problem could be a faulty output shaft speed sensor. Tech B says that the problem could be a faulty transmission front pump. Who is correct?
   A. Tech A
   B. Tech B
   C. Both A and B
   D. Neither A nor B

_____ 10. Tech A says that automatic transmissions will change gear ratios based upon winter weather conditions. Tech B says that on a vehicle with traction control, the transmission will change shift programming to assist. Who is correct?
   A. Tech A
   B. Tech B
   C. Both A and B
   D. Neither A nor B

CHAPTER 20
Servicing the Automatic Transmission/Transaxle

Tire Tread:
© AbleStock

Chapter Review

The following activities have been designed to help you refresh your knowledge of this chapter. Your instructor may require you to complete some or all of these activities as a regular part of your training program. You are encouraged to complete any activity that your instructor does not assign as a way to enhance your learning.

Matching

Match the following terms with the correct description or example.

A. Bidirectional control
B. Cooler flow test
C. Hydraulic pressure test
D. Morse-style chain
E. Noise, vibration, and harshness (NVH) test
F. Power train mount

____E____ 1. A test to measure for any audible noises, vibrations, and harsh operation. It can be completed by the technician with or without the aid of a tester. The tester is used to pinpoint the exact frequencies of the noise and vibrations.

____F____ 2. A rubber or metal bracket used to secure the engine and transmission into the vehicle. Some vehicles use hydraulic or electrohydraulic power train mounts.

____A____ 3. The ability to command different solenoids and actuators "on" and "off" to check their operation.

____D____ 4. A heavy-duty chain constructed of many links and held together with pins. It is used in some transaxles and transfer cases to transmit torque from one component to another.

____B____ 5. The placement of a specialty-measuring device into the transmission cooler line to measure fluid flow to the cooler.

____C____ 6. The use of a hydraulic pressure gauge to measure the amount of hydraulic pressure produced in each gear range.

Multiple Choice

Read each item carefully, and then select the best response.

____C____ 1. How much fluid does it take to raise the level from the bottom of the crosshatched area or add mark to the full mark on most transmission dipsticks?
   A. Half-pint
   B. Quart
   C. Pint
   D. Gallon

____D____ 2. When raised in the air on a hoist, a vehicle's speed should never be allowed to exceed _____ mph on the speedometer.
   A. 10
   B. 25
   C. 35
   D. 45

Chapter 20 Servicing the Automatic Transmission/Transaxle

___C___ 3. What type of test checks the pump operation, the pressure regulator, and the seals and gaskets inside the transmission?
   A. Air check
   B. NVH test
   C. Hydraulic pressure test
   D. Cooler flow test

___C___ 4. What feature makes a high-quality aftermarket scan tool or factory scan tool better than a code reader?
   A. Live data
   B. Reads DTCs
   C. Bidirectional control
   D. Both A and C

___B___ 5. During which test is the vehicle placed in a gear with the emergency brake fully applied and the brake pedal firmly held?
   A. Hydraulic pressure test
   B. Stall test
   C. NVH test
   D. Preload test

___D___ 6. Which of the following repairs can be performed without the removal of the transmission?
   A. Replace shift solenoid
   B. Replace speed sensor
   C. Replace power train mounts
   D. All of the above

___B___ 7. Replacement of which of the following requires the removal of the driveshaft and the extension housing?
   A. Speed sensor seal
   B. Extension housing bushing
   C. Pan gasket
   D. All of the above

___A___ 8. A loud thump when the accelerator is applied in drive or reverse and again when the brake pedal is applied may indicate a _____.
   A. broken power train mount
   B. dry extension-housing bushing
   C. transmission fluid leak
   D. bad shift sensor

___B___ 9. If the _____ is not adjusted properly, the transmission position indicator will not be set correctly and the customer will not know which gear the vehicle is in.
   A. speed sensor
   B. shift sensor
   C. linkage
   D. lock-up torque converter

___B___ 10. Which tool is often used to check torque converter end play?
   A. Feeler gauge
   B. Dial indicator
   C. Outside caliper
   D. Factory scan tool

___A___ 11. What term refers to the amount that a bearing is able to move before hitting the bearing race?
   A. Free play
   B. Preload
   C. Torsion
   D. End play

___D___ 12. The amount of force placed on a bearing before any load is on the bearing is referred to as _____.
   A. loading
   B. torquing
   C. tightening
   D. preload

13. _D_ Which of the following components is generally very durable and does not require service when rebuilding a transmission?
   A. Front pump
   B. Planetary gear set
   C. Torque converter
   D. Extension housing bushing

14. _C_ Damage found in the race from high bearing load is called _____.
   A. scoring
   B. pitting
   C. brinelling
   D. sprag

15. _A_ What causes wear to the valve body or the separator plate?
   A. Metal check balls
   B. One-way rollers
   C. Sprags
   D. Sealing rings

16. _A_ What tool should you use to measure the amount of warpage on the drum?
   A. Straightedge
   B. Dial indicator
   C. Feeler gauge
   D. Both A and C

17. _C_ During rebuilding, the valve body to transmission case surface should be checked for _____.
   A. brinelling
   B. scoring
   C. warpage
   D. clearance

18. _B_ A transmission with an integrated final drive unit such as that found in the differential of a rear-wheel drive vehicle is called a _____.
   A. Continuously variable transmission
   B. Transaxle
   C. Crossover
   D. Morse-style transmission

19. _A_ When assembling the final drive unit, what must be checked in order to prevent seizing of the gears or excessive noise?
   A. Output shaft end play
   B. Chain deflection
   C. Side gear backlash
   D. Both A and C

20. _C_ After the transmission is installed properly into the vehicle, add about _____ of the total transmission fluid to the transmission.
   A. 25%
   B. 50%
   C. 75%
   D. 90%

True/False

If you believe the statement to be more true than false, write the letter "T" in the space provided. If you believe the statement to be more false than true, write the letter "F".

_T_ **1.** Some vehicles do not have a transmission dipstick.

_T_ **2.** All transmissions are checked with the engine at operating temperature, idling, and the transmission in park.

_F_ **3.** Some modern transmission fluids are a darker red when they are brand new and even have a slightly burnt smell to them.

Chapter 20 Servicing the Automatic Transmission/Transaxle

___T___ 4. The presence of a small amount of clutch material or metal in the bottom of the pan indicates a serious transmission problem.

___F___ 5. Scanning the power train control module for trouble codes will indicate the specific transmission problem.

___T___ 6. Transmissions operate much differently when they are placed under a load, such as driving up a hill, than they do sitting in a shop on a hoist.

___T___ 7. A hydraulic pressure test puts maximum load on the engine, transmission, and brake system.

___T___ 8. A stall test should not be performed for more than 5 seconds, as severe transmission damage can occur.

___F___ 9. Low line pressure can be a sign of a stuck pressure regulator valve or failed EPC solenoid.

___F___ 10. Removal of the transmission should be carried out only after a thorough diagnosis of the transmission in the vehicle.

___T___ 11. If a transmission suffers catastrophic failure, metal particles and old clutch material can become stuck inside the fluid cooler or lines.

___F___ 12. The front pump bushing and the extension housing bushing do not require replacement when rebuilding a transmission.

___T___ 13. Sprags are easy to disassemble, but one-way rollers cannot easily be completely disassembled.

___F___ 14. Nearly every transmission has a vent that needs to be checked during rebuilding.

___F___ 15. When rebuilding an automatic transmission, all clutch assemblies should be rebuilt.

___T___ 16. You always should air check a clutch before the clutch pack is assembled and the snap ring is engaged.

___T___ 17. To properly rebuild an automatic transmission, the valve body must be completely disassembled.

___F___ 18. In the gear-style transaxle assembly, power is transferred from the transmission planetary gears to the differential through a set of helically cut gears.

___T___ 19. When assembling the final drive unit, bearing preload must be measured following the manufacturer's procedure.

___F___ 20. Sometimes it is difficult to get the torque converter fully seated when installing it into the transmission.

Fill in the Blank

Read each item carefully, and then complete the statement by filling in the missing word(s).

1. When fluids leak, they travel __downward__ due to gravity and typically toward the __Back__ of the vehicle.

2. If a transmission leak is hard to locate, place a leak detection __dye__ in the transmission fluid.

3. The new way to check the quality of transmission fluid is to take a few drops and place them on a clean __paper towel__.

4. During the __test-drive__, it is critical to place the transmission in the same operating conditions as the customer stated the problem occurred in.

5. Most manufacturers do not recommend __stall__ testing on late-model vehicles, as it can damage the transmission.

6. Generally, line pressure should be highest in the __reverse__ gear range.

7. With the __bidirectional control__ bound, the driver might be able to start a vehicle while it is in gear or have the vehicle in the wrong gear.

8. Some hydraulic __pressure test__ mounts use a computer-controlled electronic control valve that can open an alternate passageway or vary the size of the orifice.

9. The lock-up torque converter contains a __torque converter clutch__ that forces the turbine and impeller to match speeds.

10. __flex plates__ can crack, resulting in knocking noises and front pump damage.

11. Before reusing a torque converter, the __end play__ of the converter must be checked.

12. A transmission __Fluid Cooler__ that is leaking could cause the new transmission to fail from low fluid level or could cause anti-freeze to be drawn into the cooler from the radiator, contaminating the fluid.
13. The __front pump__ is the heart of the transmission.
14. Always use a __bushing driver__ to install new bushings into a transmission.
15. A __micrometer__ should be used to check the thickness of a thrust washer, and the thickness should be compared to the manufacturer's specifications.
16. A transmission __case diagram__ is used to identify the fluid passages through the case.
17. A __gasket__ is designed to fill in small imperfections in the sealing surface, but if the surface has deep gouges or scratches it will be unable to do its job.
18. Transmission __band__ can become burned during service or glazed like a brake lining.
19. After rebuilding and reassembling a clutch pack, it is necessary to measure the amount of __clearance__.
20. The valve __bore__ should be measured to check for valve clearance.

Labeling

Identify the correct component as shown in the following illustrations.

1. A bearing race showing brinelling:

   A. __Brinetting Marks__

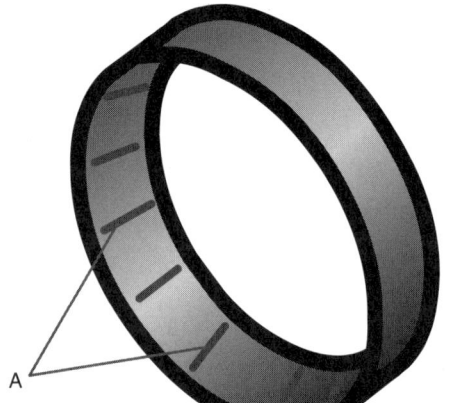

2. Power transfer:

A. __Gear Style__

B. __Chain Style__

Skill Drills

Place the skill drill steps in the correct order.

1. Draining and Replacing Fluid and Filter:

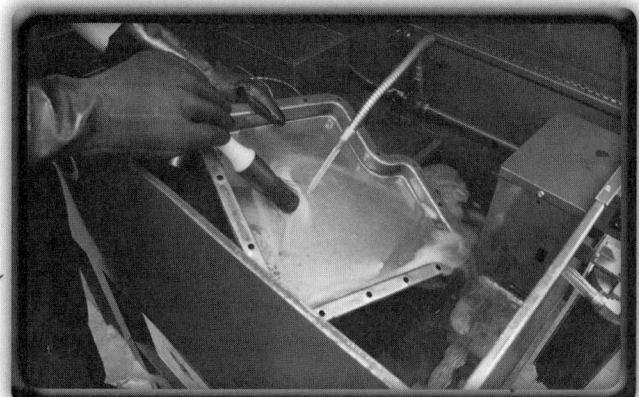

___3___ **A.** Clean the pan and the magnet thoroughly in a parts washer. Dry the pan with a lint-free shop towel or compressed air.

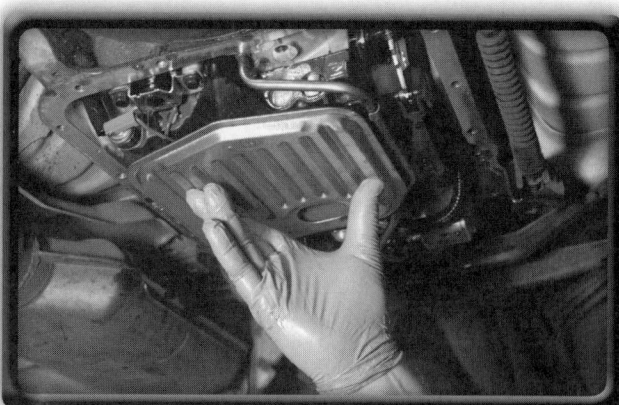

_____ **B.** Put the new gasket on the transmission pan, clean the transmission sealing surface, and place the pan onto the transmission. Start all of the bolts before tightening.

___2___ **C.** Hold the pan against the transmission with one hand and loosen the remaining bolts a few turns. Allow the transmission pan to tilt downward and the transmission fluid to flow into the drain pan. Once most of the fluid has been drained, push the pan up against the transmission again and remove the remaining bolts.

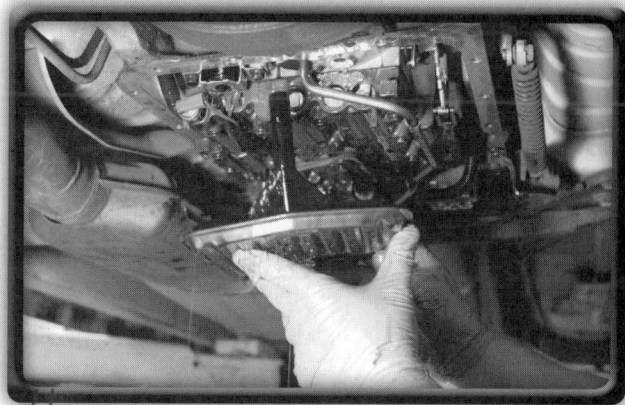

_____ **D.** Remove any bolts or clips holding the transmission filter in place, and lower the filter.

____8____ **E.** Torque the bolts to specifications. Lower the vehicle, and install 75% of the new fluid. Start the vehicle and check the fluid level. Add enough fluid to bring it to the bottom of the safe or add mark. With your foot firmly depressing the brake pedal, place the gear selector in each of the gear ranges. Check the fluid level with the engine at operating temperature, and top off as necessary. Raise the vehicle, and check for any leaks.

____1____ **F.** Identify the correct type of transmission fluid and capacity. Run the vehicle to bring the engine and transmission up to operating temperature. Shut off the engine, and raise the vehicle on the hoist. Place a drain pan with a large funnel under the transmission pan. If the transmission has a drain plug, remove it. The transmission fluid will be hot. If the transmission does not have a drain plug, loosen and remove all of the bolts except for three that are at a corner and next to each other.

____6____ **G.** Compare the new filter and transmission gasket to the old one to make sure they are correct. If the filter kit came with a new filter gasket or grommet, install the new one. Install the new filter and torque any retaining bolts to specifications.

____3____ **H.** Lower the pan and inspect the inside of the pan for debris. Check for nonferrous metal and old clutch material. Inspect the magnet in the pan (if equipped) for the amount of ferrous metal present.

2. Stall Testing:

____2____ **A.** Check engine and transmission fluid levels. If the vehicle does not have a tachometer, place a tachometer on the engine. Place a large fan in front of the vehicle's radiator.

____5____ **B.** Release the throttle and place the vehicle into neutral. Record the stall speed. Raise the engine speed to 1500 rpm and allow the vehicle to cool down 30 seconds before testing the next gear range. Continue testing the remaining gear ranges.

Chapter 20   Servicing the Automatic Transmission/Transaxle

_____ **C.** Compare your results with the manufacturer's specifications to determine which components are faulty, if any. **(There is no image associated with this step.)**

_____ **D.** Start the vehicle and allow it to reach operating temperature. Make sure the test area is clear of bystanders. Firmly apply the emergency brake, depress the brake pedal and place the vehicle into a gear. You will eventually test all gear ranges (R, D4, D3, D2, D1).

_____ **E.** Look up the service information to determine whether a stall test is recommended; if so, follow the procedure. Place wheel chocks in front of and behind the wheels.

_____ **F.** Increase the engine rpm to the stall speed for a maximum of 5 seconds while firmly holding the brake pedal. Note the maximum engine rpm.

3. Performing Pressure Tests:

_____ **A.** Install a transmission pressure tester(s) into the test port(s) on the transmission.

_____ **B.** Clean off any transmission fluid that dripped onto the transmission, restart the vehicle to check for leaks, and top off, if necessary.

___3___ **C.** Start the vehicle and place it in the correct operating conditions to monitor the pressure(s) according to the manufacturer. Record the pressure(s).

___4___ **D.** Shut off the vehicle. Remove the transmission pressure tester(s), seal the threads, and reinstall the test port plug(s).

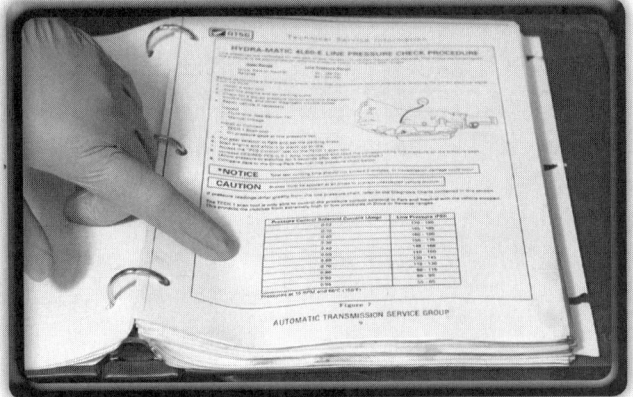

___1___ **E.** Use the service information to find the procedure to test the transmission's hydraulic pressures. Verify the correct transmission fluid level. Place a drain pan under the transmission and remove the correct pressure test port plug(s).

4. Removing and Reinstalling the Transmission:

___11___ **A.** Remove the transmission power train mount. Remove the transmission frame cross member.

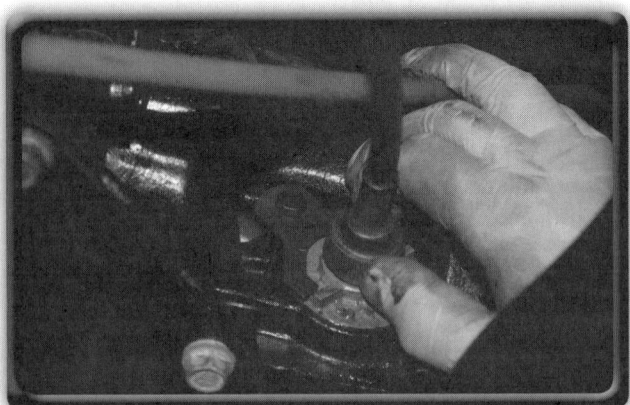

___5___ **B.** Remove the speedometer cable/vehicle speed sensor harness and vacuum line to the vacuum modulator (if equipped) from the transmission.

Chapter 20 Servicing the Automatic Transmission/Transaxle   189

_____8_____ **C.** Remove the dust cover from the torque converter (if equipped) and the bolts connecting the torque converter to the flywheel.

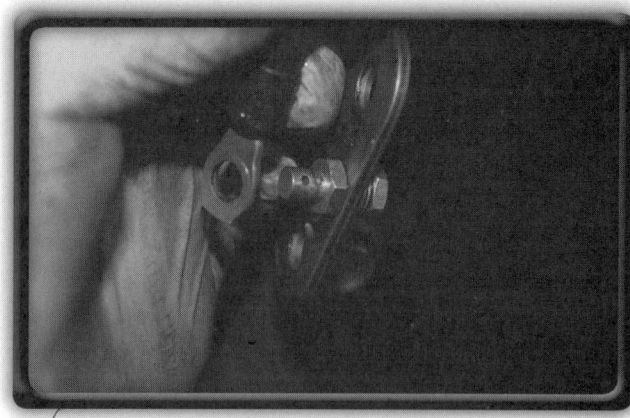

_____6_____ **D.** Remove the TV cable (if equipped), the shift linkage, and any electrical connectors.

_____12_____ **E.** Lower the transmission slightly. Loosen and remove the bolts connecting the bell housing to the engine. Pull the transmission away from the engine until it clears the flex plate and the alignment dowels. Lower the transmission, and move it to where it will be repaired. Inspect the crankshaft rear main seal, engine core plugs, dowel pins, dowel pin holes, and mating surfaces. To reinstall, reverse the order of removal.

_____3_____ **F.** Remove the driveshaft from the vehicle.

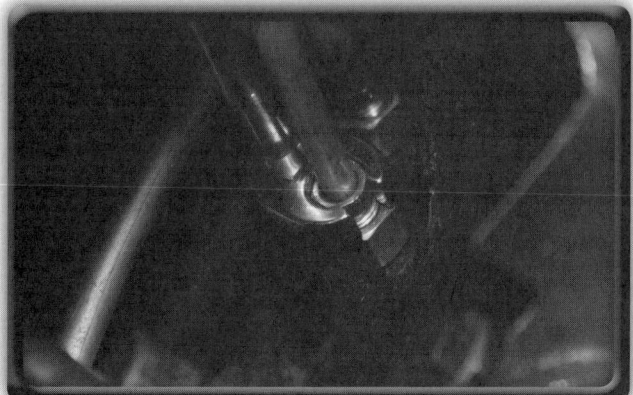

_____9_____ **G.** Disconnect and plug the transmission cooler lines and the transmission vent hose (if equipped).

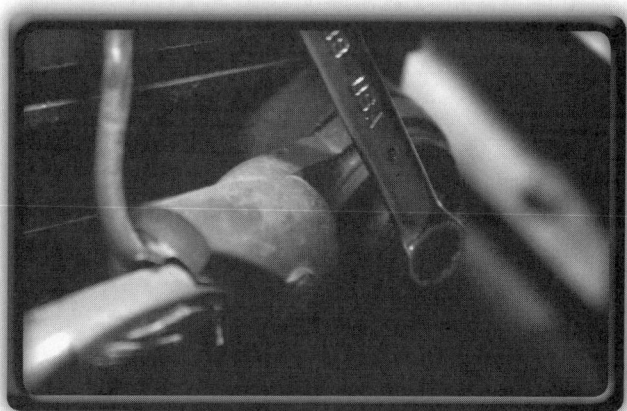

_____1_____ **H.** Look up the procedure for removal of the transmission. Disconnect the negative battery cable. Place tape around the cable end to prevent it from accidentally touching the battery terminal.

___7___ **I.** Remove the throttle valve cable. Remove the starter motor and the exhaust and catalytic converter if necessary to gain access.

___13___ **J.** Set the vehicle on the hoist, taking into consideration the center of gravity of the vehicle with the transmission both in and out of the vehicle. **(There is no image associated with this step.)**

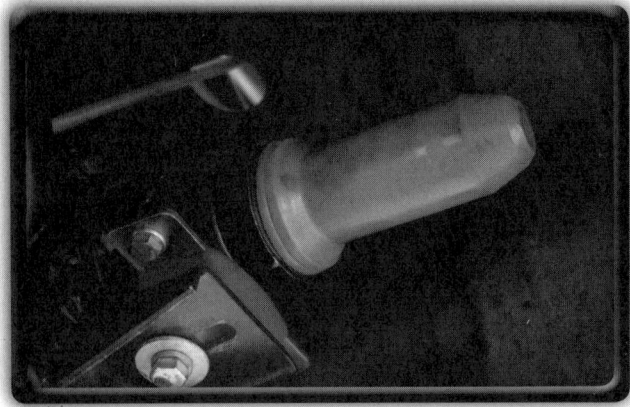

___4___ **K.** Install an output shaft cap over the end of the transmission output shaft to prevent fluid from leaking out while the transmission is removed.

___10___ **L.** Remove the transmission dipstick tube. Install the transmission jack onto the transmission. Secure the safety chain around the transmission. Raise the transmission jack slightly to support the transmission.

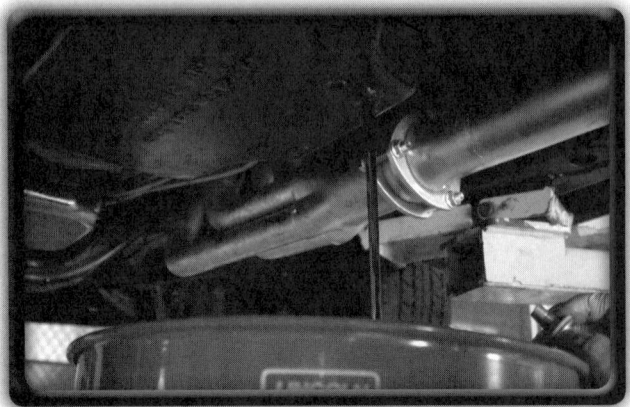

___2___ **M.** Drain the transmission fluid. Reinstall the transmission pan after the transmission has been drained.

5. Inspecting the Converter and Flex Plate:

___1___ **A.** Inspect the torque converter pilot and crankshaft pilot bore for signs of damage.

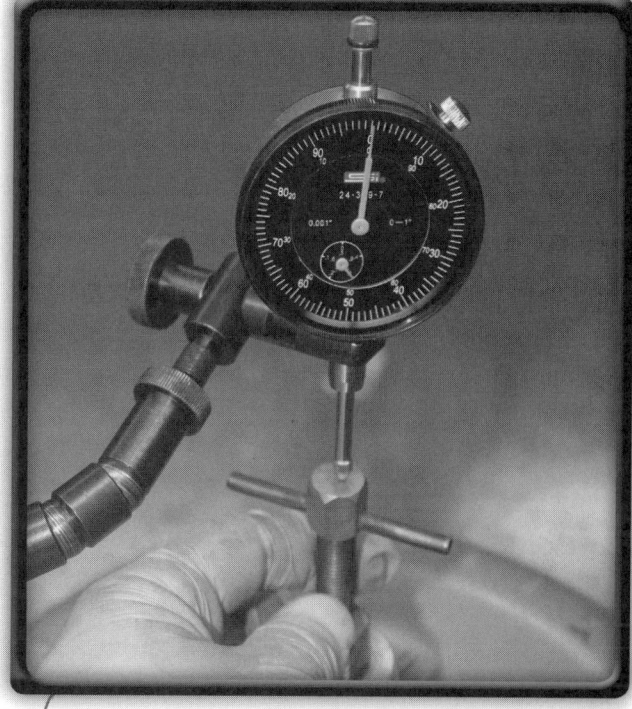

___6___ **B.** Install the required tool into the turbine to check for torque converter end play. Use a dial indicator to measure the amount of turbine movement.

___4___ **C.** Inspect the torque converter mounting pads for damage.

___3___ **D.** Remove the bolts securing the flex plate to the crankshaft.

___2___ **E.** Inspect the bolts and bolt holes for the flex plate to crankshaft and flex plate to torque converter.

___5___ **F.** Inspect the pump drive tangs for damage.

Crossword Puzzle

Use the clues to complete the puzzle.

Across

1. Further pressure applied to bearing-supported parts after all the free play is taken up.
3. The use of compressed air to check clutch and servo operation on transmissions during and after assembly.
4. A test that involves raising the engine rpm to wide-open throttle while the brake is firmly applied and the transmission is in gear. The test is used to check torque convertor and transmission operation on some vehicles.

Down

1. A rubber or metal bracket used to secure the engine and transmission into the vehicle. Some vehicles use hydraulic or electrohydraulic power train mounts.
2. The amount of movement between two mating parts.

ASE-Type Questions

Read each item carefully, and then select the best response.

____ 1. Tech A says that you should check fluid level in many automatic transmissions with the engine idling and the transmission in park. Tech B says that most manufacturers use selectively fit thrust washers to adjust transmission end play. Who is correct?
  A. Tech A
  B. Tech B
  C. Both A and B
  D. Neither A nor B

____ 2. Tech A says that increasing the transmission fluid level from the add mark to the full mark will require a quart of fluid, just like with engine oil. Tech B says that transmission solenoids should be tested electrically and mechanically. Who is correct?
  A. Tech A
  B. Tech B
  C. Both A and B
  D. Neither A nor B

Chapter 20   Servicing the Automatic Transmission/Transaxle

___A___ 3. Tech A says that when checking the fluid level on the dipstick, always read the highest level on either side. Tech B says that when air checking a clutch, the clutch plates should be removed so you can see if the clutch seals are leaking. Who is correct?
   A. Tech A
   B. Tech B
   C. Both A and B
   D. Neither A nor B

___A___ 4. Tech A says that transmission pan gaskets are always replaced. Tech B says that some newer transmissions have reusable pan gaskets. Who is correct?
   A. Tech A
   B. Tech B
   C. Both A and B
   D. Neither A nor B

___B___ 5. Tech A says that the cooler and cooler lines should be flushed as part of a transmission rebuild. Tech B says that the torque converter can be fully seated in the transmission by tightening the bell housing bolts to the engine. Who is correct?
   A. Tech A
   B. Tech B
   C. Both A and B
   D. Neither A nor B

___C___ 6. Tech A says that performing a stall test can reveal slippage within the torque converter or transmission. Tech B says that a stall test can damage some transmissions and should not be performed on all transmissions. Who is correct?
   A. Tech A
   B. Tech B
   C. Both A and B
   D. Neither A nor B

___A___ 7. Tech A says that using a scan tool to activate the TCC with the engine idling in drive and the brakes applied is a typical troubleshooting task. Tech B says that diagnosis is a waste of time on a faulty transmission since it will be rebuilt anyway. Who is correct?
   A. Tech A
   B. Tech B
   C. Both A and B
   D. Neither A nor B

___B___ 8. Tech A says that when checking line pressure, the engine should be off. Tech B says that when checking line pressure, the engine should be running. Who is correct?
   A. Tech A
   B. Tech B
   C. Both A and B
   D. Neither A nor B

___C___ 9. Tech A says that low line pressure can be caused by a plugged transmission vent. Tech B says that high line pressure is caused by a restricted filter. Who is correct?
   A. Tech A
   B. Tech B
   C. Both A and B
   D. Neither A nor B

___A___ 10. Tech A says that some transmissions are made without dipsticks. Tech B says that transmission end play should be measured after the front pump is installed. Who is correct?
   A. Tech A
   B. Tech B
   C. Both A and B
   D. Neither A nor B

CHAPTER 21
Hybrid and Continuously Variable Transmissions

Tire Tread:
© AbleStock

Chapter Review

The following activities have been designed to help you refresh your knowledge of this chapter. Your instructor may require you to complete some or all of these activities as a regular part of your training program. You are encouraged to complete any activity that your instructor does not assign as a way to enhance your learning.

Matching

Match the following terms with the correct description or example.

- **A.** Belt alternator starter (BAS)
- **B.** Continuously variable transmission (CVT)
- **C.** Electronic continuously variable transmission (ECVT)
- **D.** Integrated motor assist (IMA)
- **E.** Toroidal CVT
- **F.** Variable-diameter pulley (VDP)

_____ **1.** A type of CVT that uses two pulleys with moveable sheaves, allowing the effective diameter of the pulleys to change, resulting in variable gear ratios.

_____ **2.** A type of hybrid drive system that uses a belt-driven alternator/starter that operates on 42 volts.

_____ **3.** A type of hybrid transmission that often uses two electric motors in combination with an ICE. The two electric motors and the ICE transfer power through a planetary gear set, allowing an infinite amount of gear ratios.

_____ **4.** A type of transmission that has no fixed gears, as in a conventional transmission, but rather can adjust gear ratios infinitely within the design of the transmission.

_____ **5.** A Honda hybrid drive system that uses a moderate-sized electric motor installed between the engine and the transmission.

_____ **6.** A type of CVT that uses moveable rollers in contact with input and output drive discs. The rollers transfer power from one drive disc to the other. Their position determines the effective gear ratio.

Multiple Choice

Read each item carefully, and then select the best response.

_____ **1.** When the power train control module shuts off the hybrid engine at a stoplight, it is an example of _____.
- **A.** torque smoothing
- **B.** regenerative braking
- **C.** torque assist
- **D.** idle stop

_____ **2.** What ability can a hybrid employ to smooth out the power pulses of the internal combustion engine and create a flatter torque output curve?
- **A.** Torque assist
- **B.** Torque smoothing
- **C.** Idle stop
- **D.** Regenerative braking

Chapter 21   Hybrid and Continuously Variable Transmissions

_____ 3. When a conventional vehicle is being stopped, most of the kinetic energy of the vehicle's movement is converted into _____.
   A. heat
   B. electricity
   C. electromagnetic energy
   D. all of the above

_____ 4. What ability of a hybrid engine reduces the need to use the internal combustion engine at lower rpms by using an electric motor to help propel the vehicle from a stop?
   A. Torque smoothing
   B. Kinetic braking
   C. Torque assist
   D. Integrated propulsion

_____ 5. What hybrid system is classified as a parallel hybrid because the electric motor is operating at the same time as the gasoline engine?
   A. Two-mode
   B. Integrated motor assist
   C. Belt alternator starter
   D. Continuously variable

_____ 6. What type of transaxle can be used in combination with the integrated motor assist system?
   A. Automatic transmission
   B. Manual transmission
   C. Continuously variable transmission
   D. All of the above

_____ 7. What type of hybrid uses a system voltage of 300 volts to power two electric motor/generators housed inside the transmission case?
   A. Belt alternator starter
   B. Integrated motor assist
   C. Two-mode
   D. None of the above

_____ 8. The Toyota and Lexus hybrids are classified as _____ because either the internal combustion engine or the electric motor/generator can propel the vehicle, or both can be used together.
   A. series-parallel hybrid
   B. parallel hybrid
   C. series hybrid
   D. integrated hybrid

_____ 9. All of the following are basic types of CVTs commonly used in production vehicles, *except*:
   A. Variable-diameter pulley
   B. Integrated motor assist
   C. Toroidal
   D. Electronic continuously variable

_____ 10. Which continuously variable transmission design is currently limited in production due to its high manufacturing costs?
   A. Electronic continuously variable
   B. Variable-diameter pulley
   C. Toroidal
   D. All of the above

True/False

If you believe the statement to be more true than false, write the letter "T" in the space provided. If you believe the statement to be more false than true, write the letter "F".

_____ 1. Automobile manufacturers have had hybrid vehicles and vehicles with continuously variable transmissions in mass production for more than 20 years.

_____ 2. Mechanical brakes can be used to recapture kinetic energy while braking.

_____ 3. Hybrid vehicles are able to maintain a higher gas mileage rating during highway driving than in stop-and-go traffic.

_____ 4. An electric motor is capable of creating its maximum torque as soon as it begins spinning.

_____ 5. On many full hybrids, the vehicle can operate at low speeds using the electric motor only.

_____ 6. Currently, Honda integrated motor assist hybrids cannot drive using the electric motor only.

_____ 7. The gasoline engine does not run during the second mode of a two-mode hybrid.

_____ 8. Some of the early CVTs in Europe used a stiff rubber belt, as do many snowmobiles and all-terrain vehicles.

_____ 9. As the vehicle gains speed, the input pulley diameter of the VDP decreases, while the output pulley diameter increases.

_____ 10. Continuously variable transmissions require special transmission fluid.

Fill in the Blank

Read each item carefully, and then complete the statement by filling in the missing word(s).

1. Many hybrid vehicles that use idle stop have a small, electric transmission _____ _____ used to prevent a delay in the engagement of the transmission when the vehicle is restarted.

2. On a hybrid vehicle, when the driver initiates a stop, the electric motor becomes a _____.

3. The _____ _____ _____ system uses a 42-volt battery.

4. The _____-_____ hybrid was a joint venture by the General Motors, Chrysler, and BMW companies to create a hybrid system that could be used on trucks and larger luxury vehicles.

5. Toyota and Lexus hybrids use voltages from _____ to _____ volts, depending on the application.

6. The only major difference between the Ford hybrid and the Toyota hybrid is that rather than having the two electric motor/generators directly connected to the ring gear and the sun gear, they are attached through a set of _____ _____.

7. Each of the VDP pulleys has two movable drive faces called _____.

8. The _____ _____ is made up of hundreds of transversely mounted steel plates that are held in place with several steel bands running longitudinally around the edge of the plates.

9. The _____ CVT design uses two curved discs—an input disc and an output disc.

10. Most manufacturers use a limited number of _____ _____ _____ and a set of planetary gears that allow the continuously variable transmission to operate in reverse.

Chapter 21 Hybrid and Continuously Variable Transmissions

Labeling

Label the following diagrams with the correct terms. For diagram 1, also include engine status.

1. Planetary gear operation as the engine is started:

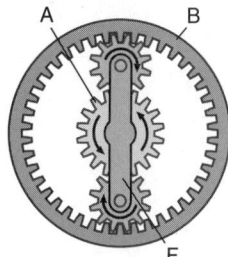 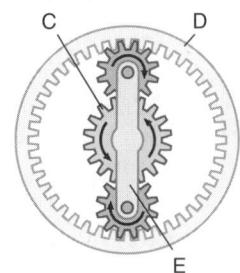

A. _____ Status: _____

B. _____ Status: _____

C. _____ Status: _____

D. _____ Status: _____

E. _____ Status: _____

F. _____ Status: _____

2. Cutaway of a Ford hybrid transmission:

A. _____

B. _____

C. _____

D. _____

E. _____

F. _____

3. A VDP CVT—two sheaves that are moveable:

A. _____
B. _____
C. _____
D. _____

4. The changing sizes of the input and output pulleys:

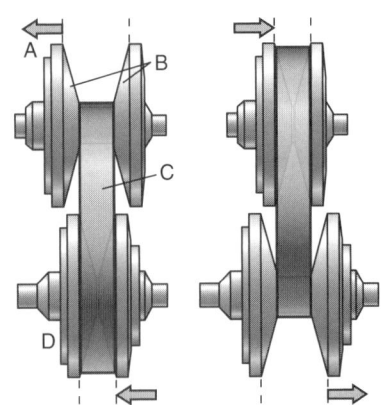

A. _____
B. _____
C. _____
D. _____

5. A toroidal CVT (the transmission on the left is in high torque multiplication ratio, the transmission on the right is in an overdrive ratio):

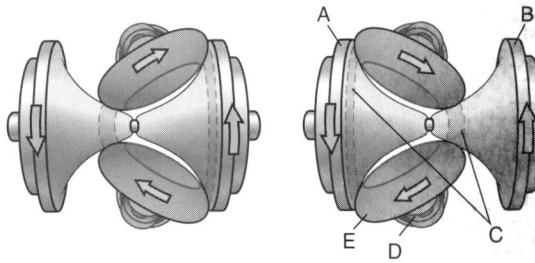

A. _____
B. _____
C. _____
D. _____
E. _____

Crossword Puzzle

Use the clues to complete the puzzle.

Across

3. A type of braking in which the kinetic energy of the vehicle's motion is captured rather than being lost to heat in a conventional braking system.
4. A process that uses an electric motor to smooth out engine power pulses when an ICE is operating at low rpm or when the vehicle is using fuel management techniques such as cylinder deactivation.
5. A drive system that uses two or more propulsion systems such as electric motors and an ICE.

Down

1. A type of hybrid drive system in which there are two distinct modes of operation. In one mode, the electric motor can propel the vehicle and be used for regenerative braking; in the second mode, the electric motors can be used to assist the engine while the engine uses fuel management techniques such as cylinder deactivation.
2. Use of an electric motor to supplement the engine's torque whenever additional torque is needed, allowing for a smaller ICE to be used.

ASE-Type Questions

Read each item carefully, and then select the best response.

_____ 1. Tech A says that hybrid vehicles have an internal combustion engine and an electric motor. Tech B says that most hybrid vehicles utilize regenerative braking to help improve fuel economy. Who is correct?
   A. Tech A
   B. Tech B
   C. Both A and B
   D. Neither A nor B

_____ 2. Tech A says that it is critical to have the proper safety equipment before working on a hybrid. Tech B says that hybrid vehicles use safety interlocks to prevent technician injury so it is not necessary to use protective gloves. Who is correct?
   A. Tech A
   B. Tech B
   C. Both A and B
   D. Neither A nor B

_____ 3. Tech A says that hybrid vehicles will typically use the same transmission as a non-hybrid vehicle. Tech B says that some hybrid transmissions have a small electric fluid pump for when the transmission is in idle/stop mode. Who is correct?
   A. Tech A
   B. Tech B
   C. Both A and B
   D. Neither A nor B

_____ 4. Tech A says that BAS vehicles use an alternator to help slow the vehicle and act as a starter. Tech B says that the BAS alternator can be used to propel the vehicle without the ICE. Who is correct?
   A. Tech A
   B. Tech B
   C. Both A and B
   D. Neither A nor B

_____ 5. Tech A says that the transmission in a Toyota hybrid has a separate gear to provide reverse. Tech B says that Toyota hybrids use a VDP CVT transmission. Who is correct?
   A. Tech A
   B. Tech B
   C. Both A and B
   D. Neither A nor B

_____ 6. Tech A says that VDP CVTs are commonly rebuilt. Tech B says that VDP CVTs require more frequent fluid changes than a conventional automatic transmission. Who is correct?
   A. Tech A
   B. Tech B
   C. Both A and B
   D. Neither A nor B

_____ 7. Tech A says that some CVTs require a heating system to warm the transmission fluid during cold weather operation. Tech B says that CVTs use conventional automatic transmission fluid. Who is correct?
   A. Tech A
   B. Tech B
   C. Both A and B
   D. Neither A nor B

_____ 8. Tech A says that VDP CVTs will often have a planetary gear set to produce reverse. Tech B says that many CVTs in vehicles use a rubber drive belt between two pulleys to transfer power. Who is correct?
   A. Tech A
   B. Tech B
   C. Both A and B
   D. Neither A nor B

_____ 9. Tech A says that two-mode hybrid transmissions use two electric motors to propel the vehicle. Tech B says that the two-mode hybrid transmission can use one motor to propel the vehicle and one to charge the battery. Who is correct?
   A. Tech A
   B. Tech B
   C. Both A and B
   D. Neither A nor B

_____ 10. Tech A says that the Honda IMA system uses a thin electric motor between the engine and the CVT transmission. Tech B says that the Honda IMA system uses voltages between 144 and 158 volts. Who is correct?
   A. Tech A
   B. Tech B
   C. Both A and B
   D. Neither A nor B

Manual Transmission/Transaxle Principles

CHAPTER 22

Tire Tread: © AbleStock

Chapter Review

The following activities have been designed to help you refresh your knowledge of this chapter. Your instructor may require you to complete some or all of these activities as a regular part of your training program. You are encouraged to complete any activity that your instructor does not assign as a way to enhance your learning.

Matching

Match the following terms with the correct description or example.

- **A.** Accelerator pedal
- **B.** Axial load
- **C.** Dead axle
- **D.** Drive axle
- **E.** Drive train
- **F.** Gear set
- **G.** Helical gears
- **H.** Independent rear axle
- **I.** Live axle
- **J.** Power flow
- **K.** Radial load
- **L.** Rotational speed
- **M.** Shaft
- **N.** Solid rear axle
- **O.** Solid axle
- **P.** Splined
- **Q.** Spur gears
- **R.** Thrust washers
- **S.** Transfer case
- **T.** Transmission

_____ 1. Gears that have teeth set on an angle to the gear face; they operate more quietly than spur gears.

_____ 2. An assembly that houses a variety of gear sets that allow the vehicle to be driven at a wider range of speeds and terrain conditions than would be possible without a transmission.

_____ 3. The load that is perpendicular to a shaft, usually controlled by bearings or bushings.

_____ 4. Gears with straight-cut gear teeth.

_____ 5. A type of axle that has a one-piece axle housing, so the action of hitting a bump with one wheel affects the other wheel.

_____ 6. Flat, washer-shaped bearings that provide a wear surface between two rotating components that are loaded axially.

_____ 7. An axle that provides power to a wheel.

_____ 8. Typically, a shaft and gear that have parallel grooves machined in them so they mate with each other and lock together rotationally.

_____ 9. An axle that provides power to the wheels.

_____ 10. The load applied in line with a shaft. It can be controlled with thrust bearings.

_____ 11. An assembly used in four-wheel drive vehicles to transmit power to either two wheels only or all four wheels.

_____ 12. The component assemblies that transmit power from the engine all the way to the drive wheels.

_____ 13. The path that power takes from the beginning of an assembly to the end. In a transmission, power flow changes as different gears are selected by the driver.

_____ 14. The foot-operated pedal used by the driver to increase and decrease the amount of power the engine develops.

_____ 15. The speed at which an object rotates, measured in revolutions per minute (rpm).

_____ 16. An axle that supplies no power to the wheels.

_____ 17. A type of axle that is not flexible, with splines on one end to fit the final drive unit and a flange on the other end to power the wheel.

_____ 18. Two or more gears that are in mesh with each other.

_____ 19. A type of rear suspension system that allows each wheel on the axle to move independently of the other.

_____ 20. The long, narrow component that carries one or more gears or has gears machined into it.

Multiple Choice

Read each item carefully, and then select the best response.

_____ 1. In 1894, Louis René Panhard and Émile Levassor, designed a(n) _____.
   A. rear-wheel drive carriage
   B. synchromesh transmission
   C. automatic transmission
   D. multi-gear manual transmission

_____ 2. Which of the following people connected an engine to a transmission and created a live rear axle by using a metal axle shaft supported by bushings in 1898?
   A. C. E. Duryea
   B. Louis Renault
   C. Henry Ford
   D. Ferdinand Porsche

_____ 3. In 1928, Cadillac introduced the first _____.
   A. synchromesh transmission
   B. front-wheel drive automobile
   C. planetary gear set
   D. multi-gear transmission

_____ 4. The amplification of the input force by trading distance moved for greater output force is the definition of _____.
   A. ratio
   B. mechanical advantage
   C. power flow
   D. rotational speed

_____ 5. Given a 4:1 mechanical advantage, if a person pushes the lever down with a force of 100 pounds, the force the lever generates against the rock is _____.
   A. 140 lb.
   B. 40 lb.
   C. 400 lb.
   D. 4 lb.

_____ 6. If the drive gear has 15 teeth and the driven gear has 30 teeth, then the gear ratio is _____.
   A. 2:1
   B. 800 lb.
   C. 1:4
   D. 4 lb.

_____ 7. The path in which power is transmitted through a series of components is called _____.
   A. linkage
   B. power flow
   C. drive train
   D. transmission

_____ 8. Which term relates to layouts where the transmission and final drive are integrated into a common assembly and is usually used on front-wheel drive vehicles?
   A. Differential
   B. Drive train
   C. Torque converter
   D. Transaxle

_____ 9. What transmits power from the transmission to the final drive assembly on rear-wheel drive vehicles?
   A. Clutch plate
   B. Transaxle
   C. Drive shaft
   D. Live axle

_____ 10. Flexible drive axles are called _____.
   A. radial axles
   B. dead axles
   C. half-shafts
   D. shafts

_____ 11. What provides front-wheel drive axles the ability to change length as the vehicle goes over bumps and dips and allows for the wheels to be steered?
   A. Constant velocity joints
   B. Universal joints
   C. Drive shafts
   D. Synchronizers

_____ 12. Grooves may be cut into the shafts to accommodate _____, which are used for holding the gears in the proper position on the shaft once the gears are installed.
   A. thrust washers
   B. spurs
   C. snap rings
   D. either A or C

_____ 13. What component(s) allow the axles to turn at different speeds when the vehicle is cornering or turning?
   A. Constant velocity joints
   B. Universal joints
   C. Differential assembly
   D. Transfer case

_____ 14. If one wheel is stuck in the snow or ice a(n) _____ allows both of the rear wheels to supply power to the ground in order to continue forward motion.
   A. transfer case
   B. limited slip differential
   C. transaxle
   D. open differential

_____ 15. Which term refers to grease's ability to be a solid but while under stress to flow or become thin in order to lubricate properly?
   A. Thixotropy
   B. Viscosity
   C. Viscidity
   D. Glutinousness

True/False

If you believe the statement to be more true than false, write the letter "T" in the space provided. If you believe the statement to be more false than true, write the letter "F".

_____ 1. Prior to 1898, vehicles were either belt or chain driven.
_____ 2. If the output force is four times greater than the input force, the input distance moved is two times greater than the output distance.
_____ 3. As the gear ratio decreases, the output speed increases.
_____ 4. The clutch pedal is used for acceleration or deceleration of the engine.
_____ 5. The final drive assembly incorporates a set of differential gears arranged so they sit between the two axles.
_____ 6. Shafts are used to support gears and are machined precisely to accommodate bearings and individual gears.
_____ 7. Splines allow shafts and gears to have greater rotational speeds, while at the same time minimizing metal-to-metal friction, which contributes to increased fuel economy and longer transmission life.
_____ 8. The only difference between the transmission and the transaxle is that the transaxle incorporates the final drive assembly in its construction.
_____ 9. Transmissions/transaxles are rated by the manufacturers for how much twisting force, measured in foot-pounds, they can handle.
_____ 10. Every component in the manual transmission requires lubrication.

Fill in the Blank

Read each item carefully, and then complete the statement by filling in the missing word(s).

1. The _____ can be operated by the driver to disconnect and connect the transmission from the engine.
2. A(n) _____ _____ applies a friction device between the gear and the shaft to match the gear speed to the shaft speed.
3. _____ are normally expressed as the equivalent of the first number to 1.
4. A gear set in which the _____ gear has half as many teeth as the _____ gear has a gear ratio of 2:1.
5. The _____ system is the medium by which the driver can connect and disconnect the engine from the transmission, resulting in the vehicle's forward or rearward movement.
6. The final _____ _____ gives the final gear reduction to the drive train and powers the drive wheels through axles.
7. The drive shaft uses _____ _____, which allow the drive shaft to change angles due to the movement of the suspension relative to the body.
8. Both front-wheel drive and rear-wheel drive vehicles use _____ _____ to power the wheels.
9. Depressing the clutch pedal closes the contacts of the _____ _____ switch.
10. _____ are round parts with teeth cut on the outside perimeter.
11. In a manual transmission vehicle, the component that locks the engine and transmission together is the _____ system.
12. The _____ _____ takes the power from the transmission and directs it to one or both axles, depending on the mode selected.
13. With the _____ _____ assembly, if one wheel is stuck in the snow or ice, the other wheel cannot supply power.
14. The _____ _____ drive axle uses one half-shaft axle for each of the two wheels.
15. A(n) _____ _____ allows the wheels to freely rotate on the axle assembly and to not drive the wheels.

Chapter 22   Manual Transmission/Transaxle Principles

Labeling
Label the following diagrams with the correct terms.

1. Power flow in a transmission:

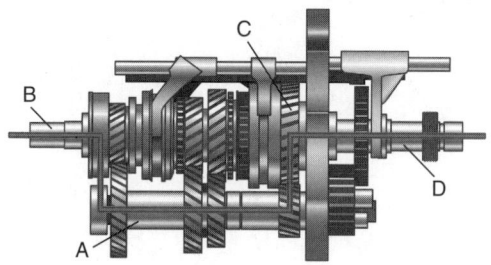

A. _____
B. _____
C. _____
D. _____

2. Typical final drive assembly:

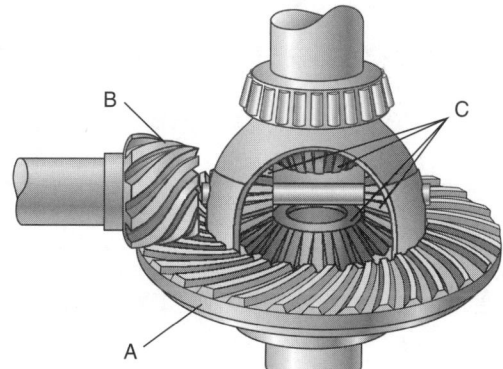

A. _____
B. _____
C. _____

3. Typical drive shaft:

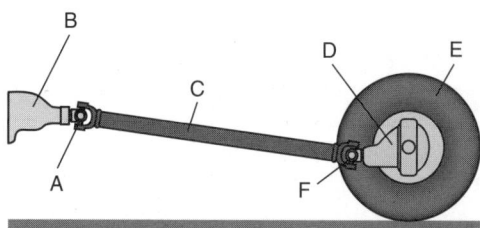

A. _____
B. _____
C. _____
D. _____
E. _____
F. _____

4. Axial-loaded thrust washer:

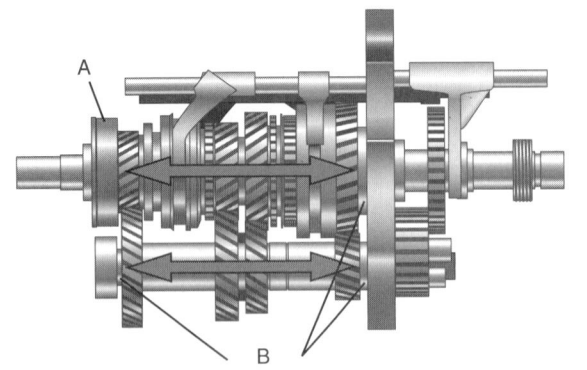

A. _____

B. _____

5. Clutch components designed to connect and disconnect power to the transmission:

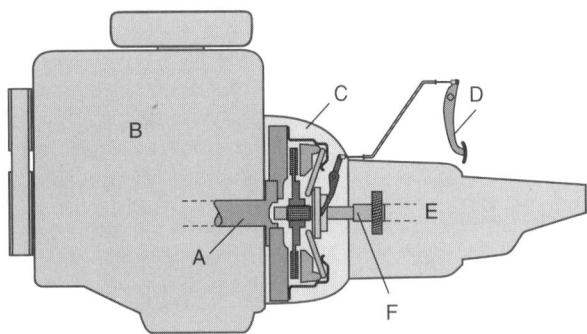

A. _____

B. _____

C. _____

D. _____

E. _____

F. _____

Crossword Puzzle

Use the clues to complete the puzzle.

Across

1. The process of using a device to get more output force than the amount of input force, with the trade-off being that the input distance is proportionately longer than the output distance.
7. The description of the difference in speed between gears in mesh, determined by comparing the number of teeth on each gear (drive/driven).
8. A differential assembly that allows both axles to turn at their own speed when turning a corner, but is dependent on the traction of the tires to deliver torque to the ground. If one wheel has no traction, all of the engine's torque will be used at that wheel, causing it to simply spin.
13. Joints commonly used in front-wheel drive vehicles to allow flexibility of the axle while turning.
15. The spring-steel C-shaped ring that is fitted in a groove and holds gears, bearings, and shafts in place.

Down

2. The components that make up the drive axle including the axles, final drive assembly, bearings, and axle housing.
3. An electrical switch that is operated by the clutch pedal and keeps the starter motor from cranking the engine over until the clutch is fully depressed.
4. An assembly used to power the drive wheels and allow the wheels to rotate at different speeds as the vehicle turns.

5. A cross-shaped joint with bearings on each leg where one set of parallel legs is connected to the end of one shaft and the other set of parallel legs is connected to the end of a second shaft. This arrangement allows the shafts to operate at shallow angles to each other.
6. Gears situated in the final drive assembly that are meshed together and with both axles, allowing the wheels to rotate at different speeds when turning a corner.
9. The ability of a semisolid grease to flow when agitated or stressed.
10. A mechanically operated assembly that connects and disconnects the engine from the transmission.
11. An axle that has CV joints on each end and that fits between the transaxle and wheel. Typically, one is used on each side of a vehicle.
12. The foot-operated pedal used by the driver to engage and disengage the clutch.
14. A relatively round, rotating part with internal or external teeth that are designed to mesh with another gear for the purpose of transmitting torque.

ASE-Type Questions

Read each item carefully, and then select the best response.

_____ 1. Tech A says that friction bearings are made up of balls and rollers. Tech B says that non-friction bearings are in sliding contact between moving surfaces. Who is correct?
   A. Tech A
   B. Tech B
   C. Both A and B
   D. Neither A nor B

_____ 2. Tech A says that gear ratios are all the same in any five-speed transmission. Tech B says that gear ratios vary from transmission to transmission. Who is correct?
   A. Tech A
   B. Tech B
   C. Both A and B
   D. Neither A nor B

_____ 3. Tech A says that a road test is helpful in diagnosing transmission issues. Tech B says that having a thorough understanding of the customer concern is important when diagnosing transmission issues. Who is correct?
   A. Tech A
   B. Tech B
   C. Both A and B
   D. Neither A nor B

_____ 4. Tech A says that gear lube can be used in all manual transmissions. Tech B says that some manual transmissions use engine oil as a lubricant. Who is correct?
   A. Tech A
   B. Tech B
   C. Both A and B
   D. Neither A nor B

_____ 5. Tech A says that the final drive is used to provide an increase in the rotational speed of the axles. Tech B says that the final drive is used to provide an increase of twisting force to the axles. Who is correct?
   A. Tech A
   B. Tech B
   C. Both A and B
   D. Neither A nor B

_____ 6. Tech A says that a gear set that has a drive gear with 9 teeth and a driven gear with 27 teeth has a gear ratio of 3:1. Tech B says that the drive gear is also called the output gear. Who is correct?
   A. Tech A
   B. Tech B
   C. Both A and B
   D. Neither A nor B

Chapter 22 Manual Transmission/Transaxle Principles

_____ 7. Tech A says that transmission fluid levels are critical to the life of the transmission. Tech B says that transmission oil should be changed when changing the engine oil. Who is correct?
   A. Tech A
   B. Tech B
   C. Both A and B
   D. Neither A nor B

_____ 8. Tech A says that a road test is part of a good process for diagnosing a customer drive train complaint. Tech B says that the ring gear and pinion gear are part of the final drive assembly. Who is correct?
   A. Tech A
   B. Tech B
   C. Both A and B
   D. Neither A nor B

_____ 9. Tech A says that the clutch uses a cone-style synchronizer to match the speed of the engine to the manual transmission. Tech B says that the fill plug hole is where to check the fluid level in most manual transmissions. Who is correct?
   A. Tech A
   B. Tech B
   C. Both A and B
   D. Neither A nor B

_____ 10. Tech A says that the differential assembly provides a means for the inside and outside wheels to turn at different speeds when going around a corner. Tech B says that the differential assembly provides smooth shifts and reduces gear "grinding" by matching gear speeds. Who is correct?
   A. Tech A
   B. Tech B
   C. Both A and B
   D. Neither A nor B

CHAPTER 23
The Clutch System

Tire Tread:
© AbleStock

Chapter Review
The following activities have been designed to help you refresh your knowledge of this chapter. Your instructor may require you to complete some or all of these activities as a regular part of your training program. You are encouraged to complete any activity that your instructor does not assign as a way to enhance your learning.

Matching
Match the following terms with the correct description or example.

- **A.** Carrier
- **B.** Clutch binding
- **C.** Clutch disc
- **D.** Clutch fork
- **E.** Coefficient of friction
- **F.** Coil spring pressure plate
- **G.** Crankshaft end play
- **H.** Diaphragm pressure plate
- **I.** Driven center plate
- **J.** Free-play
- **K.** Friction facing
- **L.** Fulcrum ring
- **M.** Pilot bearing
- **N.** Push-type clutch
- **O.** Pressure plate
- **P.** Quadrant ratchet
- **Q.** Release mechanisms
- **R.** Single-plate clutch
- **S.** Slave cylinder
- **T.** Slippage

_____ 1. The bearing or bushing that supports the front of the transmission input shaft.

_____ 2. The friction disc that is held firmly against the flywheel by a pressure plate and transfers power from the flywheel to the transmission input shaft.

_____ 3. A type of pressure plate that uses coil springs to provide the clamping force.

_____ 4. A condition in which two surfaces in firm contact with each other slide.

_____ 5. The material riveted to each side of the clutch disc that mates to the flywheel and pressure plate. Used to provide friction and a wear surface for the clutch assembly.

_____ 6. The amount of forward and rearward movement of the crankshaft in the main bearings.

_____ 7. The device used in some cable-operated clutches to provide self-adjustment as the clutch disc wears.

_____ 8. The component in a hydraulically operated clutch that converts hydraulic pressure to mechanical movement at the clutch fork.

_____ 9. The center component of the clutch assembly, with friction material riveted on each side. Also called a clutch plate or friction disc.

_____ 10. Components that operate the clutch. Usually included are the throw-out bearing and the clutch fork. Some manufacturers include the operating system.

_____ 11. A slightly conical, spring steel plate used to provide the clamping force for the clutch assembly.

_____ 12. The part of the throw-out bearing assembly that holds the bearing.

_____ 13. A clutch assembly that uses only one plate to transfer torque from the engine to the transmission. This is the most common type of light vehicle clutch.

_____ 14. The part of the clutch linkage that operates the throw-out bearing.

_____ 15. The amount of clearance in the clutch release mechanism as measured at the clutch pedal.
_____ 16. A condition in which the clutch disc is dragging, leading to grinding gears during gear shifts and possibly clutch chatter.
_____ 17. A steel ring that is used as a pivot point for the diaphragm spring in the pressure plate.
_____ 18. The amount of resistance to movement between any two surfaces that are in contact with each other.
_____ 19. The assembly that applies and removes the clamping force on the clutch disc.
_____ 20. A typical clutch system used in modern vehicles where the clutch fork pushes the release bearing forward to release the friction facing from the pressure plate.

Multiple Choice

Read each item carefully, and then select the best response.

_____ 1. The amount of torque a clutch can transmit is dependent upon which of the following?
   A. Diameter of the clutch
   B. Coefficient of friction
   C. Total spring force
   D. All of the above

_____ 2. The _____ is able to connect to and disconnect from the engine's flywheel through the operation of the clutch.
   A. crankshaft
   B. input shaft
   C. pilot bearing
   D. output shaft

_____ 3. The main purpose of the _____ is to smooth out the power pulses from the pistons during the power strokes.
   A. pressure plate
   B. throw-out bearing
   C. flywheel
   D. clutch disc

_____ 4. Which of the following is a type of flywheel?
   A. One-piece
   B. Harmonic
   C. Dual mass
   D. Both A and C

_____ 5. Refinishing the flywheel moves the pressure plate toward the engine and away from the throw-out bearing, increasing _____.
   A. free-play
   B. clutch chatter
   C. runout
   D. fuel economy

_____ 6. The snout of the input shaft rides on the _____.
   A. throw-out bearing
   B. hub
   C. pilot bearing
   D. flywheel

_____ 7. Which of the following is a type of clutch operating mechanism?
   A. Linkage style
   B. Hydraulic
   C. Cable style
   D. All of the above

___ 8. Heavier-duty or high-performance vehicles may use a ___ clutch assembly due to the higher torque-carrying ability.
   A. single-plate
   B. multiplate
   C. splined
   D. hydraulic

___ 9. What condition can usually be identified while driving the vehicle in fourth gear while accelerating?
   A. Clutch vibration
   B. Slippage
   C. Worn pilot bearing
   D. Clutch binding

___ 10. A shuddering feeling as the clutch pedal is being released is known as ___.
   A. clutch chatter
   B. torsional vibration
   C. clutch binding
   D. driveline vibration

True/False

If you believe the statement to be more true than false, write the letter "T" in the space provided. If you believe the statement to be more false than true, write the letter "F".

___ 1. Automotive manual transmission clutches are wet clutches.
___ 2. Both output shaft torque and speed can be increased at the same time.
___ 3. The output shaft is connected directly to the wheels through the drive train components and cannot be disconnected.
___ 4. The function of the dual mass flywheel is to absorb torsional crankshaft vibrations.
___ 5. Coil spring pressure plate clutches require less pedal effort to operate.
___ 6. The clutch throw-out bearing and clutch fork work together to compress the pressure plate springs when the clutch pedal is pressed.
___ 7. Some vehicles do not require a pilot bearing.
___ 8. In a cable-style clutch control system the slave cylinder is located directly behind the throw-out bearing and pushes directly on it.
___ 9. Excessive slippage from a misadjusted clutch can contribute to overheating of the clutch components and pedal pulsations.
___ 10. Use compressed air to blow off clutch parts such as the flywheel, pressure plate, clutch plate, transmission, or engine bell housings.

Fill in the Blank

Read each item carefully, and then complete the statement by filling in the missing word(s).

1. The _____ allows the driver to engage and disengage the engine from the transmission while operating the vehicle.
2. Two or more clutch plates can be used to form a _____ clutch, increasing the number of facings and the torque capacity.
3. The _____ bolts onto the flywheel.
4. The flywheel _____ enables the starter motor drive gear to crank the engine over.
5. The _____ flywheel improves the engine's fuel economy by smoothing out the power pulses and focusing them in the direction of engine rotation.
6. The diaphragm pressure plate is located inside the clutch cover on two _____, held in place by a number of rivets passing through the diaphragm.

7. The clutch disc is also called a _____ _____ _____ or a friction disc.
8. The throw-out bearing carrier slides on the sleeve of the front _____ _____ that extends from the front of the transmission.
9. The _____ is generally screwed into the bell housing and is usually replaceable.
10. While the vehicle is in motion, the transmission of engine _____ must be interrupted for shifting of the gears.
11. Older technology clutch operating systems used a series of levers with an equalizing mechanism called a(n) _____ _____.
12. A worn _____ _____ can make a howling or squeaking sound.
13. _____ _____ may come from a loose pressure plate along with motor mounts that are worn and are allowing the engine to contact the frame or cross member.
14. A _____ _____ tool is used to center the clutch plate between the flywheel and the pressure plate during the installation of the pressure plate.
15. As the clutch wears, the _____ _____ becomes thinner.

Labeling

Label the following diagrams with the correct terms.

1. Standard light vehicle clutch:

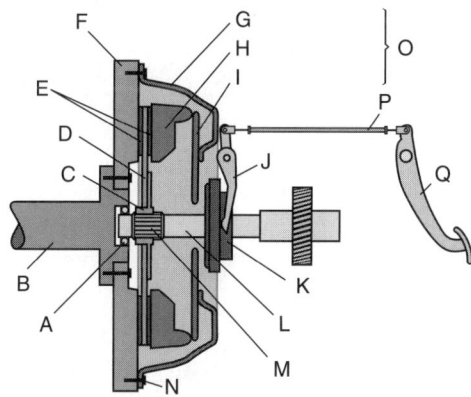

A. _____
B. _____
C. _____
D. _____
E. _____
F. _____
G. _____
H. _____
I. _____
J. _____
K. _____
L. _____
M. _____
N. _____
O. _____
P. _____
Q. _____

2. Releasing the clutch pedal:

Driven unit disengaged

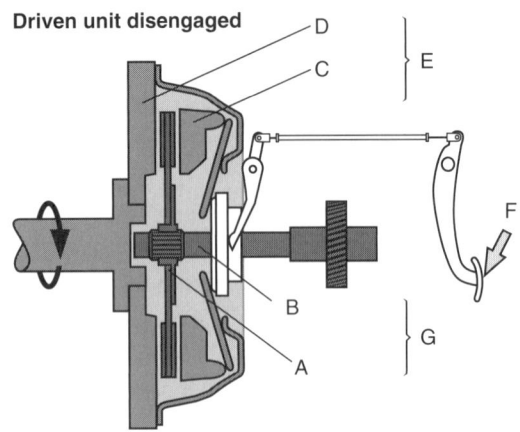

Driven unit engaged

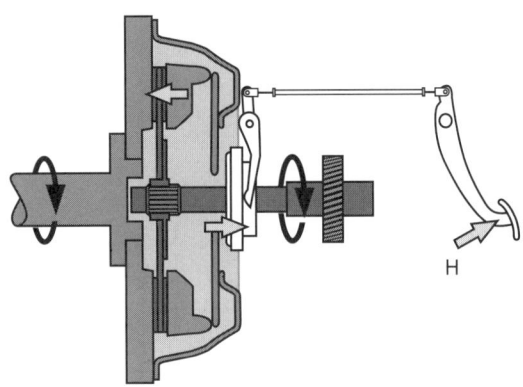

A. _____
B. _____
C. _____
D. _____
E. _____
F. _____
G. _____
H. _____

3. Two types of dual mass flywheels:

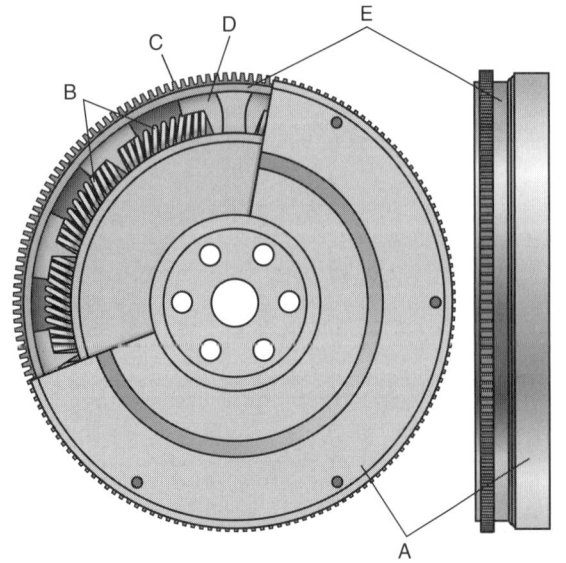

A

A. _____
B. _____
C. _____
D. _____
E. _____

F. _____

G. _____

H. _____

I. _____

J. _____

K. _____

4. Diaphragm pressure plate:

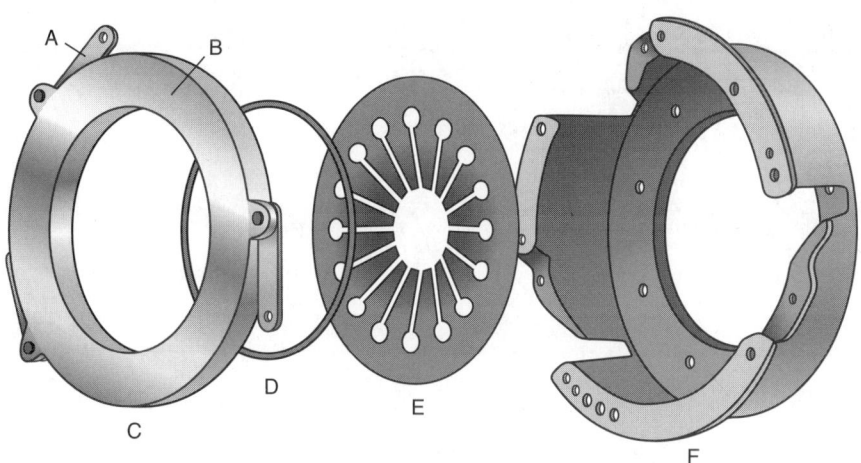

A. _____

B. _____

C. _____

D. _____

E. _____

F. _____

5. Coil spring pressure plate:

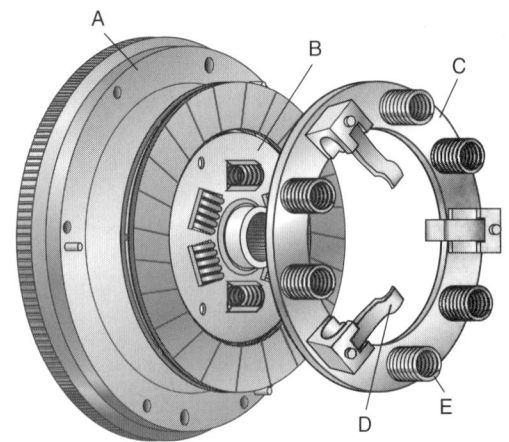

A. _____
B. _____
C. _____
D. _____
E. _____

6. Clutch disc components:

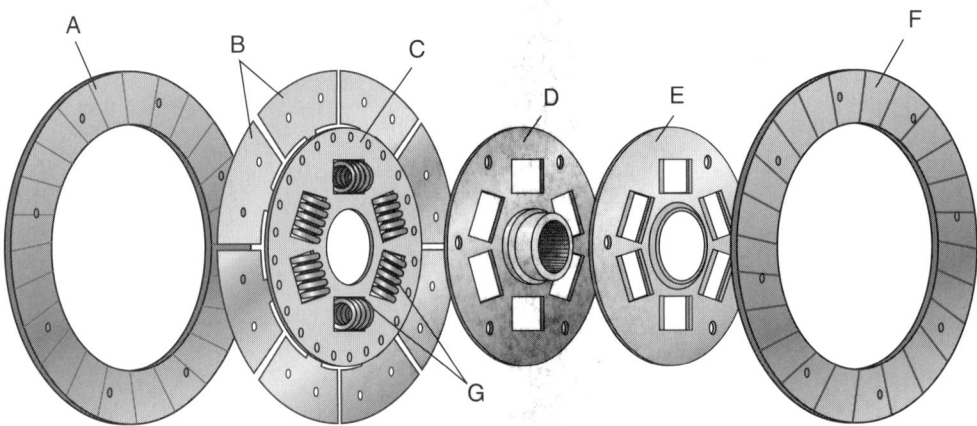

A. _____
B. _____
C. _____
D. _____
E. _____
F. _____
G. _____

Chapter 23 The Clutch System

7. Waved springs:

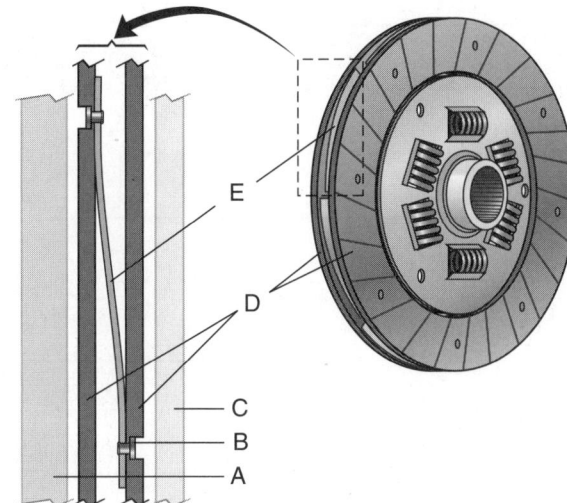

A. _____
B. _____
C. _____
D. _____
E. _____

8. Cable-operated clutch:

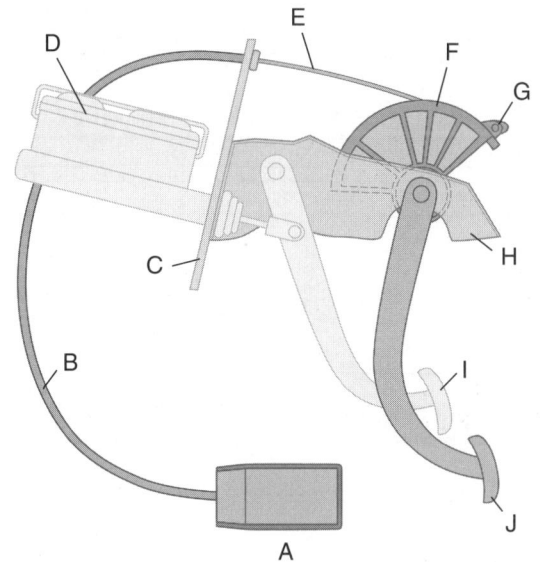

A. _____
B. _____
C. _____
D. _____
E. _____
F. _____
G. _____
H. _____
I. _____
J. _____

9. Hydraulic clutch control:

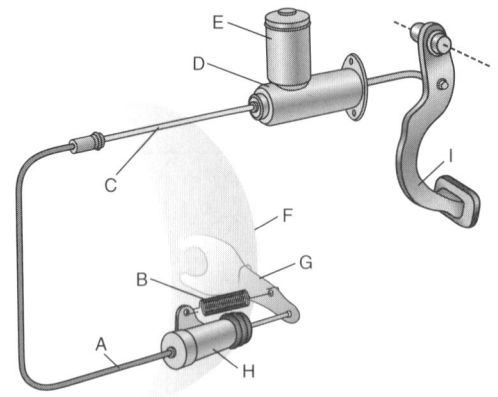

A. _____
B. _____
C. _____
D. _____
E. _____
F. _____
G. _____
H. _____
I. _____

10. Lever-operated system:

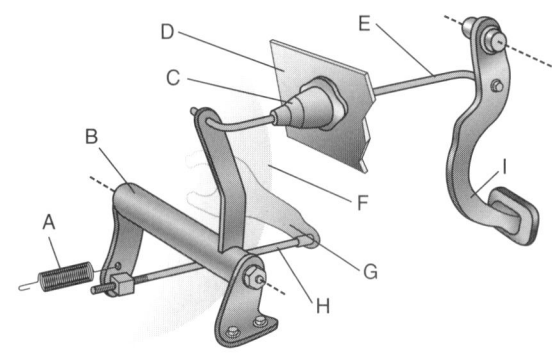

A. _____
B. _____
C. _____
D. _____
E. _____
F. _____
G. _____
H. _____
I. _____

Skill Drills

Place the skill drill steps in the correct order.

1. Checking and Adjusting a Mechanical Clutch:

_____ **A.** Check the clutch linkage components under the hood for the same signs of wear or damage as the components under the dash.

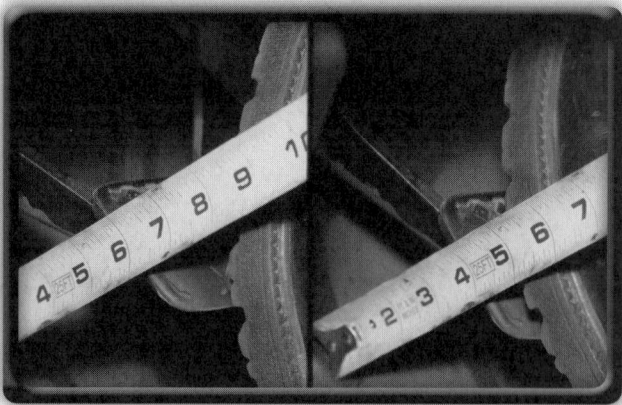

_____ **B.** Measure the clutch pedal free-play. Perform any adjustments as necessary, following the manufacturer's procedure.

_____ **C.** Following the specified procedure, inspect the clutch linkage parts for damaged, worn, bent, or missing components. Look for signs of binding, looseness, and excessive wear. Start with the clutch pedal assembly and inspect all components under the dash. Operate the clutch pedal while you are inspecting the components to observe looseness or binding.

_____ **D.** Start the vehicle and depress the clutch. The clutch should engage at the proper height and have the proper free-play. Make a gear selection to ensure the gears do not clash going into mesh. While in gear, slowly release the clutch and see how far the clutch pedal must travel before the clutch starts to engage in forward motion. If it is not within the manufacturer's specifications, discuss this with your supervisor. **(There is no image associated with this step.)**

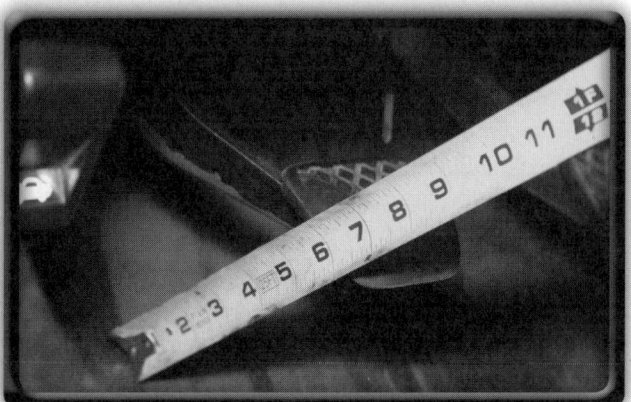

_____ **E.** Measure the clutch pedal height. Compare your reading to the specifications and determine any necessary actions to correct any fault.

2. Checking and Adjusting a Hydraulic Clutch:

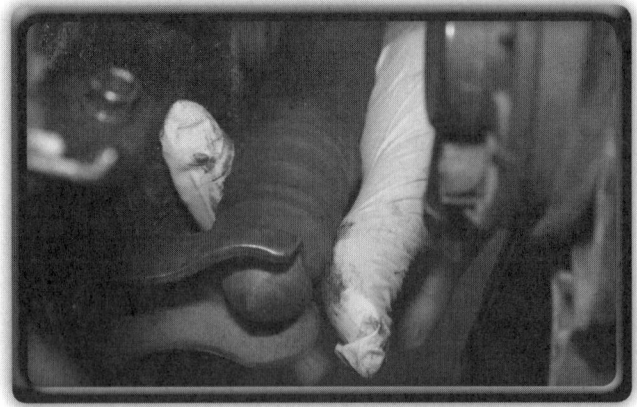

_____ **A.** Check the boot on the slave cylinder for seepage, which may indicate a leaking slave cylinder piston seal.

_____ **B.** Inspect the clutch master cylinder for correct fluid level, and test the quality of the fluid. Inspect all line connections to the master cylinder.

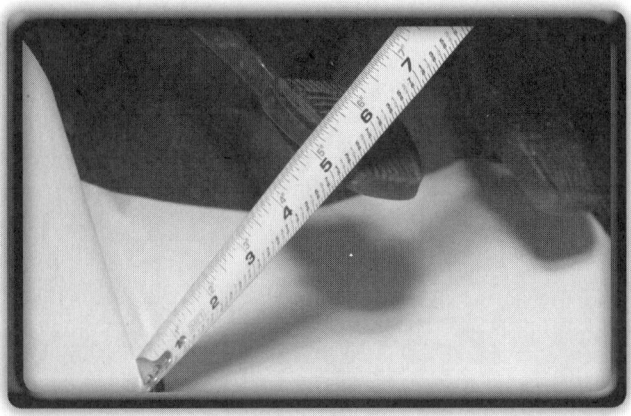

_____ **C.** Check clutch pedal height. Measure clutch pedal free-play using a tape measure. Compare your readings to the specifications, and determine any necessary actions to correct any fault.

_____ **D.** Check that all hydraulic lines are not kinked or leaking at their connections. This will require that the system be repaired and bled of any air. Check all rubber hoses for dry rot, bulges, or leaks. Make sure all hydraulic components are secure in their mountings.

3. Removing and Reinstalling a Transmission/Transaxle:

_____ **A.** If the vehicle is equipped with a hydraulic clutch, reinstall the slave cylinder and bleed and/or adjust if necessary.

_____ **B.** Check that all electrical wires, connectors, linkages, and other removed components are properly installed and adjusted. Place exhaust hose(s) over the exhaust pipe(s) and set the parking brake.

_____ **C.** Lightly lubricate input shaft splines to ensure that there is no binding upon entering the clutch disc hub. Position all wires, hoses, and tubes out of the way and in their specified positions. Secure them temporarily if necessary.

_____ **D.** Disconnect the drive shaft or axles and secure them from hanging. Disconnect all wires, tubes, and hoses that may be present and inspect for damage. Disconnect the clutch and shifter linkage and inspect for wear or damage. Secure the transmission with a transmission jack and remove any transmission mounts.

_____ **E.** Prepare the transmission to be reinstalled. Ensure that the release (throw-out) bearing and clutch fork are properly installed and the clutch disc is centered in the pressure plate. If specified, lubricate the pilot bushing/bearing.

_____ **F.** Following the specified procedure for the specific type of transmission/transaxle, remove the transmission/transaxle from the vehicle. At this time, you may want to overhaul the transmission/transaxle or perform other tasks such as inspecting and replacing the clutch assembly or replacing the rear crankshaft main seal.

_____ **G.** Research the procedures and specifications. Ensure that bolts, clips, and fasteners are kept in containers. Make sure the vehicle is secure on the lift. If necessary, drain the transmission fluid to avoid a spill on the floor that may cause a safety hazard.

_____ **H.** Following the specified procedure for the specific type of transmission/transaxle, reinstall the transmission/transaxle. Tighten all fasteners to the proper torque. If the vehicle is equipped with a cable- or linkage-style clutch release system, reinstall and adjust it for the proper free-play.

_____ I. Refill the transmission/transaxle to the proper level with the specified fluid.

4. Inspecting the Flywheel and Ring Gear:

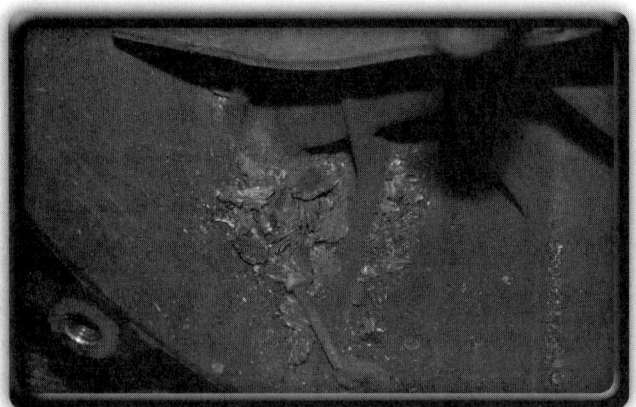

_____ A. Clean up any debris using an approved method for disposing of hazardous dust. Use approved equipment and methods for the removal and disposal of clutch dust.

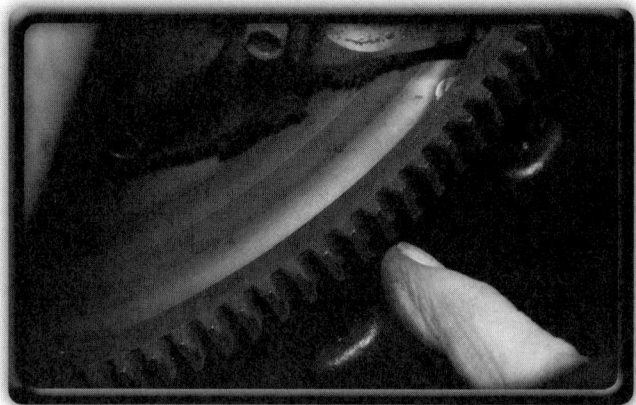

_____ B. Inspect the ring gear for wear, chipped teeth, and cracks. If teeth are worn in one area, it may also be necessary to replace the starter drive.

_____ C. Inspect the starter drive, as it may have been damaged from a faulty ring gear.

_____ D. Inspect the flywheel for wear, hot spots, bluing, and cracks. Use of a straightedge can give a preliminary check for flatness.

Chapter 23   The Clutch System   223

_____ **E.** Follow the specified procedure to remove the pressure plate and clutch disc. Remove the pressure plate bolts evenly, backing each bolt out one turn at a time to avoid warping the pressure plate. Have an assistant hold the pressure plate and clutch disc when they get close to becoming loose.

_____ **F.** Check that the ring gear is secure on the flywheel.

5. Installing a Clutch Assembly:

_____ **A.** Following the specified procedure, install the pilot bushing/bearing, if removed. Lubricate it with the specified lubricant, if applicable.

_____ **C.** Install the release (throw-out) bearing and linkage in the bell housing. **(There is no image associated with this step.)**

_____ **B.** Following the specified procedure, install the clutch disc and pressure plate. Use an appropriate clutch alignment tool to ensure proper positioning of the clutch disc.

_____ **D.** Check the release (throw-out) bearing and linkage for damage or binding. Check the pilot bearing/bushing for damage and its cavity. Determine any necessary action(s) for correcting failures, or replace any worn parts with new/remanufactured components.

_____ **E.** Research the procedure. Inspect the pressure plate assembly for worn springs or diaphragm and also for irregularities associated with any kind of failure, including hot spots. Inspect the clutch disc for loose dampening springs or rubber blocks, and check the condition of the rivets.

Fundamentals of Automotive Technology Student Workbook

_____ **F.** Tighten the pressure plate bolts down evenly by tightening each bolt a turn or two, one at a time, until the pressure plate is evenly seated on the flywheel surface. Torque each pressure plate bolt according to the specified torque and sequence. Install the release (throw-out) bearing and linkage in the bell housing.

_____ **G.** Following the specified procedure, install the flywheel. Be sure to torque the flywheel bolts in the proper sequence.

Crossword Puzzle

Use the clues to complete the puzzle.

Across

4. The part of the clutch release mechanism that imparts clutch pedal force to the rotating pressure plate levers.
5. The device used in some cable-operated clutches to provide self-adjustment as the clutch disc wears.
6. A slightly conical, spring steel plate used to provide the clamping force for the clutch assembly.
9. The part of the throw-out bearing assembly that holds the bearing.
11. The part of the clutch linkage that operates the throw-out bearing.
12. The component in a hydraulically operated clutch that converts hydraulic pressure to mechanical movement at the clutch fork.
13. The material riveted to each side of the clutch disc that mates to the flywheel and pressure plate. Used to provide friction and a wear surface for the clutch assembly.
14. Components that operate the clutch. Usually included are the throw-out bearing and the clutch fork. Some manufacturers include the operating system.

Down

1. A condition in which the clutch disc is dragging, leading to grinding gears during gear shifts and possibly clutch chatter.
2. Rotational fluctuations caused by out-of-balance, misaligned, worn, or bent driveline components.
3. The speeding up and slowing down of a shaft, which happens at a relatively high frequency. Crankshafts have torsional vibrations due to the power pulses of the pistons.
4. The shaft that brings engine torque into the transmission.
7. A condition in which two surfaces in firm contact with each other slide.
8. The assembly that applies and removes the clamping force on the clutch disc.
10. A steel ring that is used as a pivot point for the diaphragm spring in the pressure plate.

ASE-Type Questions

Read each item carefully, and then select the best response.

_____ 1. Tech A says that the pressure plate friction surface rides on the flywheel friction surface to transmit torque. Tech B says that the flywheel can either be flat or stepped. Who is correct?
   A. Tech A
   B. Tech B
   C. Both A and B
   D. Neither A nor B

_____ 2. Tech A says that insufficient clutch pedal clearance (free-play) can cause gear clashing when shifting. Tech B says that when the engine is idling and the clutch pedal is released, the friction disc should stop rotating. Who is correct?
   A. Tech A
   B. Tech B
   C. Both A and B
   D. Neither A nor B

_____ 3. Tech A says that hot spots on the flywheel are a result of excessive heat. Tech B says that a pulsation in a clutch pedal could be due to uneven clutch pressure plate levers. Who is correct?
   A. Tech A
   B. Tech B
   C. Both A and B
   D. Neither A nor B

_____ 4. Tech A says that a leaking rear main seal can cause clutch damage. Tech B says that clutch chatter is a result of a bad throw-out bearing. Who is correct?
   A. Tech A
   B. Tech B
   C. Both A and B
   D. Neither A nor B

_____ 5. Tech A says that the flywheel runout can be checked with a dial indicator. Tech B says that when checking flywheel runout, it is good practice to also check crankshaft end play. Who is correct?
   A. Tech A
   B. Tech B
   C. Both A and B
   D. Neither A nor B

_____ 6. Tech A says that clutch slippage can be a result of excessively strong pressure plate spring(s). Tech B says that when replacing the friction disc, it is good practice to also replace the pressure plate. Who is correct?
   A. Tech A
   B. Tech B
   C. Both A and B
   D. Neither A nor B

_____ 7. Tech A says that the pilot bearing can be a needle-style bearing. Tech B says that the pilot bearing can be a brass bushing style. Who is correct?
   A. Tech A
   B. Tech B
   C. Both A and B
   D. Neither A nor B

_____ 8. Tech A says that a bad pilot bearing can cause a whirring noise. Tech B says that a clutch fork can cause a whining noise. Who is correct?
   A. Tech A
   B. Tech B
   C. Both A and B
   D. Neither A nor B

_____ 9. Tech A says that most friction discs are made of asbestos. Tech B says that new friction discs can be made from ceramic materials. Who is correct?
   A. Tech A
   B. Tech B
   C. Both A and B
   D. Neither A nor B

_____ 10. Two technicians are discussing pressure plates. Tech A says that diaphragm pressure plates require more pedal effort than coil spring pressure plates. Tech B says that the two types require different free pedal adjustment procedures. Who is correct?
   A. Tech A
   B. Tech B
   C. Both A and B
   D. Neither A nor B

Manual Transmissions/Transaxles Basic Diagnosis and Maintenance

CHAPTER 24

Tire Tread:
© AbleStock

Chapter Review

The following activities have been designed to help you refresh your knowledge of this chapter. Your instructor may require you to complete some or all of these activities as a regular part of your training program. You are encouraged to complete any activity that your instructor does not assign as a way to enhance your learning.

Matching

Match the following terms with the correct description or example.

- A. Blocking ring
- B. Detent mechanism
- C. Direct drive
- D. Gear ratio
- E. Gear reduction
- F. Gradient resistance
- G. Helix
- H. Hydraulic actuator
- I. Idler gear
- J. Interlock mechanism
- K. Overdrive
- L. Paddle
- M. Pocket bearing
- N. Roller bearing
- O. Selector gate
- P. Shift fork
- Q. Spring-loaded key
- R. Synchronizer
- S. Thrust washer
- T. Torque multiplication

_____ 1. A copper or brass washer that controls clearances on all transmission shafts.

_____ 2. A hydraulically controlled cylinder that engages or disengages the clutch pedal.

_____ 3. The increase of torque.

_____ 4. A long cylindrical roller held in position by a cage.

_____ 5. The curve created by a smooth spiral and used in the angle of gear teeth and coil springs.

_____ 6. A part of the synchronizer that helps hold the synchronizer collar in position.

_____ 7. A synchronizer part that increases or decreases a gear's speed to match shaft speed so that the synchronizer sleeve can lock the gear to the shaft.

_____ 8. A gear used in between two gears to change the direction of the rotation of the driveshaft or drive axles in the transmission.

_____ 9. The ratio of the number of turns that a drive gear must complete to turn the driven gear one turn.

_____ 10. The use of a small gear to drive a large gear. The result is an increase in torque but a decrease in speed.

_____ 11. Resistance encountered when a vehicle travels up a hill, requiring torque to be applied to overcome it.

_____ 12. A roller in the rear of the input shaft that supports the front of the main shaft.

_____ 13. A condition in which the engine and transmission output are turning at the same rate of speed.

_____ 14. A gear in which the output speed is faster than the engine speed.

_____ 15. A mechanism that moves the synchronizer sleeve to lock the gear to the main shaft.

_____ 16. A mechanical device that prevents engagement of two different gears at the same time.

_____ 17. An assembly that allows for the selection of gears without grinding by matching the speed of the two assemblies.

_____ 18. A shifting mechanism or electronic control usually attached to the steering wheel.

_____ 19. The mechanism that holds or helps hold the shift rail into position to ensure that the gear does not pop out when selected and to let the driver feel when a shift is completed.

_____ 20. The U-shaped cutaway in shift shafts that the shifter lever fits into.

Multiple Choice

Read each item carefully, and then select the best response.

_____ 1. If the driven gear has 20 teeth and the drive gear has 5 teeth, the gear ratio is _____.
   A. 1:4
   B. 5:1
   C. 4:1
   D. none of the above

_____ 2. What type of gear simply transfers motion and does not have an effect on the gear ratio of the input to output gears?
   A. Drive gear
   B. Driven gear
   C. Overdrive gear
   D. Idler gear

_____ 3. The set of shafts that are connected to the shift forks are called the _____.
   A. synchronizer
   B. selector shift rail
   C. selector gate
   D. gear shift lever

_____ 4. The _____ engages the dog teeth on the selected gear and transmits torque from the gear to the output shaft.
   A. selector gate
   B. shift fork
   C. synchronizer sleeve
   D. detent mechanism

_____ 5. The fifth gear/_____ engages in a groove in the synchronizer sleeve of the reverse gear synchronizer.
   A. detent mechanism
   B. reverse shift fork
   C. interlock mechanism
   D. selector gate

_____ 6. What device does the power train control module engage to prevent the manual transmission from being shifted into reverse gear at speeds above 5 mph?
   A. Lockout solenoid
   B. Blocking ring
   C. Interlock mechanism
   D. Hydraulic actuator

_____ 7. Which type of loads are applied along the length of the components?
   A. Radial
   B. Linear
   C. Thrust
   D. Horizontal

_____ 8. Washers that can be used to provide for adjustment of end play or preload are called _____.
   A. thrust washers
   B. selective thickness shims
   C. flanges
   D. tapered washers

_____ 9. What type of seals are designed with helical flutes molded into the sealing lip and must be installed with the correct direction of shaft rotation?
   A. Hydrodynamic
   B. O-ring
   C. Fiber
   D. RTV

_____ 10. What type of gear design reduces gear noise and distributes the load more evenly?
   A. Spur
   B. Pinion
   C. Helical
   D. Spline

_____ 11. A(n) _____ is used to synchronize the speeds of the gear and shaft before engagement.
   A. baulk-ring synchromesh unit
   B. vehicle speed sensor
   C. blocker ring synchromesh unit
   D. either A or C

_____ 12. The tapered cut on the teeth of the gear, blocker ring, and sleeve is called a _____.
   A. spur
   B. chamfer
   C. pinion
   D. mesh

_____ 13. A _____ is sometimes referred to as a semiautomatic or automated manual transmission.
   A. dual-clutch transmission
   B. transaxle synchromesh unit
   C. baulk-ring synchromesh unit
   D. direct drive transmission

_____ 14. When the shifter will not move smoothly into the desired gear and requires excessive force by the driver to force it into gear it is called _____.
   A. grinding
   B. detent
   C. hard shifting
   D. slippage

_____ 15. What tool is used to check input or output shaft end play?
   A. Feeler gauge
   B. Dial indicator
   C. Micrometer
   D. Outside caliper

True/False

If you believe the statement to be more true than false, write the letter "T" in the space provided. If you believe the statement to be more false than true, write the letter "F".

_____ 1. When two gears are in mesh, one is a drive (output) gear. The other, providing the turning action, is the driven (input) gear.

_____ 2. The terms *output gear* and *drive gear* have the same meaning.

_____ 3. Fourth gear is normally a gear ratio of 1:1, or direct drive.

_____ 4. The overall gear ratio is the gearbox ratio multiplied by the final drive ratio.

_____ 5. Selector gates are the parts that actually move the components that change the gears selected inside the transmission.

_____ 6. The interlock mechanism's function is ensuring that two gears cannot be selected at the same time.
_____ 7. Thrust loads try to force the gears and shafts apart.
_____ 8. Thrust washers are plain bearings, meaning they have no rollers.
_____ 9. In transaxle designs, the drive is transferred through the clutch unit to a primary or input shaft.
_____ 10. Fifth gear is always the lowest of the gear ratios in forward.
_____ 11. A dual-clutch transmission uses two clutch pedals to engage five separate gear ratios.
_____ 12. Grinding or clashing noises during upshifts and downshifts are fairly common on manual transmissions.
_____ 13. Clicking noises in any gear selection are typically related to a tooth of a gear that is damaged on the main shaft, input shaft, or countershaft.
_____ 14. A visual inspection should always be performed when diagnosing the manual transmission.
_____ 15. All transmissions use the same fluid.

Fill in the Blank

Read each item carefully, and then complete the statement by filling in the missing word(s).

1. _____ resistance is the resistance caused by the tires scrubbing on the road surface due to front end alignment defects or road crown.
2. If the gears have the same number of teeth, they will create a one-to-one gear ratio, commonly known as _____ _____.
3. Different gear _____ are used inside the transmission to achieve varied torque to accelerate the vehicle.
4. _____ _____ trains have two or more pairs of gears in constant mesh so that they rotate together.
5. A(n) _____ _____ _____ allows the driver to manually select gears via a gear shift mechanism.
6. Each selector shift rail has a _____ mechanism, usually in the form of a spring-loaded steel ball held in the casing.
7. Many newer manual transmissions are equipped with electronic _____ _____ systems, which use an electronic solenoid to prevent the transmission from being shifted into reverse while the vehicle is in forward motion.
8. _____ roller bearings can be caged, or they can be a loose number assembled to operate directly on a hardened gear or shaft.
9. _____ roller bearings are normally used in pairs and can sustain heavy radial and thrust loads in either direction.
10. _____ gears use straight-cut teeth for more coverage and heavy-duty operation.
11. Vehicle _____ _____ are mounted in the transmission case and are used to generate a signal based on the vehicle's speed that is sent to the PCM and/or speedometer.
12. A(n) _____-_____ transaxle has dual input shafts that are connected by gears to two main shafts containing the synchronizers.
13. The _____ is used to ensure proper engagement of gears without grinding.
14. Service consultants may use a _____ _____ to help the customer express all of the conditions related to the concern.
15. A _____ _____ is used to press bearings from the transmission shafts or to remove synchronizer sleeves.

Chapter 24 Manual Transmissions/Transaxles Basic Diagnosis and Maintenance

Labeling

Label the following diagrams with the correct terms.

1. The idler gear is a gear used to change the direction of the rotation of shafts and is used to provide a reverse:

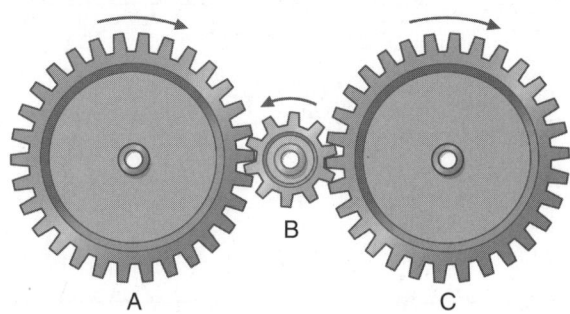

A. _____

B. _____

C. _____

2. The interlock mechanism:

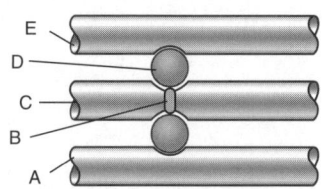

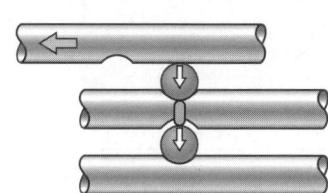

A. _____

B. _____

C. _____

D. _____

E. _____

3. Single-row ball bearings:

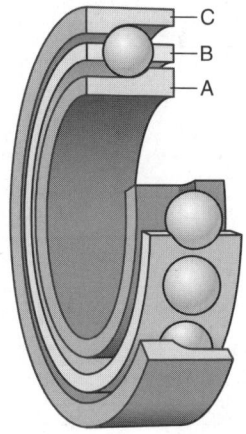

A. _____

B. _____

C. _____

4. Roller bearings:

A. _____
B. _____

5. Caged roller bearings:

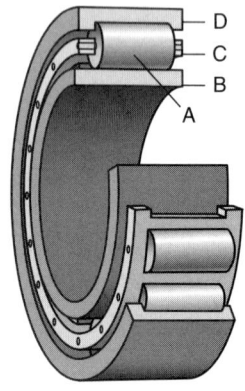

A. _____
B. _____
C. _____
D. _____

6. Tapered roller bearings:

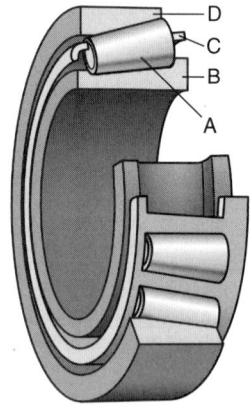

A. _____
B. _____
C. _____
D. _____

7. Pinion gear:

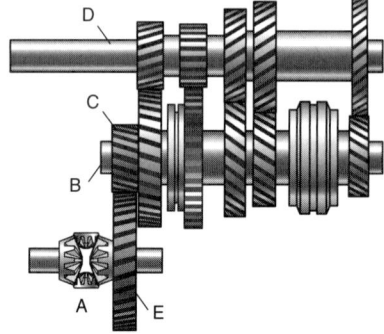

A. _____
B. _____
C. _____
D. _____
E. _____

8. Power flow through a transmission:

First gear:

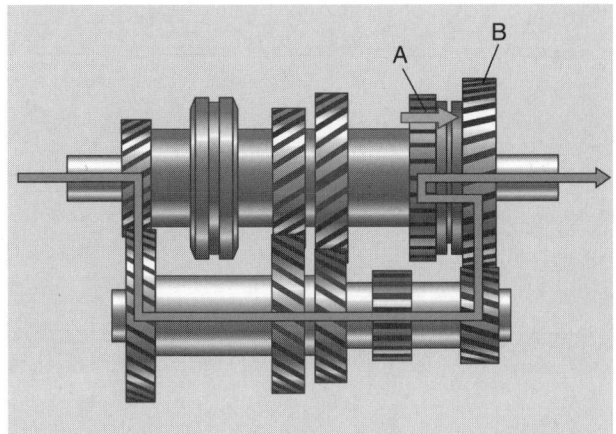

A. _____
B. _____

Second gear:

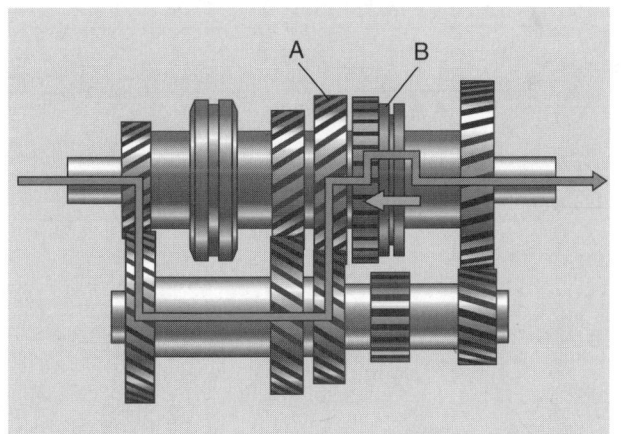

A. _____
B. _____

Third gear:

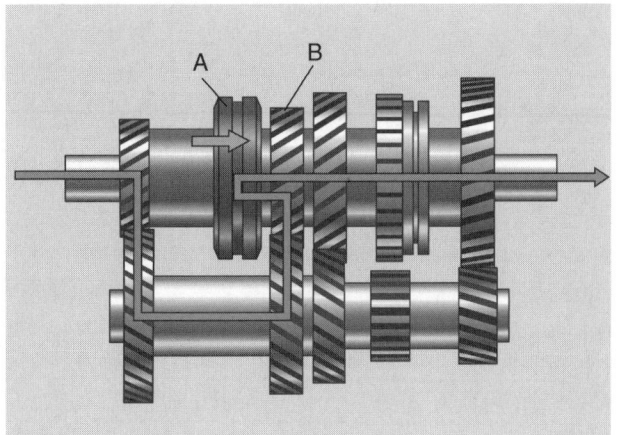

A. _____
B. _____

Fourth gear:

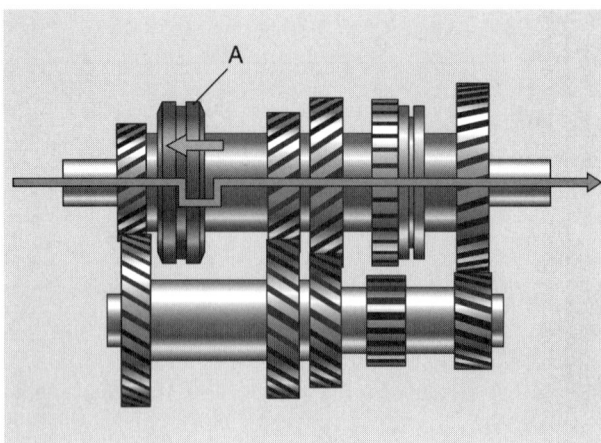

A. _____

9. Baulk-ring syncromesh unit:
   Blocker ring teeth line up with the dog teeth to allow smooth engagement of the sleeve:

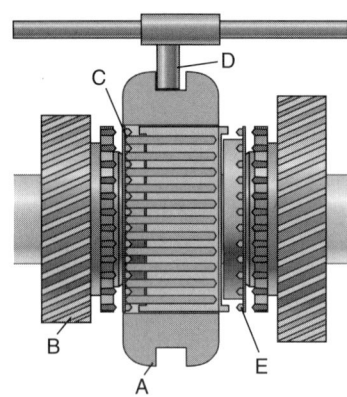

A. _____

B. _____

C. _____

D. _____

E. _____

Blocker ring teeth out of alignment, which allows the blocker ring to bring the gear to the same speed:

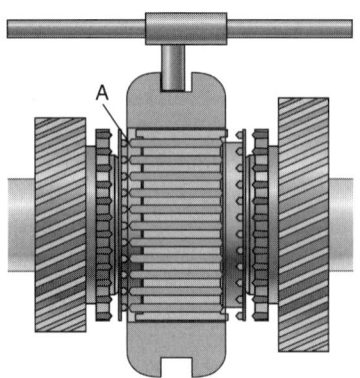

A. _____

Chapter 24 Manual Transmissions/Transaxles Basic Diagnosis and Maintenance

10. Dual-clutch transmission layout:

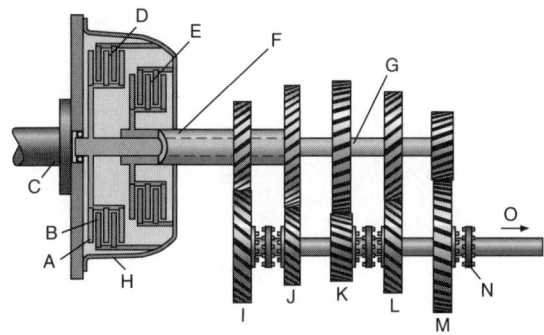

A. _____
B. _____
C. _____
D. _____
E. _____
F. _____
G. _____
H. _____
I. _____
J. _____
K. _____
L. _____
M. _____
N. _____
O. _____

Skill Drills

Test your knowledge of skill drills by filling in the correct words in the photo captions.

1. Checking the Fluid Level of a Manual Transmission:

Step 1: Raise the vehicle on the lift. Inspect the _____ for leaks. Remove the filler plug using the proper _____. Inspect the filler plug and fill hole for _____ _____, and replace/repair if necessary.

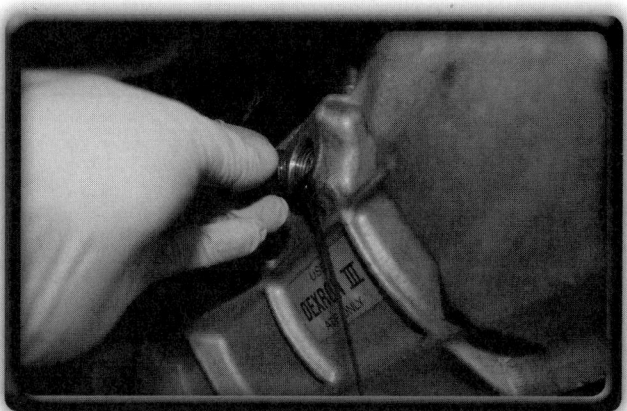

Step 2: If the gearbox fluid begins to _____ _____ as the filler plug is removed, let the gearbox fluid seek its own level before _____ the filler plug. The gearbox fluid level should be at the _____ of the filler plug hole.

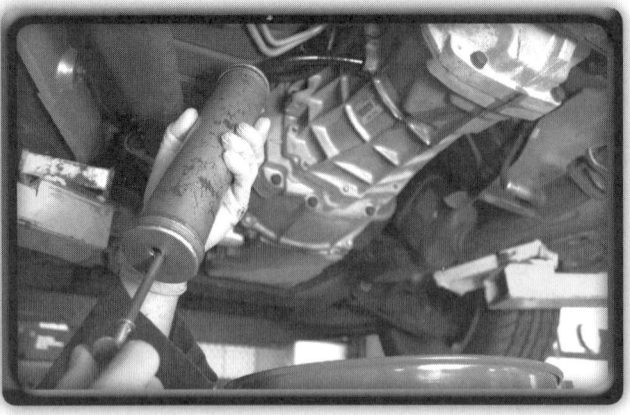

Step 3: If the fluid level is low, refill with the specified fluid, reinstall the filler plug, and wipe the area around the filler plug hole with a clean _____ _____. Tighten the filler plug to the specified _____.

2. Checking and Adjusting the Differential/Transfer Case Fluid Level:

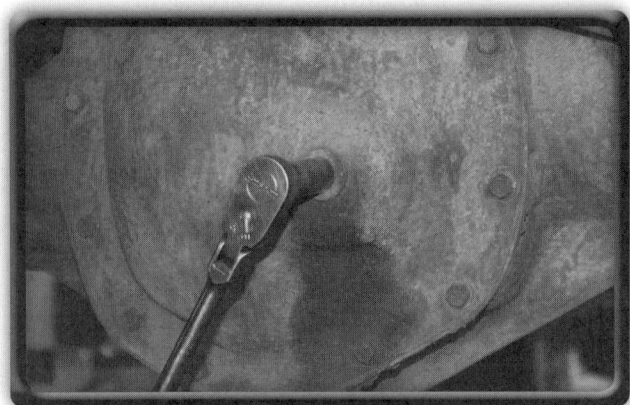

Step 1: Raise the vehicle on an approved lift. Inspect the _____ and _____ _____ before checking the fluid level. Obtain a clean _____ _____ before removal of the filler plug, as fluid may spill out. Remove the filler plug using the proper wrench. Inspect the filler plug for _____ _____, and replace if necessary. Inspect the threads in the differential and transfer case fill holes for damage also.

Step 2: If the fluid begins to run out as the filler plug is removed, let the fluid seek its own _____ before reinstalling the filler plug. The fluid level should be at the bottom of the _____ _____ _____.

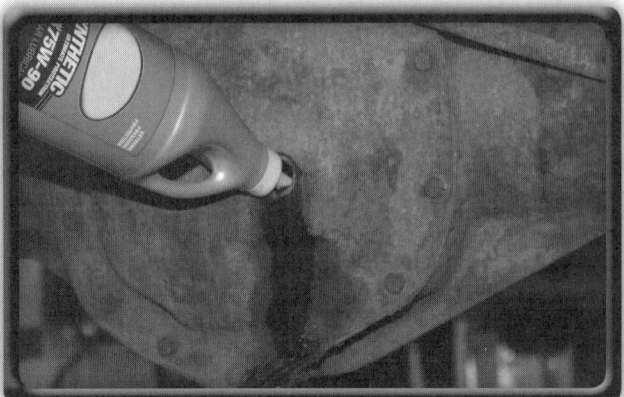

Step 3: If the _____ _____ is low, refill with the specified fluid, reinstall the filler plug, and _____ the area around the filler plug hole with a clean shop towel. _____ the filler plug to the specified torque.

3. Changing the Gearbox Fluid:

Step 1: _____ the vehicle using an approved lift. Obtain a clean drain pan to put the used _____ in.

Step 2: Inspect the _____ for leaks.

Step 3: Remove the drain plug from the _____ of the transmission, being careful of the _____ gearbox fluid. Let gearbox fluid drain until it has stopped running. If necessary, drain the _____ _____ assembly in the same manner.

Step 4: Replace the _____ _____(s) and tighten to specification, and _____ the fill plug(s).

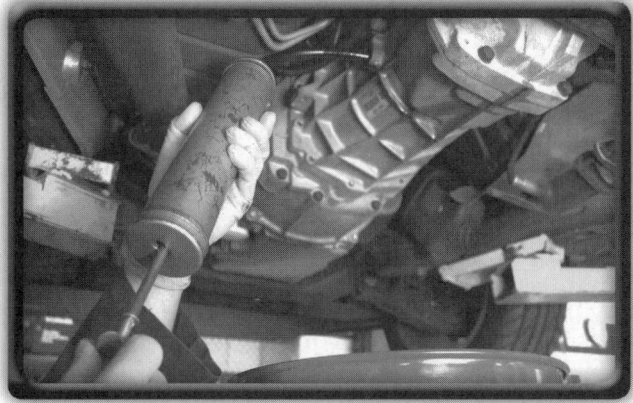

Step 5: Refill the transmission and final drive to the proper level using _____-approved gearbox fluid. _____ the fill plug(s) and tighten. Use a shop towel to wipe away any spillage. _____ _____ the vehicle. If necessary, put the vehicle back on the lift and check for any leaks that may have resulted from the service.

4. Identifying the Cause of Fluid Loss in a Transmission:

Step 1: Raise the vehicle on a _____ or jack stands. Look for leaks in the transmission _____ _____ to the engine block (front seal).

Step 2: Look for leaks in the transmission _____ outlet. Look for leaks in all case _____ areas.

Step 3: Look for leaks in the _____ tail shaft _____.

Step 4: Inspect for transmission case _____ such as _____ and _____.

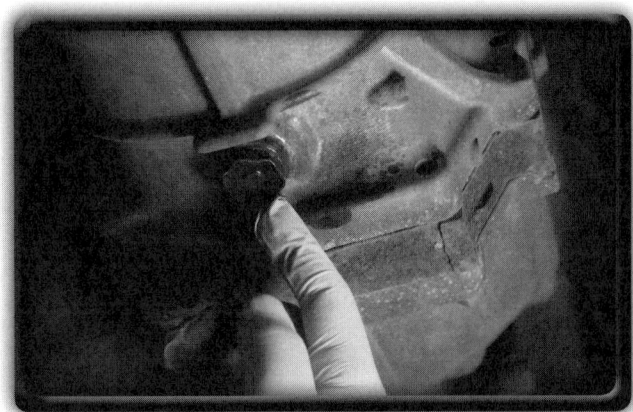

Step 5: Look for _____ at the drain and fill plugs.

Step 6: Check the gear fluid level as you are removing the drain plug, and look for evidence of excess fluid. If found, let the excess fluid drain into a container. If necessary, use your _____ _____ to check the level. If the fluid is more than half an inch below the bottom of the threaded hole, add new fluid of the correct _____ and type.

Crossword Puzzle

Use the clues to complete the puzzle.

Across

1. Typically a small-diameter pin rolling bearing that can be held in a cage or placed into the inside of a diameter of a hole.
6. The sleeve that slides to lock the selected gear to the main shaft. The synchronizer sleeve is part of the synchronizer assembly.
8. An electronic sensor that reads the driveshaft speed and sends either an analog or a digital signal to the computer that corresponds to the road speed of the vehicle.
10. A shifting problem in which the shifter will not move smoothly into the desired gear, requiring excessive force by the driver to set it into gear.
11. An electronic solenoid used to prevent accidental engagement into reverse gear while the vehicle is moving forward.

Down

2. Resistance that is present from tires contacting the road and wind resistance against the vehicle while rolling down the highway.
3. The cam and lock pin operated by an electronic solenoid that keeps the reverse gear in lockout until it is selected.
4. A technique used to shift gears when a nonsynchronized transmission is used. The driver must push the clutch in multiple times to change gears.
5. A seal that has a spring to maintain pressure on the seal's surface to prevent leakage.
7. A shift fork used to engage the reverse gear.
9. A rounded or angled edge located on the blocking ring, with ridges machined into it to help grab the cone surface of the gear.

ASE-Type Questions

Read each item carefully, and then select the best response.

_____ 1. Tech A says that in a conventional transmission, the gears freewheel around the main shaft until they are locked to it by the synchronizer. Tech B says that when not engaged, main shaft gears slide away from the counter gears until they are no longer in mesh. Who is correct?
   A. Tech A
   B. Tech B
   C. Both A and B
   D. Neither A nor B

_____ 2. Tech A says that the countershaft assembly can be made up of several gears machined out of a single piece of steel. Tech B says that the countershaft assembly is driven by the gear on the input shaft, and drives the gears on the main shaft. Who is correct?
   A. Tech A
   B. Tech B
   C. Both A and B
   D. Neither A nor B

_____ 3. Tech A says that the speeds at which two meshed gears turn depend on the number of teeth on each gear. Tech B says that if the driving gear has fewer teeth than the driven gear, the speed of the driven gear increases. Who is correct?
   A. Tech A
   B. Tech B
   C. Both A and B
   D. Neither A nor B

_____ 4. Tech A says that an overdrive gear ratio means the input gear turns faster than the output gear. Tech B says that overdrive ratios provide less torque output than underdrive ratios. Who is correct?
   A. Tech A
   B. Tech B
   C. Both A and B
   D. Neither A nor B

_____ 5. Tech A says that a benefit of helical gears is that they do not create radial or thrust loads. Tech B says that spur gears run quieter than helical gears. Who is correct?
   A. Tech A
   B. Tech B
   C. Both A and B
   D. Neither A nor B

_____ 6. Tech A says that some manual transmissions use two input shafts and two clutches. Tech B says that many manual transmissions now use planetary gear sets. Who is correct?
   A. Tech A
   B. Tech B
   C. Both A and B
   D. Neither A nor B

_____ 7. Tech A says that blocker rings prevent the vehicle from rolling when parked. Tech B says that engine torque is transferred from the dog teeth through the synchronizer sleeves. Who is correct?
   A. Tech A
   B. Tech B
   C. Both A and B
   D. Neither A nor B

_____ 8. Tech A says that shift forks slide the gears into mesh with each other when a shift occurs. Tech B says that blocker rings have fine grooves that cut through the oil on the tapered cone of the matching gear. Who is correct?
   A. Tech A
   B. Tech B
   C. Both A and B
   D. Neither A nor B

Chapter 24 Manual Transmissions/Transaxles Basic Diagnosis and Maintenance

_____ 9. Tech A says that a broken detent spring can cause the transmission to pop out of gear. Tech B says that a broken interlock pin can cause the transmission to go into two gears at the same time. Who is correct?
   A. Tech A
   B. Tech B
   C. Both A and B
   D. Neither A nor B

_____ 10. Tech A says that some manufacturers specify automatic transmission fluid to be used in their manual transmissions. Tech B says that some manufacturers specify engine oil to be used in their manual transmissions. Who is correct?
   A. Tech A
   B. Tech B
   C. Both A and B
   D. Neither A nor B

CHAPTER 25
Drive Train Components

Tire Tread:
© AbleStock

Chapter Review

The following activities have been designed to help you refresh your knowledge of this chapter. Your instructor may require you to complete some or all of these activities as a regular part of your training program. You are encouraged to complete any activity that your instructor does not assign as a way to enhance your learning.

Matching

Match the following terms with the correct description or example.

- **A.** Blocking ring
- **B.** Concentricity
- **C.** End play
- **D.** Oil slinger
- **E.** Overhauling

_____ 1. The process of refurbishing the transmission to like-new condition.
_____ 2. A device that rides partially in the transmission fluid; as it spins, it flings oil to lubricate the internal workings of the transmission.
_____ 3. The roundness of a hole. If the hole is not round it can also be referred to as out of round.
_____ 4. Referring to the input or output shaft, fore-and-aft movement in the transmission.
_____ 5. The ring on the synchronizer assembly that slows or speeds up the gear to match the main shaft speed, which allows smooth shifts.

Multiple Choice

Read each item carefully, and then select the best response.

_____ 1. All of the following are specialty tools needed for rebuilding the transmission, *except*:
- **A.** Bearing splitter
- **B.** Dial indicator
- **C.** Oscilloscope
- **D.** Bushing driver

_____ 2. When a customer is complaining of gear shifting issues or clunking sounds when accelerating or decelerating you should check the _____.
- **A.** ring gear
- **B.** power train mounts
- **C.** shift fork
- **D.** gear shift lever

_____ 3. What tool is used to check flat mating surfaces for warpage?
- **A.** Dial gauge
- **B.** Straightedge and feeler gauge
- **C.** Telescoping gauge
- **D.** Inside caliper

_____ 4. Always check the speedometer opening for any damage, and install a new _____ when assembling the transmission.
- **A.** speed sensor
- **B.** cable
- **C.** O-ring
- **D.** collar

_____ 5. To meet Environmental Protection Agency regulations, most shops use _____-based solvents.
   A. soap and water
   B. hydrocarbon
   C. mineral spirits
   D. either A or C

_____ 6. Some manual transmissions use a(n) _____ behind the input shaft that ensures oil is directed toward the front bearing and returned to the transmission case.
   A. oil slinger
   B. front pump
   C. retainer
   D. hydrodynamic seal

_____ 7. All main shift control shafts should be lubricated with _____ before installation.
   A. transmission fluid
   B. petroleum jelly
   C. motor oil
   D. mineral oil

_____ 8. The pinion gear is part of the _____ of the transaxle.
   A. input shaft
   B. synchronizer
   C. output shaft
   D. none of the above

_____ 9. The faulty part of a transmission that will be sent to a rebuilder for refurbishing is called the _____.
   A. core
   B. shell
   C. salvage
   D. hull

_____ 10. Most manufacturers require that certain _____ adjustments be done upon assembly of transmissions.
   A. end play
   B. speed sensor
   C. preload
   D. both A and C

True/False

If you believe the statement to be more true than false, write the letter "T" in the space provided. If you believe the statement to be more false than true, write the letter "F".

_____ 1. Rebuilding a transmission often requires specialty tools that are specific to each transmission.
_____ 2. Among the first tasks when doing any kind of service to a manual transmission is obtaining and reading all manufacturer publications and service information.
_____ 3. Manual transmissions do not contain any electronic components.
_____ 4. Some transmission parts are held together only by the positioning of the assemblies, and when you start to remove them, these loose parts will fall off.
_____ 5. A transmission rebuild will always cost more than a remanufactured unit from a parts supplier.
_____ 6. Dog teeth are pointed to assure complete synchronous hub engagement.
_____ 7. Transmission gel or Vaseline is used to lubricate the synchronizer hub.
_____ 8. When removing the transmission it is not necessary to remove any of the shift linkage or associated parts.
_____ 9. A special bearing driver is needed to install a new countershaft pocket bearing in the transmission case.
_____ 10. Incorrect end play and bearing preload will result in failure of the transmission.

Fill in the Blank

Read each item carefully, and then complete the statement by filling in the missing word(s).

1. Top-level _____ kits include gaskets, seals, synchronizer rings, bearings, and possibly even shafts.
2. Power train mounts isolate _____ _____ created by the running engine.

3. Hook up a _____ _____ to the diagnostic link connector to read information or any codes that may be present.
4. The _____ in the bottom of the transmission case will indicate gear and bearing wear by the amount of filings found on it.
5. Check all bores for _____ or roundness and any specific defects.
6. Check the transmission case for cracks and _____.
7. Check to make sure the bolt holes are not _____ or damaged from bolts that were loose.
8. _____ assemblies allow for the smooth shifting of the transmission.
9. Gold-colored particles in drained transmission fluid may indicate that the _____ _____ are excessively worn and may need to be serviced.
10. When reassembling all components, it is a good practice to _____ parts as they are being installed.

Skill Drills

Place the skill drill steps in the correct order.

1. Disassembling a Manual Transmission:

_____ **A.** Remove the rear bearing race from the transmission case with a brass drift and hammer.

_____ **B.** Disassemble the output shaft:
Mount the output shaft in the press to remove fifth gear, which is pressed onto the shaft. This will also remove first gear and the rear output shaft bearing.

_____ **C.** Remove the shift mechanism housing bolts, and remove the shift mechanism housing from the top of the manual transmission.

_____ **D.** The front pocket bearing (which is a caged roller-type bearing) will also have to be removed in order to be replaced upon overhaul.

Chapter 25 Drive Train Components

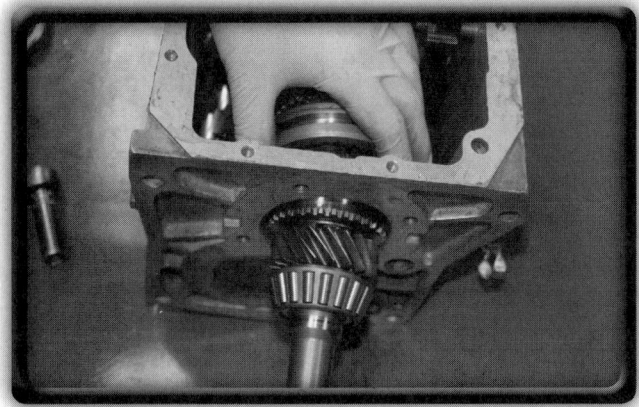

_____ **E.** Rotate the input shaft so the dog teeth cutout is lined up with the gear on the cluster shaft for easy removal. Pay attention to the roller bearings in the end of the input shaft. Now pull the input shaft from the transmission case.

_____ **F.** To disassemble the main shaft/output shaft:
- Position the shaft in a soft-jaw vice. Remove the first snap ring that holds third and fourth gears in place.
- Remove the third and fourth gear sets and roller bearing cages.
- Remove the second snap ring, and remove the second gear.
- Remove the first and second/reverse synchronizer sliding collar.

_____ **G.** Remove the fifth gear driven gear from the countershaft by removing the synchronizer and snap ring for access.

_____ **H.** Remove the retaining bolts and remove the shift lever from its seat. On some manual transmissions, it is best to place the gear shift in neutral prior to removing the shift lever.

_____ **I.** Tilt the output shaft on an upward angle to enable removal of the shaft from the transmission case. Before removal, make sure the third gear synchronizer is engaged to make clearance for removal of the output shaft. Remove the output shaft from the case.

_____ **J.** Remove the input shaft bearing retainer bolts, and remove the retainer from the transmission case.

_____ **K.** Remove the output shaft pocket bearings from the inside of the input shaft. Place the input shaft in the press, and press off the front bearing.

_____ **L.** Remove the bolts that retain the tail shaft housing, then remove the housing.

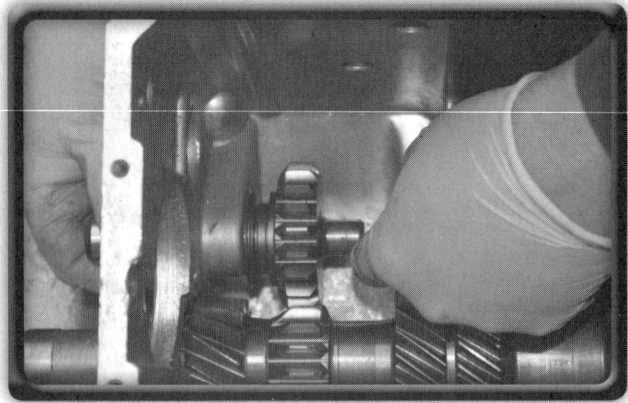

_____ **M.** Remove the reverse idler gear shaft retaining roll pin from the idler shaft. Remove the idler shaft and gear from the transmission case.

_____ **N.** Remove the speedometer gear and retainer, and slide it off of the output shaft. Some gears are pressed on, so you may have to use a gear puller to remove it.

_____ **O.** To remove the countershaft from the transmission case, it may be necessary to put the transmission case in a press to remove the rear countershaft bearing. Press out the bearing, and then remove the countershaft from the transmission.

2. Inspecting and Servicing the Shift Cover:

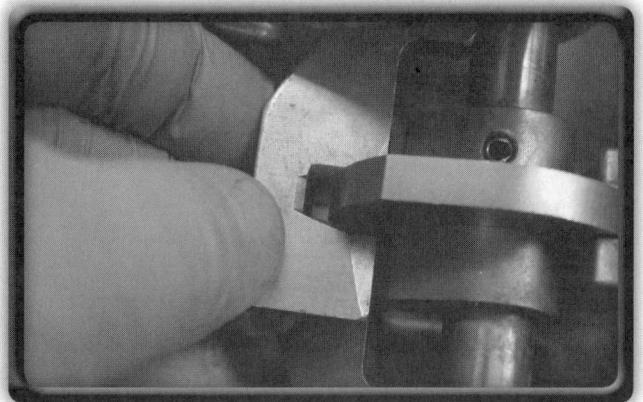

_____ **A.** Check the gear selector interlock sleeve for excessive wear and binding. If necessary, clean or replace.

_____ **B.** Install the shift lever, and check that all gears can be shifted without binding.

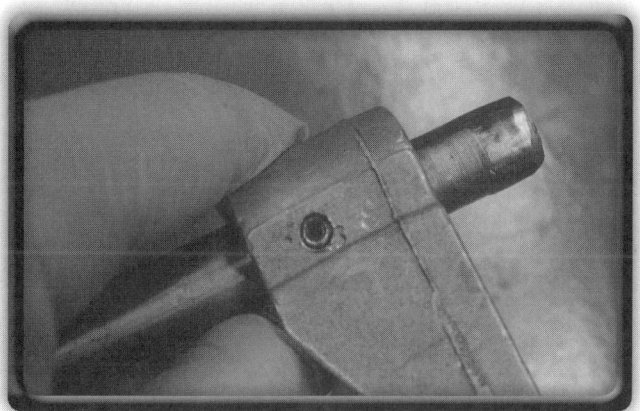

_____ **C.** If roll pins are used to secure shift forks or sleeves, inspect for tightness, and replace as necessary by driving the pin out and replacing with a new one. If the new pin fits loosely in the case, the case may be damaged and need to be replaced or repaired.

_____ **D.** Lubricate all main shift control shafts with petroleum jelly before installation.

_____ **E.** Secure a new shift cover gasket usually by using a dab of RTV sealer, and install the shift cover. Some transmissions will use RTV without an additional gasket, as does the transmission in the picture. Follow the manufacturer's directions to ensure a leak-free repair. Make sure the shift forks line up with the corresponding gears.

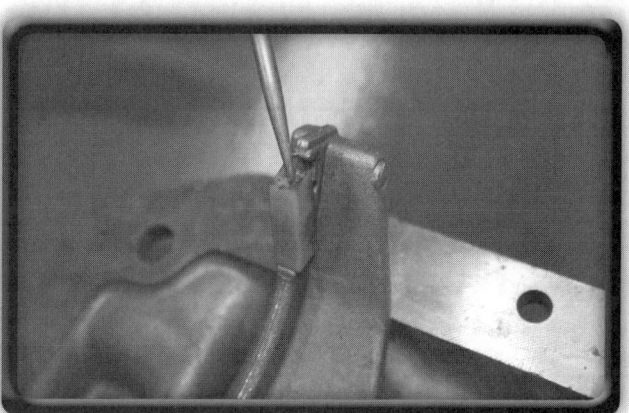

_____ **F.** Check all shifter forks for wear in the gear selector slots. If the gear shift fork inserts are worn or cracked, they should be replaced before the installation of the shift cover.

_____ **G.** As the shift rails are being installed, check all detent springs and balls for damage, and replace as necessary.

_____ **H.** If an O-ring is present that is used to seal the cover to the output shaft housing, it should be replaced.

3. Inspecting and Servicing Shift Linkages:

_____ **A.** If the vehicle has plastic or brass bushings, inspect them for excessive wear and replace as necessary.

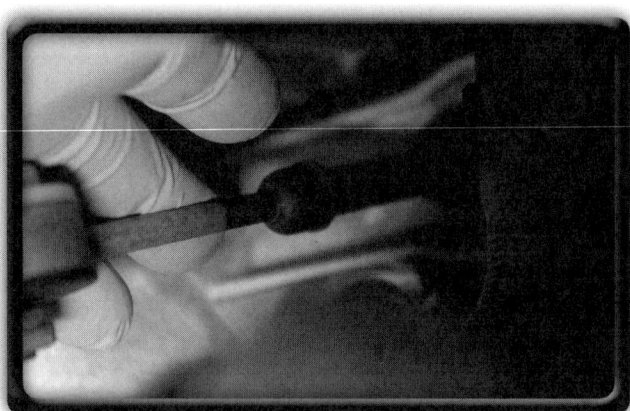

_____ **B.** Inspect the cable linkage for fraying and stretching.

_____ **C.** Inspect mounting brackets for excessive wear and looseness, and replace as necessary. Check all sensors and solenoids.

_____ **D.** Inspect the mechanical linkage for worn or elongated holes where the bushings ride.

Chapter 25 Drive Train Components

_____ **E.** Lay all the associated parts out on a bench to get a clear picture for inspection if possible. Clean off the dirt and grease in order to properly inspect.

_____ **F.** Inspect linkage rod ends for excessive wear from abuse. Inspect the rods for abnormal shape, which can occur as a result of driver abuse.

4. Removing and Replacing the Transaxle Final Drive:

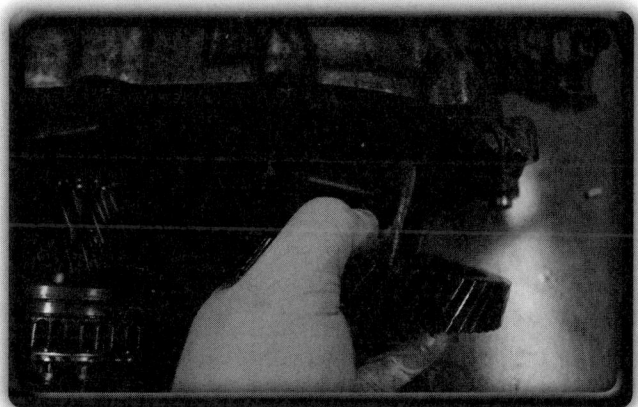

_____ **A.** Remove the output shaft from the transaxle case and inspect. Remove the differential case from the transaxle housing. Be careful of the spider gear orientation, and do not let them fall out of the differential case. In some cases, there is no retaining pin to hold them in place.

_____ **B.** Measure the preload and correct by using shims under the case bearings of the differential by using a dial indicator following the manufacturer's procedure. Reassemble the transaxle's major components.

_____ **C.** Reinstall the output shaft in the transaxle case and mate with all other shafts.

_____ **D.** Following the manufacturer's specifications, disassemble the transaxle case and shafts as necessary to gain access to the final drive assembly. Parts of the final drive include the output shaft that has the pinion gear molded on to it and the differential case.

_____ **E.** After performing any needed service, reinstall the differential case in the transaxle housing, seating the bearing in its race.

5. Measuring and Adjusting End Play/Preload:

_____ **A.** Install the correct shims to bring the end play into specifications. Reinstall the front bearing retainer and gasket. Torque the bolts to the manufacturer's specifications.

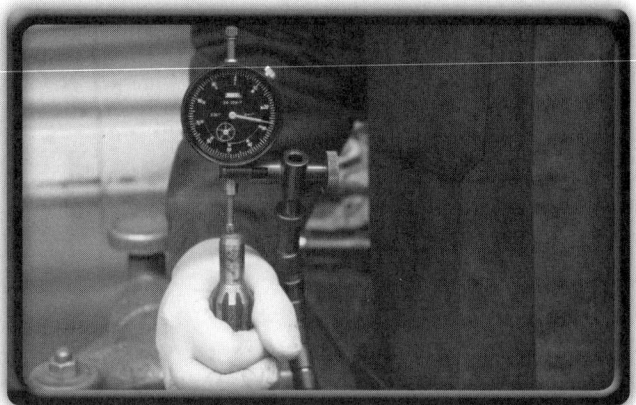

_____ **B.** Grab the output shaft, and raise it upward until the end play is totally removed; doing so will give you an accurate reading. Record the measurement, and compare it with the manufacturer's specifications. Most are usually around 0.005" to 0.025" (0.127 to 0.635 mm).

_____ **C.** Once the transmission assembly is together, turn the transmission case on its end and mount a dial indicator on the end of the output shaft. Rotate the dial indicator to zero.

_____ **D.** If an adjustment is necessary, remove the front bearing retaining cover. Remove the front bearing race, which may expose any existing shims.

Crossword Puzzle

Use the clues to complete the puzzle.

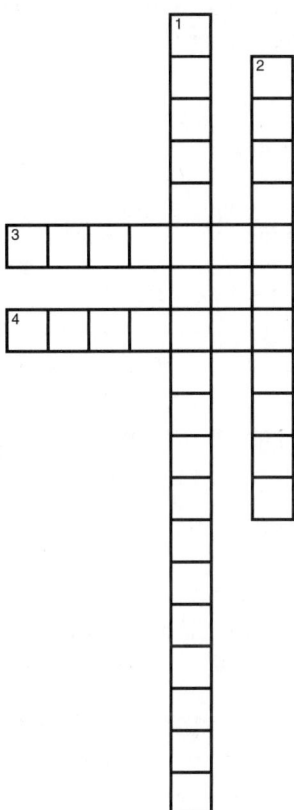

Across

3. The level of pressure placed on bearings that ensures the bearings will be held together.
4. A change in the shape of a surface due to distortion or wear.

Down

1. Pulsations in the crankshaft rotation caused by the power pulses on the crankshaft and felt as a vibration.
2. A sensor responsible for measuring the speed of the vehicle that is read by the speedometer.

ASE-Type Questions

Read each item carefully, and then select the best response.

_____ 1. Tech A says that broken or worn power train mounts can cause a transaxle to have shifting problems. Tech B says that poor shift boot alignment cannot cause a transaxle to jump out of gear. Who is correct?
   A. Tech A
   B. Tech B
   C. Both A and B
   D. Neither A nor B

_____ 2. Tech A says that a leaky transmission output shaft seal is likely to cause clutch plate contamination. Tech B says that a worn transmission rear bushing may contribute to output shaft seal leakage. Who is correct?
   A. Tech A
   B. Tech B
   C. Both A and B
   D. Neither A nor B

_____ 3. A car jumps out of gear into neutral when decelerating or going down hills. Tech A says that the shift lever and internal gearshift linkage should be checked. Tech B says that a detent spring could be broken. Who is correct?
   A. Tech A
   B. Tech B
   C. Both A and B
   D. Neither A nor B

_____ 4. Tech A says that as long as the dog (clutch) teeth of the synchronizer are rounded, the synchronizer can be reused. Tech B says that the synchronizer sleeve should slide smoothly on its splines. Who is correct?
   A. Tech A
   B. Tech B
   C. Both A and B
   D. Neither A nor B

_____ 5. A transmission is being assembled, but the input shaft does not slide all the way in. Tech A says this is normal and to use the bolts to pull the front bearing retainer into place. Tech B says that you should use a hammer to tap the bearing retainer into place. Who is correct?
   A. Tech A
   B. Tech B
   C. Both A and B
   D. Neither A nor B

_____ 6. A rough growling noise occurs when the vehicle is moving and the transmission is in the forward or reverse gear. The noise is not heard when the transmission is in neutral. Tech A says that the output shaft (main shaft) bearings may be faulty. Tech B that says the pilot bearing may be faulty. Who is correct?
   A. Tech A
   B. Tech B
   C. Both A and B
   D. Neither A nor B

_____ 7. Tech A says that the input shaft normally needs to be installed before the main shaft. Tech B says that most manual transmissions are first disassembled into subassemblies before further disassembling. Who is correct?
   A. Tech A
   B. Tech B
   C. Both A and B
   D. Neither A nor B

_____ 8. Tech A says that the condition of blocking rings should be checked by measuring the clearance between the blocking ring and the face of the gear. Tech B says that the blocking ring should be visually inspected for worn clutch teeth and smooth internal surfaces. Who is correct?
   A. Tech A
   B. Tech B
   C. Both A and B
   D. Neither A nor B

_____ 9. Transaxle inspection is being discussed. Tech A says that in some cases, a magnet is installed in the case to catch any ferrous metal particles, and it should be inspected. Tech B says that tapping on the gears with a ball-peen hammer and listening for a ringing sound is a good way to determine if a gear is reusable. Who is correct?
   A. Tech A
   B. Tech B
   C. Both A and B
   D. Neither A nor B

_____ 10. Transmission end play is being discussed. Tech A says that a dial indicator should be used to measure the inward and outward movement of the output shaft. Tech B says that an angle gauge should be used to measure output shaft rotation. Who is correct?
   A. Tech A
   B. Tech B
   C. Both A and B
   D. Neither A nor B

CHAPTER 26
Basic Drive Layouts

Tire Tread:
© AbleStock

Chapter Review

The following activities have been designed to help you refresh your knowledge of this chapter. Your instructor may require you to complete some or all of these activities as a regular part of your training program. You are encouraged to complete any activity that your instructor does not assign as a way to enhance your learning.

Matching

Match the following terms with the correct description or example.

A. Backlash clearance
B. Brinelling
C. Companion flange
D. Crush sleeve
E. Drive line angularity
F. Full-floating axle
G. Hypoid bevel gear
H. Live axle
I. Locking hub
J. Longitudinal

K. Power take-off (PTO)
L. Rzeppa joint
M. Side gear
N. Speedy sleeve
O. Torque steer
P. Transfer case
Q. Transverse
R. Tulip/tripod joint
S. Universal joint (U-joint)
T. Viscous coupling

_____ 1. An aftermarket repair kit that consists of a thin metal sleeve that fits tightly over the seal surface of the axle, providing a new, undamaged surface for the seal to ride against.

_____ 2. Four-wheel drive front axle hubs that are manually locked or unlocked by turning the knob on the hub.

_____ 3. A constant velocity joint that has three equally spaced fingers shaped like a star. This configuration allows in-and-out movement of the shaft while allowing flexing.

_____ 4. A device attached to the transmission that is gear driven and can be used to run accessories such as winches and towing equipment. It can also refer to the gears that send power to the rear axle in a predominantly front-wheel drive vehicle.

_____ 5. An axle that does not support any weight; if removed, the vehicle will still roll on its wheels.

_____ 6. A silicone clutch assembly used in all-wheel drive differentials to provide a slight amount of differential action for control of axle rotational speeds.

_____ 7. The orientation of the engine in which the front of the engine is facing the side of the vehicle.

_____ 8. Damage done to the surface of a bearing caused by excessive load, which exceeds the limit of the bearing material, typically from shock loads.

_____ 9. A cross-shaped flexible joint on which caps fit over the ends of the cross. Needle bearings fit between the ends of the cross and the caps, allowing the caps to rotate smoothly.

_____ 10. A special design of spiral bevel gear, with the centerline of the pinion below the centerline of the ring gear.

_____ 11. The relationship of the driveshaft to the component that the driveshaft attaches, measured in degrees of angle.

_____ 12. The orientation of the engine in which the front of the engine is facing the front of the vehicle. It is most commonly found in rear-wheel drive vehicles.

_____ 13. A type of fixed constant velocity joint that has an inner race, six steel ball bearings, a bearing cage, and an outer race.

_____ 14. A splined flange that transmits power from the drive shaft to the pinion gear.

Chapter 26   Basic Drive Layouts   255

_____ 15. A gear that is splined to the axle shaft and meshes with the spider gears and allows the axles to rotate at their own speeds when cornering and turning.

_____ 16. The amount of movement between the pinion teeth versus the ring teeth.

_____ 17. An axle that is powered and can move the vehicle. It is usually found on the rear of rear-wheel drive vehicles.

_____ 18. A condition in which the vehicle pulls to one side during hard acceleration.

_____ 19. A collapsible spacer between the bearings that provides a means of maintaining a preset torque on the pinion nut.

_____ 20. A component that is bolted to the back of the transmission and connects the front and rear axles via the driveshafts.

Multiple Choice

Read each item carefully, and then select the best response.

_____ 1. The _____ slides in and out on the output shaft, allowing the length changes needed as the suspension moves up and down.
   A. slip yoke
   B. universal joint
   C. half-shaft
   D. speedy sleeve

_____ 2. A two-piece driveshaft, joined in the middle, that can slide on itself to increase and decrease in length is called a _____.
   A. half-shaft
   B. full-floating driveshaft
   C. sliding spline driveshaft
   D. rzeppa joint

_____ 3. The _____ is a device that transfers torque from one component to another, such as from the transmission output shaft to the final drive assembly.
   A. half-shaft
   B. driveshaft
   C. drive axle
   D. constant velocity joint

_____ 4. The _____ does not provide any drive capabilities and is used on the rear of front-wheel drive vehicles, and the front of rear-wheel drive vehicles.
   A. dead axle
   B. half-shaft
   C. drive axle
   D. driveshaft

_____ 5. What type of rear-wheel drive solid axle has a single roller bearing between the hub and the outside of the axle housing?
   A. Dead axle
   B. Semi-floating axle
   C. Three-quarter floating axle
   D. Full-floating axle

_____ 6. A(n) _____ can be of the single-lip or double-lip design depending on the application.
   A. flange
   B. yoke
   C. speedy sleeve
   D. axle seal

_____ 7. The use of different-sized axle shafts can cause a problem known as _____.
   A. misalignment
   B. torque steer
   C. slip differential
   D. angularity

_____ 8. Which type of joint allows the shaft to transmit torque through a change of drive angle?
   A. Universal joint
   B. Ball joint
   C. Constant velocity joint
   D. Both A and C

_____ 9. Which type of constant velocity joint does not slide to allow for shaft lengthening or shortening; it simply allows for angle changes as the suspension moves?
   A. Fixed-type joint
   B. Rzeppa joint
   C. Plunge-type joint
   D. Tulip-tripod joint

_____ 10. The electronically controlled transfer case uses a(n) _____ to lock the front and rear axles automatically, as needed, and monitors wheel speed to determine the torque needed.
   A. electromagnetic clutch
   B. speed sensor
   C. viscous coupling
   D. control module

True/False

If you believe the statement to be more true than false, write the letter "T" in the space provided. If you believe the statement to be more false than true, write the letter "F".

_____ 1. Longitudinally mounted means that the front of the engine is facing the side of the vehicle.
_____ 2. In some all-wheel drive vehicles, only two of the wheels are normally powered, typically the front wheels.
_____ 3. The purpose of the drive axle is to transfer the torque that comes from the engine, transmission, driveshaft, and differential to the wheels and tires, propelling the vehicle forward or backward.
_____ 4. If a vehicle with a semi-floating axle were to hit a curb and break or snap the axle shaft, the wheel would come off of the vehicle.
_____ 5. Full-floating axles can be removed without removing the wheels.
_____ 6. A double Cardan joint greatly increases the change in velocity of a single Cardan joint.
_____ 7. Universal joints are larger than constant velocity joints.
_____ 8. If you hold the ring gear and turn the pinion, you will have a slight clearance back and forth.
_____ 9. In part-time four-wheel drive vehicles, differentials are fitted to both the front and rear axle assemblies.
_____ 10. A transfer case is used on vehicles that are primarily front-wheel drive; a power take-off is used on vehicles that are primarily rear-wheel drive.

Fill in the Blank

Read each item carefully, and then complete the statement by filling in the missing word(s).

1. _____-_____, or solid rear axle assemblies, enclose the final drive gears, differential gears, and axle shafts into one housing.
2. _____-_____ joints allow for smoother transfer of power and allow for the vehicle to turn more tightly without the joint binding.
3. The _____ can be made up of as many as four separate segments.
4. On a(n) _____-_____ axle, the axle shafts are splined to the differential side gears, and the outer bearing is between the outer end of the axle shaft and the inside of the axle housing.
5. Lug studs are pressed into the axle _____ and used to secure the wheel onto the vehicle.
6. In front-wheel drive vehicles and all-wheel drive vehicles, the driveshafts transfer the drive directly from the _____ _____ inside the transaxle to the front wheels.
7. The _____ _____ is a short section of shaft that typically has a bearing pressed onto it, used to make both half-shafts the same length from left to right.
8. The _____ joint consists of a steel cross with four hardened bearing journals, mounted on needle rollers in hardened caps, which locate the cross in the eyes of the yokes.

9. Often, a sliding spline or a _____-_____ CV joint is used as the inner half-shaft joint to accommodate for changes in shaft length when traveling on different types of terrain.
10. A CV _____ is used to keep the grease inside the joint, and to keep dirt and debris out.
11. The speed reduction gears in the final drive are called the _____ and _____ gears.
12. A(n) _____-_____ limited slip differential responds very quickly to changes in traction. They also do not bind from friction in turns and do not lose their effectiveness since there are no clutches, like a clutch-style unit.
13. Freewheeling hubs or manually _____ hubs can be installed on the front hubs to disconnect the hubs from the axle shafts and prevent rotation of the front axle components.
14. The transfer case is typically bolted to the _____ of the transmission, while the PTO is bolted to the _____ of the transaxle.
15. A _____ _____ uses two sets of clutch plates that are alternated front to back. One set of plates is splined to the outer housing, and the other set of plates is splined to the inner housing.

Labeling

Label the following diagrams with the correct terms.

1.

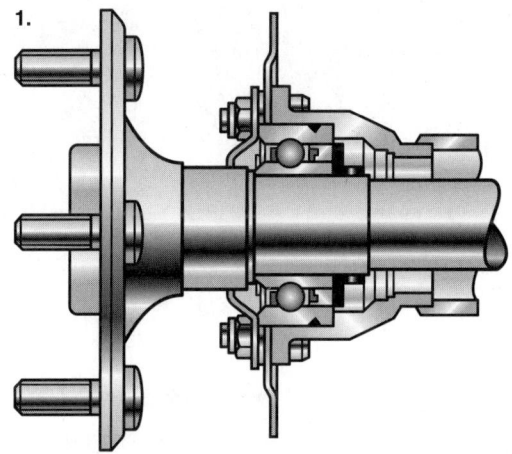

1. _____

2.

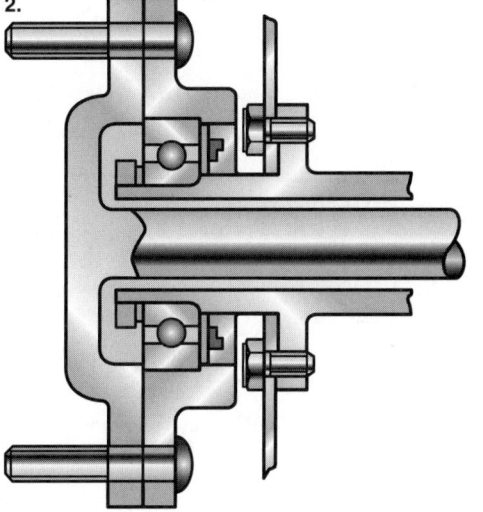

2. _____

3.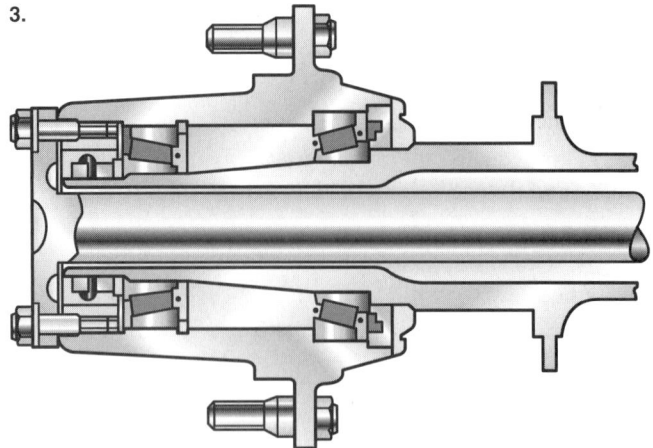

3. _____

4.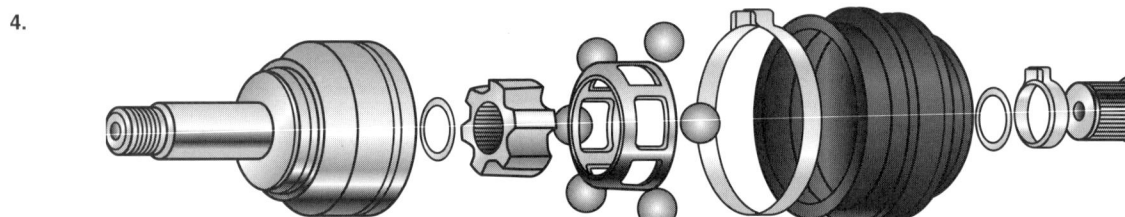

A.

A. _____

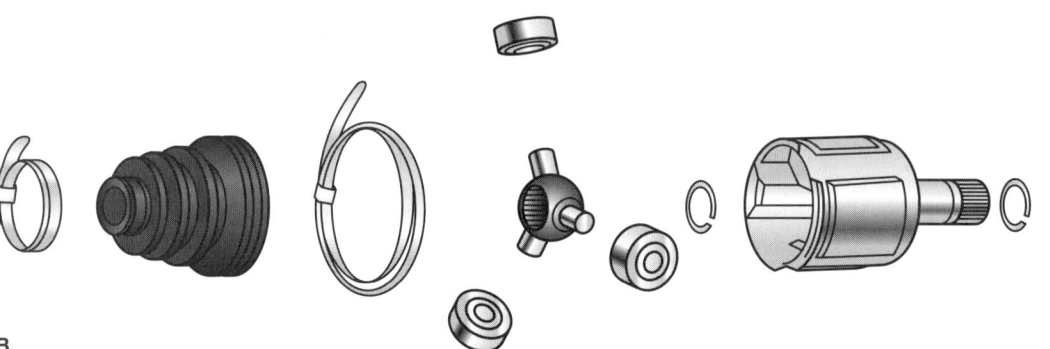

B.

B. _____

5.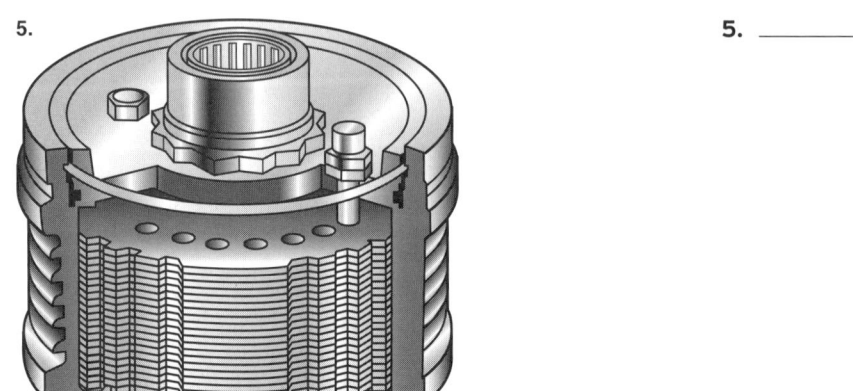

5. _____

Skill Drills

Test your knowledge of skill drills by filling in the correct words in the photo captions.

1. Replacing the Integral Axle Housing Seal and Bearing:

Step 1: Remove the _____.

Step 2: On some models, remove the bearings and seal by putting the axle in a _____. On some models, _____ the bearing out of the housing.

Step 3: Replace all _____, and inspect the seal _____ of the axle.

2. Inspecting Half-Shaft Components:

Step 1: Clamp the entire half-shaft into a soft-jawed _____, and make sure it is secure. Remove the retaining _____ from the CV boot. Slide the _____ down the shaft, paying attention to the condition of the boot.

Step 2: Wipe out as much _____ as possible to be able to access the retaining ring from the CV joint itself. If there is a retaining ring present, _____ the retaining ring with the appropriate tool. When the retaining ring is removed, remove the CV joint from the half-shaft, and inspect the _____ on the end of the half-shaft. This also applies to the other end of the half-shaft.

Step 3: Inspect the old joint to gain an accurate _____ of the failure to prevent a reoccurrence of this failure. Reinstall the new CV joint onto the _____ _____ as required.

Step 4: Apply the lubrication grease that comes with the new CV joint. Tighten the _____ _____ to ensure that grease will not be lost. This applies to both sides of the half-shaft. Reinstall the half-shaft following the _____ guidelines.

3. Inspecting Fluid Leakage:

Step 1: Put the vehicle on a _____, and make sure it is secure. Inspect around the _____ where the axle seats for any seepage.

Step 2: Inspect the _____ _____ for possible seepage.

Step 3: If necessary, remove the rear wheels and install a _____ _____ on the axle flange to check for any _____.

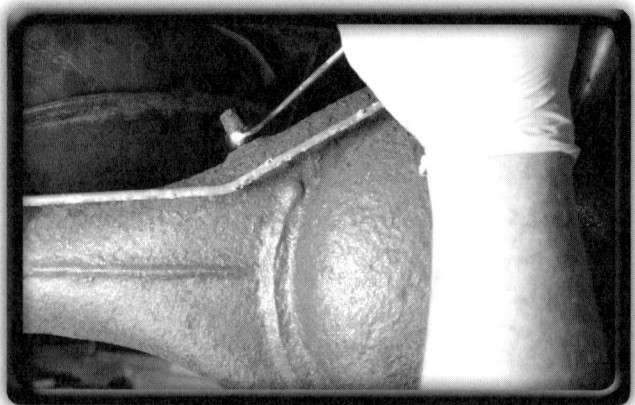

Step 4: Check and clean the _____ or _____ for any obstructions that may cause a pressure buildup to occur.

Step 5: Check the differential _____ for a leaking _____, if so equipped, and tighten if loose or replace as necessary.

Step 6: Check the fluid level for lack of fluid or _____ of the differential, as either one can indicate that a _____ is present.

4. Inspecting the Ring Gear and Measuring Runout:

Step 1: Clean off the flat side of the _____ _____ to remove any excess oil. Attach a dial indicator to the _____ housing near the area of the ring gear. Zero out the indicator so an accurate reading can be taken.

Step 2: Rotate the ring gear by installing a socket and a ratchet to the pinion companion _____ _____. Load the ring gear up by wedging a _____ or equivalent in order to get an accurate reading.

Step 3: Rotate the _____ _____ slowly, paying attention to the dial, and compare the reading with _____. Determine any necessary actions.

5. Checking Ring and Pinion Tooth Contact Patterns:

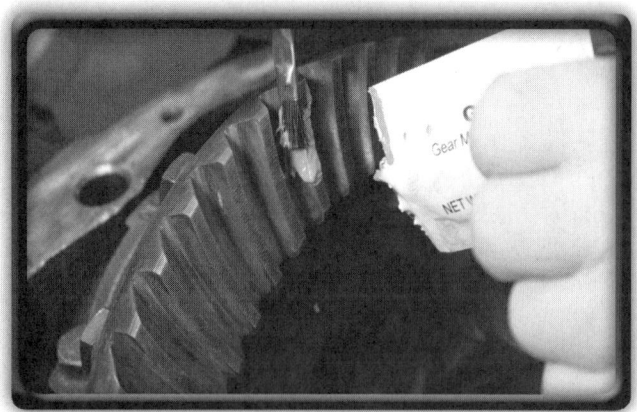

Step 1: Paint several _____ with a suitable compound to make the _____ stand out in order to read it for any corrective actions that may be necessary.

Step 2: Rotate the ring gear a full _____ degrees with a _____ and a _____ using the companion flange nut.

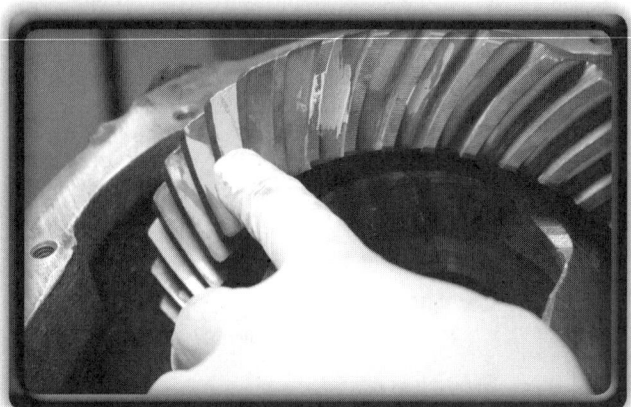

Step 3: Wedge the ring gear with a screwdriver in order to limit the _____ from affecting the pattern. Read the pattern and compare to the _____, and determine the necessary course of action if it is not satisfactory. Adjust the pattern as necessary, and _____ the components thoroughly before reassembling.

Chapter 26 Basic Drive Layouts

Crossword Puzzle

Use the clues to complete the puzzle.

Across

3. A rear-wheel drive axle assembly that has a solid tube incorporating the differential gears.
5. The inner joint on the half-shaft that allows for changes in shaft length.
6. A type of joint that uses two Cardan joints housed in a short carrier and that reduces the change in velocity of a single Cardan joint by using the second joint to cancel out the changes in velocity of the first joint.
7. A joint that consists of a steel cross with four hardened bearing journals, mounted on needle rollers in hardened caps, which locate the cross in the eyes of the yokes. The cross swivels in the yokes as the drive is transferred across the joint.
9. A joint that does not slide to allow for shaft lengthening or shortening; it simply allows for angle changes as the suspension moves.

Down

1. A joint used to transmit torque through wider angles and without the change of velocity that occurs in U-joints.
2. An axle that does not have the capability to drive the vehicle. It is usually found on the rear of front-wheel drive vehicles.
4. An axle that carries the weight of the vehicle; if removed, there is no way to connect the wheel to the vehicle.
8. Part of a two-piece driveshaft that is splined and allows for a change in length of the shaft as the suspension compresses and rebounds.

ASE-Type Questions

Read each item carefully, and then select the best response.

_____ 1. Tech A says that viscous clutch oil should generally be changed at regular intervals. Tech B says that if a viscous clutch fails, it should be replaced. Who is correct?
   A. Tech A
   B. Tech B
   C. Both A and B
   D. Neither A nor B

_____ 2. Tech A says that all-wheel drive vehicles have a differential in the middle of the drive train. Tech B says that many all-wheel drive vehicles incorporate a viscous clutch in the center differential. Who is correct?
   A. Tech A
   B. Tech B
   C. Both A and B
   D. Neither A nor B

_____ 3. Tech A says that all front wheel hubs lock automatically. Tech B says that all front wheel hubs lock manually. Who is correct?
   A. Tech A
   B. Tech B
   C. Both A and B
   D. Neither A nor B

_____ 4. Tech A says that some front differentials are operated by a vacuum motor. Tech B says that some front differentials are operated by electric motors. Who is correct?
   A. Tech A
   B. Tech B
   C. Both A and B
   D. Neither A nor B

_____ 5. Tech A says that part-time four-wheel drive does not use a transfer case. Tech B says that a viscous clutch is located in the rear axle. Who is correct?
   A. Tech A
   B. Tech B
   C. Both A and B
   D. Neither A nor B

_____ 6. Tech A says that constant-velocity joints are used on some four-wheel drive vehicles. Tech B says that a slip joint is part of a driveshaft. Who is correct?
   A. Tech A
   B. Tech B
   C. Both A and B
   D. Neither A nor B

_____ 7. Tech A says that a limited slip differential causes spring-loaded clutch plates to provide torque to both wheels. Tech B says that wrong tire sizes can affect the transfer case operation. Who is correct?
   A. Tech A
   B. Tech B
   C. Both A and B
   D. Neither A nor B

_____ 8. Tech A says that some transfer cases have a chain inside them to transfer torque to the wheels. Tech B says that the fluid level in a transfer case should be within ¼ inch of the bottom of the fill hole. Who is correct?
   A. Tech A
   B. Tech B
   C. Both A and B
   D. Neither A nor B

___ **9.** Tech A says that the drive train in some all-wheel drive vehicles is controlled by the PCM/TCM. Tech B says that some all-wheel drive vehicles can activate the wheel brake unit on a wheel that is spinning because it lost traction. Who is correct?
  A. Tech A
  B. Tech B
  C. Both A and B
  D. Neither A nor B

___ **10.** Tech A says that U-joints can make a chirping or squeaking noise when they are going bad. Tech B says that U-joints are typically used on front wheel drive axles to provide torque through all angles of steering. Who is correct?
  A. Tech A
  B. Tech B
  C. Both A and B
  D. Neither A nor B

CHAPTER 27
Servicing Wheels

Tire Tread:
© AbleStock

Chapter Review

The following activities have been designed to help you refresh your knowledge of this chapter. Your instructor may require you to complete some or all of these activities as a regular part of your training program. You are encouraged to complete any activity that your instructor does not assign as a way to enhance your learning.

Matching

Match the following terms with the correct description or example.

- A. Bar
- B. Casing plies
- C. Drop center
- D. EH2 rim
- E. Negative offset
- F. Neutral steer
- G. Ply rating
- H. Rim flanges
- I. Schrader valve
- J. Side force
- K. Slip angle
- L. Steel-disc–type rim
- M. Tapered seat
- N. Traction grade
- O. Tread wear grade
- P. Valve core
- Q. Valve stem
- R. Wheel center
- S. Wheel studs
- T. Zero offset

_____ 1. A plain steel wheel that is typically covered by a hubcap.

_____ 2. A condition in which the plane of the hub mounting surface is positioned toward the brake side or back of the wheel centerline.

_____ 3. A rubber or steel piece that attaches the tire valve to the rim.

_____ 4. A one-way valve used in a valve stem.

_____ 5. A standardized grading system that indicates how well a tire will maintain contact with the road surface when wet.

_____ 6. A rating system that denotes the number of belt layers or plies that make up the tire carcass. In radial tires, ply rating denotes the relative strength of the plies, not the actual number of plies.

_____ 7. The part of the wheel containing the holes for the lug studs.

_____ 8. The specialized rim design that is used with some run-flat tires.

_____ 9. The number imprinted on the sidewall of a tire by the manufacturer as required by the National Highway Traffic Safety Administration that indicates the tire's tread life.

_____ 10. A condition in which both the front and the rear tires of a vehicle are experiencing the same slip angle.

_____ 11. A condition in which the plane of the hub mounting surface is even with the centerline of the wheel.

_____ 12. The pressure on the wheel that pushes it toward the outside or inside of the rim as the vehicle makes a turn.

_____ 13. A wheel design with part of the center section of the wheel a smaller diameter than the rest. It is used for mounting and demounting the tire.

_____ 14. A tire's sideways distortion that makes the vehicle follow a path at an angle to the direction the road wheel is pointing.

_____ 15. The threaded fasteners that attach the wheel to the vehicle.

_____ 16. A metric unit of measure for pressure.

_____ 17. A type of lug nut with a tapered end toward the rim that helps center the wheel on the wheel studs.
_____ 18. A network of cords that give the tire shape and strength.
_____ 19. The one-way spring-loaded valve that screws into the valve stem that allows air to be pumped into a tire and prevents it from flowing out.
_____ 20. The outside edge of the wheel that helps keep the tire from popping off the wheel.

Multiple Choice

Read each item carefully, and then select the best response.

_____ 1. The force that acts between the tread and the road surface is called _____.
   A. side force
   B. road force
   C. cornering force
   D. slip angle

_____ 2. When the front slip angles are larger than the rear slip angles, the vehicle is said to be in a(n) _____ condition, which is referred to as the vehicle "pushing" in the corners.
   A. understeer
   B. oversteer
   C. neutral steer
   D. lateral steer

_____ 3. The distance across the rim flanges at the bead seat is called the _____.
   A. drop center
   B. flange width
   C. contact area
   D. rim width

_____ 4. What type of rim is designed to hold the tire bead in place on the rim in the event of inadequate tire pressure?
   A. Split rims
   B. Drop well rims
   C. Safety rims
   D. Steel-disc type

_____ 5. All of the following are standard types of wheel retaining studs or nuts, *except*:
   A. tapered seat
   B. tapered seat with washer
   C. flat seat without washer
   D. flat seat with washer

_____ 6. When the plane of the hub mounting surface is shifted from the centerline toward the outside or front side of the wheel it has a _____.
   A. positive offset
   B. negative offset
   C. positive caster
   D. negative camber

_____ 7. The measure of the innate strength of a material is called _____.
   A. ply rating
   B. tensile strength
   C. flexibility
   D. elasticity

_____ 8. The first two letters following the letters "DOT" on a tire identify the _____.
   A. tire manufacturer
   B. tire's size
   C. manufacturing plant
   D. both A and C

_____ 9. The standardized system designed to provide tire buyers with a comparative measure of a tire's tread life, traction, and temperature characteristics, is known as the _____.
   A. Uniform Tire Quality Grading System
   B. Department of Transportation compliance code
   C. International Organization for Standardization Tire Class
   D. Manufacturer's load index

_____ 10. If a tire has a four-digit Department of Transportation code, when was it manufactured?
   A. 1970s
   B. 1980s
   C. 1990s
   D. After 2000

_____ 11. What type of monitoring system utilizes the wheel speed antilock braking system to measure the difference in the rotational speed of the four wheels?
   A. Direct TPMS
   B. Centrifugal TPMS
   C. Indirect TPMS
   D. Automatic TPMS

_____ 12. What type of tires are designed to get the driver to a service center in order to have the regular tire fixed or purchase a new one?
   A. Space-saver
   B. Self-sealing
   C. Run-flat
   D. Zero-pressure

_____ 13. The amount of air pressure in the tire that provides it with load-carrying capacity and affects the overall performance of the vehicle is called the _____.
   A. maximum inflation pressure
   B. pressure index
   C. tire inflation pressure
   D. iso pressure

_____ 14. What type of rotation pattern should you use for a four-tire rotation with nondirectional tires?
   A. Forward-cross
   B. Rearward-cross
   C. X pattern
   D. Any of the above

_____ 15. What type of balancing does not take into consideration that the tire has width?
   A. Dynamic
   B. Road force
   C. Static
   D. Horizontal

True/False

If you believe the statement to be more true than false, write the letter "T" in the space provided. If you believe the statement to be more false than true, write the letter "F".

_____ 1. During cornering, centrifugal force puts more weight on the outside wheels.
_____ 2. The bead seat is the edge of the rim that creates a seal between the tire bead and the wheel.
_____ 3. Vehicles manufactured in the mid-70s through the 80s were typically built with positive offset wheels.
_____ 4. Most lug nuts and studs are left-hand threaded, which means they tighten when turned clockwise.
_____ 5. Pick-up trucks and large SUVs can have as many as 6, 8, or 10 studs.
_____ 6. The primary function of the Schrader valve is to assist in keeping debris out of the valve stem.
_____ 7. Bias-ply tires have much more flexible sidewalls because they use two or more layers of casing plies.
_____ 8. A properly inflated radial tire runs cooler than a comparable bias-ply tire.
_____ 9. Asymmetric tread patterns are designed in such a way that the tire can be mounted on the wheel for any direction of rotation.

_____ 10. Radial tires are marked with the section width in millimeters, but with the rim diameter in inches.
_____ 11. The lower a tire's aspect ratio, the wider the tire is in relation to its height.
_____ 12. The DOT tire date manufacturing code is a six-digit code.
_____ 13. It is illegal to sell a tire intended for use on a public road within the United States without a DOT stamp.
_____ 14. The use of a centrifugal switch in the indirect TPMS sensor allows the sensor to go to sleep when the vehicle stops.
_____ 15. Self-sealing tires feature standard tire construction with the addition of a flexible and malleable lining inside the tire in the tread area.
_____ 16. The proper tire pressure for the vehicle is referred to as the recommended cold inflation pressure.
_____ 17. Nitrogen-filled tires can be identified by the green valve stem caps that are placed on the valve stem when the tire is inflated.
_____ 18. Directional tires require keeping the tires on the same side of the vehicle during rotation and generally use a front-to-rear pattern.
_____ 19. Dynamic imbalance tends to cause the tire to move purely up and down.
_____ 20. A tire pressure gauge measures pressures in pounds per square inch (psi), kilopascals (kPa), or bars.

Fill in the Blank

Read each item carefully, and then complete the statement by filling in the missing word(s).

1. _____ is when the rear slip angle is larger than the front slip angle; it is referred to as the vehicle being "loose" in the corners.
2. The _____ of a wheel is the outer circular lip of the metal on which the inside edge of the tire is mounted.
3. In _____ _____ wheels, the drop center is closer to the rear of the wheel.
4. Wheels are fastened to the rims by wheel _____ and _____ nuts.
5. The _____ of a wheel is the distance from its hub mounting surface to the centerline of the wheel.
6. After the wheel assembly has been fitted, the wheel retaining nuts that hold the wheel onto the vehicle are _____ to the manufacturer's specifications.
7. The _____ _____ refers to the number and spacing of the lug nuts or wheel studs on the wheel hub on the wheel rim.
8. A(n) _____ _____ is a specially designed opening that allows a tire to be inflated and then automatically closes to prevent air from escaping.
9. A(n) _____-_____ tire is the older form of tire and is still in use on some trailers and off-road vehicles, primarily because of a slightly lower cost and their more durable construction.
10. Directional _____ _____ are designed to provide a range of attributes during particular driving conditions.
11. _____ tread patterns have the same tread pattern on both sides of the tire and are usually nondirectional tires.
12. _____ _____ is a representation of a tire's ability to resist and dissipate heat.
13. Tires with a speed rating designation of _____ may be driven up to 124 mph (200 kph).
14. The _____ _____ of a tire is the ratio of its height to its width. It is usually given as a percentage.
15. As part of US _____ _____ _____ regulations, there must be a tire manufacture date code stamped on the sidewall of every tire.
16. A(n) _____ _____ _____ system monitors the tires for low air pressure and alerts the driver when one or more tires are lower than (or in some cases, higher than) the designated thresholds.
17. The major safety benefit of _____-_____ technology is that it enables a driver to maintain vehicle control if a tire suffers a rapid pressure loss when in motion.
18. The _____ _____ is inspected to ensure that it is above the built-in wear indicators, which indicate that the tire is at the end of its legal life and should be replaced.

19. Filling the tires with _____ will increase the vehicle's fuel efficiency and tire life by reducing the effects of pressure loss due to permeation.

20. _____ balancing is performed by placing specific amounts of weight on each side of the rim to provide the exact counterbalance needed.

Labeling

Label the following diagrams with the correct terms.

1. Parts of a tire:

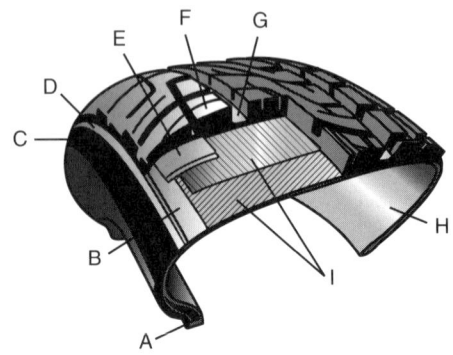

A. _____
B. _____
C. _____
D. _____
E. _____
F. _____
G. _____
H. _____
I. _____

2. Parts of a radial tire:

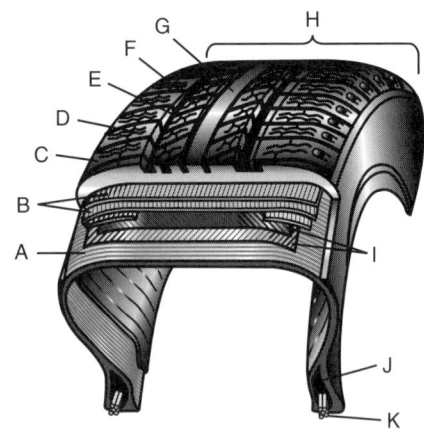

A. _____
B. _____
C. _____
D. _____
E. _____
F. _____
G. _____
H. _____
I. _____
J. _____
K. _____

Skill Drills

Place the skill drill steps in the correct order.

1. Checking Tire Wear Patterns:

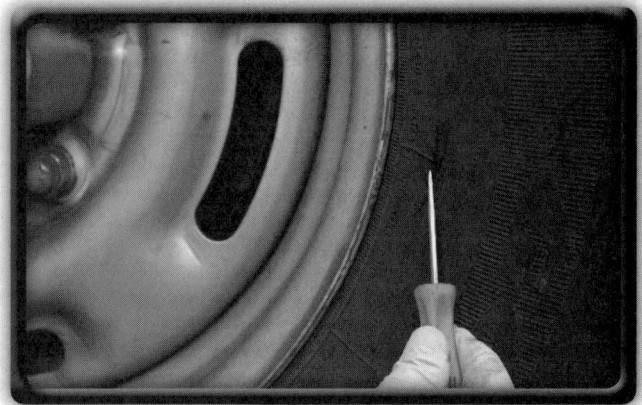

_____ **A.** Inspect the sidewalls of the tires for signs of weather cracking and gouges from impacts with blunt objects. Carefully examine the tread area for separation. This is usually identified as bubbles under the tread area.

_____ **B.** Look for signs of wear on all tires, including the spare. Check the air pressure in the tires.

_____ **C.** Spin the wheel, and see if it is running true. If it is wobbling as it rotates, report it to your supervisor.

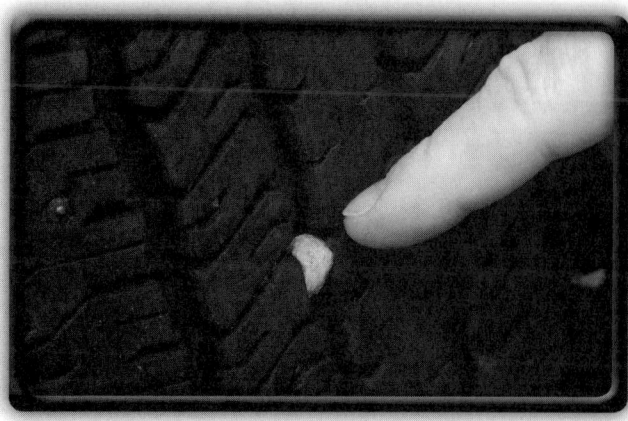

_____ **D.** Inspect the tires for embedded objects in treads and remove them. If anything penetrates the tread, mark the hole with a tire crayon.

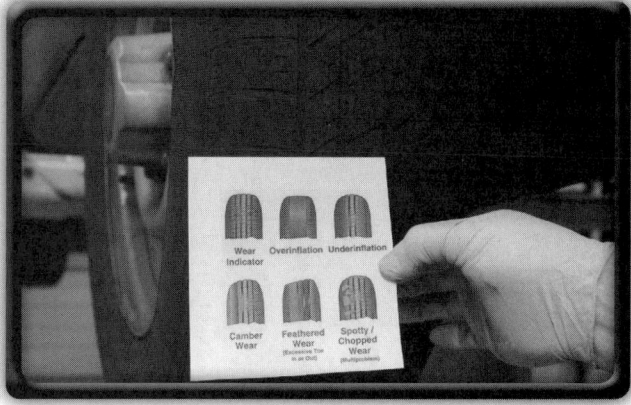

_____ **E.** Check the tread wear patterns with the vehicle's service information to indicate the types of wear that have occurred.

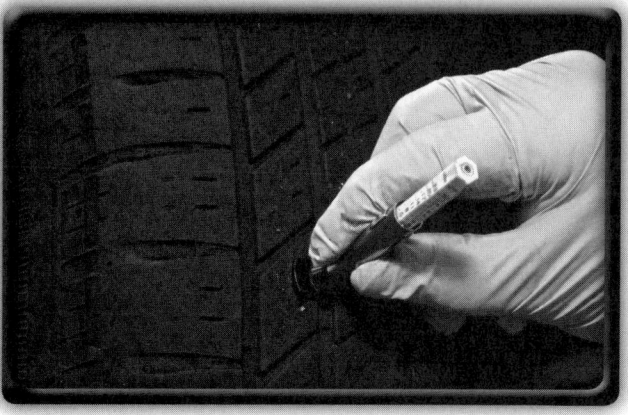

_____ **F.** Check the tread wear depth. Inspect the wear indicator bars. Tires should have at least 1/16" (2 mm) of tread remaining. If the tread is worn down to that level or below, the tires are unserviceable and must be replaced.

2. Replacing a Rubber Press-Fit Valve Stem:

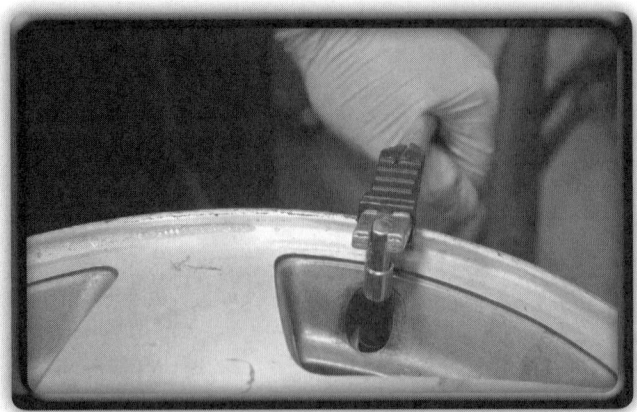

_____ A. Pry the rubber press-fit valve stem from the rim. Use one hand to hold onto the portion of the valve stem in case it breaks off.

_____ B. Insert the valve stem into the hole in the rim from the inside, remove the cap, and screw the valve stem tool onto the threaded end of the stem.

_____ C. Remove the wheel from the vehicle, deflate the tire, and break the top bead using the tire machine.

_____ D. Clean and inspect the hole in the rim. Clean any rust or corrosion with some sandpaper or other appropriate tool. Lubricate the new valve stem with tire lube.

_____ E. Use the handle of the valve stem tool as a lever to pull the retaining ridge through the hole in the rim, and verify that the valve stem is properly installed.

_____ F. Screw the valve stem tool onto the old rubber press-fit stem.

3. Mounting a Tire:

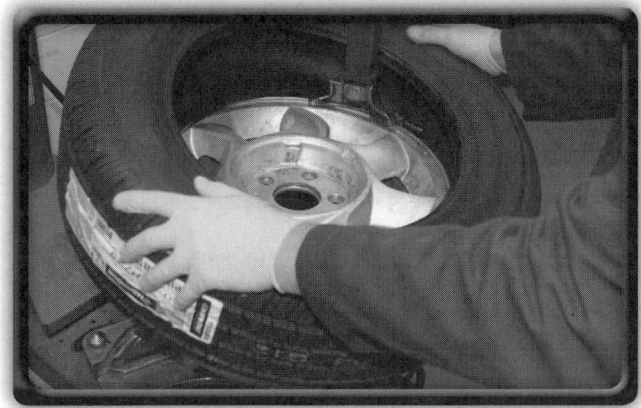

_____ A. Activate the turntable, and guide the lower tire bead onto the rim.

_____ B. When the beads are properly seated, remove the tire inflator chuck, keeping your hands and face clear of the valve stem opening. Once the tire has completely deflated, screw the valve core into the valve stem using the valve core tool.

_____ C. Once the lower bead is fitted, position the upper bead into the guide while holding the other side of the upper bead in the drop center.

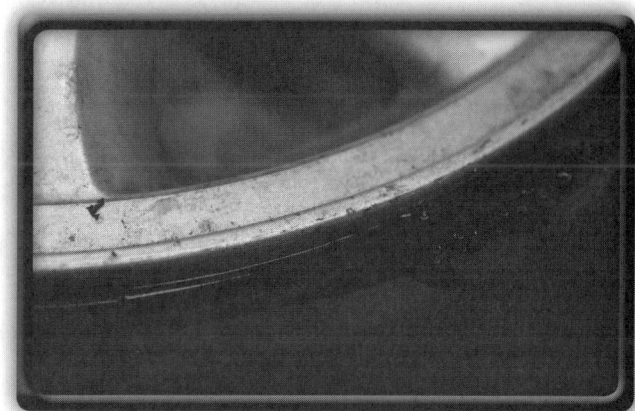

_____ D. Use a soft brush and apply a small amount of soapy water to the bead. If there are any air leaks, they will be indicated by bubbles.

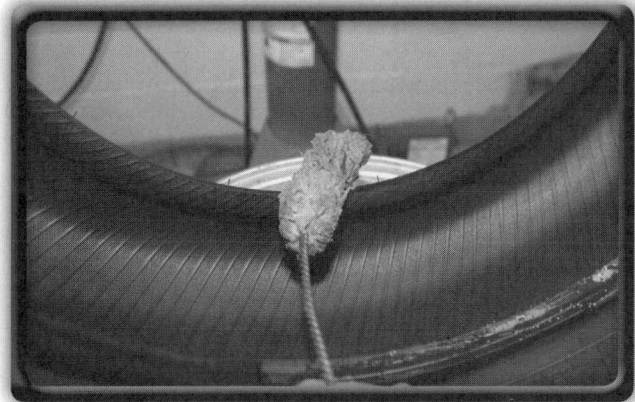

_____ E. Apply some lubricant to the tire bead and rim ridges.

_____ F. Attach the tire inflator chuck to the valve stem. Stand clear of the tire and inflate it, being careful to not exceed 30 psi (207 kPa) if both beads have not seated against the rim. If they have not seated by 30 psi, deflate the tire and inspect the rim and tire for damage. If they are OK, relube the tire and rim and reattempt to inflate the tire. If it still will not seat the beads by 30 psi, inform your supervisor.

_____ **G.** Mount the wheel to the tire machine. Examine the wheel, and remove any rust or dirt from the rim bead seat.

_____ **H.** Select the correct type of tubeless valve stem, lube it with tire lube, and insert it through the hole in the rim from the inside. Using the valve stem tool, pull the stem through until its groove locates in the hole. Use the valve core tool to unscrew the valve core from the valve stem.

_____ **I.** Check the location of the bead indicator ridge to make sure the bead is fully seated. If the rim is clamped from the outside, it will be necessary to release the clamps so the tire can inflate fully.

_____ **J.** Position the tire on top of the rim so a portion of the lower bead is positioned in the drop center while keeping the lower bead in the tire machine guide.

_____ **K.** Reattach the inflator, stand clear, and inflate the tire to the correct pressure as listed on the vehicle's tire placard or in the owner's manual. Be careful to never exceed the maximum tire pressure listed on the tire sidewall.

_____ **L.** Activate the turntable, and guide the tire onto the rim. As the turntable rotates, push the sidewall down, keeping your fingers clear of the rim, so the tire bead is guided below the safety ridge into the drop center.

4. Balancing a Tire:

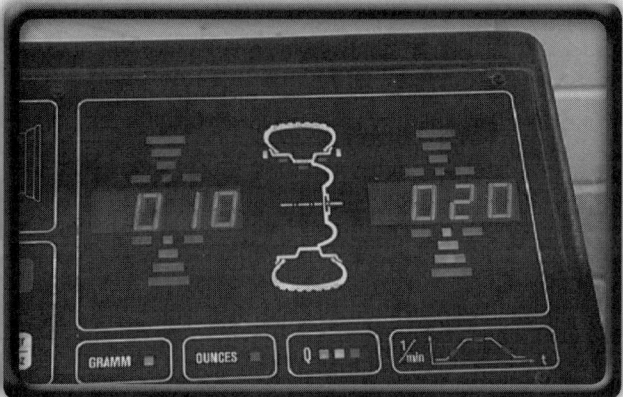

_____ **A.** Read the balancer's analysis. If the wheel is out of balance, you should remove the old weights and recheck the balance of the wheel before adding new weights.

_____ **B.** Calibrate the balancer to the wheel by measuring the width of the rim with a rim caliper, using the gauge on the balancer to determine the offset location of the flange on the wheel, and the diameter of the wheel as listed on the tire. Input these data into the balancer's computer, if fitted. If no computer is fitted, set the balancer adjustments manually according to the instruction manual.

_____ **C.** Respin the wheel to check for accuracy of the balancing job and to confirm that balance has been achieved. Repeat the process for the rest of the wheels and tires. Reinstall the wheels and tires to the vehicle.

_____ **D.** Check and adjust the tire pressure before balancing the tire. Mount the wheel and tire on the balancer, putting the inside part of the wheel toward the balancer in most cases. Secure the wheel by screwing the hub nut assembly on the balancer shaft.

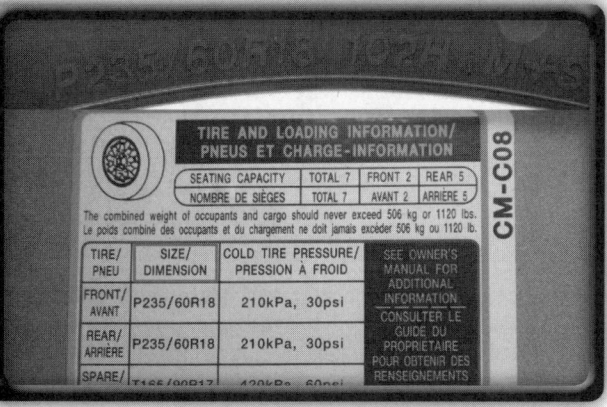

_____ **E.** If using hand tools, prepare the vehicle by loosening the lug nuts, and then raise the vehicle into a comfortable working position. Check that the tires fitted to the wheels on the vehicle are the appropriate size and rating for the vehicle.

_____ **F.** If equipped, lower the safety hood over the wheel. Spin the wheel.

_____ **G.** Install new weights as recommended by the machine's display.

_____ **H.** Mark the inside of the wheel or tire in relation to its location on the vehicle, and then remove it.

5. Patching a Tire:

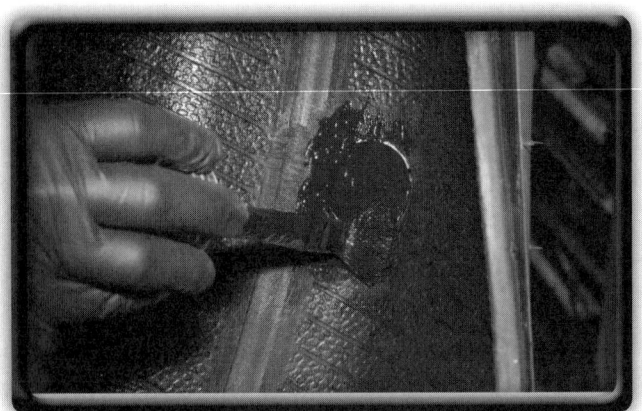

_____ **A.** After the rubber patch sealant has dried, the tire can be reassembled onto the rim in its original position and inflated to the recommended pressure. Check the wheel assembly for balance before it is reinstalled on the vehicle.

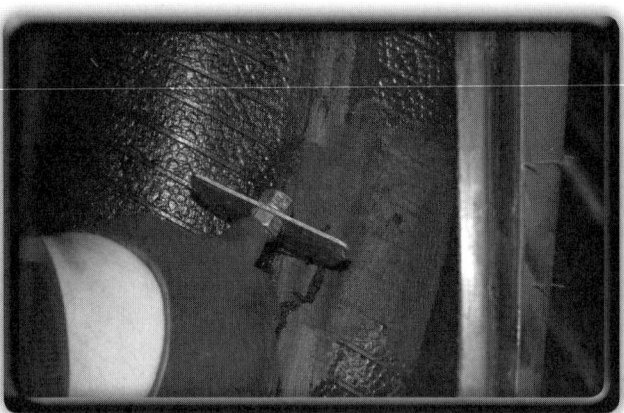

_____ **B.** Liberally apply the liquid buffing solution to a clean rag, and scrub the area just buffed. Or use cleaner and a scraper to clean the area. Repeat this step once or twice, as needed.

_____ **C.** Mount the tire in a tire spreader so the hole can be accessed from both the inside and the outside of the tire.

_____ **D.** Using a pair of pliers, grip the stem of the patch and pull it out so the disc portion of the patch comes into contact with the cemented area. Pull this pointy part of the patch away from the tire's tread. The sticky side of the patch has now been tightly pressed onto the buffed surface.

Chapter 27 Servicing Wheels

_____ **E.** Use a tire buffer to smooth the area inside the tire around the hole. Smooth the area approximately 1/2" (13 mm) beyond the expected patch area.

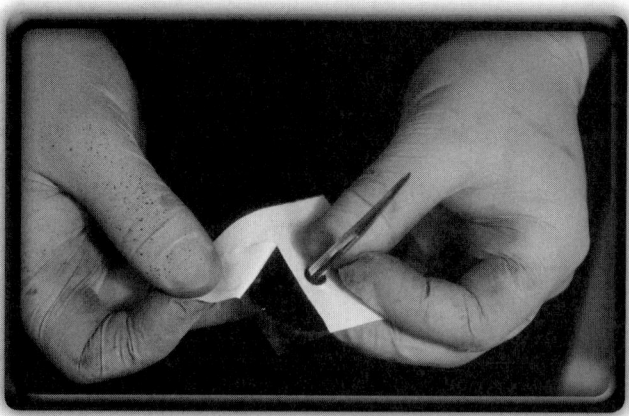

_____ **F.** After selecting the appropriate tire patch, remove the plastic protective cover that is on the sticky side of the tire patch without getting your fingerprints on the sticky side.

_____ **G.** After marking the location of the air leak on the tread of the tire and the position of the tire and weights on the rim, remove the tire from the rim assembly.

_____ **H.** After completing the buffing process, clean out all the accumulated debris with a vacuum.

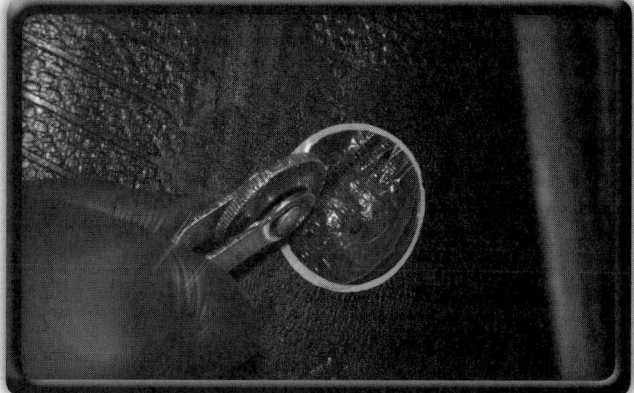

_____ **I.** Use a stitching tool and roll the inner side of the tire patch tight onto the inner surface of the tire. Start at the center of the patch, and work outward. Cover the buffed area and newly applied patch with a rubber patch sealant. Trim the plug material even with or just above the tread using a utility knife or other appropriate cutting tool.

_____ **J.** Take the pointed part of the patch and push it through the inner side of the tire's hole that was roughed previously, pushing it through to the outside of the tire.

278 Fundamentals of Automotive Technology Student Workbook

_____ **K.** Use an air die grinder with the properly sized pointy bit that matches the plug patch, and drill into the tire where the leak was located.

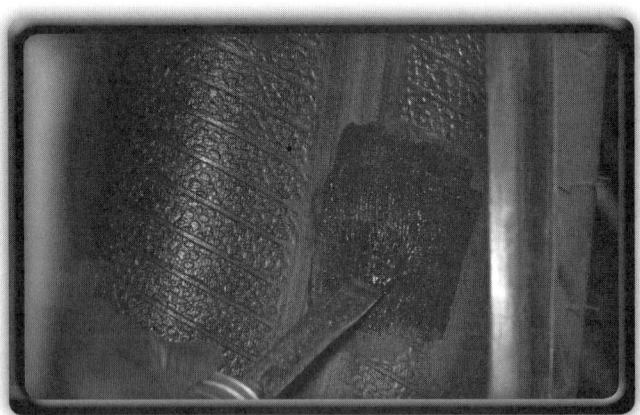

_____ **L.** Apply vulcanizing cement evenly to the inner buffed surface of the tire. The cement needs to stand until the cement is relatively dry and is only tacky to touch.

Crossword Puzzle

Use the clues to complete the puzzle.

Chapter 27 Servicing Wheels

Across

2. The process of matching the tire's lightest point with the rim's heaviest point (generally at the valve stem) for the purpose of reducing the tire's radial imbalance.
5. A condition in which a vehicle's front slip angles are larger than the rear slip angles. This vehicle is said to be "pushing" in the corners.
7. The ratio of sidewall height to section width of a tire.
9. A tire with two or more layers of casing plies and cord loops running radially from bead to bead.
12. The force between the tread and the road surface as a vehicle turns.
14. A tread pattern that differs on each side and, therefore, is usually directional.
15. Nuts that secure the wheel onto the wheel studs.

Down

1. A tread pattern designed to pump water out from under the tire; each tire must be placed in a particular spot on the vehicle.
3. A standardized grading system that indicates the extent to which heat is generated or dissipated by a tire.
4. A switch that is only activated when centrifugal forces are placed on a vehicle.
6. The process of matching the tire's highest point with the rim's lowest point for the purpose of reducing the tire's radial runout.
8. A type of automated tire pressure monitoring system that measures tire pressure and possibly temperature via a sensor installed inside each wheel.
10. A condition in which the plane of the hub mounting surface is positioned toward the outside or front of the wheel centerline.
11. A wheel with negative offset, which gives the outside of the wheel a deep dish appearance.
13. The outer circular lip of the metal on which the inside edge of the tire is mounted.

ASE-Type Questions

Read each item carefully, and then select the best response.

_____ 1. Tech A says that when turning a corner, both wheels being steered remain parallel to each other as the wheels are steered. Tech B says that on some vehicles, the rear wheels also can be steered. Who is correct?
   A. Tech A
   B. Tech B
   C. Both A and B
   D. Neither A nor B

_____ 2. Tech A says that most wheels have a drop center or deep well that is used in installing a tire on the wheel. Tech B says that the drop center or deep well is used to prevent the tire from coming off the wheel in the case of low tire pressure. Who is correct?
   A. Tech A
   B. Tech B
   C. Both A and B
   D. Neither A nor B

_____ 3. Tech A says that all wheels must be torqued to prevent wheels from loosening up and falling off. Tech B says that all wheels must be torqued to prevent overtightening, which can weaken lug studs and warp brake rotors. Who is correct?
   A. Tech A
   B. Tech B
   C. Both A and B
   D. Neither A nor B

_____ 4. Tech A says that underinflated tires reduce fuel economy. Tech B says that underinflated tires are unsafe and cause accelerated tire wear. Who is correct?
   A. Tech A
   B. Tech B
   C. Both A and B
   D. Neither A nor B

_____ 5. Tech A says that incorrect toe settings will cause feathered wear across the tire tread. Tech B says that underinflated tires have more wear in the center of the tread. Who is correct?
   A. Tech A
   B. Tech B
   C. Both A and B
   D. Neither A nor B

_____ 6. Tech A says that tires are marked with a date code indicating the date the tires should be discarded. Tech B says that any tires with a three-digit date code should be discarded. Who is correct?
   A. Tech A
   B. Tech B
   C. Both A and B
   D. Neither A nor B

_____ 7. Tech A says that a TPMS system can save fuel over time. Tech B says that a TPMS will help prevent blowouts. Who is correct?
   A. Tech A
   B. Tech B
   C. Both A and B
   D. Neither A nor B

_____ 8. Tech A says that the use of nitrogen to fill a tire will prevent the tire from blowing out. Tech B says that when mounting or dismounting a tire on a wheel with a TPMS sensor, you need to position the wheel/tire properly on the tire machine or the TPMS sensor can be easily broken. Who is correct?
   A. Tech A
   B. Tech B
   C. Both A and B
   D. Neither A nor B

_____ 9. Tech A says that using a tire plug to repair a hole in a tire is the best and fastest way to fix a tire. Tech B says that using a tire plug patch is the only approved method of repairing a tire in many states. Who is correct?
   A. Tech A
   B. Tech B
   C. Both A and B
   D. Neither A nor B

_____ 10. Tech A says that directional tires cannot be rotated. Tech B says that all old wheel weights should be removed before balancing a tire. Who is correct?
   A. Tech A
   B. Tech B
   C. Both A and B
   D. Neither A nor B

Servicing the Steering System

CHAPTER 28

Chapter Review

The following activities have been designed to help you refresh your knowledge of this chapter. Your instructor may require you to complete some or all of these activities as a regular part of your training program. You are encouraged to complete any activity that your instructor does not assign as a way to enhance your learning.

Matching

Match the following terms with the correct description or example.

- A. Active control
- B. Actuating
- C. Beam axle
- D. Chassis
- E. Drag link
- F. Gear reduction
- G. Inertia
- H. Intermediate shaft
- I. Knuckle
- J. Pitch
- K. Power section
- L. Power unit
- M. Rack
- N. Relay lever
- O. Spline
- P. Steering box
- Q. Steering linkage
- R. Steering system
- S. Tie-rod
- T. Torque sensor

_____ 1. Steel rods that connect the steering box to the steering arms on the steering knuckle.
_____ 2. A steel rod that transfers movement from the drag link to an idler arm.
_____ 3. A device used to measure the load on the steering wheel.
_____ 4. A chamber in the rack where pressurized fluid acts upon pistons that assist in steering.
_____ 5. A steel or iron rod that transfers movement of the pitman arm to a relay lever.
_____ 6. A device that converts the rotary motion of the steering wheel to the linear motion needed to steer the vehicle.
_____ 7. A suspension system in which one set of wheels is connected laterally by a single beam or shaft.
_____ 8. A steering component that transfers linear motion from the steering box to the steering arms at the front wheels.
_____ 9. A system of providing constant feedback from sensors in the vehicle to the control unit.
_____ 10. A term used to describe all of the components and parts involved in steering a vehicle.
_____ 11. The act of making something move or work.
_____ 12. The part that contains the wheel hub or spindle and attaches to the suspension components.
_____ 13. A gear ratio used to make large turns of the steering wheel into smaller turns of the tire to ease steering for the driver.
_____ 14. A belt- or gear-driven pump that produces hydraulic pressure for use in the steering box or rack.
_____ 15. The resistance to a change in motion.
_____ 16. A ridge or tooth on a driveshaft that meshes with grooves in a mating piece and transfers torque to it, maintaining the angular correspondence between them.
_____ 17. The frame of a vehicle, to which the suspension pieces attach.
_____ 18. On a helix, the distance moved in one full revolution of the cylinder.

_____ 19. A steel rod positioned at an angle from the steering column to the steering gear that functions in transferring movement from one to the other.

_____ 20. A steel rod driven by the pinion with tie-rods on each end or tie-rods connected to the center of the rack.

Multiple Choice

Read each item carefully, and then select the best response.

_____ 1. Parallelogram steering uses a _____ gearbox, which changes the direction of steering wheel rotation 90 degrees.
   A. pitman
   B. rack and pinion
   C. worm
   D. linear

_____ 2. What type of steering system is used on the majority of front-wheel drive vehicles due to the restriction of space under the hood?
   A. Worm and sector
   B. Rack and pinion
   C. Recirculating ball
   D. Direct linkage

_____ 3. As the driver turns the steering wheel the forces are transferred to the _____, causing the rack to move in either direction.
   A. worm gear
   B. pitman arm
   C. tie-rods
   D. pinion

_____ 4. The _____ is attached between the tie-rod shaft and the steering arm; it pivots as the rack is extended or retracted when the vehicle is negotiating turns.
   A. inner tie-rod
   B. outer tie-rod
   C. pitman arm
   D. idler arm

_____ 5. The _____ connects the pitman arm to the idler arm.
   A. center link
   B. tie-rod
   C. steering knuckle
   D. adjustment sleeve

_____ 6. The _____ connects the tie-rod to the tie-rod end and provides the adjustment point for setting toe-in or toe-out.
   A. adjustment sleeve
   B. idler arm
   C. pitman arm
   D. center link

_____ 7. The rotary electrical connector located between the steering wheel and the steering column that maintains a constant electrical connection with the wiring while the vehicle's steering wheel is being turned is called the _____.
   A. swivel switch
   B. air bag relay
   C. slot switch
   D. clock spring

_____ 8. In the rack-and-pinion steering system, the steering rack is supported at the pinion end by being sandwiched between the pinion and a(n) _____.
   A. flexible rubber bellows
   B. spring-loaded rack guide yoke
   C. idler arm
   D. pitman arm

_____ 9. Nylon bushings enable the flexibility needed to accommodate slight radial or lateral movement to assist in _____, or release when something gets stuck.
   A. springback
   B. antibinding
   C. slippage
   D. torsion

_____ 10. The worm-and-nut steering gear is also known as a _____ steering gear.
   A. recirculating ball
   B. box-type
   C. worm and sector
   D. rack and pinion

_____ 11. When the rear wheels can be steered independently of or in conjunction with the front wheels it is known as _____.
   A. independent steering
   B. four-wheel drive
   C. four-wheel steering
   D. worm and roller steering

_____ 12. All of the following are types of power steering, *except*:
   A. hydraulically assisted power steering
   B. compressed air-assisted power steering
   C. electrically powered hydraulic steering
   D. fully electric power steering

_____ 13. The spring-loaded piece of steel connected to the pinion gear or worm at its bottom end and the input shaft at its top end is called a _____.
   A. clock spring
   B. helix
   C. leaf spring
   D. torsion bar

_____ 14. All power steering pumps have a(n) _____ to vary fluid flow and power steering system pressures.
   A. pressure relief valve
   B. spool valve
   C. actuator
   D. flow-control valve

_____ 15. What type of steering system still uses fluid and a pump, but the pump is driven by an electric motor to reduce power drawn from the engine?
   A. Electrically assisted steering system
   B. Electric power steering system
   C. Electrically powered hydraulic steering system
   D. Hybrid steering system

True/False

If you believe the statement to be more true than false, write the letter "T" in the space provided. If you believe the statement to be more false than true, write the letter "F".

_____ 1. The steering box converts the rotary motion of the steering wheel to the linear motion needed to pivot the wheels.

_____ 2. The parallelogram steering system gets its name because the center link and axle, along with the pitman arm and idler arm, always move parallel to each other.

_____ 3. The outer tie-rod is attached to the end of the rack and allows for suspension movement and slight changes in steering angles.

_____ 4. Toe-out is a condition where the fronts of the wheels, as seen from above, are closer together than the rears of the wheels.

_____ 5. The toe-setting is the symmetric angle that each wheel makes with the longitudinal axis of the vehicle.

284 Fundamentals of Automotive Technology Student Workbook

_____ **6.** To reduce serious injury, all steering columns are now fitted with collapsible sections that help protect the driver.

_____ **7.** When a technician needs to carry out a servicing procedure on the steering column, it is good practice to disarm the triggering system for the driver's side airbag.

_____ **8.** Both the pinion and the rack teeth are worm gears.

_____ **9.** A disadvantage of the rack-and-pinion steering system is that it typically is not manually adjustable.

_____ **10.** Worm gear steering boxes made the process of turning the front wheels an easier task for drivers of the early automobile.

_____ **11.** A ball-return guide is a special passage or metal tube through which the balls move in recirculating ball steering boxes.

_____ **12.** Power steering fluid gets contaminated with rubber and metallic particles from internal wear in the system.

_____ **13.** Direct drive steering is a completely electrically powered power-assist system that eliminates all hydraulic components and fluid.

_____ **14.** In a pinion-assist type EPS steering system, the power assist unit, controller, and torque sensor are attached to the steering column.

_____ **15.** In an EPS system, sensor inputs are compared to determine how much power assistance is required according to the forces capability map data stored in the ECU's memory.

Fill in the Blank

Read each item carefully, and then complete the statement by filling in the missing word(s).

1. The _____ _____ transmits the driver's steering effort from the steering wheel down to the steering box.

2. The steering linkage transfers the linear steering effort to the wheels by connecting the steering box to the _____ _____ on each of the steering knuckles, which pivot on the ball joints, allowing the wheels to steer the vehicle.

3. The steering knuckle pivots on one or two _____ _____, depending on the type of suspension.

4. The relationships between the steering system, the wheel positions, and the suspension system form what is called the _____ _____.

5. The rack slides in the housing and is moved by the action of the _____ pinion.

6. The _____ _____ protects the inner joints from dirt and contaminants and retains the grease lubricant inside the rack-and-pinion housing.

7. The _____ _____ transfers movement from the steering box to the center link.

8. The _____ _____ is attached to the chassis (the frame of the vehicle) and is positioned parallel to the pitman arm.

9. _____ _____ is the undesired condition produced when, upon hitting a bump, the vehicle darts to one side as the steering linkage is pushed or pulled as a result of the travel of the suspension.

10. To compensate for variations in driving positions, many manufacturers have included a steering column _____ and/or _____ mechanism to their vehicles.

11. Mechanical advantage is gained in the steering system by the _____ _____ between the turns of the steering wheel and the angle turned of the wheel.

12. The rack is typically supported at both ends of the _____ _____ or tube by a nylon bushing.

13. The pinion is supported by two bearings in the rack housing that must be _____ to ensure the pinion is in the correct position, relative to the rack, and to eliminate free play.

14. The steering arm, the stub axle knuckle, and the stub-axle carrier can be forged as one piece, and can be referred to as a(n) _____ _____.

15. The _____ _____ can read both torque and rotation from the steering wheel and convert them into voltage signals for the ECU to monitor.

Labeling

Label the following diagrams with the correct terms.

1. The components of a basic steering system:

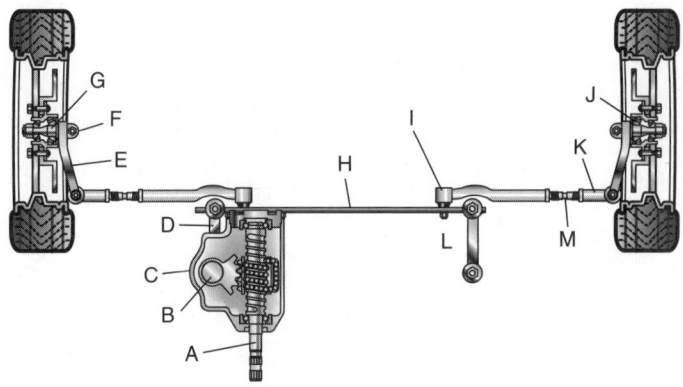

A. _____
B. _____
C. _____
D. _____
E. _____
F. _____
G. _____
H. _____
I. _____
J. _____
K. _____
L. _____
M. _____

2. Rack-and-pinion steering system:

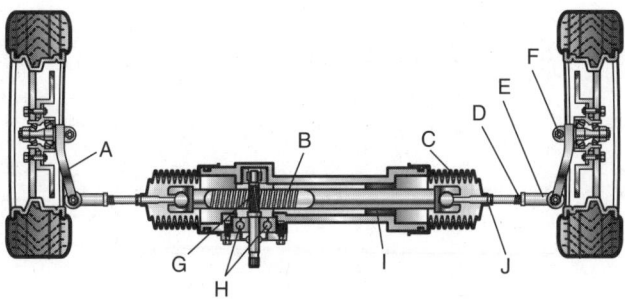

A. _____
B. _____
C. _____
D. _____
E. _____
F. _____
G. _____
H. _____
I. _____
J. _____

3. Recirculating ball steering system:

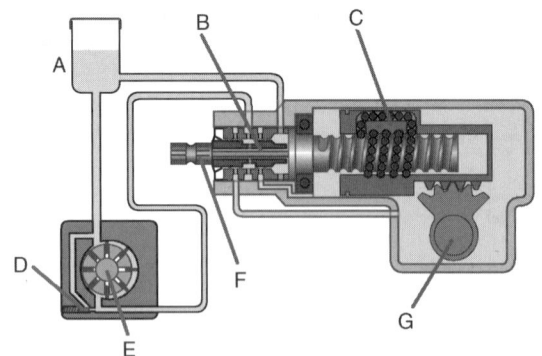

A. _____
B. _____
C. _____
D. _____
E. _____
F. _____
G. _____

4. Power-assisted rack-and-pinion system:

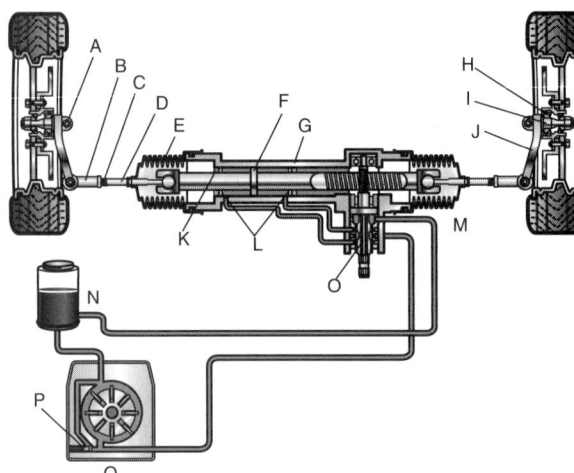

A. _____
B. _____
C. _____
D. _____
E. _____
F. _____
G. _____
H. _____
I. _____
J. _____
K. _____
L. _____
M. _____
N. _____
O. _____
P. _____
Q. _____

5. Power-assisted recirculating gearbox:

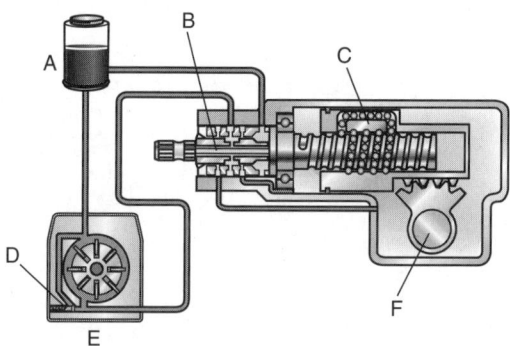

A. _____
B. _____
C. _____
D. _____
E. _____
F. _____

Skill Drills

Place the skill drill steps in the correct order.

1. Removing and Reinstalling the Power Steering Pump:

_____ **A.** To reinstall the new or overhauled power steering pump, reinstall in the reverse order. Fill and bleed the system with the specified power steering fluid.

_____ **B.** After researching the procedure for removing the power steering pump, raise the vehicle on a lift or support it on safety stands. Drain the power steering system's fluid by disconnecting the power steering return line, and dispose of fluid in accordance with environmental legislation.

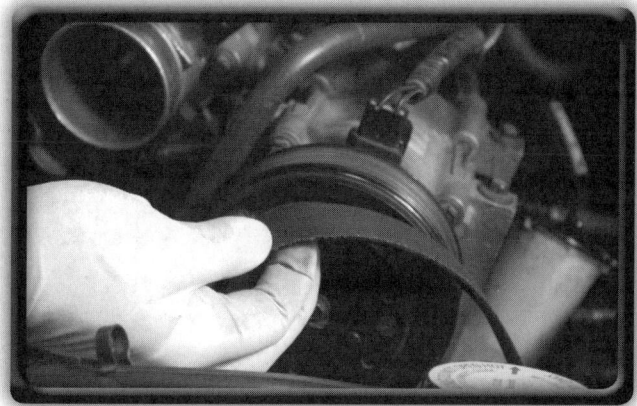

_____ **C.** Disconnect the drive belt(s), and inspect for cracking or groove depth with a belt groove depth gauge. Replace any belt that is cracked or worn beyond specifications.

_____ **D.** Disconnect the power steering hoses from the pump and check for damage or leaks. Replace hoses if needed. New O-rings should be used on hose ends on reassembly, if equipped.

_____ **E.** Remove the pump retaining bolts, and remove the power steering pump.

2. Removing and Replacing the Rack-and-Pinion Steering Gear:

_____ **A.** Remove cotter pins and nuts from the outer tie-rod ends, and separate tie-rod ends with an approved tie-rod end puller, hammer method, or pickle fork.

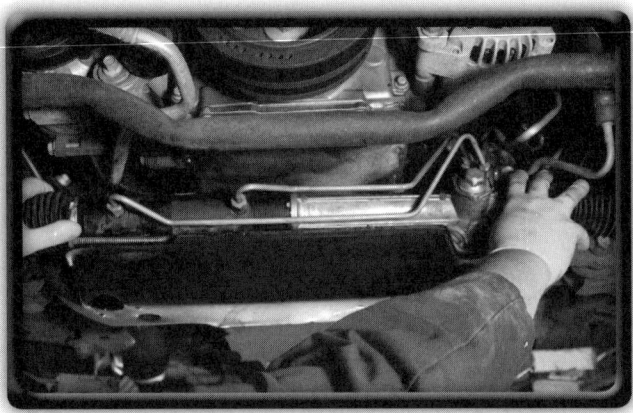

_____ **B.** To install the rack-and-pinion steering gear, reverse the removal steps. Be careful not to damage any lines, tubes, or fittings. Also, make sure the rack is centered in its travel so the clock spring will be indexed to the rack.

_____ **C.** After researching the service information, start the engine and straighten the wheels. Turn the engine off, and lock the steering column. This will keep the clock spring centered. Remove the front tires.

_____ **D.** Check the fluid level in the power steering reservoir. Top off with the proper fluid, and check for leaks from the rack and pinion. Follow the manufacturer's instructions to bleed any air from the power steering system. Top off fluid as necessary.

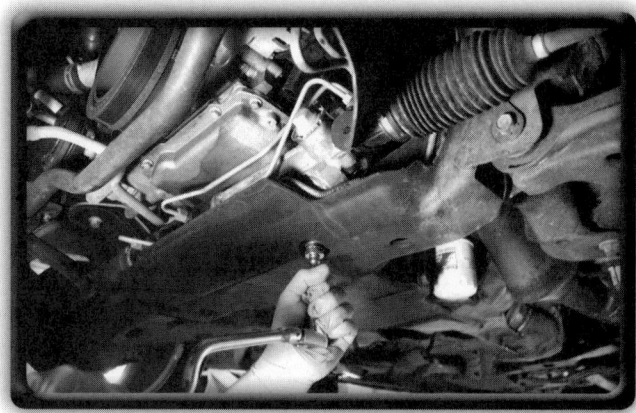

_____ **E.** Remove bolts from the rack-and-pinion bushing brackets, and remove brackets, if equipped.

_____ **F.** Place an oil catch container under the power steering lines. Use flare nut wrenches to disconnect the power steering lines from the rack, and plug them to keep fluid from leaking.

_____ **G.** With the engine running, turn the steering wheel from lock to lock. Check for binding, excessive steering effort, and uneven steering effort. After a major repair like this, it is recommended to perform an alignment to ensure all alignment angles are correct.

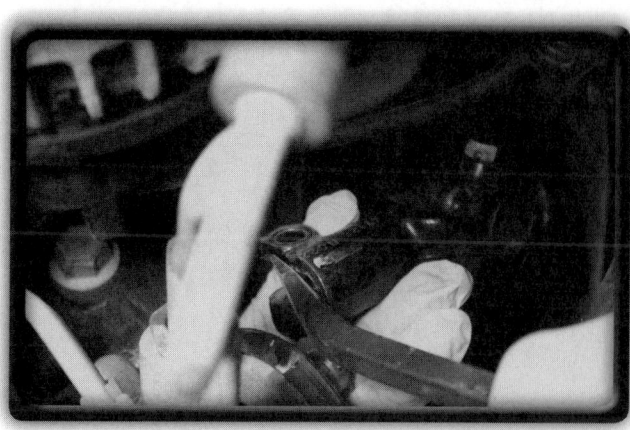

_____ **H.** Pry the universal joint coupler off the pinion shaft.

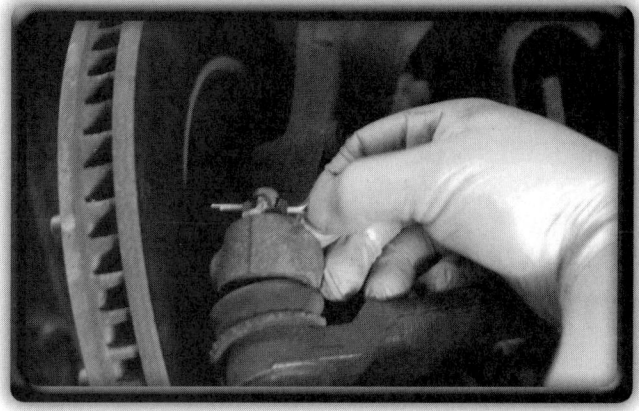

_____ **I.** Torque all fasteners and install new cotter pins.

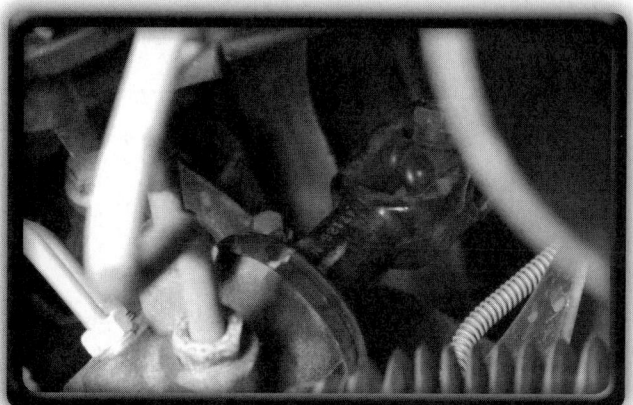

_____ **J.** Locate the universal joint coupler from the intermediate shaft to the pinion shaft of the rack-and-pinion assembly. Remove the bolt from this coupler. Place a paint line on the coupler and the pinion shaft if reinstalling the same rack.

_____ **K.** Slide the rack-and-pinion assembly out through the vehicle's wheel well.

3. Inspecting and Replacing the Pitman Arm:

_____ **A.** Install the new pitman arm and center link in reverse order of removal.

_____ **B.** Remove the pitman arm nut holding the arm to the sector shaft of the steering gear.

_____ **C.** Push and pull side to side on the front driver's side tire while watching for looseness in the pitman arm joint. If the movement is out of specifications, the joint will need to be replaced. This joint can be located on the pitman arm or the center link.

_____ **D.** If the pitman arm needs to be replaced, place an alignment mark on the pitman arm to the sector shaft with white paint or a punch to ensure correct positioning on reassembly.

Chapter 28 Servicing the Steering System

_____ **E.** If the center link needs to be replaced, remove the cotter pin and nut holding the pitman arm taper stud to the center link.

_____ **F.** Using a pitman arm puller, pull the pitman arm free from the sector shaft.

4. Disabling and Enabling the SRS:

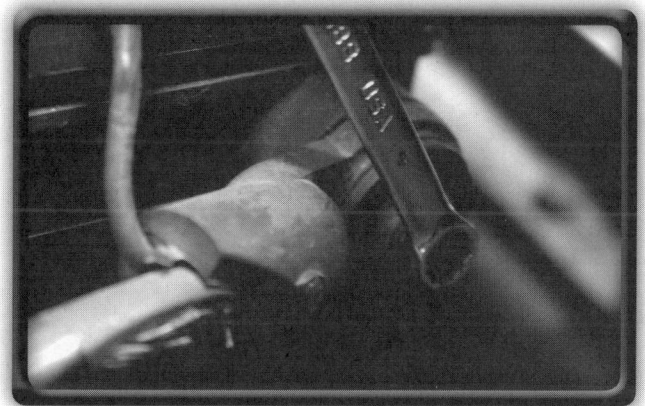

_____ **A.** Remove the negative battery cable, and allow a minimum of 15 minutes to pass to let the SRS system capacitors discharge. Note any radio presets or other memory features of the vehicle that will be erased when the battery is disconnected. Do not use a memory minder or auxiliary power source!

_____ **B.** Without being in front of or reaching across the driver's side airbag, turn on the ignition switch and observe the SRS light. It should illuminate briefly and then go out, and stay out. If so, the SRS system should be ready to be placed back into service.

_____ **C.** To enable the SRS system, verify that all SRS modules, components, and connectors are installed and connected properly.

_____ **D.** Reconnect the negative battery terminal and tighten properly.

_____ **E.** Find and remove the SRS fuse. Verify by turning the key on and observing that the SRS light remains lit for at least 30 seconds. If it goes out, you did not remove the correct fuse or all of the required fuses. Make sure the wheels are steered straight ahead. Turn the key off.

_____ **F.** Reinstall the SRS fuse.

5. Inspecting the Steering Shaft U-Joint(s), Flexible Coupling(s), Collapsible Column, Lock Cylinder Mechanism, and Steering Wheel:

_____ **A.** To inspect the collapsible column, push and pull on the steering wheel to check for any looseness of the steering shaft or column assembly.

_____ **B.** Place the key in the lock position, and check that the steering wheel is locked. If any faults are found, follow the service information for the repair procedures.

_____ **C.** Inspect the steering shaft flexible couplings for obvious damage or wear. Have an assistant wiggle the steering wheel back and forth as you watch for play or improper movement. If any is noted, the joint will need to be replaced.

_____ **D.** To inspect the lock cylinder mechanism, insert and remove the key from the lock cylinder, checking for any resistance or difficulties.

Chapter 28   Servicing the Steering System

_____ **E.** Look under the dashboard at the steering column, and inspect the mounting bolts. Any broken bolts indicate a collapsed steering column.

_____ **F.** Look on the column for the plastic inserts that are typically used to connect the upper and lower halves of the steering shaft. Any damage indicates a collapsed steering column.

_____ **G.** Place the key inside the lock cylinder, and turn the key to each position, checking for smooth and proper operation.

Crossword Puzzle

Use the clues to complete the puzzle.

Across

1. A special rotary electrical connector located between the steering wheel and the steering column that maintains a constant electrical connection with the wiring system while the vehicle's steering wheel is being turned.
4. The release of something when it gets stuck.
5. The protrusion of the worm gear that serves as the point of attachment to the steering column.
9. The undesired condition produced when hitting a bump where the vehicle darts to one side as the steering linkage is pushed or pulled as a result of the travel of suspension.
11. Setting of the toe-in or toe-out of the tires to the centerline of the vehicle.
12. A pinion when it is mated with the rack.
14. A brace or nylon part that pushes against the rack to adjust the mesh of the rack teeth to the pinion teeth.
15. A special passage or metal tube through which the balls move in recirculating ball steering boxes.

Down

2. Any device that controls another object such as a computer.
3. A gear located at the end of the steering shaft connected to the rack. It moves from side to side as the pinion rotates, controlling the direction of the wheels.

6. A component that connects the tie-rods together and to the center link on some applications, providing the adjustment point for toe-in or toe-out, depending on the manufacture's specifications.
7. The electric motor in electric power assist steering systems.
8. A term used to describe two or more parts, such as gears, that are in constant contact with each other.
10. The outer shell of the rack-and-pinion steering system that is mounted to the chassis.
13. Unwanted lateral movements of the worm shaft.

ASE-Type Questions

Read each item carefully, and then select the best response.

_____ 1. Tech A says that the steering column uses one or more flexible joints to connect to the steering gearbox. Tech B says that the pitman arm is bolted to the frame and relays the steering linkage movement to the opposite wheel from the steering gearbox. Who is correct?
   A. Tech A
   B. Tech B
   C. Both A and B
   D. Neither A nor B

_____ 2. Tech A says that the clock spring assists the turning of the steering wheel on vehicles with EPS. Tech B says that the clock spring is used to transmit an electrical signal to the driver's side airbag. Who is correct?
   A. Tech A
   B. Tech B
   C. Both A and B
   D. Neither A nor B

_____ 3. Tech A says that power steering fluid is universal and can be used in virtually any vehicle's power steering system. Tech B says that in a power steering system, the force needed to turn the wheels is created by the hydraulic pump or electric motor. Who is correct?
   A. Tech A
   B. Tech B
   C. Both A and B
   D. Neither A nor B

_____ 4. Tech A says that rack-and-pinion steering systems generally do not use power steering due to their lighter duty construction. Tech B says that rack-and-pinion steering systems use a worm gear arrangement to move the rack. Who is correct?
   A. Tech A
   B. Tech B
   C. Both A and B
   D. Neither A nor B

_____ 5. Tech A says that to properly check a tie-rod end, the technician should twist the tie-rod end, and any rotational movement means the joint is bad. Tech B says that to properly check tie-rod ends, the vehicle's weight should be on the wheels, and as an assistant wiggles the steering wheel, you can check for movement in the tie-rod ends. Who is correct?
   A. Tech A
   B. Tech B
   C. Both A and B
   D. Neither A nor B

_____ 6. Tech A says that worn rack-and-pinion mount bushings can cause excessive play in the steering system. Tech B says that a worn idler arm can cause excessive play in the steering system. Who is correct?
   A. Tech A
   B. Tech B
   C. Both A and B
   D. Neither A nor B

_____ 7. Tech A says that when the vehicle is being driven straight ahead, a belt-driven power steering pump will pump fluid continuously, placing a minimal load on the engine. Tech B says that a belt-driven power steering pump is activated by an electromagnetic clutch, so it only pumps fluid when the wheels are being steered. Who is correct?
  A. Tech A
  B. Tech B
  C. Both A and B
  D. Neither A nor B

_____ 8. Tech A says that the pressure relief valve maintains a preset minimum pressure in the system. Tech B says that the pressure relief valve prevents excessive pressure. Who is correct?
  A. Tech A
  B. Tech B
  C. Both A and B
  D. Neither A nor B

_____ 9. Tech A says that worn tie-rod ends can cause a steering wandering complaint. Tech B says that a hard steering complaint could be caused by a worn power steering pump. Who is correct?
  A. Tech A
  B. Tech B
  C. Both A and B
  D. Neither A nor B

_____ 10. Tech A says that a pickle fork is used to hold a tie-rod while it is being tightened. Tech B says that power steering pump pulleys are typically bolted onto the pump shaft. Who is correct?
  A. Tech A
  B. Tech B
  C. Both A and B
  D. Neither A nor B

Servicing the Suspension System

CHAPTER 29

Tire Tread:
© AbleStock

Chapter Review

The following activities have been designed to help you refresh your knowledge of this chapter. Your instructor may require you to complete some or all of these activities as a regular part of your training program. You are encouraged to complete any activity that your instructor does not assign as a way to enhance your learning.

Matching

Match the following terms with the correct description or example.

- A. Axle
- B. Caster
- C. Deflecting force
- D. Elasticity
- E. Independent suspension
- F. Negative camber
- G. Overshoot
- H. Positive caster
- I. Rebound clip
- J. Scrub radius
- K. Shroud
- L. Sintering
- M. Static toe
- N. Suspension strut
- O. Suspension system
- P. Thrust line
- Q. Torsion bar
- R. Trailing arm suspension
- S. Unibody construction
- T. Yaw

_____ 1. A system within a vehicle designed to isolate the vehicle body from road bumps and vibrations.

_____ 2. A steel or plastic cover placed over the shock rod.

_____ 3. A construction style that uses the body sheet metal as the frame of the vehicle. Typically this type of construction uses small frame sections called subframes.

_____ 4. The angle formed through the wheel pivot points when viewed from the side in comparison to a vertical line through the wheel.

_____ 5. Movement of a vehicle around its z-axis (vertical axis) felt when the vehicle deviates from its straight path, as when skidding sideways and the rear comes around.

_____ 6. Backward tilt of the wheel pivot points from the vertical line.

_____ 7. The shaft of the suspension system to which the tires and wheels are attached; they are used to drive or support the wheels.

_____ 8. A metal strap that is warped around the leaf spring to prevent excessive flexing of the main leaf during rebound.

_____ 9. The ability to deform and reform into the same shape.

_____ 10. The process of using pressure and heat to bond metal particles.

_____ 11. The amount a spring extends (springs back) past its original length following compression.

_____ 12. A setting designed to compensate for slight wear in steering components that may cause the wheels to turn outward or inward while the vehicle is in motion.

_____ 13. Tilt of the top of the tire toward the centerline of the vehicle.

_____ 14. The imaginary line drawn through the center of the rear axle.

_____ 15. A force that moves an object in a different direction or into a different shape.

_____ 16. A bar made of a steel alloy that is fixed rigidly to the chassis at one end and the suspension control arm at the other to support the weight of a vehicle.

_____ 17. The distance between two imaginary points on the road surface—the point of center contact between the road surface and the tire, and the intersecting point where the steering axis centerline and the tire centerline contact the road surface.

_____ 18. A type of suspension system that uses upper and lower control arms.

_____ 19. A shock absorber designed to reduce spring oscillations.

_____ 20. A system for allowing the up and down movement of one tire without affecting the other tire on that axle.

Multiple Choice

Read each item carefully, and then select the best response.

_____ 1. A force that moves an object in a different direction or into a different shape is known as a(n) _____.
   A. applied force
   B. reaction force
   C. deflecting force
   D. suspension force

_____ 2. Parts of a vehicle that are not supported by the suspension system, including the wheels, tires, brakes, axles, and steering and suspension parts not supported by springs are considered _____.
   A. unsprung mass
   B. free weight
   C. unsprung weight
   D. either A or C

_____ 3. The force transferred from the tire contact patch through the axle housing is called _____.
   A. cornering force
   B. driving thrust
   C. reaction force
   D. roll

_____ 4. Movement around the vehicle's y-axis, commonly felt during hard braking or fast acceleration, when the front of the vehicle noses down or rises up slightly, is called _____.
   A. yaw
   B. pitch
   C. roll
   D. oscillation

_____ 5. When a spring deflects easily under a light load, but its resistance increases as the load increases, it is said to have a(n) _____.
   A. progressive rate of deflection
   B. constant rate of deflection
   C. uniform pitch
   D. progressive pitch

_____ 6. The rear of the multileaf is connected to the frame by a(n) _____, which provides a link between the spring eye and a bracket on the frame.
   A. rigid spring hanger
   B. intermediate hook
   C. rebound clip
   D. swinging shackle

_____ 7. The twisting force that is applied by anchoring one end of an object and then applying a twisting force to the other end is called _____.
   A. torsional load
   B. compression
   C. pitch
   D. deflection

Chapter 29   Servicing the Suspension System

_____ 8. A(n) _____ consists of a flexible rubber bladder, which seals the outside of the upper and lower halves of the shock absorber.
   A. strut-type shock absorber
   B. air spring
   C. gas-pressurized shock absorber
   D. suspension strut

_____ 9. The newest style of electronic adjustable shock uses a special type of fluid called _____ fluid that has the unique characteristic of changing viscosity when exposed to a magnetic field.
   A. anisotropic fluid
   B. magnetic hydraulic
   C. magneto-rheological
   D. frequency

_____ 10. In control arm applications, particularly at the rear of a vehicle, a rubber bushing may be molded with a voided section; this bushing is called a _____.
   A. rubber-bonded bushing
   B. spring shackle bushing
   C. gap-tooth bushing
   D. compliance bushing

_____ 11. What is another name for a rigid-axle coil-spring suspension that uses two bars similar to a panhard rod and a pivot point on the axle to keep the axle from moving in turns?
   A. Watt's linkage
   B. Rigid nondrive axle suspension
   C. Trailing arm suspension
   D. Rear-wheel independent suspension

_____ 12. Which of the following is an example of a computerized suspension system?
   A. Stepper motor actuated
   B. Solenoid valve actuated
   C. Electromagnetic rheological
   D. All of the above

_____ 13. The side-to-side vertical tilt of the wheel, viewed from the front of the vehicle and measured in degrees is called _____.
   A. caster
   B. camber
   C. alignment
   D. steering axis

_____ 14. When a vehicle has _____, a line drawn through the steering axis centerline meets the road surface behind the vertical centerline of the wheel.
   A. positive camber
   B. negative camber
   C. positive caster
   D. negative caster

_____ 15. The distance between two imaginary points on the road surface is called _____.
   A. scrub radius
   B. steering offset
   C. scrub geometry
   D. all of the above

_____ 16. If the camber line is outside of the steering axis inclination line, then it has _____.
   A. negative offset
   B. zero scrub radius
   C. positive scrub radius
   D. toe-out on turns

_____ 17. The condition in which the fronts of the wheels are closer together than the rears of the wheels is called _____.
   A. toe-in
   B. toe-out
   C. static toe
   D. none of the above

_____ 18. The _____ refers to the relationship between the centerline of the vehicle and the angle of the rear tires.
   A. thrust line
   B. thrust angle
   C. turning radius
   D. Ackermann angle

_____ 19. The amount of distance between the ground and a specified part of the vehicle such as the fender well, upper control arm, or rocker panel is called _____.
   A. clearance
   B. setback
   C. ride height
   D. camber

_____ 20. What type of alignment is outdated and almost never performed on modern vehicles?
   A. Front-end two-wheel alignment
   B. Thrust-angle alignment
   C. Four-wheel alignment
   D. None of the above

True/False

If you believe the statement to be more true than false, write the letter "T" in the space provided. If you believe the statement to be more false than true, write the letter "F".

_____ 1. Preventing or reducing oscillations is called dampening.
_____ 2. Pitch is vehicular movement along its x-axis. It is the rolling motion you feel when making a sharp corner and is generally what causes rollovers.
_____ 3. All front-wheel drive vehicles are of the full-frame chassis design.
_____ 4. The pitch of a spring is the distance from the center of one coil to the center of the adjacent coil; if they are evenly spaced it is called uniform pitch.
_____ 5. The longest leaf is called the main leaf and is rolled at both ends to form spring eyes; they are used to mount the spring to the frame of the vehicle.
_____ 6. Leaf springs can be used across the chassis frame in a trailing arm suspension, or as part of the connecting link between two axle assemblies on a semi-rigid axle beam.
_____ 7. In full-floating axles, the axle bearing is placed on the outside of the axle housing.
_____ 8. The most widely used hydraulic shock absorber is the direct-acting telescopic type.
_____ 9. Electronic adjustable-rate shock absorbers are also called self-leveling shock absorbers.
_____ 10. Typically straight pieces of steel used to either transfer motion or prevent motion within a vehicle's suspension system are called control arms.
_____ 11. Rubber-bonded bushings are normally used for the front eye of the spring at the fixed shackle point, and also in control arm applications.
_____ 12. A solid axle is a nonindependent suspension because the wheels on both sides of the axle are connected together.
_____ 13. MacPherson strut suspension can be used on either the front or the rear of the vehicle.
_____ 14. The primary reason that the short-/long-suspension system was designed was to ensure correct wheel alignment as the vehicle corners.
_____ 15. Most twin I-beam systems use a coil spring to support the weight of the vehicle.
_____ 16. Rear-wheel suspension systems can be of the independent or nonindependent design.

_____ 17. A wheel that leans away from the center of the vehicle at the top is said to have negative camber.

_____ 18. If the camber line is outside of the SAI line, then it has positive offset or positive scrub radius.

_____ 19. The Ackermann angle is the angle the steering arms make with the steering axis, projected toward the center of the rear axle.

_____ 20. Setback is an alignment angle that is not adjustable but that allows the technician to diagnose the vehicle.

Fill in the Blank

Read each item carefully, and then complete the statement by filling in the missing word(s).

1. When a tire hits an obstruction, there is a _____ _____, meaning the tire will move in response to the force applied by the obstruction.

2. _____ refers to the fluctuating of an object between two states, basically meaning the spring compresses and rebounds over and over again.

3. _____ force refers to the lateral movement of the axle housing during turning.

4. In _____-_____ construction, the frame and vehicle body are separately constructed pieces.

5. A _____ _____ is made from a single length of special wire, which is heated and wound on a former to produce the required shape.

6. A bar similar to the torsion bar is the _____ _____, or antiroll bar.

7. If the suspension reaches its limit of travel, rubber _____ prevent direct metal-to-metal contact, thereby reducing jarring of the suspension components.

8. A manual adjustable-rate shock absorber has a manual, external _____ _____ adjustment.

9. The A-arm style control arm is sometimes referred to as a _____ control arm, a relatively flat triangular part that mounts to the frame or subframe at each leg of the A.

10. _____ _____ bushings can be molded to form two halves, to fit into each side of the spring eye on the swinging shackle, which is located on the vehicle frame.

11. A(n) _____ _____ allows the wheel on each side of the axle to move up and down independently of each other.

12. In strut suspension systems, the _____ _____ is contained inside the strut.

13. Also referred to as a track bar, a(n) _____ _____ sits parallel with the axle. One end connects to the frame of the vehicle, and the other end connects to the axle.

14. On rear-wheel drive vehicles with independent rear suspension, drive is transmitted to each wheel by external _____ _____.

15. _____ _____ suspension is an electronically controlled air suspension system at all four wheels with a continuously adaptive dampening system.

16. Wheels are positioned on the suspension at certain angles to provide for easy driving of the vehicle. These angles, taken together, determine the vehicle's _____ _____.

17. The angle formed between the steering axis inclination and the camber line is called the _____ _____.

18. _____-_____ on turns is the relative toe setting of the front wheels as the vehicle turns.

19. The _____ _____ is a geometric alignment of linkages in a vehicle's steering such that the wheels on the inside of a turn are able to move in a different circle radius than the wheels on the outside.

20. _____ _____ is a measure of how small a circle the vehicle can turn in when the steering wheel is turned to the limit.

Labeling

Label the following diagrams with the correct terms.

1. The suspension system:

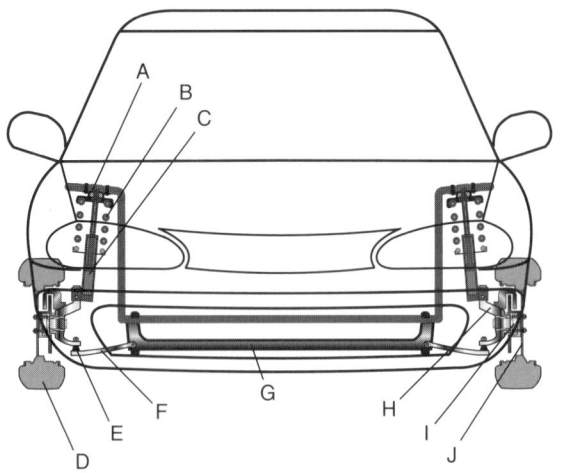

A. _____
B. _____
C. _____
D. _____
E. _____
F. _____
G. _____
H. _____
I. _____
J. _____

2. Yaw, pitch, and roll:

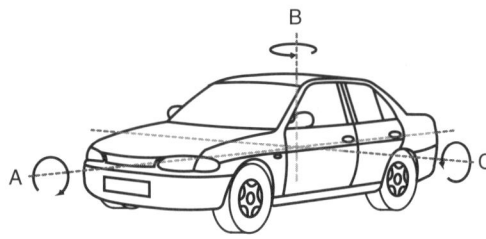

A. _____
B. _____
C. _____

3. Gas-pressurized shock absorber:

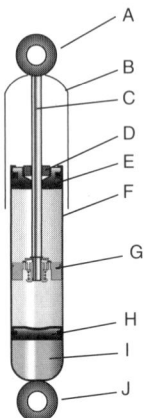

A. _____
B. _____
C. _____
D. _____
E. _____
F. _____
G. _____
H. _____
I. _____
J. _____

Chapter 29 Servicing the Suspension System

4. Electromagnetic shock absorber:

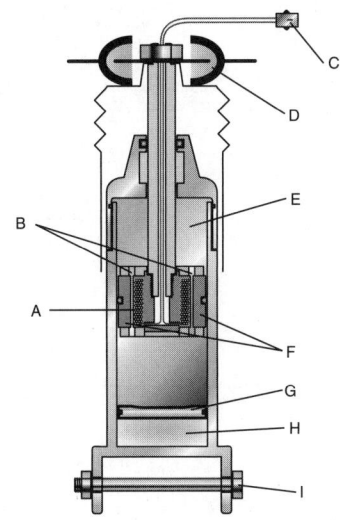

A. _____
B. _____
C. _____
D. _____
E. _____
F. _____
G. _____
H. _____
I. _____

5. Automatic load adjustable suspension system:

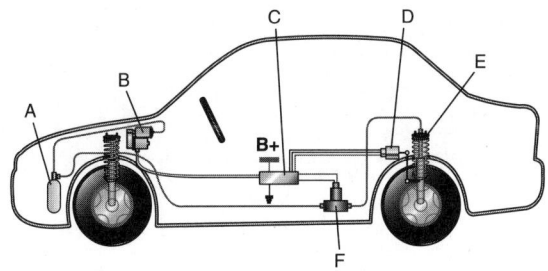

A. _____
B. _____
C. _____
D. _____
E. _____
F. _____

6. Strut suspension:

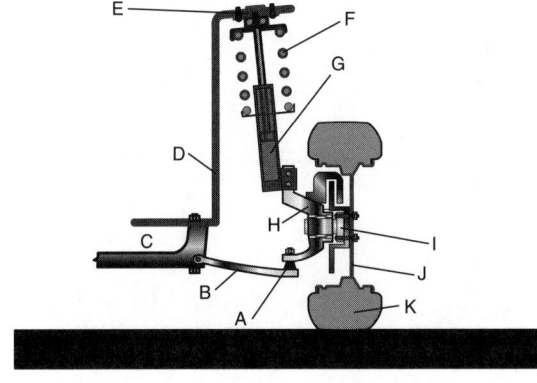

A. _____
B. _____
C. _____
D. _____
E. _____
F. _____
G. _____
H. _____
I. _____
J. _____
K. _____

7. Adaptive shock absorber:

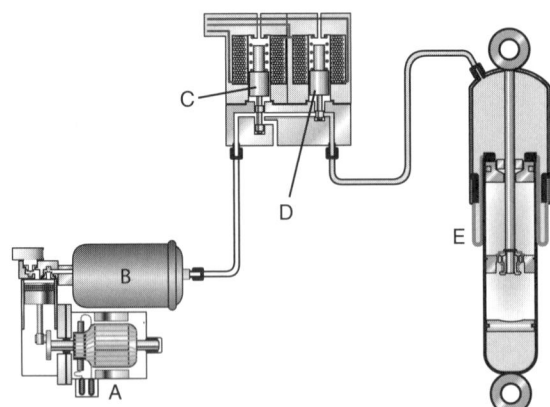

A. _____
B. _____
C. _____
D. _____
E. _____

8. Camber:

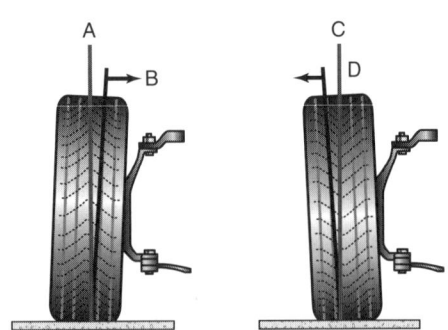

A. _____
B. _____
C. _____
D. _____

9. Caster:

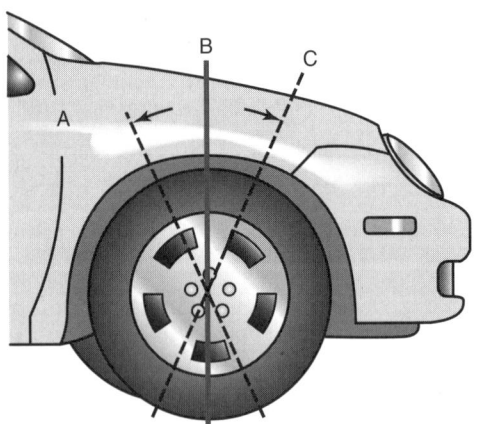

A. _____
B. _____
C. _____

Chapter 29 Servicing the Suspension System

10. Steering axis and scrub radius:

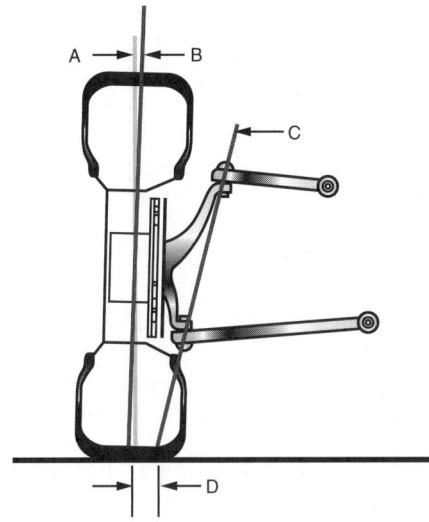

A. _____

B. _____

C. _____

D. _____

Skill Drills

Test your knowledge of skill drills by filling in the correct words in the photo captions.

1. Ride Height:

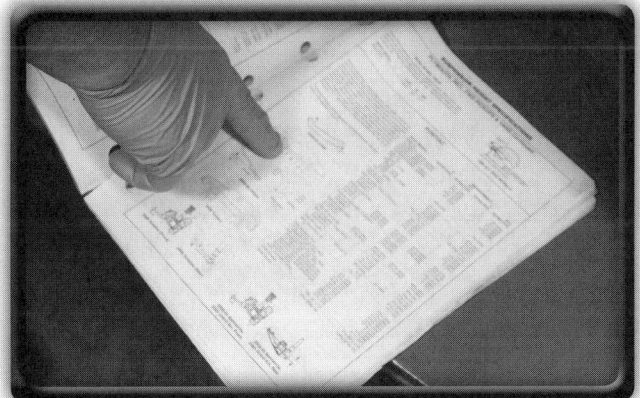

Step 1: Refer to the manufacturer's service information for correct _____ points and _____.

Step 2: Check for properly _____, _____, and _____ tires. Correct any issues found.

Step 3: Check the vehicle for any _____ _____ in the trunk or luggage area. Remove them temporarily while measuring ride height.

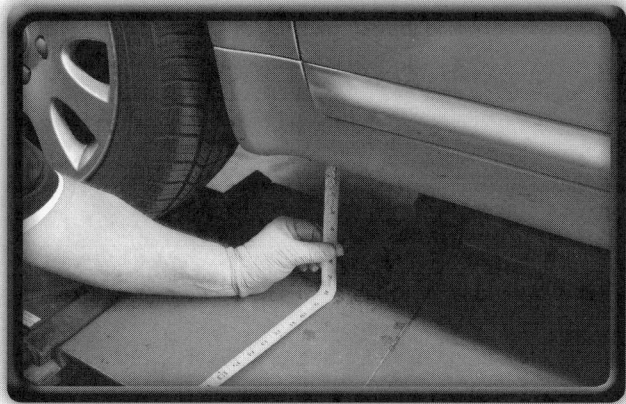

Step 4: Measure from points specified, such as from _____ to _____ on all four corners of the vehicle, and _____ measurements to specifications.

Step 5: Inspect for _____ components or a weak or broken spring if any measurements are not correct. If working with a _____ bar suspension, adjustment of ride height may need to be performed to correct the condition.

Step 6: Refer to the manufacturer's service information for proper adjustment of ride height. Remember that anytime ride height is changed, a(n) _____ _____ will need to be performed also. **(There is no image associated with this step.)**

2. Replacing a Shock Absorber:

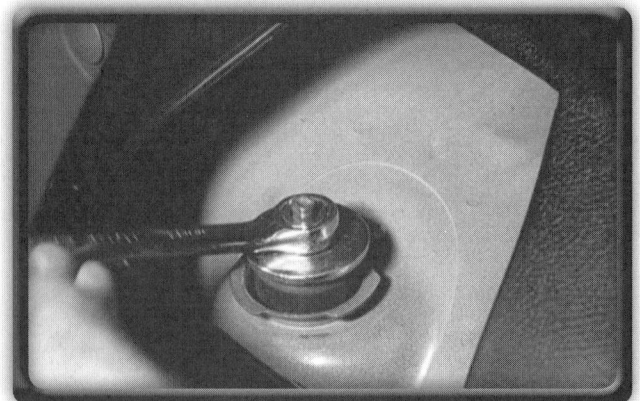

Step 1: Raise the vehicle on a lift, or raise it with a jack and support it with _____ _____ under the frame. Remove the upper bolts holding the shocks in place with a _____ and box-end wrench.

Step 2: Remove the _____ _____ holding the shock in place.

Step 3: Pull the _____ out by hand. _____ on the other side.

3. Lubricating Suspension and Steering Systems:

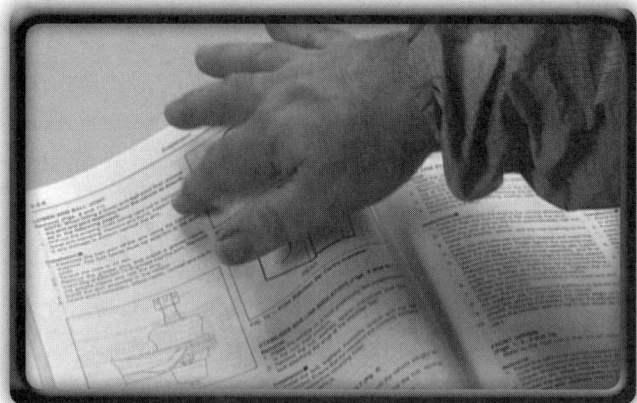

Step 1: Determine the location of lubricating points. Check the shop _____ _____ to determine where the grease points are and the type of grease required. Also look to see if any _____ grease fittings are installed. If so, grease them.

Step 2: Clean each of the lubrication fittings and the grease gun _____ by wiping them with a clean rag. You may need to remove a component's plugs and temporarily install a _____ _____. After the component has been lubricated, reinstall the original plug.

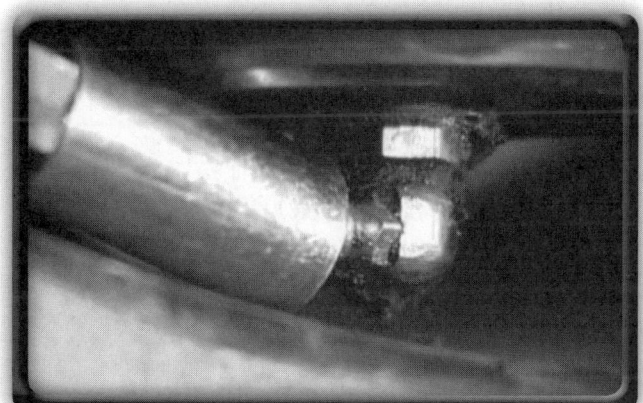

Step 3: Push the _____ _____ nozzle fully over the fitting. It should snap into place. Add enough grease to see the seal or rubber boot rise slightly. Do not _____ a lubricated joint with grease.

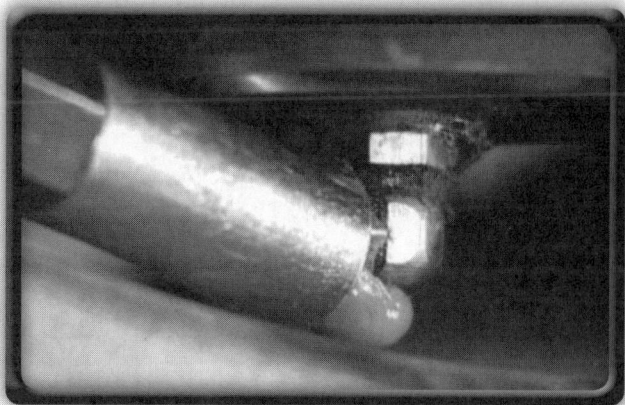

Step 4: If the fitting is clean and will not take grease, remove the grease _____ and check for blockage. If found, the fitting must be replaced with a new fitting of the same _____ and _____, and the joint relubricated.

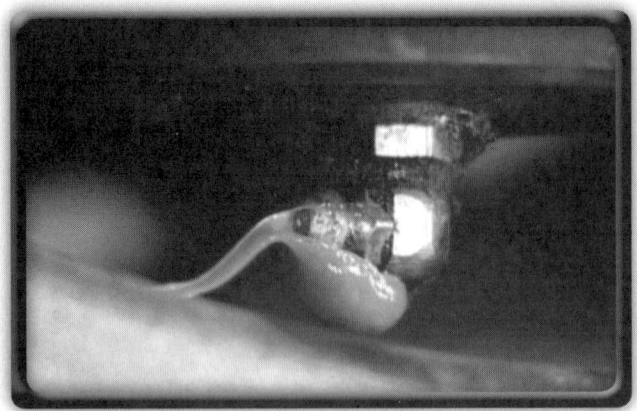

Step 5: Remove the _____ from the _____, and wipe away any excess _____ from it. Repeat the procedure until all the appropriate joints have been lubricated.

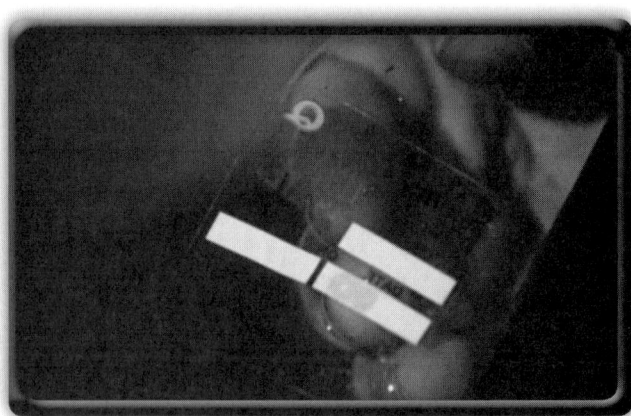

Step 6: After you have completed lubricating all the appropriate joints and cleaned off any excess grease, attach a _____ _____ _____ to the windshield, or reset the maintenance reminder system. Lower the vehicle, and remove it from the lifting device.

4. Removing, Inspecting, and Replacing Leaf Springs:

Step 1: Raise the vehicle on a lift, and support the _____ _____ with tall screw jack stands. Check to see if any of the _____ are cracked or broken and that the noise _____ inserts are positioned correctly between the leaves.

Step 2: Test the security of the _____ _____ _____, and make sure the U-bolts are tight.

Step 3: Check the condition of the bushings or mountings and the _____ _____ by placing a _____ between the frame and the eye of the spring and levering against the spring.

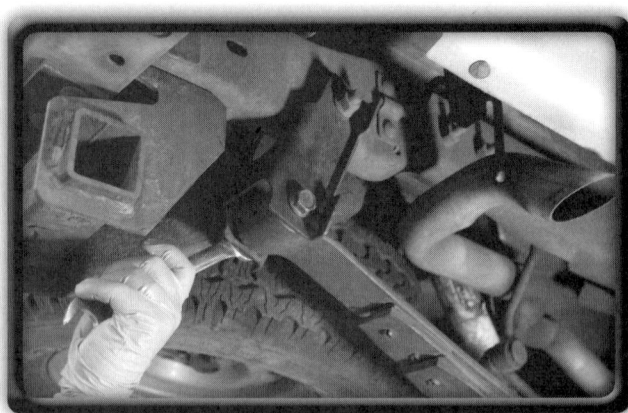

Step 4: Remove the _____ holding the leaf spring to the shackle and _____.

Chapter 29   Servicing the Suspension System   309

Step 5: Have an assistant help hold the _____ while removing the _____, holding the leaf spring to the axle.

Step 6: To install a new leaf spring, _____ the new leaf spring into position by inserting _____ into the hanger and shackle. Use new _____ whenever specified by service information.

Step 7: Reinstall U-bolts to the axle, and _____ them down a little at a time. Tighten all bolts to the manufacturer's _____.

5. Preparing a Vehicle for a Wheel Alignment:

Step 1: Remove any _____ items from the trunk and _____ compartments.

Step 2: Check the size and _____ of all four tires. Adjust the _____ _____ to specifications.

Step 3: Measure the vehicle's _____ _____.

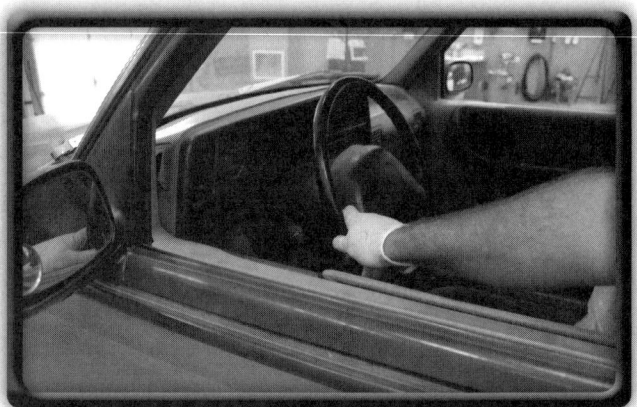

Step 4: Check the _____ of the steering wheel. Correct any _____ _____ before undertaking the wheel alignment.

Chapter 29 Servicing the Suspension System 311

Step 5: _____ each corner of the vehicle to check the correct functioning of the _____ _____.

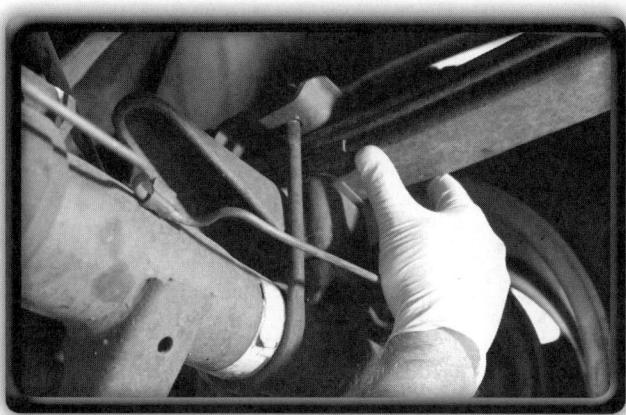

Step 6: With the vehicle raised, inspect all suspension and steering components, including the _____ _____. Repair or replace all damaged or worn _____ _____.

Step 7: Position the vehicle on the wheel alignment _____, making sure the front tires are positioned correctly on the _____.

Step 8: Position the _____ wheels on the _____ _____ or rear turntables.

Step 9: Attach the wheel _____ of the wheel _____ machine.

Crossword Puzzle

Use the clues to complete the puzzle.

Across

3. The primary load-bearing element of a vehicle's suspension system, commonly known as control arm. Can also mean the steering arm, which applies the driver's turning effort to the steering knuckle.
5. The distance between two imaginary points on the road surface—the point of center contact between the road surface and tire, and the intersecting point where the steering axis centerline and the tire centerline contact the road surface.
8. The rigid part typically welded to the body or frame of the vehicle to which the front of the leaf spring is attached.
9. The fluctuation of an object between two states. With regard to suspension springs, it refers to the uncontrolled compression and decompression of the spring following overshoot.
10. A part that provides the springing action or auxiliary spring. It is typically used in air suspension systems or heavy truck applications.
11. A force that acts in the opposite direction to another force.
13. The imaginary line drawn down the exact center of the vehicle from front to back.

Down

1. A spring made of one or more flat, tempered steel springs bracketed together that is used in the suspension system to support the weight of the vehicle.
2. The amount of ground clearance a vehicle has, measured from a point on the body or frame depending on the manufacturer. Also known as ride height.

3. The angle the steering arms make with the steering axis, projected toward the center of the rear axle.
4. A strut used on an independent suspension where the spring and shock are joined together. Used on most front-wheel drive vehicles.
6. Pressure placed on something.
7. The force applied to an axle that shifts the axle in relation to the position of the body when turning.
12. The side-to-side vertical tilt of the wheel. It is viewed from the front of the vehicle and measured in degrees.

ASE-Type Questions
Read each item carefully, and then select the best response.

_____ 1. Tech A says that the wheel and tire are examples of unsprung weight. Tech B says that the exhaust system is an example of sprung weight. Who is correct?
  A. Tech A
  B. Tech B
  C. Both A and B
  D. Neither A nor B

_____ 2. Tech A says that thrust angle refers to the direction the front wheels are pointing. Tech B says that scrub radius refers to the vertical centerline of the tire in relation to an imaginary line through the steering knuckle pivots. Who is correct?
  A. Tech A
  B. Tech B
  C. Both A and B
  D. Neither A nor B

_____ 3. Tech A says that yaw is when a vehicle deviates from its straight path. Tech B says that roll is felt during hard braking. Who is correct?
  A. Tech A
  B. Tech B
  C. Both A and B
  D. Neither A nor B

_____ 4. Tech A says that a vehicle will tend to pull toward the side with the most positive camber. Tech B says that a progressive-rate spring offers a soft ride but can also carry a heavier load. Who is correct?
  A. Tech A
  B. Tech B
  C. Both A and B
  D. Neither A nor B

_____ 5. Tech A says that a shock dampens movement of the suspension in both upward and downward movement. Tech B says that a shock only dampens upward movement. Who is correct?
  A. Tech A
  B. Tech B
  C. Both A and B
  D. Neither A nor B

_____ 6. Tech A says that ball joints must be unloaded when checking them for wear. Tech B says that ball joints must be loaded when checking them for wear. Who is correct?
  A. Tech A
  B. Tech B
  C. Both A and B
  D. Neither A nor B

_____ 7. Tech A says that positive toe is when the fronts of the tires are farther apart than the rears of the tires. Tech B says that a ball joint in a MacPherson strut suspension is a follower joint (not loaded). Who is correct?
  A. Tech A
  B. Tech B
  C. Both A and B
  D. Neither A nor B

_____ 8. Tech A says that a dead axle is designed to carry the weight of the vehicle with no drive capability. Tech B says that checking toe-out on turns is a prealignment check. Who is correct?
   A. Tech A
   B. Tech B
   C. Both A and B
   D. Neither A nor B

_____ 9. Tech A says that loose ball joints can cause the vehicle to wander. Tech B says that loose control arm bushings can affect alignment angles. Who is correct?
   A. Tech A
   B. Tech B
   C. Both A and B
   D. Neither A nor B

_____ 10. Tech A says that the front wheels should be aligned before the rear wheels. Tech B says that front wheel toe should be adjusted after front wheel camber and caster. Who is correct?
   A. Tech A
   B. Tech B
   C. Both A and B
   D. Neither A nor B

Principles of Braking

CHAPTER 30

Tire Tread:
© AbleStock

Chapter Review
The following activities have been designed to help you refresh your knowledge of this chapter. Your instructor may require you to complete some or all of these activities as a regular part of your training program. You are encouraged to complete any activity that your instructor does not assign as a way to enhance your learning.

Matching
Match the following terms with the correct description or example.

- A. Asbestos
- B. Band brake
- C. Brake assist
- D. Brake fade
- E. Brakes
- F. Conservation of energy
- G. Disc brakes
- H. Drum brakes
- I. Friction
- J. Fulcrum
- K. Jake brake
- L. Kinetic energy
- M. Master cylinder
- N. Mechanical advantage
- O. Mechanical disadvantage
- P. Parking brake
- Q. Regenerative braking
- R. Scrub brakes
- S. Top hat parking brake
- T. Weight transfer

_____ 1. A type of brake system that forces stationary brake pads against the outside of a rotating brake rotor.

_____ 2. A brake system that uses leverage to force a friction block against one or more wheels.

_____ 3. When the load distance on a lever is greater than the effort distance, which means the effort required to move the load is greater than the load itself.

_____ 4. A physical law that states that energy cannot be created or destroyed.

_____ 5. A drum brake that is located inside a disc brake rotor in order to act as a parking brake.

_____ 6. A type of brake system that forces brake shoes against the inside of a brake drum.

_____ 7. An enhanced safety system built in to some ABS systems that anticipates a panic stop and applies maximum braking force to slow the vehicle as quickly as possible.

_____ 8. Technology used in vehicles that allows the vehicle to recapture and store part of the vehicle's kinetic energy in a reusable form when braking. Primarily used in hybrid and electric vehicles.

_____ 9. The reduction in stopping power caused by a change in the brake system such as overheating, water, or overheated brake fluid.

_____ 10. The ratio of load and effort for any simple machine such as a lever.

_____ 11. A braking system that uses a metal band lined with friction material to clamp around the outside of a wheel or drum.

_____ 12. The point around which a lever rotates and that supports the lever and the load.

_____ 13. A mineral with needle-like fibers that can become embedded in lung tissue and cause cancer.

_____ 14. Weight moving from one set of wheels to the other set of wheels during braking, acceleration, or cornering.

_____ 15. A brake system that consists of an extra exhaust valve on a diesel engine, which releases compressed gases from the combustion chamber at the top of the compression stroke; also called a compression brake.

_____ 16. The resistance created by surfaces in contact.

_____ 17. Converts the brake pedal force into hydraulic pressure, which is then transmitted via brake lines and hoses to one or more pistons at each wheel brake unit.

_____ 18. A brake system used for holding the vehicle when it is stationary.

_____ 19. A system made up of hydraulic and mechanical components designed to slow or stop a vehicle.

_____ 20. The energy of an object in motion; it increases by the square of the speed.

Multiple Choice

Read each item carefully, and then select the best response.

_____ 1. The old _____ braking systems caused the vehicle to veer dangerously to one side when braking and required frequent adjustment.
   A. scrub
   B. band
   C. disc
   D. drum

_____ 2. What type of system uses a computer to monitor each wheel's speed and either hold, decrease, or apply hydraulic pressure to each wheel to prevent wheel lock-up and maintain the maximum amount of braking power just short of brake lock-up?
   A. Brake assist
   B. Antilock brake system
   C. Brake-by-wire
   D. Brake pedal emulator

_____ 3. In a brake-by-wire system a _____ tells the computer how firmly the driver intends to brake which then sends control signals to the appropriate brake actuators.
   A. brake pedal emulator
   B. service brake
   C. hydraulic actuator
   D. kinetic sensor

_____ 4. Which of the following factors influence a vehicle's ability to brake effectively?
   A. Weight of the vehicle
   B. Height of the vehicle
   C. Tire composition
   D. All of the above

_____ 5. What law states that an object will stay at rest or at uniform speed unless it is acted on by an outside force?
   A. Euler's first law of motion
   B. The law of conservation of energy
   C. Newton's first law of motion
   D. Kirchhoff's law

_____ 6. The amount of friction between two moving surfaces in contact with each other is expressed as a ratio and is called the _____.
   A. law of conservation of energy
   B. coefficient of friction
   C. rate of heat transfer
   D. second law of motion

_____ 7. When brakes are operated on a moving vehicle, a _____ force is generated.
   A. cornering
   B. directional
   C. deflecting
   D. rotational

_____ 8. Brake pedals are usually a _____. They pivot at the top end (fulcrum). The foot pressure (effort) is applied to the bottom end. And the master cylinder (load) is applied between the two.
   A. lever of the first order
   B. lever of the second order
   C. lever of the third order
   D. lever of the fourth order

_____ 9. What type of braking system is used on trailers towed by light vehicles if their gross weight exceeds a certain value?
   A. Jake brake
   B. Exhaust braking system
   C. Electric braking system
   D. Air-operated braking system

_____ 10. What type of parking brake uses a small drum brake to prevent the drive shaft from turning?
   A. Drum-style parking brake
   B. Transmission-mounted parking brake
   C. Top hat parking brake
   D. Electric parking brake

True/False

If you believe the statement to be more true than false, write the letter "T" in the space provided. If you believe the statement to be more false than true, write the letter "F".

_____ 1. The band braking system was used for more than 2,000 years with virtually no change.
_____ 2. Giving greater control of the braking system to the computer increases driving safety.
_____ 3. Aggressive driving causes tires to become hot and possibly overheated, thus reducing the tire's ability to obtain maximum traction.
_____ 4. The service brake is usually operated by hand, but some vehicles use a foot-activated pedal.
_____ 5. Faster-moving objects have more kinetic energy than slower-moving objects of the same weight.
_____ 6. An outside force needs to act upon a vehicle to cause it to decelerate, that force comes from the mass of the Earth.
_____ 7. Static friction is resistance between moving surfaces and is present in standard brakes.
_____ 8. Most of the heat generated by the braking process radiates into the atmosphere.
_____ 9. Water fade is caused by water-soaked brake linings acting like a lubricant and lowering the coefficient of friction between the braking surfaces.
_____ 10. Modern drum and disc brake systems are regularly fitted with an ABS that monitors the speed of each wheel and prevents wheel lock-up or skidding, no matter how hard brakes are applied or how slippery the road surface.
_____ 11. Air-operated braking systems use an extra lobe on the camshaft to control an auxiliary exhaust valve at the top of each cylinder.
_____ 12. When engaged, the jake brake releases the compression stroke pressure before it can be transmitted back to the power stroke of the piston.
_____ 13. Parking brake systems incorporate an automatic method of adjustment.
_____ 14. Regenerative braking is accomplished by causing the hybrid vehicle's electric motor to act as a battery.
_____ 15. Some current hybrid vehicles use brake-by-wire technology during regeneration but with a hydraulic backup brake system.

Fill in the Blank

Read each item carefully, and then complete the statement by filling in the missing word(s).

1. A(n) _____-_____-_____ system does away with the hydraulic portion of the brake system and replaces it with sensors, wires, an electronic control unit, and electrically actuated motors to apply individual brake units at each wheel.

2. In a(n) _____ braking system the amount of stopping power is controlled by how much electricity is being generated.

3. There are two brake systems on all vehicles—a _____ brake and a _____ brake.

4. By using _____ of different sizes, hydraulic forces can be increased or reduced, allowing designers to obtain the desired braking force for each wheel.

5. Heavier objects have more _____ energy than lighter objects moving at the same speed.

6. In an automobile, _____, or an increase in kinetic energy, is caused by the power from the engine.

7. The energy used to cause a vehicle to accelerate and decelerate must be _____ from one form of energy to another.

8. _____ _____ is caused by the buildup of heat in the braking surfaces, which get so hot they cannot create any additional heat, leading to a loss of friction.
9. _____ fade is caused by the brake fluid becoming so hot that it boils.
10. Brake systems use _____ and mechanical advantage to apply service and parking brakes.
11. Some vehicles come equipped with _____ pedal assemblies that allow the driver to raise or lower the brake and throttle pedal assembly for personal comfort.
12. On articulated vehicles, such as tractor/trailers, any delays in applying the trailer brakes is minimized by using a relay valve and a separate _____ _____ on the trailer.
13. A(n) _____ brake works by restricting the flow of exhaust gases through the engine by closing a butterfly valve located in the exhaust manifold.
14. On drum brakes, a drum-style parking brake _____ applies the brake shoes against the drum.
15. Part of the _____ _____ cable is inside a wound steel housing, which allows it to be somewhat flexible, yet noncompressible to guide the cable and hold everything in place.

Labeling

Label the following diagrams with the correct terms.

1. Weight transfer during braking:

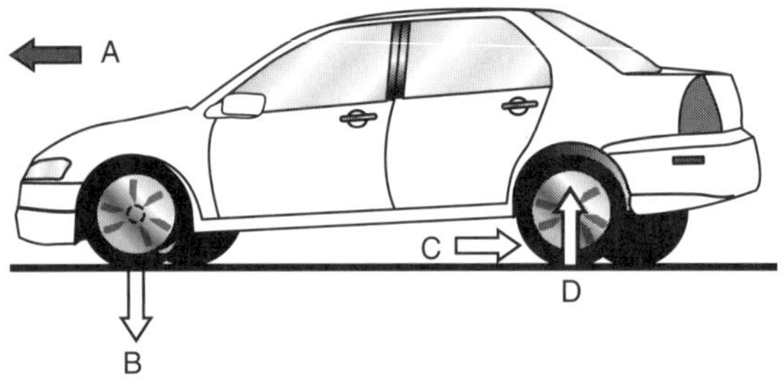

   A. _____
   B. _____
   C. _____
   D. _____

2. Basic types of levers:

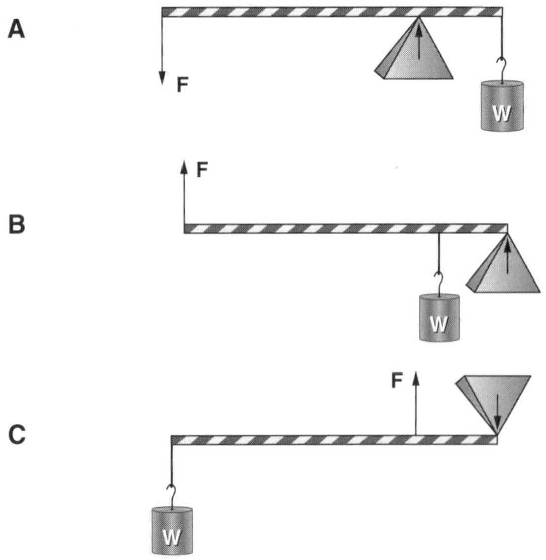

   A. _____
   B. _____
   C. _____

3. The hydraulic brake system:

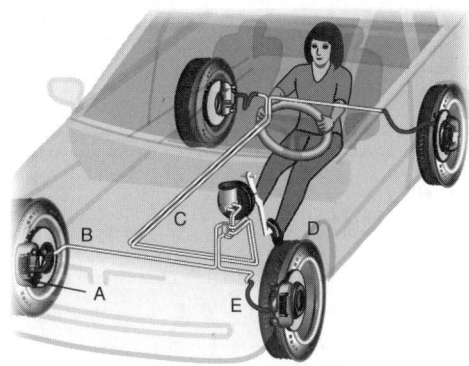

A. _____
B. _____
C. _____
D. _____
E. _____

4. Power brake booster:

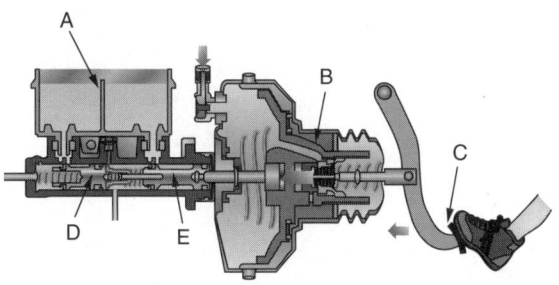

A. _____
B. _____
C. _____
D. _____
E. _____

5. The air brake system:

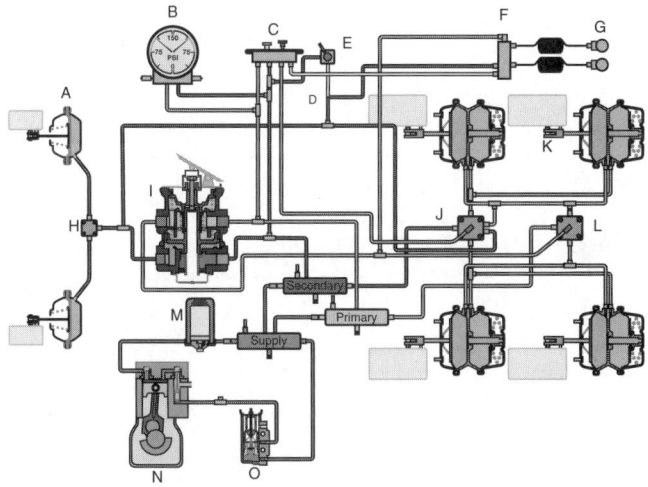

A. _____
B. _____
C. _____
D. _____
E. _____
F. _____
G. _____
H. _____
I. _____
J. _____
K. _____
L. _____
M. _____
N. _____
O. _____

6. Air brake canister:

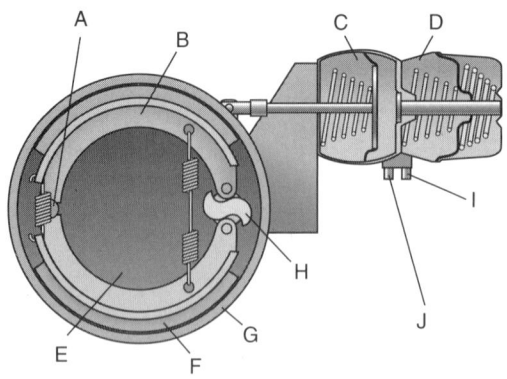

A. _____
B. _____
C. _____
D. _____
E. _____
F. _____
G. _____
H. _____
I. _____
J. _____

7. Electric braking system:

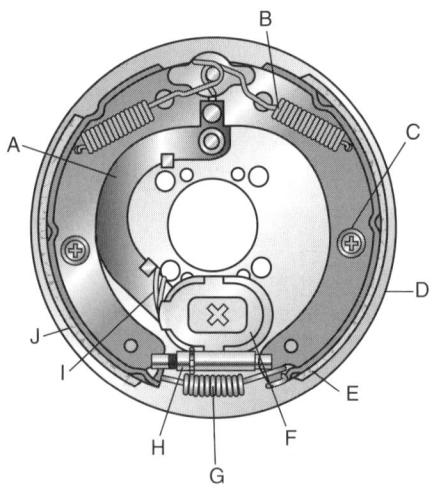

A. _____
B. _____
C. _____
D. _____
E. _____
F. _____
G. _____
H. _____
I. _____
J. _____

8. Top hat design parking brake:

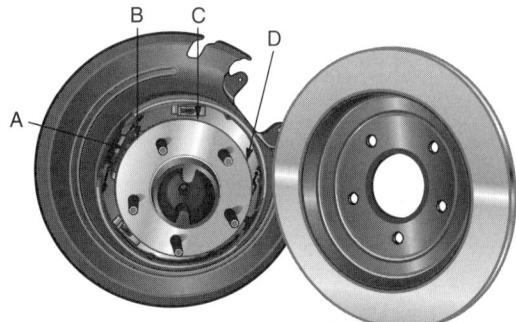

A. _____
B. _____
C. _____
D. _____

Chapter 30　Principles of Braking

9. Drum-style design parking brake:

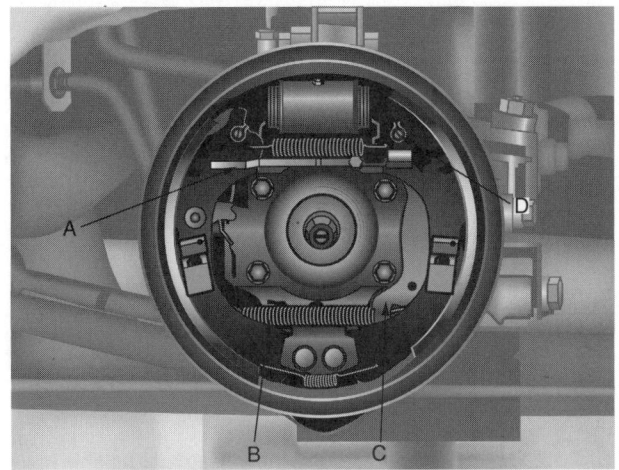

A. _____
B. _____
C. _____
D. _____

10. Electric/hydraulic brake-by-wire system:

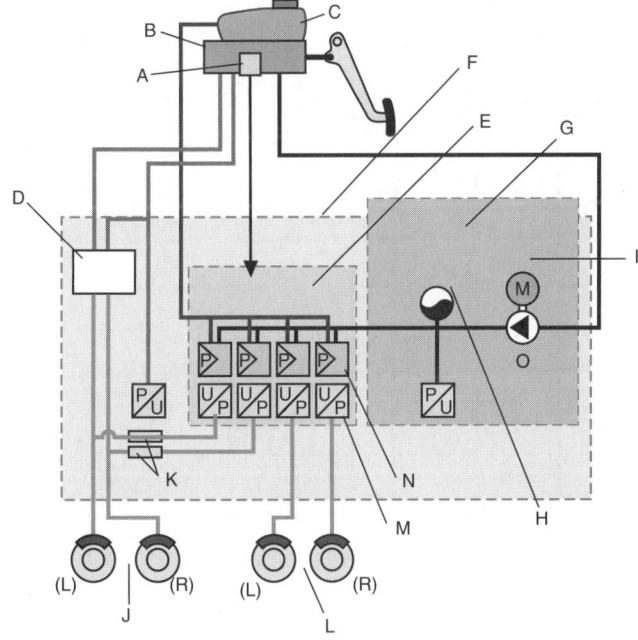

A. _____
B. _____
C. _____
D. _____
E. _____
F. _____
G. _____
H. _____
I. _____
J. _____
K. _____
L. _____
M. _____
N. _____
O. _____

Crossword Puzzle

Use the clues to complete the puzzle.

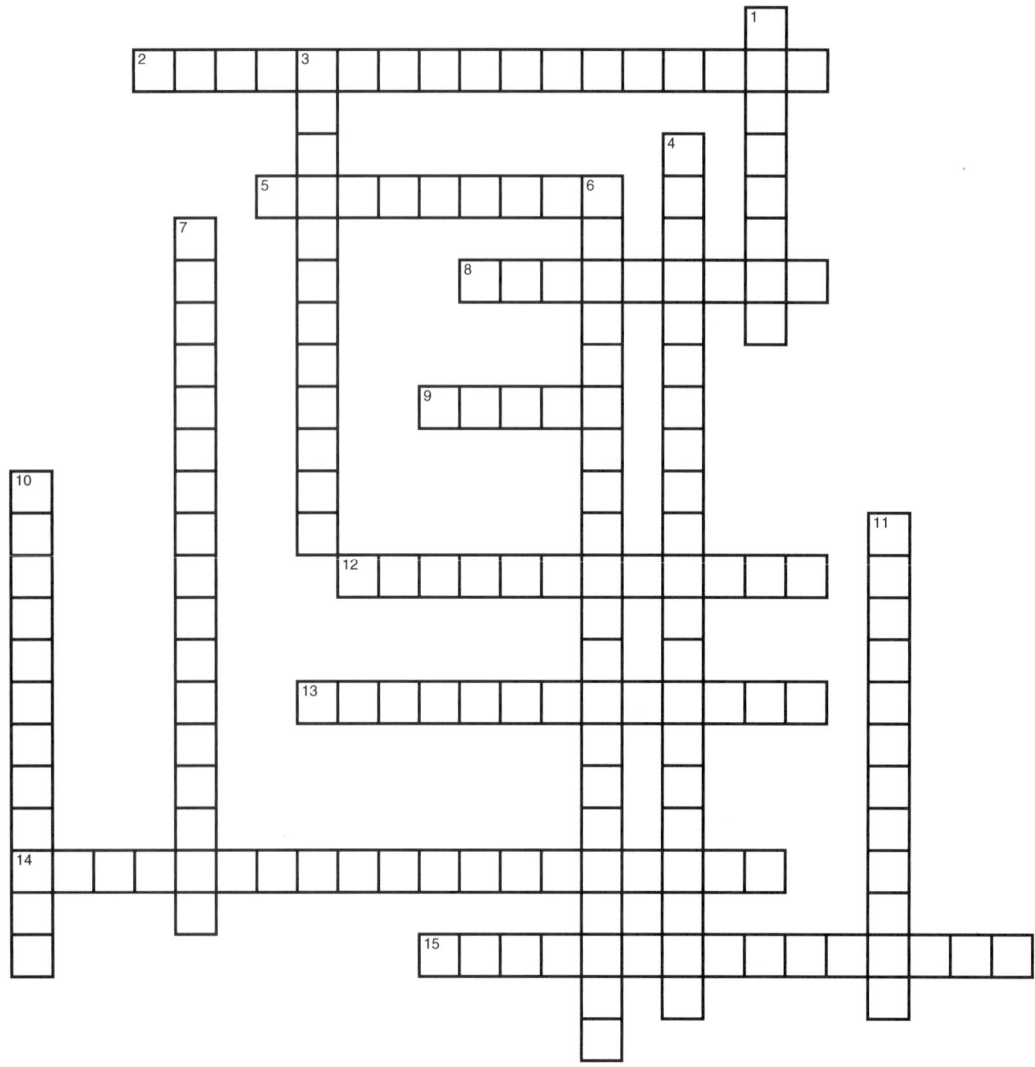

Across

2. A braking system that uses no mechanical connection between the brake pedal and each brake unit. This system uses electrically actuated motors to apply brake force.
5. A braking system that uses a metal band lined with friction material to clamp around the outside of a wheel or drum.
8. Brake fade caused by water-soaked brake linings.
9. A tool that allows the user to move a large load over a small distance at one end by applying a small force over a greater distance from the other end.
12. An increase in a vehicle's speed.
13. Brake fade caused by boiling brake fluid.
14. A braking system that uses compressed air operation on large-diameter diaphragms to provide force to the braking assembly; also called air brakes.
15. The force created by the rotating wheel when the brakes are applied; it causes the brake components to twist the brake support and ultimately the vehicle to the direction of wheel rotation.

Down

1. Brake fade caused by the buildup of heat in braking surfaces, which get so hot they cannot create any additional heat, leading to a loss of friction.
3. A brake system that restricts the flow of exhaust gases through the engine by closing a butterfly valve located in the exhaust manifold. Restricting the exhaust flow causes the engine speed to slow down, slowing the vehicle.
4. The amount of friction between two moving surfaces in contact with each other.
6. A braking system used to provide braking to trailers; the drum brakes are electrically activated in the trailer when the driver applies the brakes on the towing vehicle.
7. A mechanism used to transmit force from the parking brake actuating lever to the brake unit.
10. A brake system that is operated while the vehicle is moving to slow or stop the vehicle.
11. A decrease in the vehicle's speed.

ASE-Type Questions

Read each item carefully, and then select the best response.

_____ 1. Tech A says that regenerative braking converts brake heat into electricity. Tech B says that regenerative braking converts electrical energy into braking energy. Who is correct?
   A. Tech A
   B. Tech B
   C. Both A and B
   D. Neither A nor B

_____ 2. Tech A says that an ABS allows the front wheels to be steered during a panic stop. Tech B says that ABS sensor inputs control wheel speed during brake events. Who is correct?
   A. Tech A
   B. Tech B
   C. Both A and B
   D. Neither A nor B

_____ 3. Tech A says that water-soaked brake shoes will cause brake fade. Tech B says that disc brakes dissipate heat faster than drum brakes. Who is correct?
   A. Tech A
   B. Tech B
   C. Both A and B
   D. Neither A nor B

_____ 4. Tech A says that a brake-by-wire system can start to apply brakes before the driver can step on the brake pedal. Tech B says that brake-by-wire systems use heavy cables to transmit the brake pedal force to the wheel brake units. Who is correct?
   A. Tech A
   B. Tech B
   C. Both A and B
   D. Neither A nor B

_____ 5. Tech A says that tire pressure does not affect braking. Tech B says that heavy vehicle loads increase stopping distance. Who is correct?
   A. Tech A
   B. Tech B
   C. Both A and B
   D. Neither A nor B

_____ 6. Tech A says that light-duty service brakes are applied hydraulically. Tech B says that light-duty parking brakes are applied hydraulically. Who is correct?
   A. Tech A
   B. Tech B
   C. Both A and B
   D. Neither A nor B

_____ 7. Tech A says that the heavier the vehicle, the more stopping power is needed. Tech B says that the faster a vehicle is moving, the more braking power is needed. Who is correct?
   A. Tech A
   B. Tech B
   C. Both A and B
   D. Neither A nor B

_____ 8. Tech A says that kinetic energy is created during braking to stop the vehicle. Tech B says that kinetic energy is converted to heat energy during braking. Who is correct?
   A. Tech A
   B. Tech B
   C. Both A and B
   D. Neither A nor B

_____ 9. Tech A says that the brake pedal uses leverage to multiply foot pressure. Tech B says that when braking hard while moving forward, the vehicle's weight transfers to the rear wheels, increasing their tracion. Who is correct?
   A. Tech A
   B. Tech B
   C. Both A and B
   D. Neither A nor B

_____ 10. Tech A says that friction brakes can fade due to overheating of the brake lining. Tech B says that friction brakes can fade due to overheating of the brake fluid. Who is correct?
   A. Tech A
   B. Tech B
   C. Both A and B
   D. Neither A nor B

Hydraulic and Power Brakes

CHAPTER 31

Tire Tread:
© AbleStock

Chapter Review

The following activities have been designed to help you refresh your knowledge of this chapter. Your instructor may require you to complete some or all of these activities as a regular part of your training program. You are encouraged to complete any activity that your instructor does not assign as a way to enhance your learning.

Matching

Match the following terms with the correct description or example.

- A. Bleeding
- B. Brake fluid
- C. Brake hose
- D. Brake lines
- E. Compensating port
- F. Free play
- G. Inlet port
- H. Input force
- I. Load transfer
- J. Metering valve
- K. Output force
- L. Outlet port
- M. Poppet valve
- N. Primary cup
- O. Primary piston
- P. Quick take-up valve
- Q. Residual pressure valve
- R. Secondary cup
- S. Secondary piston
- T. Working pressure

_____ 1. A brake piston in the master cylinder moved directly by the pushrod or the power booster; it generates hydraulic pressure to move the secondary piston.

_____ 2. The amount of clearance between the brake pedal linkage and the master cylinder piston.

_____ 3. Force that equals the working pressure multiplied by the surface area of the output piston, expressed as pounds, newtons, or kilograms.

_____ 4. In drum brake systems, a valve that maintains pressure in the wheel cylinders slightly above atmospheric pressure so air does not enter the system through the seals in the wheel cylinders.

_____ 5. The process of removing air from a hydraulic braking system.

_____ 6. The force applied to the input piston, measured in either pounds or kilograms.

_____ 7. A valve used on vehicles equipped with older rear drum/front disc brakes to delay application of the front disc brakes until the rear drum brakes are applied. Located in line with the front disc brakes.

_____ 8. Hydraulic fluid that transfers forces under pressure through the hydraulic lines to the wheel braking units.

_____ 9. Links the cylinder to the brake lines.

_____ 10. Made of seamless, double-walled steel, and able to transmit over 1,000 psi (6895 kPa) of hydraulic pressure through the hydraulic brake system.

_____ 11. A piston that is moved by hydraulic pressure generated by the primary piston in the master cylinder.

_____ 12. Connects the reservoir with the space around the piston and between the piston cups in a brake master cylinder.

_____ 13. The pressure within a hydraulic system while the system is being operated.

_____ 14. A valve used to release excess pressure from the larger piston in a quick take-up master cylinder once the brake pads have contacted the brake rotors.

_____ 15. A valve that controls the flow of brake fluid at usually preset pressures.

_____ 16. Connects the brake fluid reservoir to the master cylinder bore when the piston is fully retracted, allowing for expansion and contraction of the brake fluid.

_____ 17. A seal that prevents loss of fluid from the rear of each piston in the master cylinder.

_____ 18. A seal that holds pressure in the master cylinder when force is applied to the piston.

_____ 19. A flexible section of the brake lines between the body and suspension to allow for steering and suspension movement.

_____ 20. Weight transfer from one set of wheels to the other set of wheels during braking, acceleration, or cornering.

Multiple Choice

Read each item carefully, and then select the best response.

_____ 1. What law states that pressure applied to a fluid in one part of a closed system will be transmitted without loss to all other areas of the system?
   A. Newton's law
   B. Pascal's law
   C. Kirchhoff's law
   D. Ohm's law

_____ 2. All of the following are variables to consider when talking about pressure and force in hydraulic systems, *except*:
   A. input force
   B. fluid type
   C. output force
   D. working pressure

_____ 3. Standard brake fluid is _____, which means it absorbs water.
   A. hydrotropic
   B. aerated
   C. hygroscopic
   D. hydraulic

_____ 4. Brake fluids are tested to ensure they meet the standards of quality for _____.
   A. stability
   B. resistance to oxidation
   C. boiling point
   D. all of the above

_____ 5. What type of brake fluid is silicone-based?
   A. DOT 2
   B. DOT 3
   C. DOT 4
   D. DOT 5

_____ 6. When the brake pedal is released quickly, the use of small holes drilled in the piston so that brake fluid from the reservoir can pass through the inlet port and past the edge of the primary cup, thus preventing a vacuum from being created is called _____.
   A. recuperation
   B. compensating
   C. aeration
   D. residual pressure release

_____ 7. The _____ master cylinder is a tandem master cylinder used in divided systems.
   A. single-piston
   B. dual-piston
   C. ABS
   D. quick take-up

_____ 8. What type of flare is sometimes called a bubble flare?
   A. Inverted double flare
   B. International Standards Organization flare
   C. Mushroom flare
   D. DIN flare

Chapter 31   Hydraulic and Power Brakes

_____ 9. Which of the following components is used to modify pressures within the hydraulic braking system?
  A. Proportioning valves
  B. Pressure differential valves
  C. Antilock hydraulic control units
  D. All of the above

_____ 10. All of the following conditions will cause the brake warning light to illuminate, *except*:
  A. Parking brake is engaged
  B. Brake fluid is low
  C. Brake pedal is pressed
  D. Prove out circuit

True/False

If you believe the statement to be more true than false, write the letter "T" in the space provided. If you believe the statement to be more false than true, write the letter "F".

_____ 1. In a closed system, hydraulic pressure is transmitted equally in all directions throughout the system.
_____ 2. Because silicone-based fluid tends to aerate when forced at high pressure through small passages, it is not to be used in any vehicle equipped with antilock brakes.
_____ 3. The outlet port adjusts for changes in the volume of the brake fluid ahead of the piston.
_____ 4. Tandem systems can be split diagonally so one front wheel is paired with the rear wheel on the opposite side of the vehicle in one brake circuit, and vice versa in the other circuit.
_____ 5. Master cylinder reservoirs can be built into the master cylinder housing or can be a separate unit.
_____ 6. Master cylinder reservoirs should be filled all the way to the top to prevent air bubbles from entering the system.
_____ 7. A front-engine rear-wheel drive car has around 40% of its load on its rear wheels and 60% on its front wheels.
_____ 8. While brake hoses are designed to be flexible, they should never be pinched, kinked, or bent tighter than a specified radius.
_____ 9. During heavy braking, master cylinder pressure can reach the poppet valve's crack point.
_____ 10. Adjustable proportioning valves are not recommended for most applications due to the amount of trial and error necessary to set them properly.
_____ 11. The combination valve can combine the pressure differential valve, metering valve, and proportioning valve(s) in one unit.
_____ 12. The brake warning light is activated by a normally closed switch located on the brake pedal assembly.
_____ 13. The vacuum-assisted power booster uses the difference between engine vacuum and atmospheric pressure to increase the force that acts on the master cylinder pistons.
_____ 14. Dual-diaphragm power boosters work on the same principle of operation as the single-diaphragm power booster but are much larger in diameter.
_____ 15. While hydraulic brake boosters use power steering fluid to operate the booster, the master cylinder portion of the system still uses brake fluid.

Fill in the Blank

Read each item carefully, and then complete the statement by filling in the missing word(s).

1. The same _____ applied over different-sized surface areas will produce different levels of force.
2. If brake fluid boils, it turns from a liquid to a _____, which is compressible.
3. Brake fluids are graded against compliance standards set by the United States _____ _____ _____.
4. _____-_____ master cylinders have one piston with two cups: a primary cup and a secondary cup.
5. _____ master cylinders combine two master cylinders within a common housing that share a common cylinder bore.

6. Tandem systems can be split _____ to _____ so the front brakes operate from one circuit and the rear brakes from the other.

7. _____ _____-_____ master cylinders are used on disc brake systems that are equipped with low-drag brake calipers.

8. The _____ _____ uses leverage to multiply the effort from the driver's foot to the master cylinder.

9. An alternative to the diagonal, or X, pattern braking system split used in front-engine, front-wheel drive vehicles is an _____ _____.

10. The _____ _____ flare is created by first flaring the end of the tube outward in a Y shape. Then about half of the flared end is folded inside of itself, leaving a double-thick section of brake line on the flared portion of the Y.

11. Many brake hoses use _____ fittings to connect the hose to the wheel unit.

12. _____ valves reduce brake pressure to the rear wheels when their load is reduced during moderate to severe braking.

13. A _____ _____ valve monitors any pressure difference between the two separate hydraulic brake circuits.

14. During hydraulic braking system bleeding, the pressure differential valve may need to be _____.

15. The less common _____ braking system gets rid of the vacuum booster and replaces it with an electrically driven hydraulic pump.

Labeling

Label the following diagrams with the correct terms.

1. Single piston master cylinder with primary and secondary pumps:

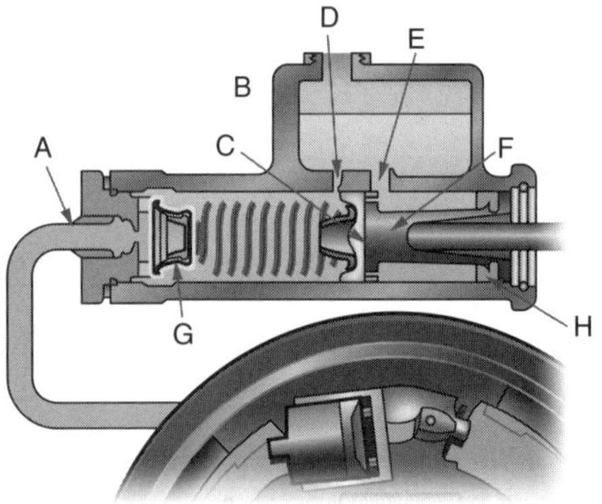

A. _____
B. _____
C. _____
D. _____
E. _____
F. _____
G. _____
H. _____

2. Single piston master cylinder with small holes in the piston to allow for recuperation:

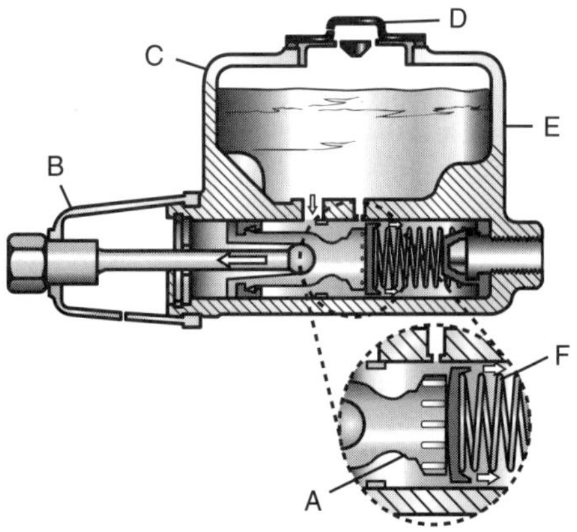

A. _____
B. _____
C. _____
D. _____
E. _____
F. _____

3. Divided hydraulic systems:

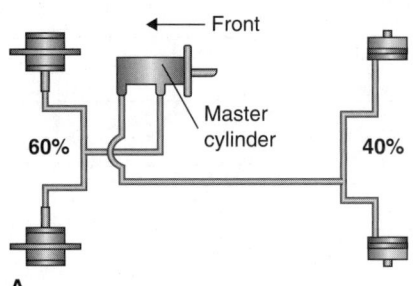

A. _____

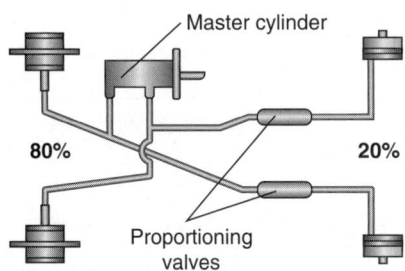

B. _____

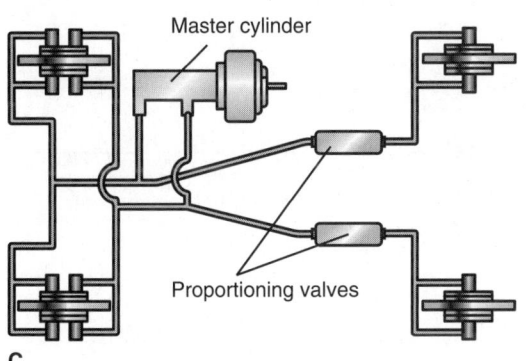

C. _____

4. Flexible brake hose construction:

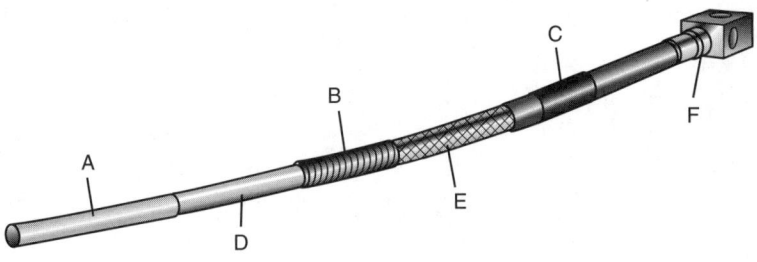

A. _____

B. _____

C. _____

D. _____

E. _____

F. _____

5. Pressure differential valve with a leak in the hydraulic braking system:

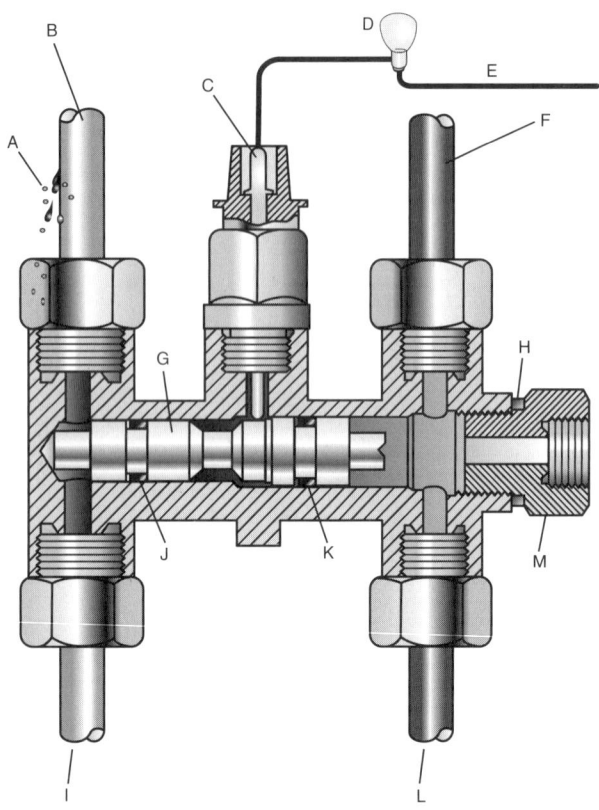

A. _____
B. _____
C. _____
D. _____
E. _____
F. _____
G. _____
H. _____
I. _____
J. _____
K. _____
L. _____
M. _____

6. Combination valve:

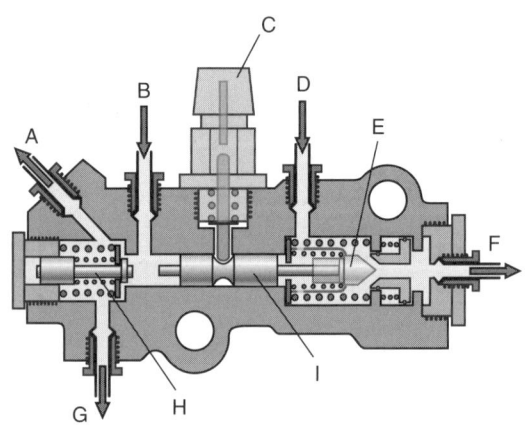

A. _____
B. _____
C. _____
D. _____
E. _____
F. _____
G. _____
H. _____
I. _____

Skill Drills

Test your knowledge of skill drills by filling in the correct words in the photo captions.

1. Selecting, Handling, Storing, and Filling Brake Fluid:

Step 1: _____ the specified type of brake fluid in the appropriate service information. Wipe around the master cylinder reservoir cover to prevent any _____ from entering the system. _____ the reservoir cover.

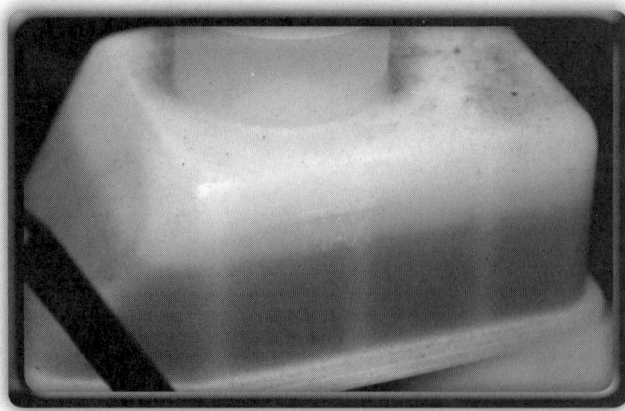

Step 2: Check the fluid level in the reservoir. The fluid should be near the full mark on the _____ of the cylinder or within half an _____ of the top of each _____ if there are no marks.

Step 3: Once you are sure there are no unresolved issues, add the manufacturer's recommended _____ _____ to bring the level to the _____ mark. Replace the cover, and check that it is properly _____. Check for any leaks around the master cylinder. _____ any brake fluid that may have been spilled with fresh clean water.

2. Performing Pressure Bleeding:

Step 1: Prepare the _____ _____, and install it on the vehicle.

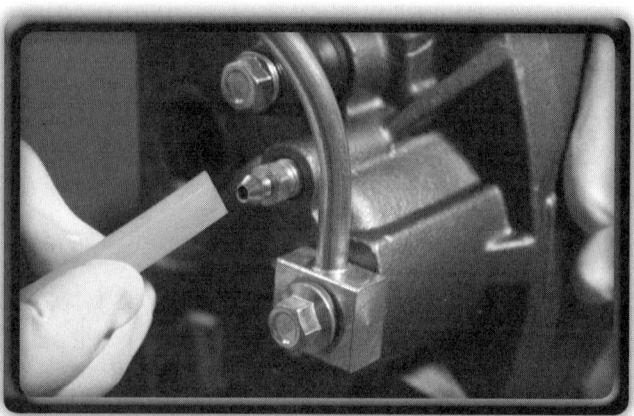

Step 2: Install a clear _____ on the farthest bleeder screw, and open it one-quarter to one-half turn. Observe any old brake fluid and _____ _____ coming out.

Step 3: Close off the _____ _____ when the brake fluid is clear and has no bubbles. Tighten it to the manufacturer's specifications. Repeat this procedure for each of the _____ _____ _____, moving closer to the master cylinder, one wheel at a time.

3. Testing Pedal Free Travel and Checking Power Assist Operation:

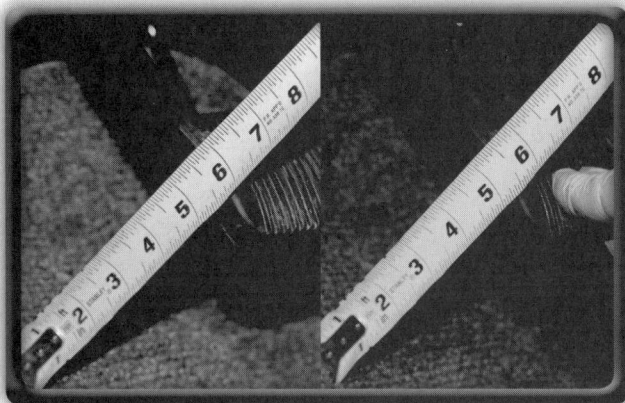

Step 1: To test brake pedal free travel, with the engine off, _____ the brake pedal several times to remove any _____ or hydraulic pressure from the _____ _____. Measure the distance of the brake pedal _____ _____ by depressing the brake pedal by hand until you just feel all of the slack taken up.

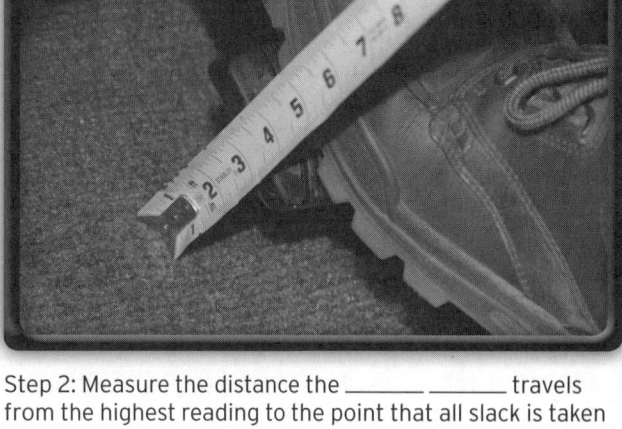

Step 2: Measure the distance the _____ _____ travels from the highest reading to the point that all slack is taken up; compare the findings to the specifications. Check _____ _____ operation by beginning with the vehicle engine off. Apply and release the brake pedal _____ or _____ times to bleed off any vacuum or hydraulic pressure in the power booster.

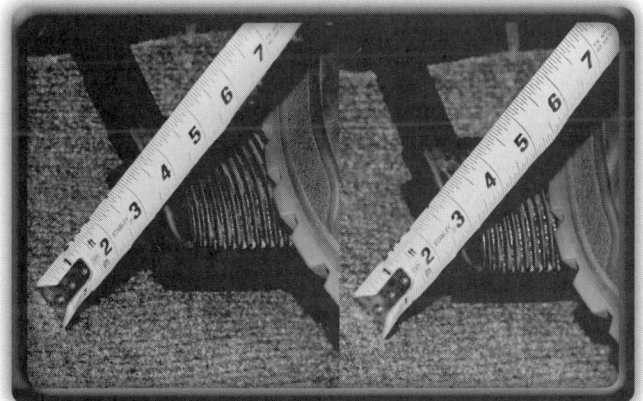

Step 3: Hold the brake pedal down with moderately firm pressure (20–30 lb [9.1–13.6 kg]). Start the engine, and _____ the brake pedal. On vacuum-assisted vehicles, if the pedal drops an _____ or two, the booster is providing boost. If it does not drop, the booster is not providing boost and the following tests will need to be performed to determine the cause of the fault. On some _____-_____ vehicles, when starting the engine with your _____ on the brake pedal, the pedal should rise or fall an inch or so (depending on the vehicle you are working on) if the booster is providing _____.

4. Checking Vacuum Supply to Vacuum-Type Power Booster:

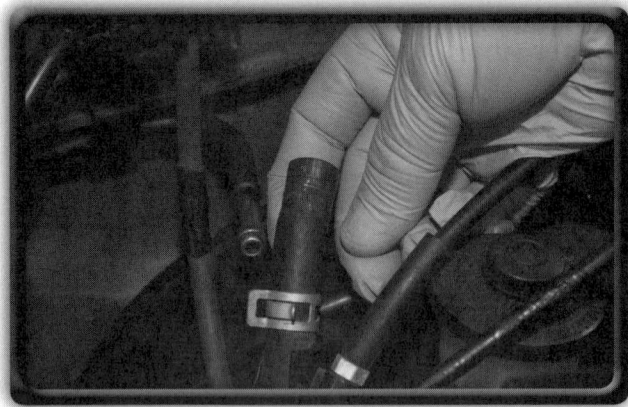

Step 1: With the engine off, remove the _____ _____ from the _____-type booster.

Step 2: Connect a _____ _____ to the vacuum supply end of the _____.

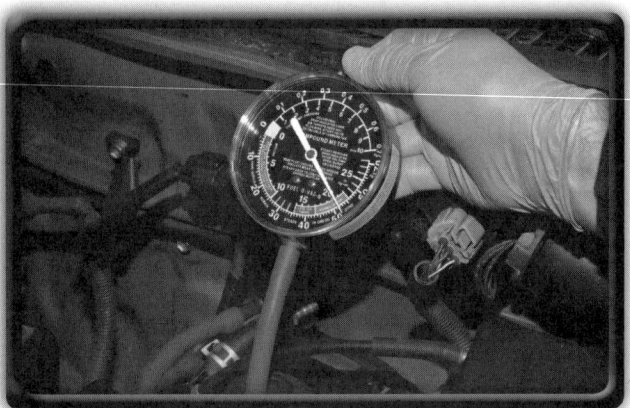

Step 3: Start the engine, and read the vacuum supply _____ to the vacuum-type booster. Vacuum should be _____ than 16 in. Hg (406 mm Hg) on most vehicles. If the reading is insufficient, check for vacuum _____ or _____ in the supply hose or for an improperly _____ engine.

5. Inspecting Brake Lines, Brake Hoses, and Associated Hardware:

Step 1: Safely raise the vehicle on a _____. Trace all brake lines from the master cylinder to each wheel's _____ _____. Inspect the steel brake lines for _____, _____, _____, _____ and _____.

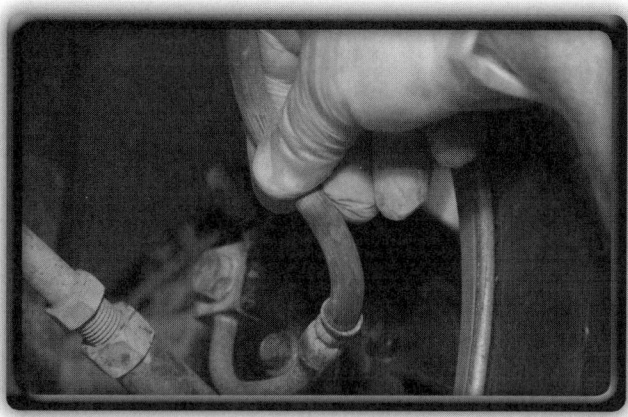

Step 2: Inspect all flexible brake hoses for _____, _____, and _____.

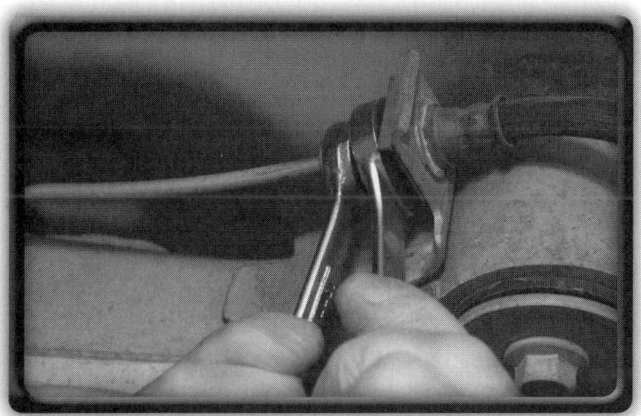

Step 3: _____ any loose _____ and supports.

Crossword Puzzle

Use the clues to complete the puzzle.

Across

3. A brake system in which the left front wheel is hydraulically paired with the right rear wheel and the right front wheel is paired with the left rear. This preserves 50% of the braking capability if one of the brake circuits begins to leak.

7. A valve used on vehicles equipped with older rear drum/front disc brakes to delay application.

9. Weight transfer from one set of wheels to the other set of wheels during braking, acceleration, or cornering.

12. Connects the reservoir with the space around the piston and between the piston cups in a brake master cylinder.

13. A seal that prevents loss of fluid from the rear of each piston in the master cylinder.

14. The force applied to the input piston, measured in either pounds or kilograms.

15. The process of removing air from a hydraulic braking system.

Down

1. A seal that holds pressure in the master cylinder when force is applied to the piston.

2. The pressure within a hydraulic system while the system is being operated.

4. A master cylinder that has two pistons that operate separate braking circuits so if a leak develops in one circuit, the other circuit can still operate.

5. Connects the brake fluid reservoir to the master cylinder bore when the piston is fully retracted, allowing for expansion and contraction of the brake fluid.

6. A bleeding method where one person manually operates the brake pedal while the other person opens and closes the bleeder screws on the wheel brake units to allow the air and old brake fluid to be pushed out.
8. The tendency to create air bubbles in a fluid.
10. A valve that controls the flow of brake fluid at usually preset pressures.
11. A substance that attracts and absorbs water (e.g., brake fluid).

ASE-Type Questions

Read each item carefully, and then select the best response.

_____ 1. Tech A says that hydraulic pressure is applied equally in all directions throughout a closed system. Tech B says that air in the hydraulic system will cause the brake pedal to be spongy. Who is correct?
   A. Tech A
   B. Tech B
   C. Both A and B
   D. Neither A nor B

_____ 2. Tech A says that a brake pedal that does not return all the way will cause the brake warning light on the instrument panel to stay on. Tech B says that a brake pedal that does not return will cause the brake lights to stay on. Who is correct?
   A. Tech A
   B. Tech B
   C. Both A and B
   D. Neither A nor B

_____ 3. Tech A says that brake fluid should periodically be checked for excessive moisture content. Tech B says that brake fluid should be replaced every 12,000 miles. Who is correct?
   A. Tech A
   B. Tech B
   C. Both A and B
   D. Neither A nor B

_____ 4. Tech A says that DOT 5 brake fluid should be used in all vehicles today because it is silicone based and will not absorb water. Tech B says that mixing DOT 4 and DOT 3 is not recommended because of different boiling points, but both are hydroscopic. Who is correct?
   A. Tech A
   B. Tech B
   C. Both A and B
   D. Neither A nor B

_____ 5. Tech A says that the secondary piston in a master cylinder is operated by hydraulic force. Tech B says that the brake pedal return spring returns the master cylinder pistons to their original position. Who is correct?
   A. Tech A
   B. Tech B
   C. Both A and B
   D. Neither A nor B

_____ 6. Tech A says that the low level brake fluid switch on a master cylinder will turn on the brake warning light when the system is low on fluid. Tech B says that the low level switch also monitors the condition of the fluid and will activate the warning light when the brake fluid needs to be replaced. Who is correct?
   A. Tech A
   B. Tech B
   C. Both A and B
   D. Neither A nor B

_____ 7. Tech A says that the metering valve controls pressure to the rear brakes. Tech B says that the proportioning valve controls pressure to the front brakes. Who is correct?
   A. Tech A
   B. Tech B
   C. Both A and B
   D. Neither A nor B

_____ 8. Tech A says that the vacuum booster uses vacuum and atmospheric pressure to multiply the driver's foot pressure applied to the master cylinder push rod. Tech B says that the vacuum booster increases the vacuum in the brake system. Who is correct?
   A. Tech A
   B. Tech B
   C. Both A and B
   D. Neither A nor B

_____ 9. Tech A says that bench bleeding a master cylinder will prevent having to bleed air from the brake lines during replacement. Tech B says that bleeding the master cylinder on the vehicle is preferred to verify brake pedal and booster operation. Who is correct?
   A. Tech A
   B. Tech B
   C. Both A and B
   D. Neither A nor B

_____ 10. Tech A says that a faulty vacuum booster can affect engine operation. Tech B says that steel brake line can be replaced with a copper line, since it is easier to bend into shape. Who is correct?
   A. Tech A
   B. Tech B
   C. Both A and B
   D. Neither A nor B

Disc Brake System

CHAPTER 32

Tire Tread:
© AbleStock

Chapter Review

The following activities have been designed to help you refresh your knowledge of this chapter. Your instructor may require you to complete some or all of these activities as a regular part of your training program. You are encouraged to complete any activity that your instructor does not assign as a way to enhance your learning.

Matching

Match the following terms with the correct description or example.

- A. Backing plate
- B. Bearing races
- C. Bonded linings
- D. Brake booster
- E. Brake wash station
- F. Caliper
- G. Dial indicator
- H. Drawing-in method
- I. Electronic control module
- J. Independent rear suspension
- K. Lateral runout
- L. Low-drag caliper
- M. Off-car brake lathe
- N. Parallelism
- O. Pushrod
- P. Riveted linings
- Q. Rotor
- R. Sliding or floating caliper
- S. Steering knuckle
- T. Ventilated rotor

_____ 1. A method for replacing wheel studs that uses the lug nut to draw the wheel stud into the hub or flange.

_____ 2. Also called warpage, the side-to-side movement of the rotor surfaces as the rotor turns.

_____ 3. A piece of equipment designed to safely clean brake dust from drum and disc brake components.

_____ 4. A type of brake caliper that only has piston(s) on the inboard side of the rotor. The caliper is free to slide or float, thus pulling the outboard brake pad into the rotor when braking force is applied.

_____ 5. A caliper designed to maintain a larger brake pad-to-rotor clearance by retracting the pistons farther than normal.

_____ 6. Brake linings riveted to the brake pad backing plate with metal rivets and used on heavier-duty or high-performance vehicles.

_____ 7. Brake linings that are essentially glued to the brake pad backing plate; more common on light-duty vehicles.

_____ 8. A type of brake rotor with passages between the rotor surfaces that are used to improve heat transfer to the atmosphere.

_____ 9. A hydraulic device that uses pressure from the master cylinder to apply the brake pads against the rotor.

_____ 10. A vacuum or hydraulically operated device that increases the driver's braking effort.

_____ 11. A type of suspension system where each rear wheel is capable of moving independently of the other.

_____ 12. A metal plate to which the brake lining is fixed.

_____ 13. A device that connects the front wheel to the suspension; it pivots on the top and bottom, thus allowing the front wheels to turn.

_____ 14. A tool used to machine (refinish) drums and rotors after they have been removed from the vehicle.

_____ 15. Tool used to measure the lateral runout of the rotor.

_____ 16. A mechanism used to transmit force from the brake pedal to the master cylinder.

_____ 17. Hardened metal surfaces that roller or ball bearings fit into when a bearing is properly assembled.

_____ 18. The main rotating part of a disc brake system.

_____ 19. Also called thickness variation; both surfaces of the rotor should be perfectly parallel to each other so brake pulsations do not occur.

_____ 20. A computer that receives signals from input sensors, compares that information with preloaded software, and sends an appropriate command signal to output devices; used to manage the antilock brake system.

Multiple Choice

Read each item carefully, and then select the best response.

_____ 1. In high-performance vehicles, the _____ are made from composite materials, ceramics, or carbon fiber; otherwise, they are usually made of cast iron.
   A. brake pads
   B. calipers
   C. rotors
   D. all of the above

_____ 2. Disc brake caliper assemblies are bolted to the _____.
   A. axle housing
   B. wheel hub
   C. steering knuckle
   D. either A or C

_____ 3. Disc brake caliper pistons are sealed by a stationary square section sealing ring, also called a _____.
   A. square cut O-ring
   B. caliper gasket
   C. square-to-round gasket
   D. square oil seal

_____ 4. Manufacturers have dealt with the corrosion issue by making caliper pistons out of _____, which does not corrode or rust.
   A. aluminum
   B. carbon fiber
   C. phenolic resin
   D. rubber

_____ 5. The backing plate has _____ that correctly positions the pad in the caliper assembly and helps the backing plate maintain the proper position to the rotor.
   A. rivets
   B. tabs
   C. bolts
   D. lugs

_____ 6. The amount of friction between two surfaces is expressed as a ratio and is called the _____.
   A. sliding resistance
   B. coefficient of friction
   C. friction ratio
   D. drag factor

_____ 7. Today, brakes are manufactured from a variety of different materials including all of the following, *except*:
   A. kevlar
   B. semimetallic materials
   C. asbestos
   D. ceramic materials

_____ 8. A spring steel _____ mounted to the brake pad may be used to notify the driver that the brake pads are worn to their minimum limit.
   A. rivet
   B. scratcher
   C. needle
   D. squealer

Chapter 32 Disc Brake System

_____ 9. Which type of rotors are less expensive and usually found on smaller vehicles?
   A. Solid
   B. Ventilated
   C. Stainless steel
   D. Phenolic resin

_____ 10. Which type of parking brake is engaged by pushing a button on the dash?
   A. Top hat
   B. Integrated mechanical
   C. Electric
   D. Automatic

True/False

If you believe the statement to be more true than false, write the letter "T" in the space provided. If you believe the statement to be more false than true, write the letter "F".

_____ 1. The hub can be part of the brake rotor or a separate assembly that the rotor slips over and is bolted to by the lug nuts.

_____ 2. Disc brake pads require much lower application pressures to operate than drum brake shoes because they are self-energizing.

_____ 3. The sliding or floating caliper has brake pads located on each side of the rotor, but all of the pistons are only on one side, usually the inside of the rotor.

_____ 4. Bonded brake linings are less susceptible to failure under high temperatures.

_____ 5. The lower the edge code letter, the less friction the material has, and the harder the brake pedal must be applied to achieve a given amount of stopping power.

_____ 6. Society of Automotive Engineers standards assure that an EE-rated lining from one manufacturer will have the same braking characteristics as an EE-rated lining from another manufacturer.

_____ 7. Technicians can apply a high-temperature liquid rubber compound to the back of the brake pad that stays flexible, absorbs brake pad vibrations, and helps reduce brake noise.

_____ 8. Lateral runout tends to move the caliper pistons in the same direction as each other, so brake fluid is not pushed back to the master cylinder.

_____ 9. Solid rotors are used to improve heat transfer to the atmosphere.

_____ 10. Most disc brake rotors are stamped with the manufacturer's minimum thickness specification.

Fill in the Blank

Read each item carefully, and then complete the statement by filling in the missing word(s).

1. The purpose of the _____ _____ system is to provide an effective means to slow the vehicle under a variety of conditions in an acceptable distance and manner.

2. Calipers use hydraulic pressure from the _____ _____ to apply the brake pads.

3. On rear-wheel drive vehicles, the _____ is mounted onto the driving axle or hub and may be held in place by the wheel.

4. A(n) _____ _____ on the top of the piston bore allows for the removal of air within the disc brake system as well as helping with performing routine brake fluid changes.

5. _____-_____ calipers are designed to maintain a larger brake pad-to-rotor clearance by retracting the pistons a little bit farther.

6. Floating calipers are mounted in place by _____ _____ and _____.

7. The Society of Automotive Engineers has adopted letter codes to rate brake lining materials' coefficient of friction. The rating is written on the edge of the friction linings and is called the _____ _____.

8. Adding brake pad _____ and _____ to the brake pads help cushion the brake pad and absorb some of the vibration.

Fundamentals of Automotive Technology Student Workbook

9. Incorporating _____ tangs on the brake pad backing plate allow technicians to crimp the tangs so they are more firmly mounted in the caliper.

10. Most _____ _____ are mechanically applied by use of a cable and ratcheting lever assembly.

Labeling

Label the following diagrams with the correct terms.

1. Disc versus drum brakes:

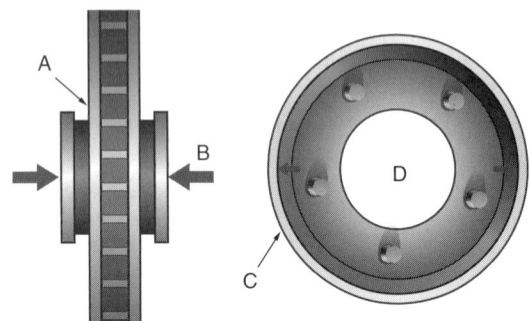

A. _____

B. _____

C. _____

D. _____

2. The master cylinder converts the pedal force into hydraulic pressure:

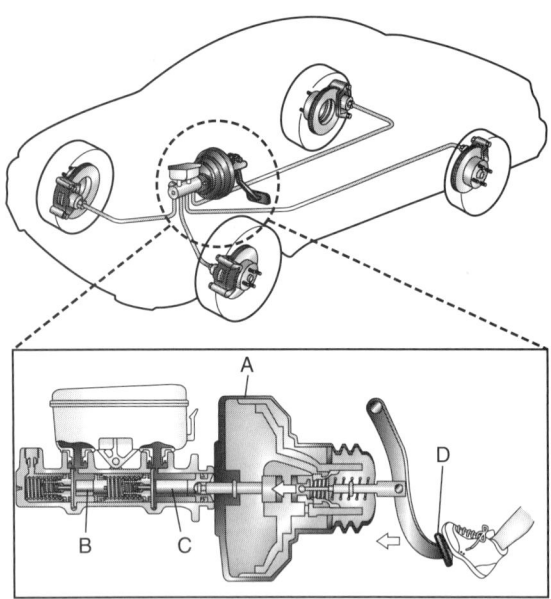

A. _____

B. _____

C. _____

D. _____

3. Identify the disc brake tools:

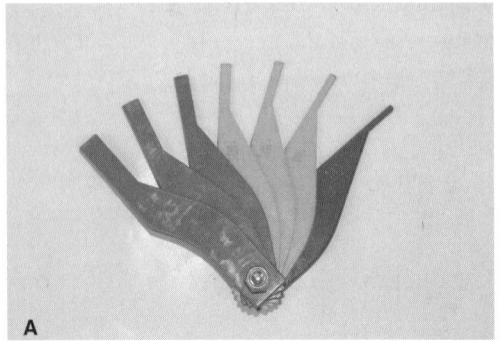

A. _____

B. _____

Chapter 32   Disc Brake System

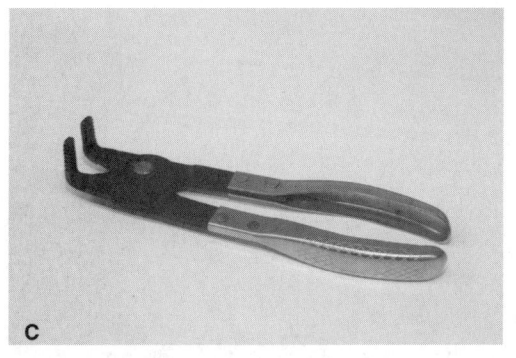

C. _____

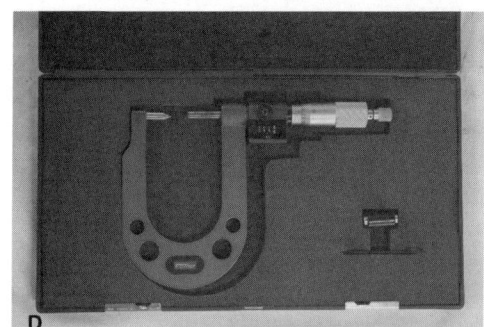

D. _____

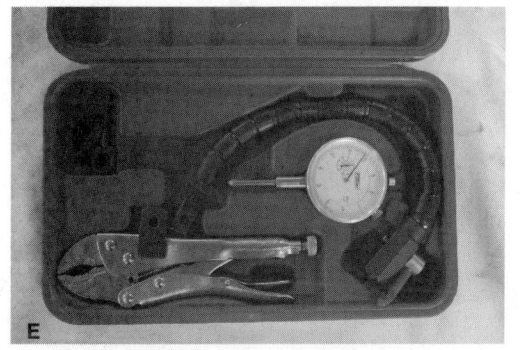

E. _____

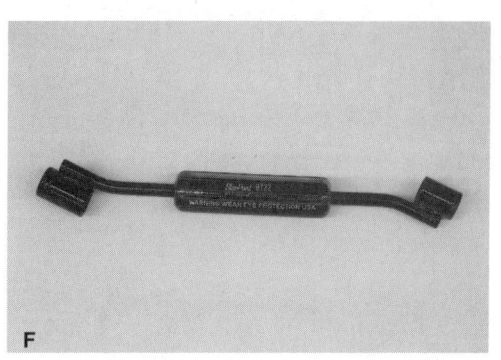

F. _____

G. _____

H. _____

I. _____

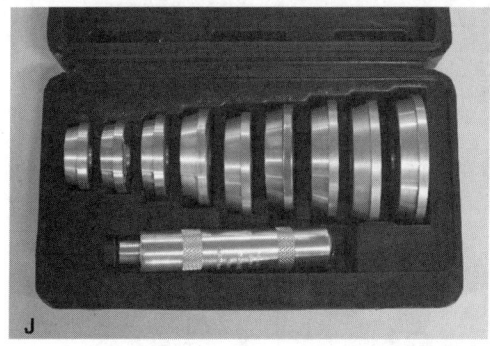

J. _____

Skill Drills

Place the skill drill steps in the correct order.

1. Removing and Inspecting Calipers:

_____ **A.** Push the caliper pistons back into their bores slightly. While many use a screwdriver as shown, a pry bar is a safer choice.

_____ **B.** Inspect the caliper for leaks or damage, including the piston dust boot. Determine any necessary actions.

_____ **C.** Use a brake pedal holding tool to slightly apply the brakes and block off the compensating ports in the master cylinder to avoid excess fluid leakage.

_____ **D.** Research the procedure for removing the caliper in the appropriate service information. Loosen the bleeder screws slightly, and then retighten them.

_____ **E.** Remove the brake line or hose from the caliper. Be careful not to lose the sealing rings.

_____ **F.** If the caliper is being rebuilt or a new caliper will be installed, it is good practice to flush the old brake fluid from the system at this time.

Chapter 32   Disc Brake System   345

_____ **G.** Remove the caliper assembly from its mountings.

2. Disassembling Calipers:

_____ **A.** Clean all of the caliper parts following the service manual procedure.

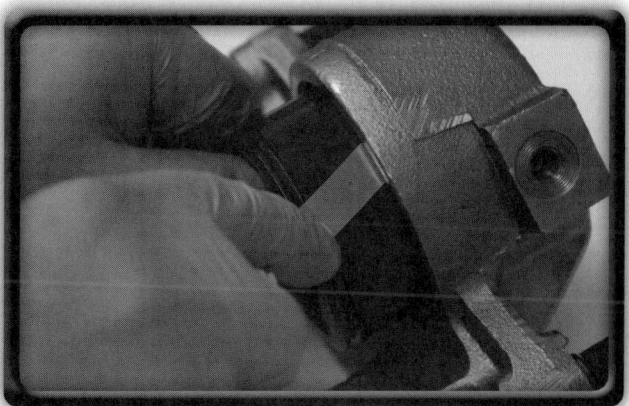

_____ **B.** Measure the caliper bore-to-piston clearance with a feeler gauge, and compare to specifications. Determine any necessary actions.

_____ **C.** Inspect each of the parts for damage, rust, and wear.

_____ **D.** Disassemble the caliper, following the service manual procedure.

3. Inspecting and Measuring Disc Brake Rotors:

_____ **A.** Measure the thickness of the rotor in a minimum of five to eight places around the face of the rotor. Calculate the maximum thickness variation, and compare to specifications.

_____ **B.** Slowly rotate the rotor to find the highest spot on the rotor. Read the dial indicator showing maximum runout.

_____ **C.** Inspect the rotor for hard spots or hot spots, scoring, cracks, and damage.

_____ **D.** Keep turning the rotor to make sure the dial indicator does not read below zero. If it does, rezero the dial caliper on the lowest spot. Keep turning the rotor to find the highest spot and reread the dial indicator. Compare all of your readings to the specifications and determine if the rotor is fit for service, is machinable, or needs to be replaced.

_____ **E.** Research the procedure and specifications for inspecting the rotor. If you have not already done so, remove the caliper assembly, brake pads, and any hardware. Clean the rotor with approved asbestos removal equipment.

_____ **F.** Set up a dial indicator to measure lateral runout. Rotate the rotor and find the lowest spot on the rotor, then zero the dial indicator.

_____ **G.** Measure the rotor thickness at the deepest groove or thinnest part of the rotor and compare to specifications.

4. Refinishing Hubless Rotors on Vehicle:

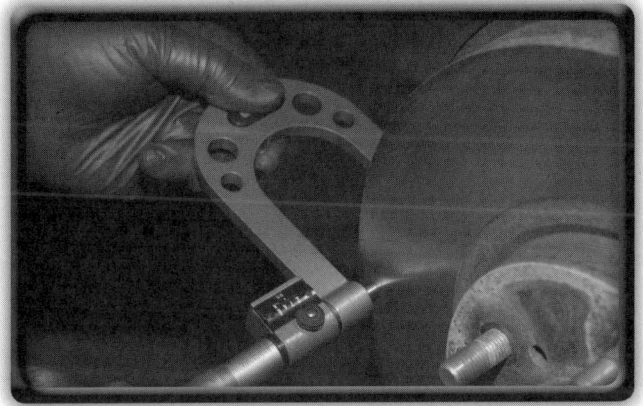

_____ **A.** Remeasure the rotor thickness to determine if the rotor is above minimum thickness specifications.

_____ **B.** Engage the automatic feed, and watch for proper machining action. If necessary, repeat this step until all damaged surface areas have been removed on both sides of the rotor.

_____ **C.** Install the antichatter device, if specified.

_____ **D.** Adjust the cutting bits, and cut off any lip at the edge of the rotor.

_____ **E.** Research the brake lathe manufacturer's procedure for properly refinishing the rotor. Mount the on-car brake lathe to the rotor after cleaning the rust and dirt from between the rotor and hub or adjusting the wheel bearing so there is no end play.

_____ **F.** Make sure the cutting bits will not contact the rotor face, and move the cutting head toward the inner diameter of the rotor face. Set the cutting bits to the proper cutting depth for machining.

_____ **G.** If necessary, perform a finish cut on the rotor.

_____ **H.** Perform the runout calibration on the brake lathe.

5. Installing Wheels, Torquing Lug Nuts, and Making Final Checks:

_____ **A.** Carefully run all of the lug nuts down so they are seated in the wheel.

_____ **B.** Once all of the lug nuts have been torqued, go around them again, this time in a circular pattern to ensure that you did not miss any in the previous pattern.

_____ **C.** Check the brake fluid level in the master cylinder reservoir. Start the vehicle, and check the brake pedal for proper feel and height. Check the parking brake for proper operation. Also inspect the system for any brake fluid leaks and loose or missing fasteners.

_____ **D.** Use a torque wrench to tighten each lug nut to the proper torque in the proper sequence.

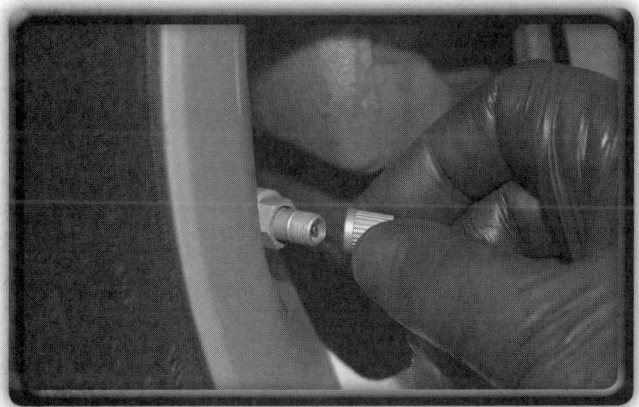

_____ **E.** If the vehicle was equipped with hubcaps and valve stem caps, reinstall them.

_____ **F.** Lower the vehicle so the tires are partially on the ground to keep them from turning while tightening the lug nuts.

_____ **G.** Start the lug nuts on the wheel studs, being careful to match up the surfaces.

Crossword Puzzle

Use the clues in the column to complete the puzzle.

Across

4. An insert with an inner bearing surface that is fitted into a hole in an object, allowing the object to rotate or slide on a pin or shaft.
5. Threaded fasteners that are pressed into the wheel hub flange and used to bolt the wheel onto the vehicle.
7. A tool used to push pistons back into the caliper bore on nonintegrated parking brakes.
8. A method for replacing wheel studs that uses a press to force the wheel stud into the flange until it bottoms out.
11. A screw that allows air and brake fluids to be bled out of a hydraulic brake system when it is loosened and seals the brake fluid in when it is tightened.
13. The main rotating part of a disc brake system.
14. A thin, spring steel wear indicator that is fixed to the backing plate of the brake pad; it emits a high-pitched squeal when the brakes are applied if the brake pads have become too thin.
15. A tool used to grip caliper pistons while removing them.

Down

1. A flange that is shaped to assist with aligning objects on other objects.
2. A type of brake bolted firmly to the steering knuckle or axle housing, having at least one piston on both sides of the rotor.
3. Pins that allow the caliper to move in and out as the brakes operate and as the brake pads wear.
6. Small tabs on the brake pad backing plate that are crimped on to the caliper, creating a secure fit and reducing noise.

9. A code printed on the edge of a friction lining that describes its coefficient of friction.
10. Tool used to measure the lateral runout of the rotor.
12. A type of brake rotor made of solid metal.

ASE-Type Questions

Read each item carefully, and then select the best response.

_____ 1. Tech A says that disc brakes operate on the principle of friction. Tech B says that disc brakes operate on the principle of regeneration. Who is correct?
   A. Tech A
   B. Tech B
   C. Both A and B
   D. Neither A nor B

_____ 2. Tech A says that some vehicles use fixed calipers. Tech B says that some vehicles use sliding/floating calipers. Who is correct?
   A. Tech A
   B. Tech B
   C. Both A and B
   D. Neither A nor B

_____ 3. Tech A says that disc brakes require higher application pressures than drum brakes. Tech B says that disc brakes are self-energizing. Who is correct?
   A. Tech A
   B. Tech B
   C. Both A and B
   D. Neither A nor B

_____ 4. Tech A says that fixed calipers use one or more pistons only on one side of the rotor. Tech B says that sliding/fixed calipers use one or more pistons on both sides of the rotor. Who is correct?
   A. Tech A
   B. Tech B
   C. Both A and B
   D. Neither A nor B

_____ 5. Tech A says that calipers use a round section O-ring to seal each piston. Tech B says that calipers use a square section O-ring to seal each piston. Who is correct?
   A. Tech A
   B. Tech B
   C. Both A and B
   D. Neither A nor B

_____ 6. Tech A says that pistons plated with chrome resist rust. Tech B says that pistons made of phenolic resin do not corrode and rust. Who is correct?
   A. Tech A
   B. Tech B
   C. Both A and B
   D. Neither A nor B

_____ 7. Tech A says that some vehicles are equipped with a spring-steel brake pad wear indicator that drags on the rotor when the lining thickness is low. Tech B says that some vehicles are equipped with an electric brake pad wear sensor that activates a warning on the dash. Who is correct?
   A. Tech A
   B. Tech B
   C. Both A and B
   D. Neither A nor B

_____ 8. Tech A says that rotors should be measured for thickness variation (parallelism). Tech B says that rotors should be measured for lateral runout. Who is correct?
   A. Tech A
   B. Tech B
   C. Both A and B
   D. Neither A nor B

_____ 9. Tech A says that rotors that are too thin cannot handle as much heat and will experience brake fade sooner. Tech B says that brake pedal pulsation is the result of air in the hydraulic system. Who is correct?
   A. Tech A
   B. Tech B
   C. Both A and B
   D. Neither A nor B

_____ 10. Tech A says that a micrometer is used to measure rotor thickness variation. Tech B says that a micrometer is used to measure rotor lateral runout. Who is correct?
   A. Tech A
   B. Tech B
   C. Both A and B
   D. Neither A nor B

Drum Brake System

CHAPTER 33

Tire Tread: © AbleStock

Chapter Review

The following activities have been designed to help you refresh your knowledge of this chapter. Your instructor may require you to complete some or all of these activities as a regular part of your training program. You are encouraged to complete any activity that your instructor does not assign as a way to enhance your learning.

Matching

Match the following terms with the correct description or example.

- A. Anchor pin
- B. Automatic brake self-adjuster
- C. Backing plate
- D. Brake drum
- E. Duo-servo drum brake system
- F. Hold-down springs
- G. Hold-down spring tool
- H. Leading/trailing shoe drum brake system
- I. Parking brake mechanism
- J. Self-energizing
- K. Servo action
- L. Specialty springs
- M. Springs and clips
- N. Twin leading shoe drum brake system
- O. Wheel cylinder

_____ 1. Type of brake shoe arrangement where one shoe is positioned in a leading manner and the other shoe in a trailing manner.

_____ 2. A drum brake design where one brake shoe, when activated, applies an increased activating force to the other brake shoe, in proportion to the initial activating force; further enhances the self-energizing feature of some drum brakes.

_____ 3. Springs that hold the brake shoes against the backing plate.

_____ 4. A mechanism that operates the brake shoes or pads to hold the vehicle stationary when the parking brake is applied.

_____ 5. Brake shoe arrangement in which both brake shoes are self-energizing in the forward direction.

_____ 6. A short, wide, hollow cylinder that is capped on one end and bolted to a vehicle's wheel; it has an inner friction surface that the brake shoe is forced against.

_____ 7. A hydraulic cylinder with one or two pistons, seals, dust boots, and a bleeder screw that pushes the brake shoes into contact with the brake drum to slow or stop the vehicle.

_____ 8. Springs used to return links and levers on the parking brake system or the self-adjuster mechanism.

_____ 9. A system on drum brakes that automatically adjusts the brakes to maintain a specified amount of running clearance between the shoes and drum.

_____ 10. A system that uses servo action in both the forward and reverse direction.

_____ 11. Various devices that hold the brake shoes in place or return them to their proper place.

_____ 12. A tool used for removing and installing hold-down springs.

_____ 13. A component of the backing plate that takes all of the braking force from the brake shoes.

_____ 14. A stamped steel plate, bolted to the steering or suspension, which supports the wheel cylinder, brake shoes, and other hardware.

_____ 15. The property of drum brakes that assists the driver in applying the brakes; when brake shoes come into contact with the moving drum, the friction tends to wedge the shoes against the drum, thus increasing the braking force.

Multiple Choice

Read each item carefully, and then select the best response.

_____ 1. All of the following are main components of the drum brake system, *except*:
   A. wheel cylinder
   B. brake shoes
   C. rotor
   D. parking brake mechanism

_____ 2. In drum brake systems, when the brake pedal is depressed, a _____ transfers the force to a hydraulic master cylinder.
   A. linkage
   B. pushrod
   C. cable
   D. brake line

_____ 3. All of the following are types of drum brake systems, *except*:
   A. twin leading shoe
   B. leading/trailing shoe
   C. twin trailing shoe
   D. duo-servo

_____ 4. Brake drums are usually made out of _____ due to its ability to withstand high temperatures, absorb a lot of heat, and maintain its shape.
   A. cast iron
   B. stainless steel
   C. chrome
   D. phenolic resin

_____ 5. The _____ must be able to take all of the braking force when the brakes are applied, so it must be strong and firmly attached to the backing plate.
   A. wheel cylinder
   B. anchor pin
   C. brake drum
   D. primary shoe

_____ 6. A reduction in the coefficient of friction capability as the heat in brake pads and linings builds up is called _____.
   A. bonding fail
   B. slippage
   C. brake fade
   D. wear

_____ 7. Which of the following can cause drum brakes to make a groaning noise?
   A. Excessive brake dust
   B. Overheating
   C. Fluid leak
   D. Worn brake pads

_____ 8. What type of brake springs are generally quite stiff, making them a challenge to install?
   A. Specialty springs
   B. Hold-down springs
   C. Torsion springs
   D. Return springs

_____ 9. What type of brake springs can be of all different shapes and sizes and can be used to push or pull components into their proper position?
   A. Torsion springs
   B. Return springs
   C. Specialty springs
   D. Hold-down springs

_____ 10. What type of self-adjuster uses two toothed pieces held in contact with each other by spring pressure that can slide over each other in one direction, but hold in the other direction?
   A. Star wheel type
   B. Ratcheting-style
   C. Servo-style
   D. Cable-style

True/False

If you believe the statement to be more true than false, write the letter "T" in the space provided. If you believe the statement to be more false than true, write the letter "F".

_____ 1. The master cylinder converts brake pedal force into hydraulic pressure.
_____ 2. Trailing shoes are self-energizing.
_____ 3. Brake drums provide the rotating friction surface that the brake lining contacts.
_____ 4. Hubless-style drums have a one-piece integrated hub/drum assembly.
_____ 5. Cylinder bores on aluminum wheel cylinders are usually honed to help them resist corrosion.
_____ 6. Drum brakes are usually designed so that the condition of the lining can only be checked once the drum has been removed.
_____ 7. In a duo-servo brake installation the matching shoes belong on the same side of the vehicle.
_____ 8. The lining on brake shoes is much thinner than on disc brake pads.
_____ 9. If you switch a self-adjuster from one side of the vehicle to the other it will retract the adjustment, causing the brake shoe clearance to increase as the brakes adjust.
_____ 10. Drum parking brake systems mechanically apply the regular service brake shoes.

Fill in the Blank

Read each item carefully, and then complete the statement by filling in the missing word(s).

1. The drum is bolted to the vehicle's axle _____ by the lug nuts.
2. Each drum brake has two brake shoes with a friction material called a(n) _____ attached.
3. Manufacturers might use linings with different _____ _____ _____ for each of the shoes to get the desired braking load between the two shoes.
4. Brake drums are machined to a specific diameter from the manufacturer, which is called its _____ diameter.
5. All of the brake unit components, except the _____ _____, are mounted on a backing plate bolted to the vehicle axle housing or suspension.
6. The cylinder bore, or inside diameter of the cylinder, is created by drawing a properly sized _____ _____ through the bore.
7. The _____ _____ is a hollow screw with a taper on the end that mates with a matching tapered seat in the wheel cylinder.
8. Most modern vehicles use _____-_____ wheel cylinders because they are simpler to design, install, and bleed.
9. The terms primary and secondary refer to the _____ _____ in a duo-servo brake system.
10. Since 1968 manufacturers have incorporated a _____-_____ mechanism into their drum brake systems that is capable of maintaining proper shoe-to-drum clearance.

Labeling

Label the following diagrams with the correct terms.

1. The main components of a drum brake system:

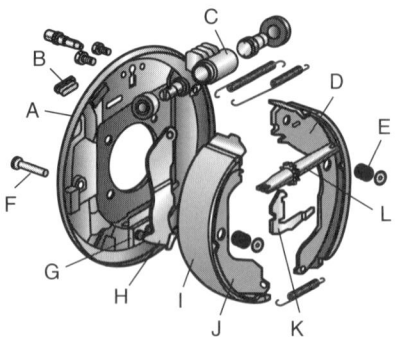

A. _____
B. _____
C. _____
D. _____
E. _____
F. _____
G. _____
H. _____
I. _____
J. _____
K. _____
L. _____

2. The leading/trailing shoe drum brake system:

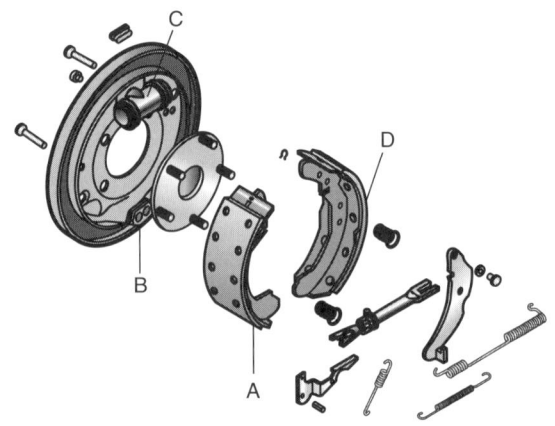

A. _____
B. _____
C. _____
D. _____

3. The duo-servo drum brake system:

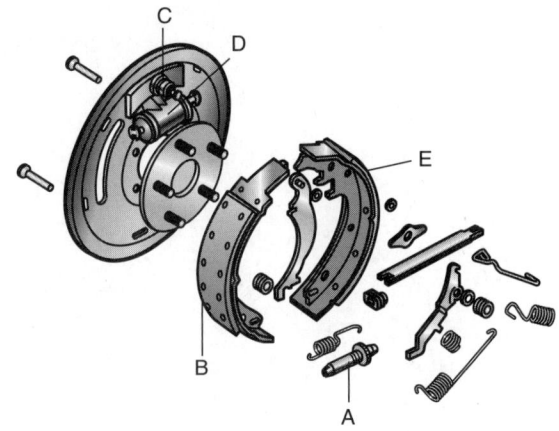

A. _____
B. _____
C. _____
D. _____
E. _____

4. Dual-acting wheel cylinder:

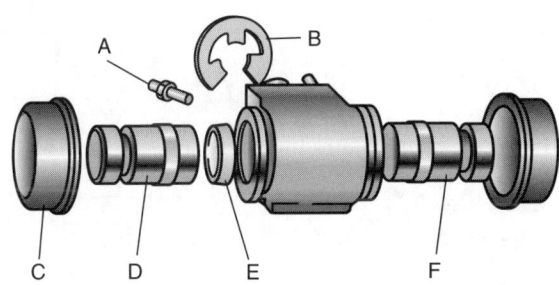

A. _____
B. _____
C. _____
D. _____
E. _____
F. _____

5. Star wheel assembly:

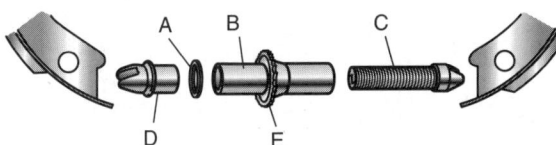

A. _____
B. _____
C. _____
D. _____
E. _____

6. Parking brake assembly for a drum brake:

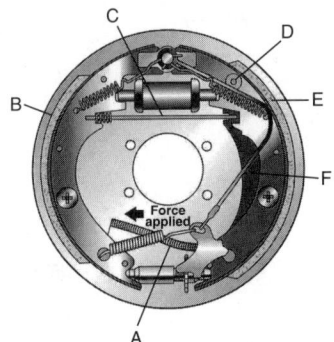

A. _____
B. _____
C. _____
D. _____
E. _____
F. _____

7. Drum brake tools:

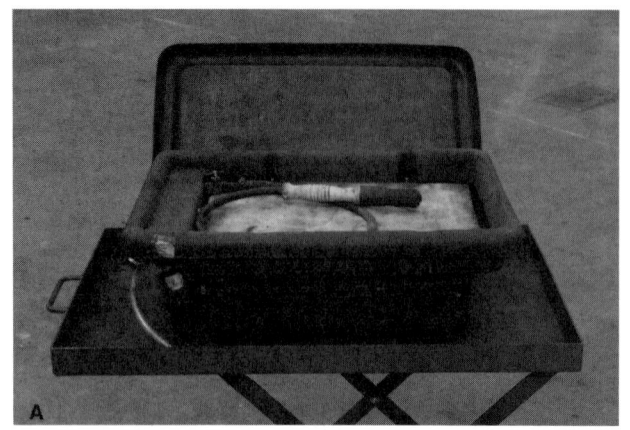

A. _____

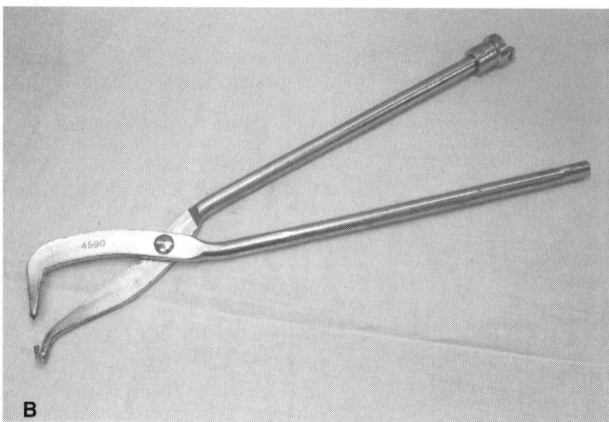

B. _____

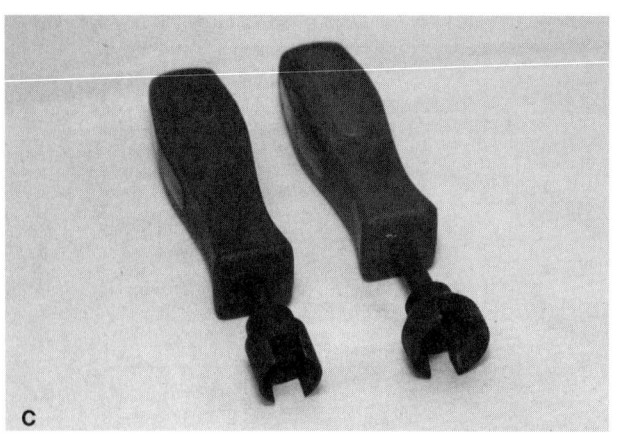

C. _____

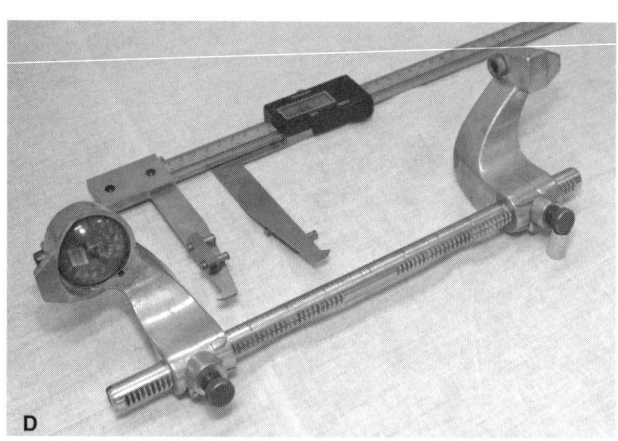

D. _____

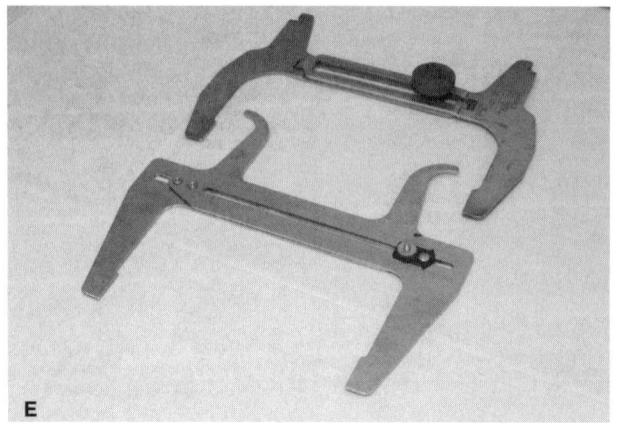

E. _____

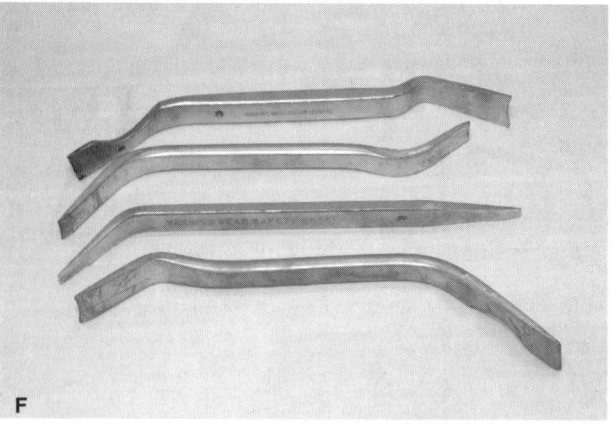

F. _____

Chapter 33   Drum Brake System

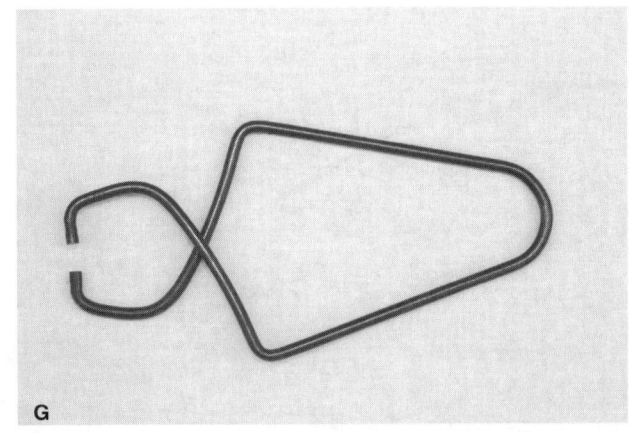

G. _____

H. _____

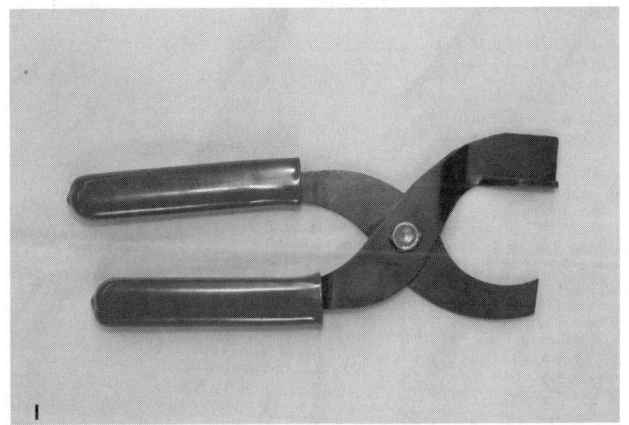

I. _____

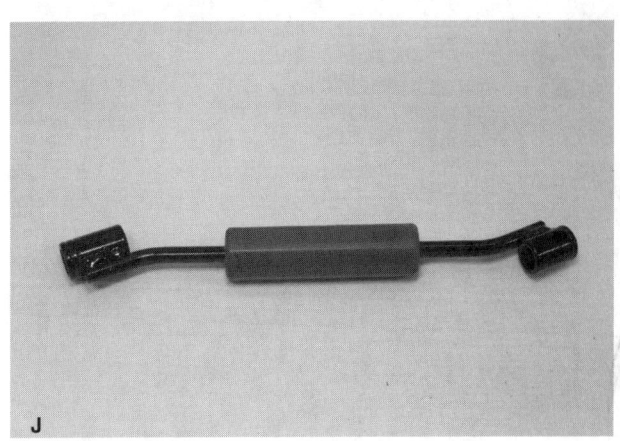

J. _____

Skill Drills

Test your knowledge of skill drills by filling in the correct words in the photo captions.

1. Refinishing Brake Drums:

Step 1: Research the brake lathe manufacturer's procedure for properly _____ the drum. Clean any nicks, burrs, or debris from the mounting surfaces of the _____. Mount the drum on the brake lathe. Check to see that the drum is running _____ on the lathe.

Step 2: Install the _____-_____ band on the drum.

Step 3: Set the position of the _____ _____ so the brake drum is close to the brake _____ when the cutting bit is in the far corner of the drum. Make sure the cutting bits will not _____ the face of the drum, and move the cutting head about 0.5" (12.7 mm) in from the _____ of the drum. Turn on the brake lathe, and set the _____ of the cutting tool so it just touches the _____ of the drum. Slowly remove the ridge.

Step 4: Run the drum all the way in so the cutting bit is in the _____ _____ of the drum.

#

Step 5: Set the cutting bit to the proper depth and _____ it in place. Engage the automatic _____ feed, and set it to the proper _____ (if equipped), lock it in place, and watch for proper machining action. Repeat this step until the worn surface areas have been _____ all the way around the surface of the drum. If the brake lathe is not a single-cut machine, perform a _____ _____ on the drum.

Step 6: Move the drum well away from the cutting bit, and use _____ to give the drum surface a nondirectional finish. Remeasure the drum _____, and discard if _____ specifications.

2. Removing, Cleaning, Inspecting, and Reassembling a Duo-Servo Brake:

Step 1: Research the procedure for _____ the brake assembly. Clean brake shoes, hardware, and _____ _____ using equipment and procedures for dealing with _____/dust.

Step 2: To disassemble a _____-_____ brake, first remove the return springs, cable guide (if installed), and _____ guide. Remove the parking brake _____ and spring. Remove the primary shoe _____-_____ _____, retainer, and pin. Remove the self-adjuster spring and _____ _____ assembly and primary shoe.

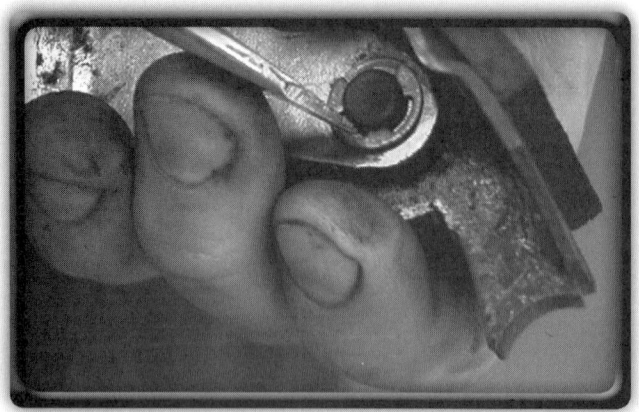

Step 3: Remove the _____ hold-down spring, retainer, pin, and secondary shoe. Disassemble the _____ _____ lever from the brake shoe and hardware from the backing plate. Finally, _____ and _____ all parts according to the manufacturer's procedure.

Step 4: To reassemble a duo-servo brake, first reassemble the parking brake lever on the _____ _____ and parking brake cable. Reassemble the star wheel assembly, _____ the floating end, and set aside. Also lubricate the _____ on the backing plate. Install both shoes to the backing plate with the hold-down spring assemblies. Install the shoe guide and self-adjuster cable over the _____ _____.

Step 5: Install the cable _____, and return spring in the secondary shoe. Also align the wheel cylinder _____ in the shoe. Position the secondary shoe in place, and use brake spring pliers to install the _____ _____ over the anchor pin. Install the parking brake strut rod onto the secondary shoe, and _____ the primary shoe engaged with the parking brake strut rod. Install the return spring in the primary shoe, and use _____ _____ _____ to stretch the return spring over the anchor pin. Install the self-adjuster link, cable, and spring into position. Install the star wheel between the _____ of the two shoes. Finally, check the _____ of all springs, clips, and levers.

3. Disassembling, Cleaning, Inspecting, and Reassembling a Non-Servo Brake:

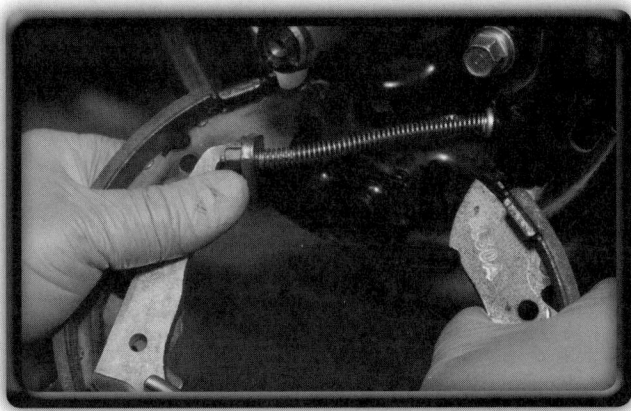

Step 1: Research the procedure for disassembling the brake assembly. Clean brake shoes, hardware, and _____ _____ using equipment and procedures for dealing with _____ /dust. To disassemble a non-servo brake disassembly, first remove the hold-down springs, _____, and pins. Spread the _____ apart, and remove the parking brake _____ and self- adjuster components.

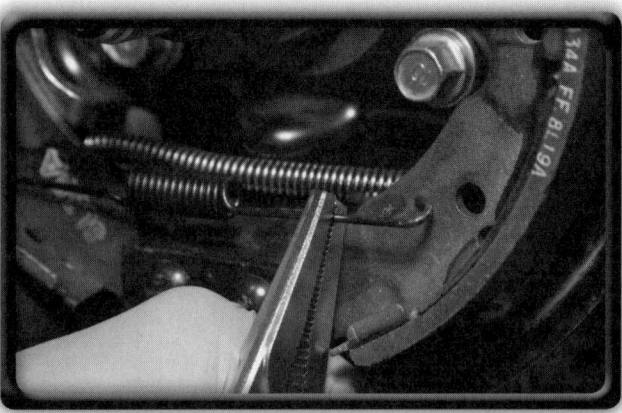

Step 2: Remove the _____ _____. Disassemble the parking brake lever from the _____ _____ and hardware from the backing plate. Finally, clean and inspect all parts.

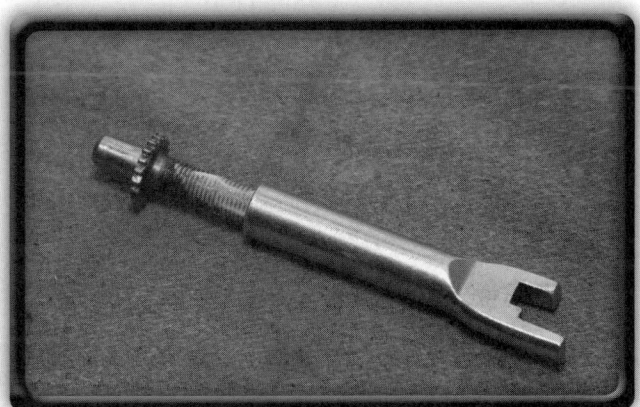

Step 3: To reassemble a _____-_____ assembly, first assemble and _____ the self-adjuster/parking brake strut assembly. Lube the backing plate _____. Place one shoe on the backing plate, and _____ the hold-down spring and pin. Place the self-adjuster/parking brake strut on the installed brake shoe.

Step 4: Place the _____ springs on both shoes, and fit the _____ shoe to the backing plate, being sure to line up the wheel cylinder _____, the self-adjuster, and the parking brake mechanism. Install the _____-_____ spring and pin. Check the fit of all springs, clips, and levers.

4. Removing, Inspecting, and Installing Wheel Cylinders:

Step 1: Use a _____ nut or line wrench to unscrew the brake line from the wheel cylinder. Remove the _____ _____ from the backing plate.

Step 2: Peel back the _____ _____, and check for _____ _____ behind them. Determine any necessary actions.

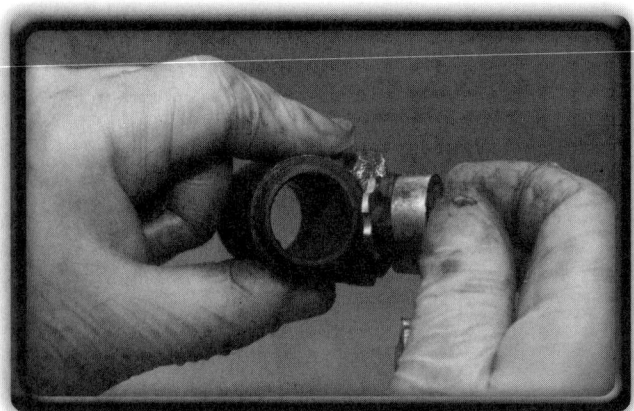

Step 3: _____ the wheel cylinder and _____ each part. Determine any necessary actions.

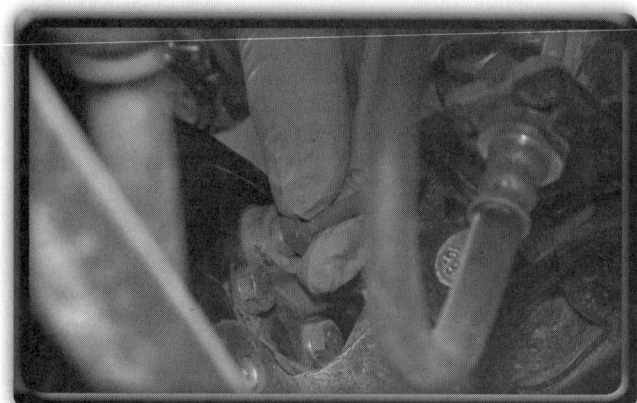

Step 4: If the cylinder can be reused, _____ it, preferably with new seals and dust boots. If not, _____ it with a new wheel cylinder. Reinstall the _____ _____ by hand. Install and tighten any mounting screws. After that, tighten the brake line with a flare nut or _____ _____.

5. Pre-Adjusting Brakes and Installing Drums and Wheel Bearings:

Step 1: Make sure the brake shoes are _____ on the backing plate.

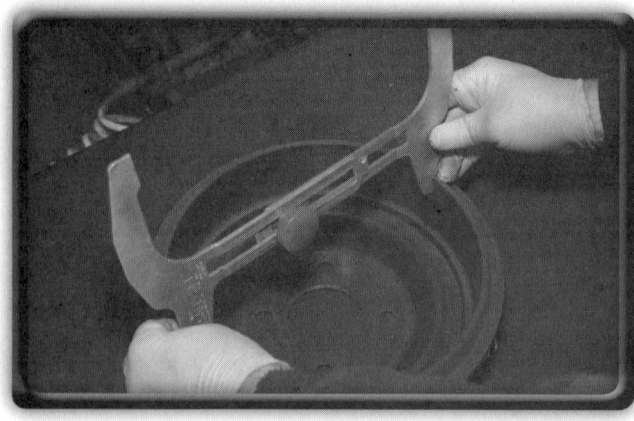

Step 2: Set the _____-_____ gauge to the drum _____, and lock it in place.

Step 3: Place the pre-adjustment gauge over the _____ of the _____ _____.

Step 4: Adjust the _____ _____ until the centers of the brake shoes just _____ the pre-adjustment gauge.

Step 5: Test install the _____ _____ to verify the drum fits. Adjust the _____ as necessary. Pre-adjust the _____ _____ according to the manufacturer's procedure. If the drum has serviceable wheel _____, repack, install, adjust, and secure them.

Crossword Puzzle

Use the clues to complete the puzzle.

Across

3. A tool used to adjust the brake lining-to-drum clearance when the drum is installed on the vehicle.
5. A tool that prevents the pistons from being pushed out of the wheel cylinders while the brake shoes are being replaced.
7. Brake shoes that are installed so they are applied in the same direction as the forward rotation of the drum and thus are self-energizing.
8. Springs that retract the brake shoes to their released position.
9. The inside diameter of a cylinder.

Down

1. A tool used for measuring the inside diameter of the brake drum.
2. A tool used to adjust the brake lining-to-drum clearance when the drum is installed on the vehicle.
3. A tool used for removing and installing brake return springs.
4. Brake shoes installed so they are applied in the opposite direction to the forward rotation of the brake drum—not self-energizing and less efficient at developing braking force.
6. A steel shoe and brake lining friction material that apply force to the brake drum during braking.

ASE-Type Questions

Read each item carefully, and then select the best response.

_____ 1. Tech A says that parking brakes on drum brake vehicles use separate brake shoes for backup in an emergency. Tech B says that parking brakes on drum brake vehicles mechanically operate the standard drum brake shoes. Who is correct?
   A. Tech A
   B. Tech B
   C. Both A and B
   D. Neither A nor B

_____ 2. Tech A says that most brake drums are designed to be machined if minor surface issues are present. Tech B says that brake drums can be reused if they are machined over specifications, as long as the surface is smooth. Who is correct?
   A. Tech A
   B. Tech B
   C. Both A and B
   D. Neither A nor B

_____ 3. Tech A says that duo-servo brake shoes are only anchored on the top. Tech B says that typically duo-servo brake shoes adjust automatically during normal brake applications in a forward direction. Who is correct?
   A. Tech A
   B. Tech B
   C. Both A and B
   D. Neither A nor B

_____ 4. Tech A says that when inspecting brake shoes, if the shoes are unequally worn, this could be caused by a stuck wheel cylinder piston. Tech B says that the lining on the primary shoe is typically shorter in length and is installed toward the front of the vehicle. Who is correct?
   A. Tech A
   B. Tech B
   C. Both A and B
   D. Neither A nor B

_____ 5. Tech A says that it is almost impossible to install self-adjusters on the wrong side of the vehicle. Tech B says that grease seals can be reused. Who is correct?
   A. Tech A
   B. Tech B
   C. Both A and B
   D. Neither A nor B

_____ 6. Tech A says to use an air hose to clean the backing plate of dust and contamination. Tech B says to use a brake cleaning solution to clean the backing plate of dust and contamination. Who is correct?
   A. Tech A
   B. Tech B
   C. Both A and B
   D. Neither A nor B

_____ 7. Tech A says that riveted lining is usually for heavy-duty or high-performance vehicles. Tech B says that you need to identify each shoe individually to help ensure proper installation. Who is correct?
   A. Tech A
   B. Tech B
   C. Both A and B
   D. Neither A nor B

_____ 8. Tech A says that a grinding noise in drum brakes generally requires replacing the brake shoes and resurfacing or replacing the drums. Tech B says that a click noise in drum brakes requires replacing the brake shoes and resurfacing or replacing the drums. Who is correct?
   A. Tech A
   B. Tech B
   C. Both A and B
   D. Neither A nor B

_____ **9.** Tech A says that brake shoe linings saturated with brake fluid from a leaky wheel cylinder can be cleaned with brake cleaner and reused as long as they aren't worn out. Tech B says that brake shoes should be replaced in axle sets. Who is correct?
   A. Tech A
   B. Tech B
   C. Both A and B
   D. Neither A nor B

_____ **10.** Tech A says that when performing a brake job on the rear axle of an older vehicle, inspection finds brake fluid under the dust boot; this suggests wheel cylinder replacement on both rear wheels. Tech B says that when a drum brake return spring has failed, springs on both rear wheels need to be replaced. Who is correct?
   A. Tech A
   B. Tech B
   C. Both A and B
   D. Neither A nor B

CHAPTER 34

Wheel Bearings

Tire Tread:
© AbleStock

Chapter Review

The following activities have been designed to help you refresh your knowledge of this chapter. Your instructor may require you to complete some or all of these activities as a regular part of your training program. You are encouraged to complete any activity that your instructor does not assign as a way to enhance your learning.

Matching

Match the following terms with the correct description or example.

- A. Antifriction bearing
- B. Ball bearings
- C. Bearing packer
- D. Castellated nut
- E. Cylindrical roller bearing assembly
- F. End play
- G. Grease seal
- H. Interference fit
- I. Lithium soap
- J. Outer race
- K. Preload
- L. Running clearance
- M. Tapered roller bearing assembly
- N. Unitized wheel bearing hub
- O. Wheel bearing

_____ 1. A component that is designed to keep grease from leaking out and contaminants from leaking in.

_____ 2. A condition where the wheel bearing components are forced together under pressure and therefore have no end play.

_____ 3. An adjusting nut with slots cut into the top such that it resembles a castle; used with a cotter pin to prevent the nut from turning.

_____ 4. An assembly consisting of the hub, wheel bearing(s), and possibly the wheel flange, which is preassembled and ready to be installed on a vehicle.

_____ 5. The outside component of a wheel bearing that has a smooth, hardened surface for rollers or balls to ride on.

_____ 6. A component that allows the wheels to rotate freely while supporting the weight of the vehicle, made up of an inner race, outer race, rollers or balls, and a cage.

_____ 7. Wheel bearing assemblies that use surfaces that are in rolling contact with each other to greatly reduce friction compared to surfaces in sliding contact.

_____ 8. A condition when two parts are held together by friction because the outside diameter of the inner component is slightly larger than the inside diameter of the outer component.

_____ 9. A tool that forces grease into the spaces between the bearing rollers.

_____ 10. A type of wheel bearing with races and rollers that are tapered in such a manner that all of the tapered angles meet at a common point, which allows them to roll freely and yet control thrust.

_____ 11. A thickening agent for grease to give it the proper consistency.

_____ 12. The in-and-out movement of the hub caused by clearance within the wheel bearing assembly.

_____ 13. The amount of space between wheel bearing components while in operation.

_____ 14. The rolling components of a wheel bearing consisting of hardened balls that roll in matching grooves in the inner and outer races.

_____ 15. A type of wheel bearing with races and rollers that are cylindrical in shape and roll between inner and outer races, which are parallel to each other.

Multiple Choice

Read each item carefully, and then select the best response.

_____ 1. What type of bearings must be serviced periodically by disassembling, cleaning, inspecting, and repacking them with the specified lubricant?
   A. Sealed bearings
   B. Serviceable bearings
   C. Ball bearings
   D. Unitized bearings

_____ 2. What type of wheel bearing assembly is used where heavier loads need to be supported and the wheel bearings are put under a side load condition?
   A. Tapered roller bearing assemblies
   B. Unitized bearings
   C. Cylindrical roller bearing assemblies
   D. Either A or C

_____ 3. Lubrication of serviceable tapered wheel bearing assemblies is usually accomplished by using _____.
   A. wheel bearing grease
   B. lithium soap
   C. gear lube
   D. either A or C

_____ 4. Many seals use a(n) _____ to help hold the lips of the seal in contact with the shaft it is sealing.
   A. O-ring
   B. garter spring
   C. rubber gasket
   D. sealing bushing

_____ 5. In a _____ the weight of the vehicle is fully carried by a pair of tapered roller bearing assemblies, which ride between the hub and axle tube.
   A. semifloating axle
   B. ¾ floating axle
   C. full-floating axle
   D. ½ floating axle

_____ 6. In a(n) _____, the wheel flange is part of the axle, which is supported by a single bearing assembly near the flange end of the axle.
   A. ¾ floating axle
   B. semifloating axle
   C. full-floating axle
   D. solid axle

_____ 7. In a _____ design, there is a single bearing assembly between the outside of the axle tube and the hub.
   A. ¾ floating axle
   B. full floating axle
   C. semifloating axle
   D. splined axle

_____ 8. The level of gear lube should normally be within _____ of the bottom of the fill plug hole.
   A. 0.25"
   B. 0.50"
   C. 0.75"
   D. none of the above

_____ 9. Grease is made out of a base oil, plus a thickening agent such as _____.
   A. lithium soap
   B. polysaccharides
   C. molybdenum thickening agents
   D. either A or C

_____ **10.** On adjustable wheel bearings, the proper clearance must be set using the _____.
  A. keyed washer
  B. adjusting nut
  C. lock cage
  D. lock nut

True/False

If you believe the statement to be more true than false, write the letter "T" in the space provided. If you believe the statement to be more false than true, write the letter "F".

_____ **1.** Sealed bearings are designed so they cannot be disassembled or adjusted.
_____ **2.** The outer bearing assembly is typically larger than the inner bearing assembly.
_____ **3.** Packing is best performed with a bearing packing tool, but it can be successfully performed by hand.
_____ **4.** Roller bearings roll easier than ball bearings since they have a smaller contact area; thus, they provide a small increase in vehicle efficiency.
_____ **5.** Sealed wheel bearings need to be adjusted for proper running clearance after installation.
_____ **6.** On some vehicles equipped with ABS brakes, the ABS sensor is integrated into the unitized wheel bearing assembly.
_____ **7.** Gear lube is somewhat thicker than bearing grease.
_____ **8.** Automotive wheel bearing grease is a thickened lubricant, designated as a plastic solid.
_____ **9.** Sealed bearings and double-row bearing assemblies come from the manufacturer with the proper clearance machined into them.
_____ **10.** On four-wheel drive vehicles, the adjustable wheel bearing locking mechanism commonly includes a keyed washer, adjusting nut, keyed lock washer, and lock nut.

Fill in the Blank

Read each item carefully, and then complete the statement by filling in the missing word(s).

1. In many instances, _____ roller bearing assemblies use the surface of the axle as the inner bearing race.
2. _____ a bearing means that the spaces between the rollers and races are completely filled with grease.
3. Virtually all wheel bearings using a ball bearing assembly are of the _____-_____ ball bearing variety, and they are commonly used in automotive light vehicle applications.
4. In some applications, the _____ _____ is press-fit into the axle housing, which is stationary, and seals against the axle shaft, which is rotating.
5. A(n) _____ _____ is a soft metal pin that can be bent into shape and is used to retain the bearing adjusting nut.
6. In many rear-wheel drive vehicles, the wheel bearing assemblies are open to the axle housing, which is partially filled with _____ _____ that also lubricates the differential assembly.
7. A(n) _____ _____ is usually threaded and can be removed to allow the level of a fluid to be checked and filled.
8. _____ refers to the thickness of the gear lube; the higher the number, the thicker the gear lube.
9. The thickness of grease is graded by the _____ _____ _____.
10. _____ refers to the absence of clearance in the bearing and the specified amount of pressure forcing the bearing components together.

Labeling

Label the following diagrams with the correct terms.

1. Components of a wheel bearing:

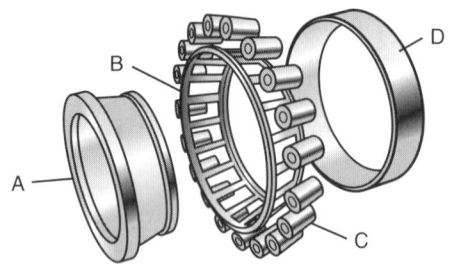

A. _____

B. _____

C. _____

D. _____

2. Identify the following components:

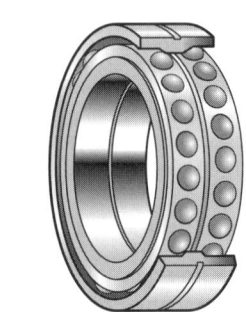

A. _____

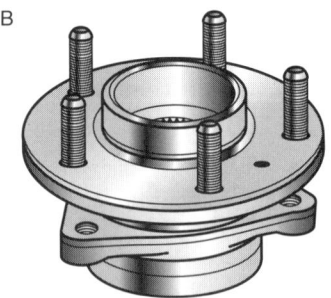

B. _____

C. _____

3. Lock nut-style wheel bearing locking mechanism:

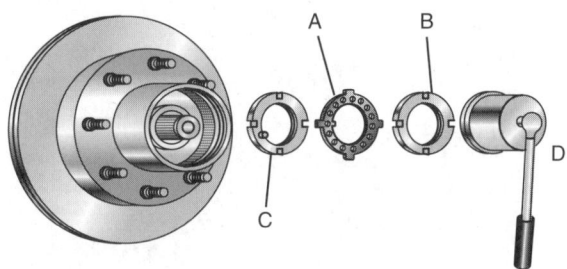

A. _____

B. _____

C. _____

D. _____

Skill Drills

Place the skill drill steps in the correct order.

1. Removing, Cleaning, and Inspecting Wheel Bearings:

_____ **A.** Wipe any old grease off of the wheel bearings, races, and spindle with a rag, and give them a quick visual inspection. Consult the bearing diagnosis chart to identify any faults.

_____ **B.** Remove the locking mechanism. Remove the adjusting nut, keyed washer, and outer bearing. Reinstall the adjusting nut approximately five turns back onto the spindle.

_____ **C.** Remove the wheel bearing dust cap with dust cap pliers or a narrow cold chisel and hammer.

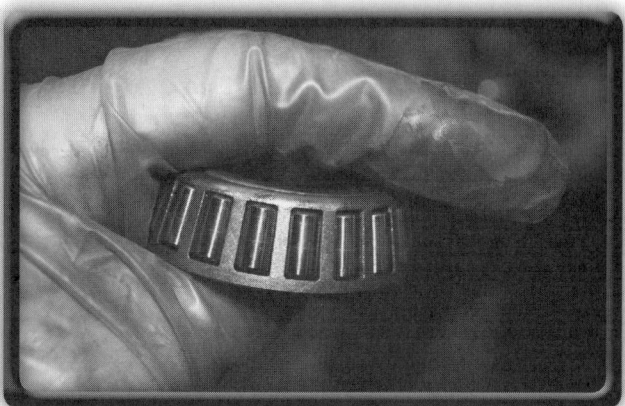

_____ **D.** Give the parts a final inspection, and consult the bearing diagnosis chart if there are any signs of damage. Using the specified grease, pack both wheel bearings, being careful to keep dirt and debris out of the grease.

_____ **E.** Grasp the drum/rotor at the 1 o'clock and 7 o'clock positions or 11 o'clock and 5 o'clock positions. While holding downward pressure, quickly pull the drum/rotor toward you. The adjusting nut should catch the inner bearing race and pop the grease seal and bearing out of the hub, leaving them sitting on the spindle.

_____ **F.** If the wheel bearings are in serviceable condition, completely clean the wheel bearings, races, and hub. If using solvent to clean any of the components, make sure there is no solvent-contaminated grease left on the parts.

2. Packing Grease by Hand:

_____ **A.** Smear some grease around the outside of the bearing. Repeat this process on the other bearing.

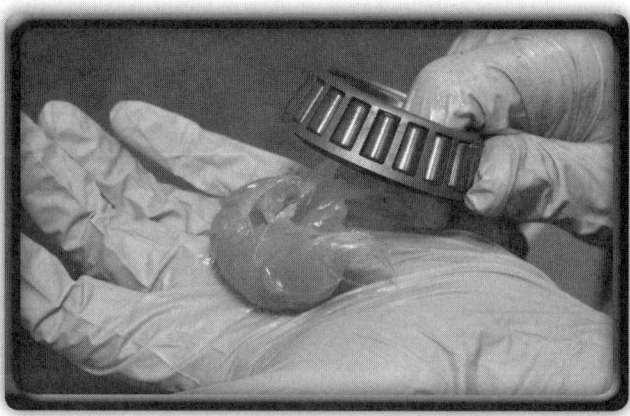

_____ **B.** Using a pair of latex or nitrile (nitro) gloves, place a small glob of grease in the palm of your nondominant hand. Place the index finger of your other hand through the bearing center hole with the larger diameter facing down.

_____ **C.** Carefully turn the bearing as a unit to a new space, and keep forcing grease between the bearings. Do this until all of the spaces are full.

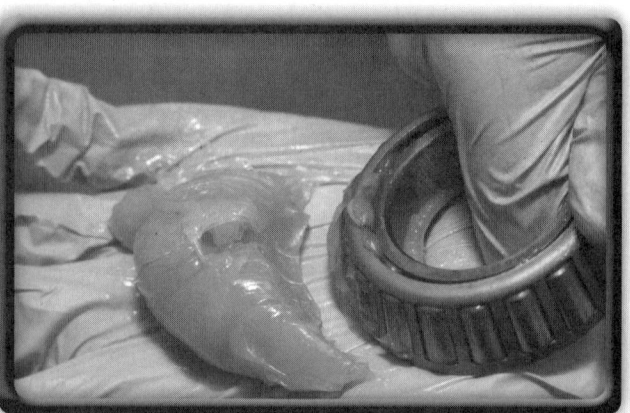

_____ **D.** Push the large diameter of the bearing down the edge of the grease into your palm. This should force grease into the space between the bearings and races. Continue this process until grease comes out of the top of the bearing.

3. Installing the Locking Mechanism:

_____ **A.** If it is a locking nut style, then tighten the lock nut to the specified torque. This is usually a substantial torque of 50 ft-lb (67.79 Nm) or more.

_____ **B.** If it is a bendable tang locking style, then place the tang washer against the adjusting nut and thread the locking nut up against it. Torque the locking nut to the specified torque, and bend the appropriate tang out toward you against the flat side of the locking nut with a small pry bar to lock the adjustment in place.

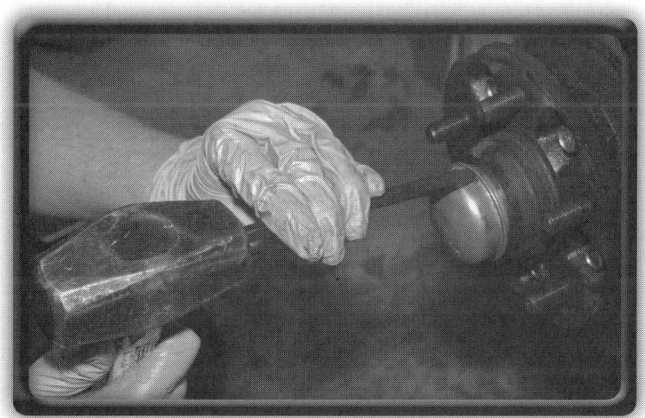

_____ **C.** Install the dust cap, being sure it is fully seated in the hub. Make sure the drum/rotor turns freely without binding or making any unusual noises.

_____ **D.** Install the locking mechanism. If it is a cotter pin, insert the new cotter pin through the castellated nut or locking cage and spindle. The short leg of the cotter pin should be against the castellated nut and the long leg should be toward you. With the cotter pin fully engaged in the notch, bend the outer leg toward you and up over the end of the spindle. Cut it off just short of the spindle. Also cut the short leg off so it does not extend beyond the nut or cage. Make sure the cotter pin will not hit the inside of the dust cap.

4. Replacing Wheel Bearings and Races:

_____ **A.** Using a hydraulic press or a hammer and bearing race installer, carefully drive the race until it is fully seated in the hub. When using a hammer and punch, a distinct sharp metallic sound should be produced when it seats. Inspect the race to verify that it is fully seated. Also check for any damage caused by installation. If everything is good, pack the new bearing and install it.

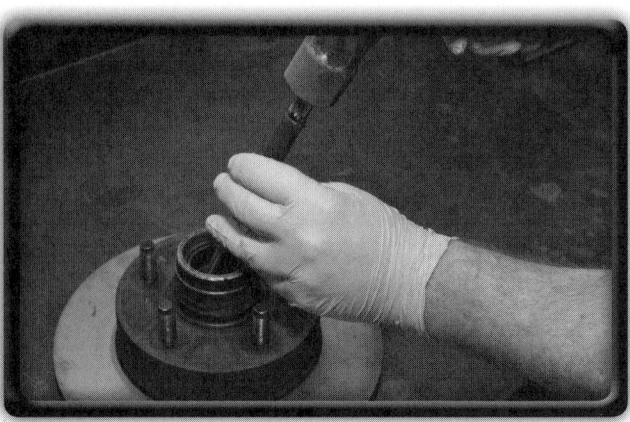

_____ **B.** With the wheel bearings removed from the wheel hub, clean and inspect the bearing and race for damage. Determine which bearing and race need to be replaced. Using a hydraulic press or a hammer and punch from the opposite side of the hub, carefully force the race from the hub. Keep it as straight as possible while removing it.

_____ **C.** Clean the inside diameter of the hub in a parts washer. Remove any burrs with a fine file or Dremel™, and remove any debris from the seat. Lightly lubricate the outside surface of the new race, and set it thick side down in the hub.

5. Removing and Reinstalling Sealed Wheel Bearings Using the Unitized Wheel Bearing Hub Style:

_____ **A.** If the wheel you are working on is a drive wheel, remove the axle hub nut and tap the drive axle loose with a dead blow hammer.

_____ **B.** Carefully compare the new hub to the old one, then fit the new hub assembly (over the axle shaft, if equipped) to the knuckle, making sure it is fully seated in place, and torque the mounting bolts to the specified torque. Reassemble the brake assembly and ABS sensor following the specified procedure, install the wheel, and torque the lug nuts. Install the drive axle nut, if equipped. Use a new hub nut if called for by the manufacturer, and torque to specifications.

_____ **C.** Loosen the axle hub nut, if equipped, while the tire is still on the ground. Remove the wheel and brake assembly following the specified procedure. Also disconnect the ABS connector and/or sensor if mounted to the hub.

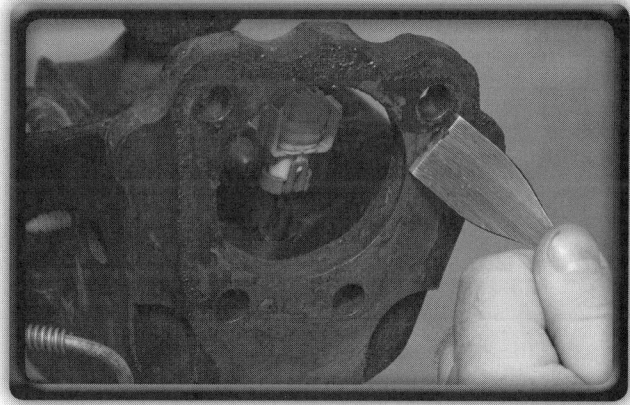

_____ **D.** Unbolt and remove the hub assembly from the steering knuckle. Clean the knuckle assembly, and check the hub seat for nicks, burrs, or other damage.

Crossword Puzzle

Use the clues to complete the puzzle.

Across

3. A component that is designed to keep grease from leaking out and contaminants from leaking in.
7. The nut that holds the adjusting nut from turning; usually tightened much tighter than the adjusting nut.
8. The inside component of a wheel bearing that has a smooth, hardened surface for rollers or balls to ride on.
9. A one-use soft metal pin that can be bent into shape and is used to retain bearing adjusting nuts.
10. A bearing that uses sliding motion between components, such as a clutch pilot bearing.
12. The component in a wheel bearing that maintains the proper spacing between the roller bearings or ball bearings.
13. A lubricating liquid thickened to make it suitable for use with many wheel bearings.

Down

1. Wheel bearings that are assembled by the manufacturer with the proper lubrication and sealed for life; cannot normally be disassembled.
2. The washer that fits between the adjusting nut and the wheel bearing and that has the center hole keyed to fit a slot on the spindle or axle tube.
4. The measurement of the thickness of a liquid.

5. A type of lubricant primarily used to lubricate transmission and differential gears but also used to lubricate some wheel bearings.
6. A sleeve type bearing made of metal or plastic bearing material used to support a rotating shaft.
7. The stamped sheet metal cap that fits over the bearing adjustment nut and is secured by a cotter pin going through it and the spindle/axle.
10. Usually a threaded plug that can be removed to allow the level of fluid to be checked and filled. This could also be a rubber snap fit plug.
11. The nut used to adjust the end play or preload of a wheel bearing.

ASE-Type Questions

Read each item carefully, and then select the best response.

1. Tech A says that cylindrical roller bearings can carry more weight than similarly sized ball bearings. Tech B says that tapered roller bearings, used in opposing pairs, control side thrust. Who is correct?
   A. Tech A
   B. Tech B
   C. Both A and B
   D. Neither A nor B

2. Tech A says that a tapered roller bearing assembly has less rolling resistance than a similarly sized ball bearing assembly. Tech B says that the bearing assembly in a unitized wheel bearing assembly can normally be disassembled, cleaned, and repacked. Who is correct?
   A. Tech A
   B. Tech B
   C. Both A and B
   D. Neither A nor B

3. Tech A says that over time a grease seal can wear a groove in the sealing surface of the axle or shaft. Tech B says that grease seals need to be replaced every time the bearing is removed. Who is correct?
   A. Tech A
   B. Tech B
   C. Both A and B
   D. Neither A nor B

4. Tech A says that in a full-floating axle, the axle does not support the weight of the vehicle. Tech B says that when installing tapered wheel bearings, the final torque should be about 20 ft/lbs. Who is correct?
   A. Tech A
   B. Tech B
   C. Both A and B
   D. Neither A nor B

5. Tech A says that serviceable wheel bearings can be repacked by removing the dust cap, filling it with grease, and reinstalling it. Tech B says that the cotter pin must be replaced with a new one every time it is removed. Who is correct?
   A. Tech A
   B. Tech B
   C. Both A and B
   D. Neither A nor B

6. Tech A says that the grease level in the final drive is okay as long as you can touch the level with your finger. Tech B says that the grease level should normally be no more than ¼" below the threads on the fill plug hole. Who is correct?
   A. Tech A
   B. Tech B
   C. Both A and B
   D. Neither A nor B

_____ 7. Tech A says that wheel bearings need to be replaced as a set, bearing and race. Tech B says that the wheel bearings and races on both sides of the vehicle must be replaced if one side fails. Who is correct?
   A. Tech A
   B. Tech B
   C. Both A and B
   D. Neither A nor B

_____ 8. Tech A says that unitized hubs have a wheel nut with a higher installation torque than serviceable wheel bearings. Tech B says that unitized hubs have the proper bearing end play designed into the assembly once they are torqued properly. Who is correct?
   A. Tech A
   B. Tech B
   C. Both A and B
   D. Neither A nor B

_____ 9. Tech A says when installing a bearing race you should use a 3-lb hammer and a brass drift. Tech B says that when installing a bearing race you should use a 3-lb hammer and a 3/8" drive extension. Who is correct?
   A. Tech A
   B. Tech B
   C. Both A and B
   D. Neither A nor B

_____ 10. Tech A says that when a race is fully seated, a sharp metallic sound will be produced when installation is complete. Tech B says that to be sure a race is fully seated, the wheel bearing adjusting nut should be tightened to at least 100 ft-lbs of torque, which will finish seating it. Who is correct?
   A. Tech A
   B. Tech B
   C. Both A and B
   D. Neither A nor B

Electronic Brake Control

CHAPTER 35

Tire Tread:
© AbleStock

Chapter Review

The following activities have been designed to help you refresh your knowledge of this chapter. Your instructor may require you to complete some or all of these activities as a regular part of your training program. You are encouraged to complete any activity that your instructor does not assign as a way to enhance your learning.

Matching

Match the following terms with the correct description or example.

- A. ABS proportioning valve depressor
- B. Accumulator
- C. Body control module (BCM)
- D. Channel
- E. Common bore
- F. Fault codes
- G. High-pressure accumulator
- H. Integral ABS system
- I. Isolation valve
- J. Magneto-resistive sensor
- K. Nonintegral ABS systems
- L. Oversteer
- M. Roll-rate sensor
- N. Steering angle sensor
- O. Steering wheel position sensor
- P. Tone wheel
- Q. Understeer
- R. Variable reluctance sensor
- S. Vehicle speed sensor
- T. Wheel speed sensor

_____ 1. A type of wheel speed sensor that uses the principle of magnetic induction to create its signal.

_____ 2. A brake system in which the master cylinder, power booster, and HCU are all combined in a common unit.

_____ 3. A sensor that measures the amount of turning a driver desires. This information is used by the ESC system to know the driver's directional intent.

_____ 4. An alphanumeric code system used to identify potential problems in a vehicle system.

_____ 5. A storage container that holds pressurized brake fluid.

_____ 6. The valve in the HCU that either allows or blocks brake fluid that comes from the master cylinder from entering the HCU hydraulic circuit.

_____ 7. A brake system in which the master cylinder, power booster, and HCU are all separate units.

_____ 8. The computer that controls the electrical system in the body of a vehicle, including power windows, door locks, heating and A/C systems, and in some cases the EBC system.

_____ 9. The component that creates an electrical signal based on the speed of the vehicle, which is sent to the EBCM.

_____ 10. A sensor that measures the amount of roll around the vehicle's horizontal axis that a vehicle is experiencing.

_____ 11. A device that creates an analog or digital signal according to the speed of the wheel.

_____ 12. A storage container designed to contain high-pressure liquids such as brake fluid.

_____ 13. A sensor that signals to the EBCM both the position and speed of the steering wheel.

_____ 14. A tool used to hold the proportioning valve open on some ABS HCUs.

_____ 15. The number of wheel speed sensor circuits and hydraulic circuits the EBCM monitors and controls.

_____ 16. A condition in which the front wheels are turned further than the direction the vehicle is moving and the front tires are slipping sideways toward the outside of the turn.

_____ 17. A condition in which the rear wheels are slipping sideways toward the outside of the turn.

_____ 18. The part of the wheel speed sensor that has ribs and valleys used to create an electrical signal inside of the pick-up assembly.

_____ 19. When a single cylinder is used for two pistons. A tandem master cylinder would be an example of two pistons in one bore.

_____ 20. A type of wheel speed sensor that uses an effect similar to a Hall effect sensor to create its signal.

Multiple Choice

Read each item carefully, and then select the best response.

_____ 1. ABS systems use a computer that sends electrical signals to the _____ that momentarily hold or release hydraulic pressure to that wheel until it speeds up and starts rolling again.
   A. wheel speed sensors
   B. master cylinder
   C. solenoid valves
   D. spool valves

_____ 2. Which component of the ABS system contains electric solenoid valves controlled by the EBCM to modify hydraulic pressure in each hydraulic circuit?
   A. Master cylinder
   B. Hydraulic control unit
   C. Power booster
   D. Accumulator

_____ 3. Either separate or combined into a common assembly, the isolation valve and dump valve are controlled by a(n) _____.
   A. electric solenoid
   B. spool valve
   C. check valve
   D. brake pedal sensor

_____ 4. A _____ system uses separate speed sensors and hydraulic control circuits for each of the four wheels.
   A. single-channel
   B. two-channel
   C. three-channel
   D. four-channel

_____ 5. If the high-pressure pump fails for any reason, the _____ holds enough brake fluid at high pressure to apply the brakes 10 to 20 times before the boost is used up.
   A. hydraulic control unit
   B. accumulator
   C. master cylinder
   D. reservoir

_____ 6. A wheel sensor assembly consists of a toothed tone wheel (or tone ring) that rotates with the wheels and a stationary _____ attached to the hub or axle housing.
   A. pick-up assembly
   B. resistor
   C. tooth
   D. relay

_____ 7. The height of the sine wave is called its _____.
   A. peak
   B. amplitude
   C. phase
   D. duration

_____ 8. What style of speed sensor is sometimes called a passive system since it is self-contained and needs no outside power to function?
   A. Variable reluctance
   B. Magneto-resistive
   C. Hall effect
   D. Either A or C

_____ **9.** Some ABS systems provide _____ through the ABS warning lamp when a specific terminal is grounded or two specific terminals are shorted together.
   A. blink codes
   B. morse codes
   C. flash codes
   D. either A or C

_____ **10.** An electronic stability control system may integrate a(n) _____ into the basic ABS and TCS systems.
   A. yaw sensor
   B. steering angle sensor
   C. roll-rate sensor
   D. all of the above

True/False

If you believe the statement to be more true than false, write the letter "T" in the space provided. If you believe the statement to be more false than true, write the letter "F".

_____ **1.** Electronic stability control systems take the ABS and TCS systems to the next level by adding sensor information regarding the driver's directional intent.

_____ **2.** Understeer occurs when the vehicle is turning more sharply than the front wheels are being steered.

_____ **3.** Applying the brakes too hard or on a slippery surface can cause the wheels to lock.

_____ **4.** Maximum braking traction occurs when the wheels are rotating 20–25% slower than the vehicle speed.

_____ **5.** Nonintegral ABS systems use a fairly standard tandem master cylinder and a typical vacuum or hydraulic power booster.

_____ **6.** Some hydraulic control units use a single, three-position solenoid valve per circuit, while others use dual, two-position valves per hydraulic circuit.

_____ **7.** The pick-up assembly and tone wheel do not touch each other; a small gap, called an air gap, must be maintained at the specified clearance.

_____ **8.** The magneto-resistive sensor does not function effectively below vehicle speeds of around 5 mph.

_____ **9.** The brake switch is a normally open switch, meaning that if the switch is not affected by any outside force, electrical current will flow through it.

_____ **10.** Many electronic stability control system-equipped vehicles also monitor signals from the throttle position sensor, vehicle speed sensor, and brake pedal position sensor to help prevent a loss of control of the vehicle.

Fill in the Blank

Read each item carefully, and then complete the statement by filling in the missing word(s).

1. In the quest for increased safety, manufacturers developed a series of electronic _____ _____ systems; the first-generation was the antilock brake system.

2. The _____ _____ system applies brake pressure to the slipping tire, which causes more of the engine's torque to be transmitted to the wheel or wheels with the most traction.

3. The _____-_____ _____ system is designed to prevent wheels locking or skidding, no matter how hard the brakes are applied or how slippery the road surface, and to maintain steering control of the vehicle.

4. When the ignition switch is turned on, the ABS controller illuminates the yellow ABS warning lamp and performs an automatic _____-_____ of the system.

5. Drivers need to be taught to expect ABS brake pedal _____ when in a panic stop.

6. The ABS control module (or EBCM) sends commands in the form of _____ _____ to the hydraulic control unit.

7. A(n) _____-_____ system uses one sensor circuit with the speed sensor typically located in the differential and one hydraulic control circuit to control both rear wheels.

8. _____-_____ accumulators hold brake fluid in a spring-loaded chamber when it is released by the dump valves during an EBC event.

Fundamentals of Automotive Technology Student Workbook

9. The electronic brake control module supplies the magneto-resistive or Hall effect sensor systems with a _____ _____ of between 5 and 12 volts, depending on the manufacturer.

10. On a traction control system _____ _____ direct hydraulic pressure from the accumulator to the ABS solenoid valves so individual wheel brake units can be applied independently.

Labeling

Label the following diagrams with the correct terms.

1. HCU solenoid valve arrangement:

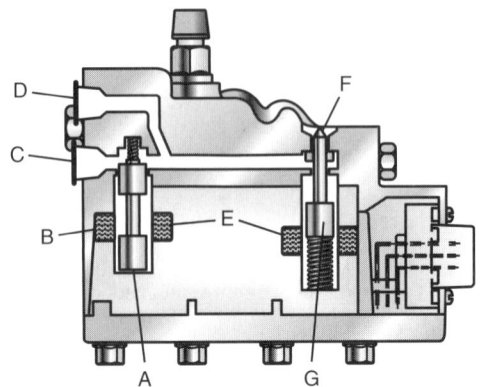

A. _____
B. _____
C. _____
D. _____
E. _____
F. _____
G. _____

2. Portless ABS master cylinder:

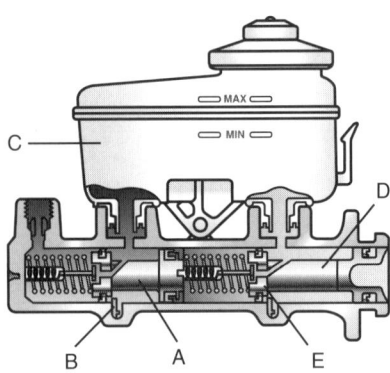

A. _____
B. _____
C. _____
D. _____
E. _____

3. The four types of ABS channels:

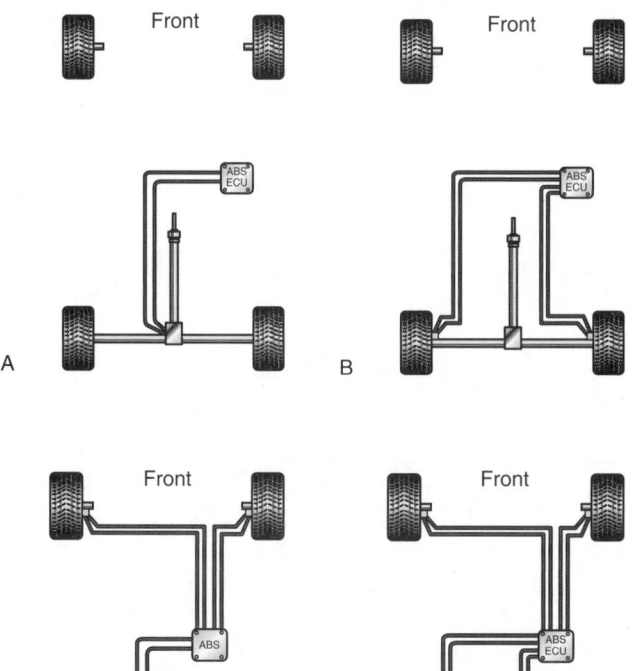

A. _____
B. _____
C. _____
D. _____

4. High-pressure accumulator:

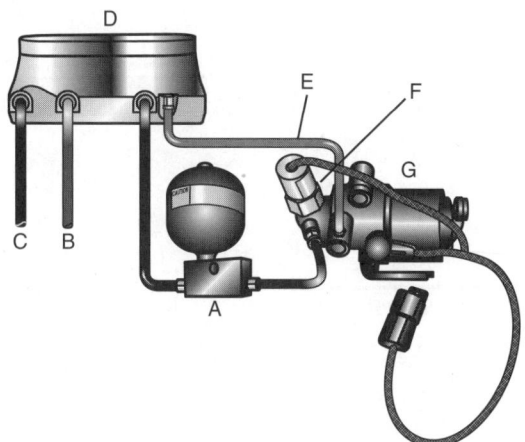

A. _____
B. _____
C. _____
D. _____
E. _____
F. _____
G. _____

5. Wheel speed sensor sine wave:

A. _____

B. _____

C. _____

6. An HCU with boost valves:

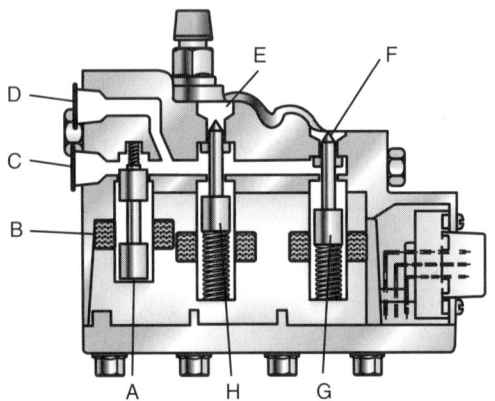

A. _____

B. _____

C. _____

D. _____

E. _____

F. _____

G. _____

H. _____

7. Tools used to diagnose electronic brake systems:

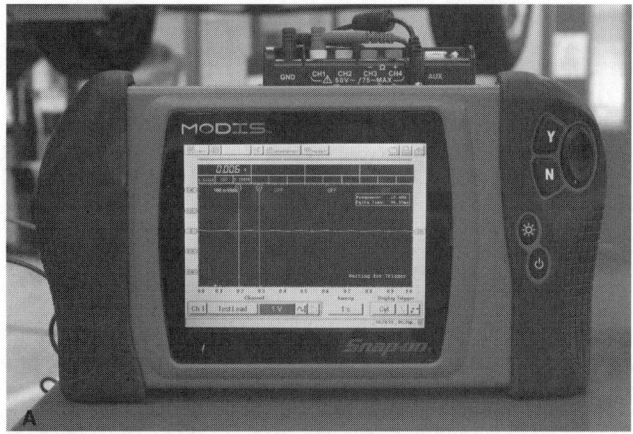

A. _____

B. _____

C. _____

Skill Drills

Test your knowledge of skill drills by filling in the correct words in the photo captions.

1. Testing Variable Reluctance Sensors:

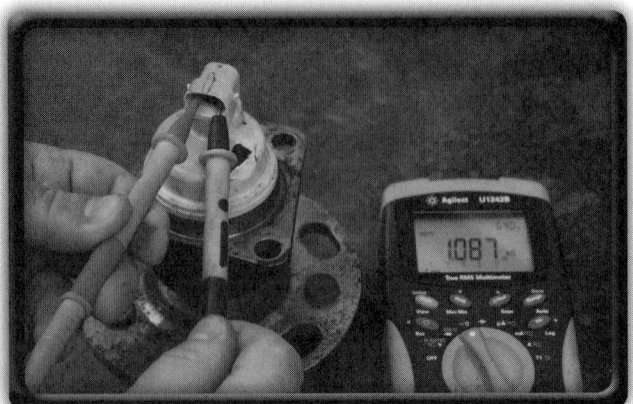

Step 1: Disconnect the suspect sensor, measure its _____, and _____ the reading to the specifications. If the resistance does not meet the manufacturer's specifications, _____ the sensor.

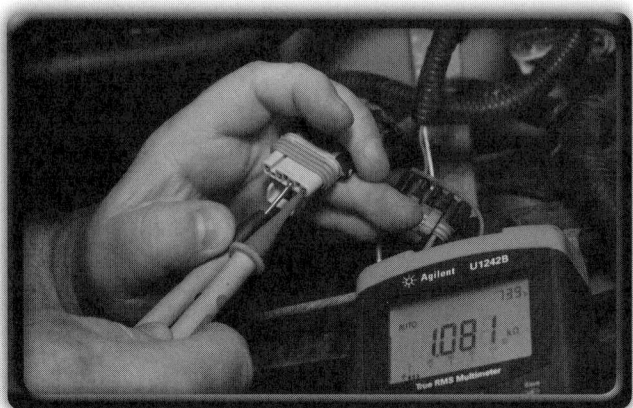

Step 2: If it is within specifications, test the two-wire circuit back to the EBCM for _____, _____, _____ _____, and _____. Repair as necessary.

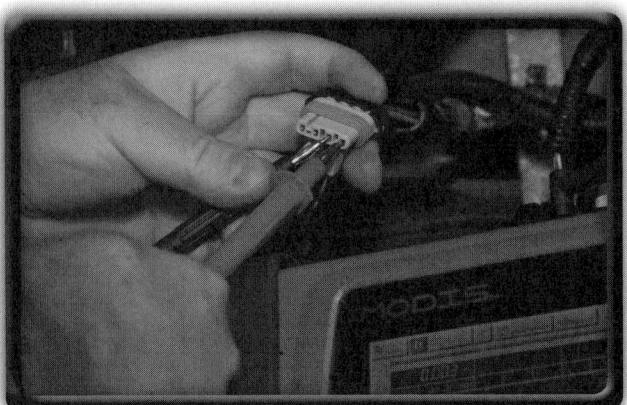

Step 3: If the circuit is good, reconnect the _____ _____ connector and attach a GMM or DSO to one of the sensor _____.

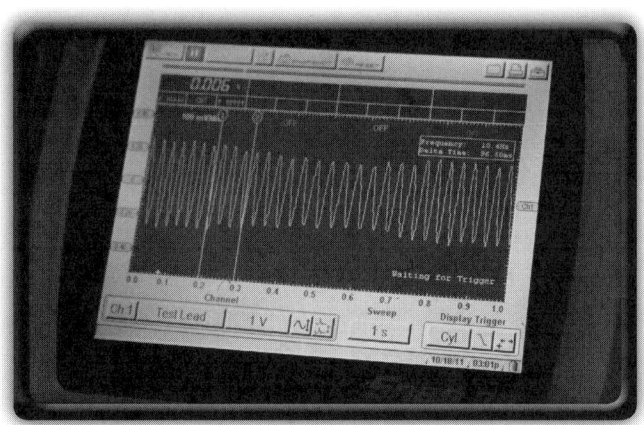

Step 4: Spin the _____ _____, and observe the pattern. It should be a clean _____ analog pattern of sufficient _____ (voltage).

Step 5: If the _____ is not correct, _____ the tone wheel for _____, and replace as necessary.

Step 6: If the tone wheel is good, replace the _____ _____ _____. If the pattern looks OK, you may have to compare it to the other wheel speed sensor _____ while _____ the vehicle. After repair, clear the _____ _____, if directed by the service information.

2. Testing Variable Reluctance Sensors:

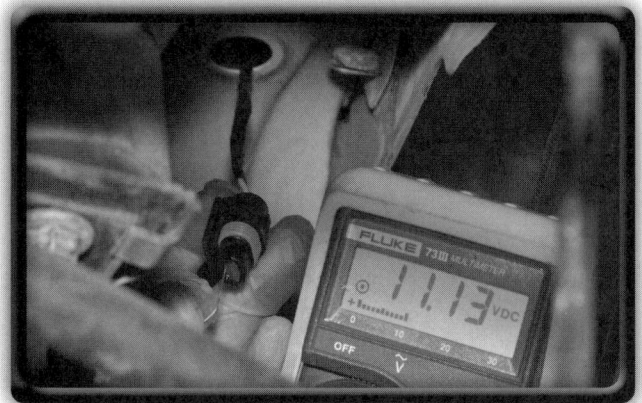

Step 1: Make sure the _____ _____ is in the "run" position, and measure the available _____ at the suspect sensor. Also check the _____ for voltage drop.

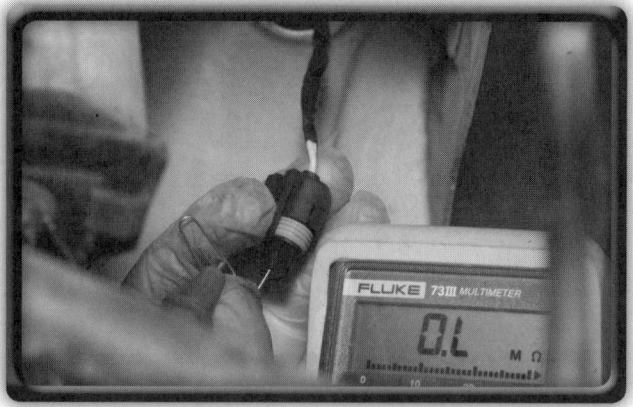

Step 2: If the _____ does not meet the manufacturer's specifications, test the circuit for opens, shorts, _____ _____, or grounds. _____ as necessary.

Step 3: If the voltage to the _____ is within the manufacturer's specifications, connect a _____ or DSO to the _____ _____.

Step 4: With the key in the "run" position, _____ the tone wheel and _____ the pattern. It should be a clean digital _____ _____ signal of the appropriate height and shape. If the pattern is not correct, _____ the tone wheel for _____ and replace as necessary. If the tone wheel is _____, replace the wheel speed sensor. After repair, clear the _____, if directed by the service information.

Crossword Puzzle

Use the clues to complete the puzzle.

Across

5. A valve located in the HCU that is controlled by the EBCM; it allows brake fluid under high pressure to flow into the HCI hydraulic circuits to apply the brakes when commanded.

8. Codes used to communicate DTCs; they are given by the EBCM as a series of blinks illuminated by the ABS warning lamp.

10. A tool used to read codes, access live data, and communicate with the vehicle's computers.

11. A component with a wire coil wrapped around a ferrous metal core; it is used to generate an electrical signal when a magnetic field passes through it.

13. An assembly that houses electrically operated solenoid valves used in electronic braking systems; also called a modulator.

14. An electrically operated valve that when used in a brake system is designed to control the flow of brake fluid in the hydraulic system.

15. A brake system in which the master cylinder, power booster, and HCU are all combined in a common unit.

Down

1. A sensor that measures the amount a vehicle is turning around its vertical axis. This information is used by the ESC system to know how much a vehicle is turning.
2. A computer-controlled system added to ABS to help prevent loss of traction while the vehicle is accelerating.
3. Electrically operated valves, which in brake systems are used to control the flow of brake fluid in the hydraulic system.
4. A high-pressure gauge designed to connect to the HCU and used to measure high hydraulic pressures in the system.
6. The space or clearance between two components, such as the space between the tone wheel and the pick-up coil in a wheel speed sender.
7. An electrical effect where electrons tend to flow on one side of a special material when exposed to a magnetic field, causing a difference in voltage across the special material. When the magnetic field is removed, the electrons flow normally and there is no difference of voltage across the special material. This effect can be used to determine the position or speed of an object.
9. A tool that shows graphically what is happening to voltage over a period of time; it is used to diagnose electrical faults.
12. The electrical switch that is activated by the brake pedal; it turns on the brake lights and signals the EBCM that the brakes are being applied.

ASE-Type Questions

Read each item carefully, and then select the best response.

_____ 1. Tech A says that an anti-lock brake system (ABS) helps shorten the stopping distance during a panic stop. Tech B says that antilock brake systems work by increasing the hydraulic pressure in the brake system so the brakes can be applied harder. Who is correct?
  A. Tech A
  B. Tech B
  C. Both A and B
  D. Neither A nor B

_____ 2. Tech A says that traction control can reduce the power output of the engine to increase traction. Tech B says that electronic stability control increases the risk of rollover. Who is correct?
  A. Tech A
  B. Tech B
  C. Both A and B
  D. Neither A nor B

_____ 3. Tech A says that an electronic braking system has sensors that monitor wheel speed. Tech B says that under steer is generally easier to recover from than over steer. Who is correct?
  A. Tech A
  B. Tech B
  C. Both A and B
  D. Neither A nor B

_____ 4. Tech A says that ABS controls braking every time the brakes are used. Tech B says that during an ABS event, it is normal for the brake pedal to pulsate. Who is correct?
  A. Tech A
  B. Tech B
  C. Both A and B
  D. Neither A nor B

_____ 5. Tech A says that during anti-lock braking, brake fluid may be returned to the master cylinder. Tech B says that solenoid valves in the hydraulic control unit will isolate the master cylinder from the brake circuit when it is in the "hold" mode. Who is correct?
  A. Tech A
  B. Tech B
  C. Both A and B
  D. Neither A nor B

_____ 6. Tech A says that mismatched tires may cause the ABS system to register a fault code. Tech B says that a TCS may automatically apply brake pressure to a wheel brake unit even if the vehicle is not being braked. Who is correct?
   A. Tech A
   B. Tech B
   C. Both A and B
   D. Neither A nor B

_____ 7. Tech A says that an ABS key-on system test checks for faults in the vehicle's base brake system. Tech B says that on most vehicles ABS DTCs are stored in memory for later retrieval. Who is correct?
   A. Tech A
   B. Tech B
   C. Both A and B
   D. Neither A nor B

_____ 8. Tech A says that a yaw sensor tells the computer the vehicle's actual direction. Tech B says that raising a vehicle's curb height has no effect on the ESC system. Who is correct?
   A. Tech A
   B. Tech B
   C. Both A and B
   D. Neither A nor B

_____ 9. Tech A says that a scan tool may be required to bleed air from the ABS. Tech B says that some vehicles have a high-pressure accumulator that may need to have the pressure bled off before hydraulic brake repairs are made. Who is correct?
   A. Tech A
   B. Tech B
   C. Both A and B
   D. Neither A nor B

_____ 10. Tech A says that some ABS wheel speed sensors create a square wave digital pattern. Tech B says that some wheel speed sensors create an AC sine wave pattern. Who is correct?
   A. Tech A
   B. Tech B
   C. Both A and B
   D. Neither A nor B

Principles of Electrical Systems

CHAPTER 36

Tire Tread:
© AbleStock

Chapter Review

The following activities have been designed to help you refresh your knowledge of this chapter. Your instructor may require you to complete some or all of these activities as a regular part of your training program. You are encouraged to complete any activity that your instructor does not assign as a way to enhance your learning.

Matching

Match the following terms with the correct description or example.

- A. Alternating current (AC)
- B. Amp
- C. Capacitor
- D. Circuit breaker
- E. Diode
- F. Electromagnet
- G. Energy
- H. Ground
- I. Insulator
- J. Invertor
- K. Ohm
- L. Phase
- M. Polarity
- N. Semiconductor
- O. Short
- P. Silicon
- Q. Solenoid
- R. Switch
- S. Transistor
- T. Volt

_____ 1. A term used to describe one set of windings from an alternator or alternating current electric motor.

_____ 2. A material that has properties that prevent the easy flow of electricity. These materials are made up of atoms with five to eight electrons in the valance ring.

_____ 3. An electromagnet with a moving iron core that is used to cause mechanical motion.

_____ 4. A conductor wound in a coil that produces a magnetic field when current flows through it.

_____ 5. The state of charge, positive or negative.

_____ 6. A two-lead electronic component that allows current flow in one direction only.

_____ 7. A material commonly used to make semiconductors.

_____ 8. A type of current flow that flows back and forth.

_____ 9. A semiconductor device that allows a small current in the base lead to control a larger current through the emitter collector leads.

_____ 10. The ability to do work.

_____ 11. Also called a short circuit, the flow of current along an unintended route.

_____ 12. A device that can quickly store a small amount of electrical energy, at which point it is charged.

_____ 13. An electrical device with contacts that turns current flow on and off.

_____ 14. A device that changes direct current into alternating current.

_____ 15. An abbreviation for amperes, the unit for current measurement.

_____ 16. The unit for measuring electrical resistance.

_____ 17. The unit used to measure potential difference or electrical pressure.

_____ 18. A material used to make microchips, transistors, and diodes.

_____ 19. A device that trips and opens a circuit, preventing excessive current flow in a circuit. It is resettable to allow for reuse.

_____ 20. The return path for electrical current in a vehicle chassis, other metal of the vehicle, or dedicated wire.

Multiple Choice

Read each item carefully, and then select the best response.

_____ 1. Electromotive force is also referred to as _____.
   A. voltage
   B. resistance
   C. amperage
   D. electrons

_____ 2. The number of charge carriers, either an excess of electrons or deficiency of electrons can be altered by _____, or adding very small quantities of impurities to a semiconductor material.
   A. spiking
   B. doping
   C. diluting
   D. charging

_____ 3. The point where P-type and N-type semiconductors join is called the _____.
   A. depletion area
   B. hole
   C. pn junction
   D. carrier

_____ 4. All of the following are materials used to make semiconductors, *except*:
   A. germanium
   B. mica
   C. silicon carbide
   D. gallium-arsenide

_____ 5. Which type of current flow is produced by a battery?
   A. Alternating current
   B. Three phase
   C. Direct current
   D. Sine wave

_____ 6. _____ refers to the number of separate staggered power windings in the motor or alternator.
   A. Phase
   B. Sine wave
   C. Hertz
   D. Current

_____ 7. The term _____ describes a low-voltage circuit that does not have a complete circuit and therefore cannot conduct current.
   A. closed
   B. open
   C. short
   D. grounded

_____ 8. When two dissimilar metals are immersed in an electrolyte, the breakdown of chemicals into charged particles that results in a flow of electricity is called _____.
   A. photovoltaic effect
   B. induction
   C. electrolysis
   D. electrostatic energy

_____ 9. All of the following are effects of the flow of electricity, except:
   A. Chemical reactions
   B. Mechanical action
   C. Heat
   D. Magnetism

_____ 10. Which of the following Ohm's law formulas is correct?
   A. $A = V/R$
   B. $V = A \times R$
   C. $R = V \div A$
   D. All of the above

Chapter 36 Principles of Electrical Systems

_____ 11. The rate of transforming energy is also known as _____.
   A. power
   B. voltage
   C. work
   D. magnetism

_____ 12. In a(n) _____ circuit, all components are connected directly to the voltage supply.
   A. series
   B. parallel
   C. series/parallel
   D. all of the above

_____ 13. Which law states that current entering any junction is equal to the sum of the current flowing out of the junction?
   A. Ohm's law
   B. Kirchhoff's current law
   C. The law of conservation of energy
   D. Newton's first law of energy

_____ 14. A _____ is made up of an electromagnet, a set of switch contacts, terminals, and the case.
   A. thermocouple
   B. solenoid
   C. diode
   D. relay

_____ 15. An ignition coil can be described as a _____.
   A. solenoid
   B. relay
   C. step-up transformer
   D. step-down transformer

_____ 16. Mechanical variable resistors with three connections, two fixed and one moveable, are called _____.
   A. thermistors
   B. rheostats
   C. potentiometers
   D. transistor

_____ 17. A _____ can be thought of as the electronic version of a one-way check valve.
   A. transistor
   B. diode
   C. resistor
   D. capacitor

_____ 18. A _____ is a semiconductor device used as a switch and to amplify currents.
   A. diode
   B. capacitor
   C. resistor
   D. transistor

_____ 19. All of the following are examples of the type of shielding used in shielded wiring harnesses, except:
   A. Twisted pair
   B. Drain lines
   C. Seamless plastic
   D. Mylar tape

_____ 20. Which of the following is an example of a crimp type terminal?
   A. Push-on spade
   B. Butt connector
   C. Eye ring
   D. All of the above

True/False

If you believe the statement to be more true than false, write the letter "T" in the space provided. If you believe the statement to be more false than true, write the letter "F".

_____ 1. A deficiency of electrons gives an atom an overall positive charge.
_____ 2. Most wiring diagrams are written from the conventional theory perspective, while electronic circuits are typically designed and operate on the electron theory perspective.
_____ 3. Gallium-arsenide has been used to create blue light-emitting diodes, and it can withstand high operating temperatures.
_____ 4. Hertz is the measurement of frequency and indicates the number of cycles per second.
_____ 5. Voltage drop testing is the best way of finding high resistance in the feed side or ground side of the circuit.
_____ 6. A potential electrical difference across a crystal that will physically distort the crystal is called electromagnetic induction.
_____ 7. When negative ions in a solution are attracted to the negative plate and positive ions to the positive plate, a chemical reaction can occur.
_____ 8. Ohm's law is a relationship between volts, amps, and ohms, and because they must always balance out, if we know any two of the values, then we can calculate the third.
_____ 9. A light bulb uses a certain amount of electrical power, but the power used is not an indication of brightness.
_____ 10. Some capacitors, and most semiconductor components, are polarity sensitive.
_____ 11. Flasher cans are electronic devices, while flasher controls are mechanical devices.
_____ 12. A solenoid is an electromechanical device that converts electrical energy into mechanical linear movement.
_____ 13. In a motor, the armature and brushes act as switches to control the current flow through the windings of the commutator.
_____ 14. A thermistor is a mechanical variable resistor with two connections.
_____ 15. When a capacitor is charged, one surface is positively charged and the other is negatively charged.
_____ 16. A PNP transistor has an N-type semiconductor between two P-types.
_____ 17. A control module or unit is a generic term that identifies an electronic unit that controls one or more electrical systems in the vehicle.
_____ 18. Most speed control systems use pulse-width modulation to control motor speed.
_____ 19. Printed circuits are essentially a map of all of the electrical components and their connections.
_____ 20. A controlled area network bus uses two thin wires to connect, or multiplex, many of the control units and their sensors to each other.

Fill in the Blank

Read each item carefully, and then complete the statement by filling in the missing word(s).

1. _____ _____ are only loosely held by the nucleus and are free to move from one atom to another when an electrical potential (pressure) is applied.
2. Electrical _____, measured in ohms, affects the current flow in a circuit.
3. _____ _____ tells us that if we increase current flow through a resistance, the voltage used by that resistance will increase.
4. A doped semiconductor always has an excess of one type of charge carrier; electrons in excess make it an _____-_____ semiconductor, and holes in excess make it a _____-_____ semiconductor.
5. Within the PN junction some electrons and holes cancel each other out and few charge carriers are present; this very thin _____ _____ acts like an insulator.
6. Voltage can be measured by hooking a voltmeter or _____, set to read voltage across two parts of a circuit where you want to measure the difference in volts.

7. Devices are available to change DC into AC and are called _____.
8. _____ is achieved when an electrical circuit has a continuous and uninterrupted electrical connection and is thereby capable of conducting current and working as designed.
9. In its purest definition, the term _____ describes a circuit fault in which current takes a shorter path, resistance-wise, through an accidental or unintended route.
10. The process of converting light to electricity is called the _____ effect.
11. In electrical systems, _____ occurs when a current passes through a conductor and a field is created around it.
12. One _____ is produced when 1 volt causes 1 amp of current to flow.
13. The _____ _____ across each resistor can be found by subtracting the voltage after a resistor from the voltage before it, or the difference can be measured.
14. A(n) _____-_____ relay acts like a mechanical relay but does not have any moving parts.
15. The metal core of a solenoid, used by the electromagnet to strengthen the magnetic field, is referred to as a(n) _____.
16. As _____ voltage is reached, the Zener diode's resistance suddenly collapses.
17. _____ _____ are added to vehicles to build in specific time delays in turning on or off electrical devices.
18. There are many different types of sensors installed in the modern vehicle; they are used to provide information to the _____ on the vehicle.
19. There are two scales used to measure the sizes of wires: the metric wire gauge and the _____ _____.
20. Wiring _____, also known as wiring looms or cable harnesses, are used throughout the vehicle to group two or more wires together within a sheath of either insulating tape or tubing.

Labeling

Label the following diagrams with the correct terms.

1. Parts of an atom (including charge):

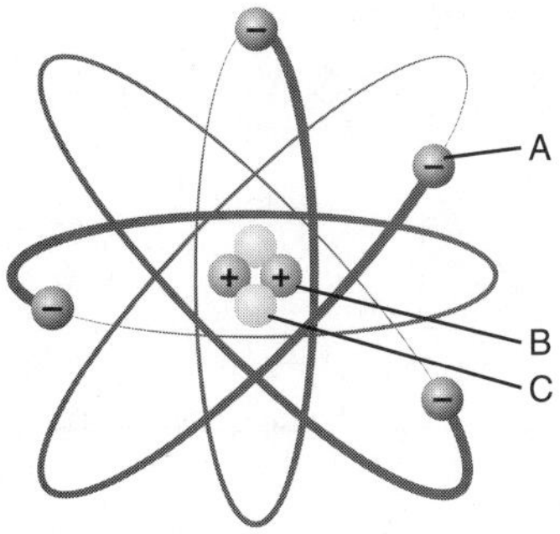

A. _____ Charge _____
B. _____ Charge _____
C. _____ Charge _____

2. Ions:

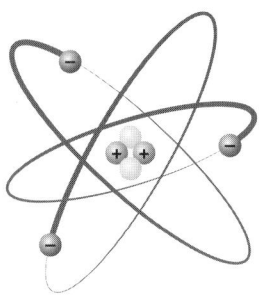 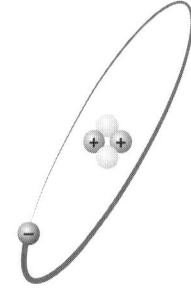

A   B

A. _____

B. _____

3. Insulators:

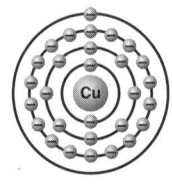 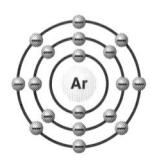 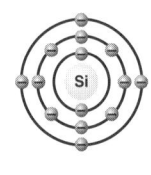

A   B   C

A. _____

B. _____

C. _____

4. Short circuit:

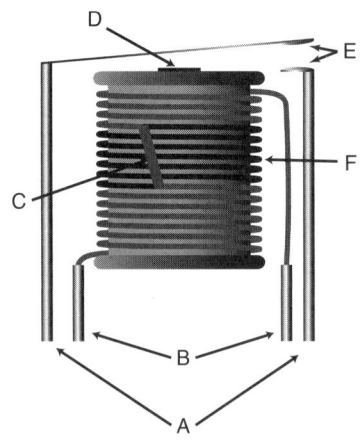

A. _____

B. _____

C. _____

D. _____

E. _____

F. _____

5. Diode symbols:

A   B

A. _____

B. _____

6. Wiring diagram symbols:

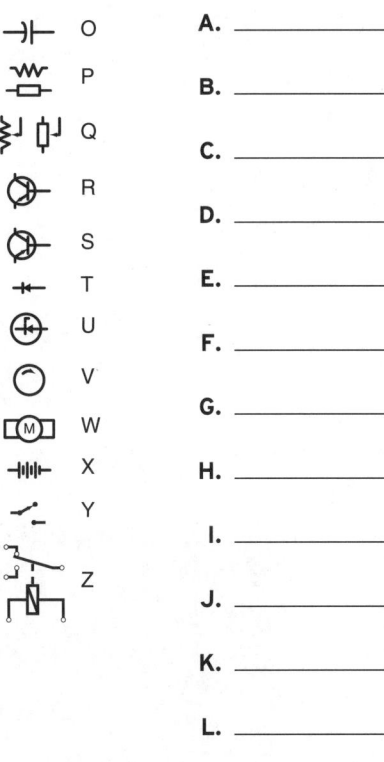

A. _____
B. _____
C. _____
D. _____
E. _____
F. _____
G. _____
H. _____
I. _____
J. _____
K. _____
L. _____
M. _____
N. _____
O. _____
P. _____
Q. _____
R. _____
S. _____
T. _____
U. _____
V. _____
W. _____
X. _____
Y. _____
Z. _____

7. Transistors:

A

| N | P | N |

B

| P | N | P |

A. _____

B. _____

Skill Drills

Test your knowledge of skill drills by filling in the correct words in the photo captions.

1. Stripping Wire Insulation:

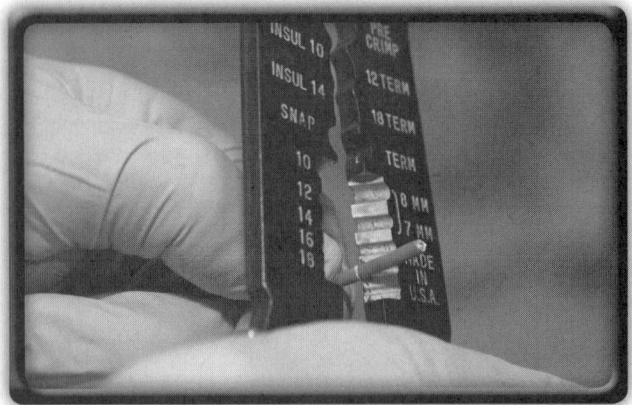

Step 1: Choose the correct _____ _____.

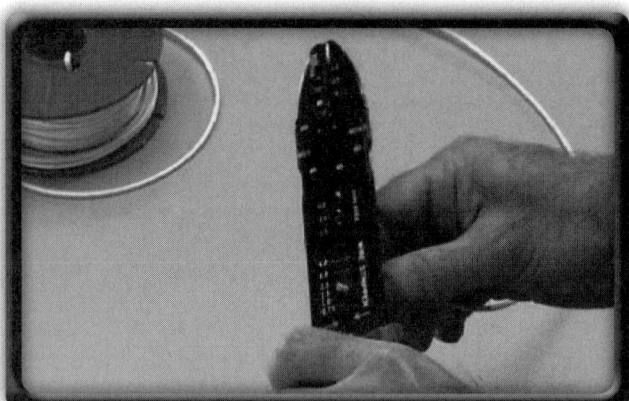

Step 2: Select the hole that matches the _____ of the wire to be stripped. Place the wire in the hole, and close the _____ firmly around it to cut the _____.

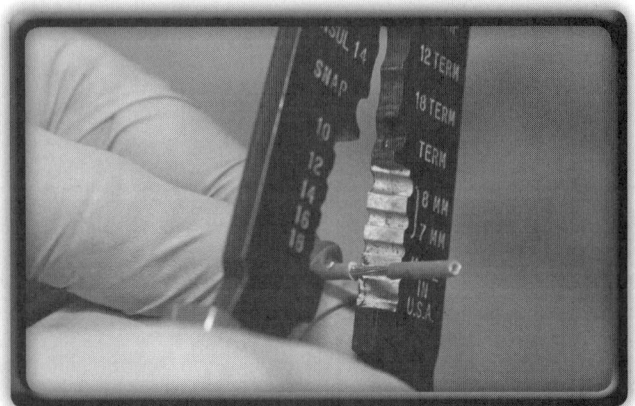

Step 3: Remove the insulation. To keep the _____ together, give them a light _____.

2. Installing a Solderless Terminal:

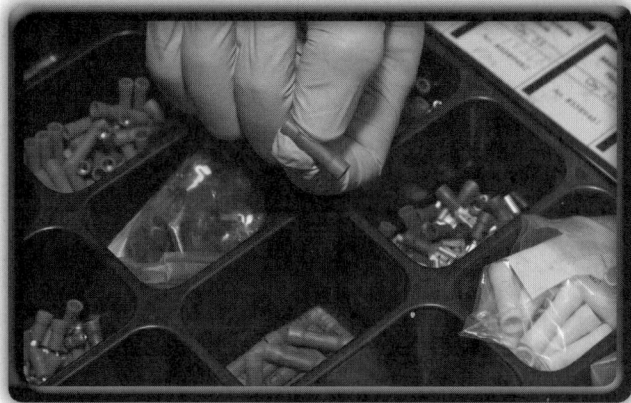

Step 1: Make sure you have the correct size of _____ for the wire to be terminated and the terminal has the correct volt/amp _____. Remove an appropriate amount of the protective _____ from the wire.

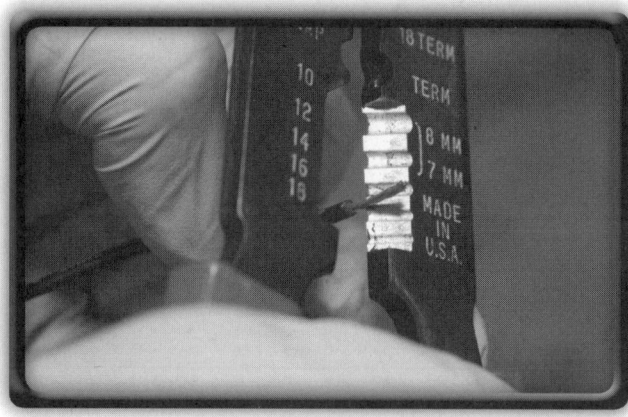

Step 2: Lightly _____ the wire _____ and place the terminal onto the _____.

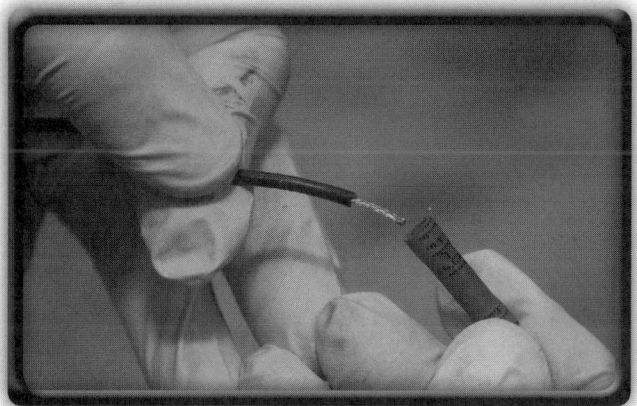

Step 3: Use a proper _____ _____ for the terminal you are crimping. Do not use _____, as they have a tendency to cut through the connection. Select the proper _____.

Step 4: Crimp the _____ section first. Use firm _____ so a good electrical contact will be made, but not _____ force, as this can _____ the pin or terminal.

Step 5: If crimping an _____ terminal, lightly crimp the insulation _____ so they hold the _____ firmly.

3. Soldering Wires and Connectors:

Step 1: Safely position the soldering iron while it is _____ _____. While the soldering iron is heating, _____ an appropriate amount of the protective _____ from the wires with _____ _____.

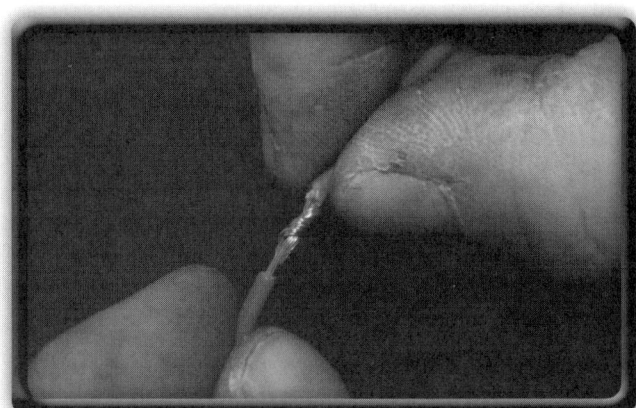

Step 2: _____ the wires together to make a good _____ _____ between them.

Chapter 36 Principles of Electrical Systems

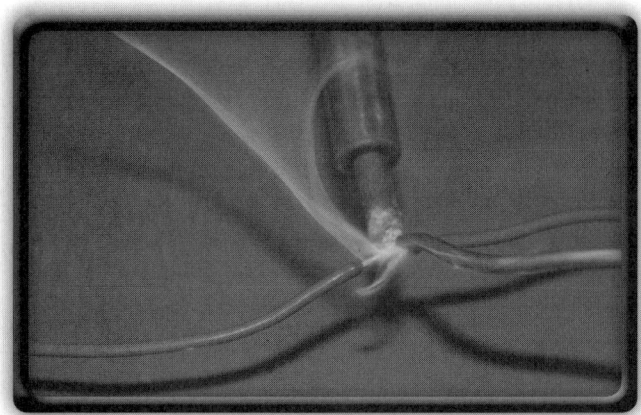

Step 3: _____ the soldering iron tip, and gently heat up the _____ while placing the _____ opposite of the soldering iron. Allow the solder to be _____ into the joint.

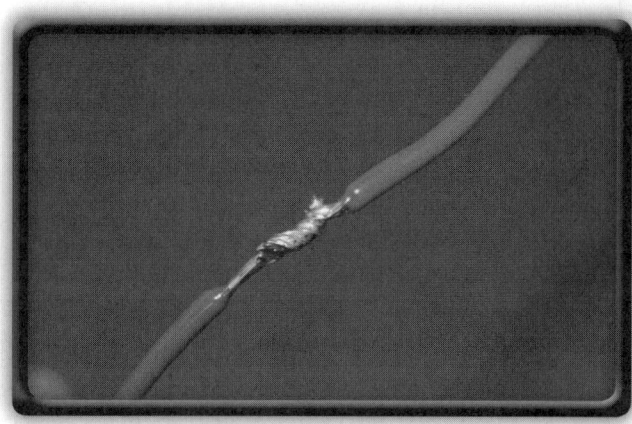

Step 4: A good _____ _____ is where the solder has been _____ in.

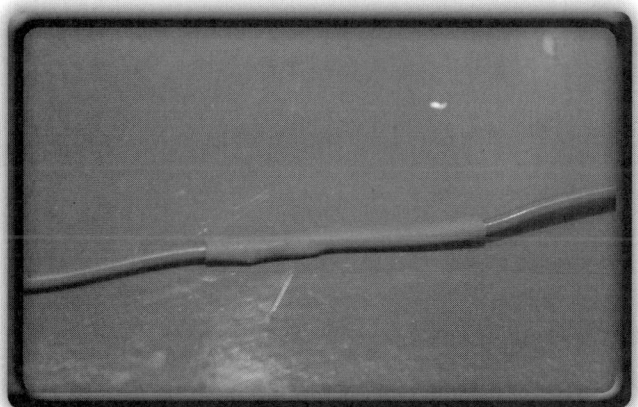

Step 5: Once the electrical connection has been made and it has _____ enough for you to handle it, slide the insulator _____ cover over the joint and use a _____ _____ to shrink the tubing around the _____.

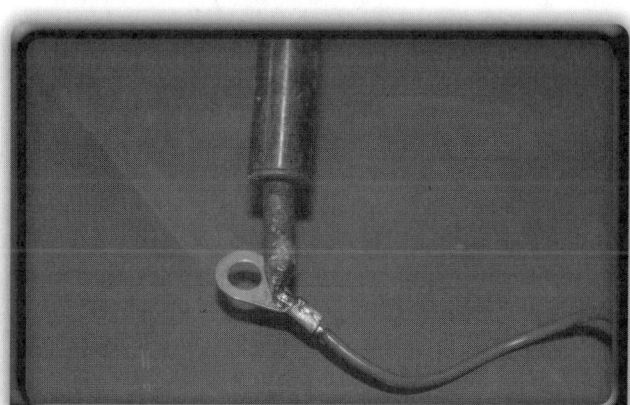

Step 6: To solder a wire to a terminal _____, it is best to _____ it in place as before and use the solder to "glue" the joint together. Place the heated iron onto the _____ to get it hot enough to _____ the solder applied to the end of the crimped wire tabs. Some solder will be _____ between the terminal and the wire. Cover the terminal with _____-_____ tubing.

Crossword Puzzle

Use the clues to complete the puzzle.

Across

3. A safety device that self-destructs to prevent excessive current flowing in a circuit in the event of a fault.
4. A method of using electrical current to create a chemical reaction.
5. The loss of electrical insulation properties.
7. A mobile particle that has a positive or negative electrical charge.
9. The rotating wire coils in motors and generators. It is also the moving part of a solenoid or relay, and the pole piece in a permanent magnet generator.
12. The resistance of a component or circuit to a low resistance. It can also refer to a faulty circuit where a section or component has excess unwanted resistance.
13. A type of electricity in which a material such as a quartz crystal produces voltage when mechanical pressure distorts it.
14. A semiconductor used in high-frequency circuits.

Down

1. The introduction of impurities to pure semiconductor materials to provide N- and P-type semiconductors.
2. A mechanism that turns the vehicle's turn signal and hazard flasher bulbs on and off.
3. Fine glass fibers through which light is transmitted.
6. A measurement of the rate at which electricity is consumed or created.

8. The unit for electrical frequency measurement.
10. An electron located on the outer ring, called the valence ring, that is only loosely held by the nucleus and that is free to move from one atom to another when an electrical potential (pressure) is applied.
11. A device used to measure current flow.

ASE-Type Questions

Read each item carefully, and then select the best response.

_____ 1. Tech A says that if the specified fuse keeps blowing, it is generally OK to replace it with a larger fuse. Tech B says that a fusible link is one type of circuit protection device. Who is correct?
   A. Tech A
   B. Tech B
   C. Both A and B
   D. Neither A nor B

_____ 2. Tech A says that the movement of electrons in a circuit is called current flow. Tech B says that the movement of electrons in a circuit is measured in amps. Who is correct?
   A. Tech A
   B. Tech B
   C. Both A and B
   D. Neither A nor B

_____ 3. Tech A says that electromotive force is also known as voltage. Tech B says that when electrons flow in one direction only, this is DC. Who is correct?
   A. Tech A
   B. Tech B
   C. Both A and B
   D. Neither A nor B

_____ 4. Tech A says that 18 AWG wire is larger than 12 AWG wire. Tech B says that the larger the diameter of the conductor, the more electrical resistance it has. Who is correct?
   A. Tech A
   B. Tech B
   C. Both A and B
   D. Neither A nor B

_____ 5. Two technicians are discussing electron flow. Tech A says that in "conventional theory" current is believed to flow from positive to negative. Tech B says that in "electron theory" current is believed to flow from negative to positive. Who is correct?
   A. Tech A
   B. Tech B
   C. Both A and B
   D. Neither A nor B

_____ 6. Tech A says that hertz is the number of cycles per second. Tech B says that hertz is the amount of current flow produced by an alternator. Who is correct?
   A. Tech A
   B. Tech B
   C. Both A and B
   D. Neither A nor B

_____ 7. Two technicians are discussing a series circuit with four resistors of various resistances. Tech A says that current flow will be different in each resistor. Tech B says that current flow will be the same in each resistor. Who is correct?
   A. Tech A
   B. Tech B
   C. Both A and B
   D. Neither A nor B

_____ 8. Tech A says that a voltage drop is typically used to find excessive resistance in a circuit. Tech B says that high resistance creates heat at the point of resistance in the circuit. Who is correct?
   A. Tech A
   B. Tech B
   C. Both A and B
   D. Neither A nor B

_____ 9. Tech A says that in a series circuit with two resistors of 120 ohms each, the total circuit resistance is 240 ohms. Tech B says that in a parallel circuit with two resistors of 120 ohms each, the total circuit resistance is 240 ohms. Who is correct?
   A. Tech A
   B. Tech B
   C. Both A and B
   D. Neither A nor B

_____ 10. Tech A says that a relay is a one-way electrical check valve used in alternators to change AC into DC. Tech B says that a relay uses electromagnetism to open or close a switch. Who is correct?
   A. Tech A
   B. Tech B
   C. Both A and B
   D. Neither A nor B

Meter Usage and Circuit Diagnosis

CHAPTER 37

Tire Tread:
© AbleStock

Chapter Review

The following activities have been designed to help you refresh your knowledge of this chapter. Your instructor may require you to complete some or all of these activities as a regular part of your training program. You are encouraged to complete any activity that your instructor does not assign as a way to enhance your learning.

Matching

Match the following terms with the correct description or example.

- **A.** Digital volt-ohmmeter (DVOM)
- **B.** High resistance
- **C.** Min/max setting
- **D.** Oscilloscope
- **E.** Probing technique
- **F.** Short to power

_____ 1. A setting on a DVOM to display the maximum and minimum readings.
_____ 2. A term that describes a circuit or components with more resistance than designed.
_____ 3. A condition in which current flows from one circuit into another.
_____ 4. A test instrument that graphs voltage over time and displays the results on a screen.
_____ 5. The way in which test probes are connected to a circuit.
_____ 6. A test instrument with a digital display for measuring voltage, resistance, and current.

Multiple Choice

Read each item carefully, and then select the best response.

_____ 1. A digital volt-ohmmeter is used to measure _____.
   - **A.** voltage
   - **B.** amperage
   - **C.** resistance
   - **D.** all of the above

_____ 2. A digital volt-ohmmeter for three-phase fixed equipment and single-phase commercial lighting would have a _____ rating.
   - **A.** CAT I
   - **B.** CAT II
   - **C.** CAT III
   - **D.** CAT IV

_____ 3. The _____ setting is often used to measure vehicle battery voltage while the engine is cranking or the battery is charging.
   - **A.** hold
   - **B.** sample
   - **C.** min/max
   - **D.** save

_____ 4. The red lead for the DVOM is labeled with the _____ symbol.
   - **A.** +
   - **B.** V/Ω
   - **C.** –
   - **D.** ±

_____ 5. All of the following are examples of common types of DVOM probes, *except*:
   A. ground probes
   B. alligator clips
   C. fine-pin probes
   D. insulation piercing probes

_____ 6. The sum of all the _____ in a series circuit equals the supply voltage.
   A. resistors
   B. batteries
   C. voltage drops
   D. loads

_____ 7. The _____ is the same in all parts of a properly working series circuit.
   A. voltage
   B. current
   C. resistance
   D. all of the above

_____ 8. Ideally the voltage drop across a fuse attached in series with a 12V battery should be _____.
   A. 24 volts
   B. 12 volts
   C. 6 volts
   D. 0 volts

_____ 9. As the position of the wiper of a potentiometer is changed, so is the _____ to a load connected to the potentiometer.
   A. voltage
   B. resistance
   C. current flow
   D. both A and C

_____ 10. _____ can be fitted to the probe leads to reduce the maximum voltage to safe levels for the oscilloscope to measure.
   A. Attenuators
   B. Cuffs
   C. Alligator clips
   D. Current clamps

True/False

If you believe the statement to be more true than false, write the letter "T" in the space provided. If you believe the statement to be more false than true, write the letter "F".

_____ 1. Hybrid vehicles typically require meters and test leads rated as CAT III or CAT IV.

_____ 2. 5208 MV is the same as 5208 millivolts. It could also be called 52.08 volts.

_____ 3. The red lead is the positive lead, and the black lead is the negative lead.

_____ 4. After back-probing always reinsulate the hole that the probe makes with room temperature vulcanizing silicone to prevent any corrosion.

_____ 5. To accurately measure the resistance of a component, you should remove or isolate the component from the circuit.

_____ 6. A DVOM reading of OL means overload and indicates the voltage being read is higher than the maximum allowed for the range.

_____ 7. Voltage drop can be measured across components, connectors, or cables, but current has to be flowing to get an accurate measurement.

_____ 8. Parallel circuits are commonly used in the vehicle's electrical system, especially for lights.

_____ 9. Oscilloscopes display time on the vertical axis, and voltage is displayed on the horizontal axis, or from left to right across the screen.

_____ 10. If the resistance and voltage of a circuit are known, then the theoretical current can be calculated using Ohm's law.

Chapter 37 Meter Usage and Circuit Diagnosis

Fill in the Blank
Read each item carefully, and then complete the statement by filling in the missing word(s).

1. A basic digital _____-_____ can measure alternating current (AC) and direct current (DC) voltage, AC and DC amperage, and resistance.
2. _____ _____ are used to connect the DVOM to or into the circuit being tested and come in pairs: one red, the other black.
3. When the _____ function is activated, the display will hold the value on the display until the function or DVOM is turned off.
4. _____-_____ occurs when the probe is pushed in from the back of a connector to make a connection.
5. The _____ _____ fastens around the conductor and measures the strength of the magnetic field produced from current flowing through the conductor.
6. Unwanted voltage _____ in vehicle circuits can cause real problems and faults.
7. Current flow is _____ proportional to resistance.
8. The sum of the current flow in each branch is equal to the total _____ circuit current flow.
9. As a _____ charges, the voltage drop across it increases and the current flow decreases.
10. _____ _____ can occur anywhere in the circuit and can be difficult to locate, especially if it is intermittent.

Labeling
Label the following diagrams with the correct terms.

1. Measuring voltage and current flow in a circuit with a single resistor and a 12-volt DC supply:

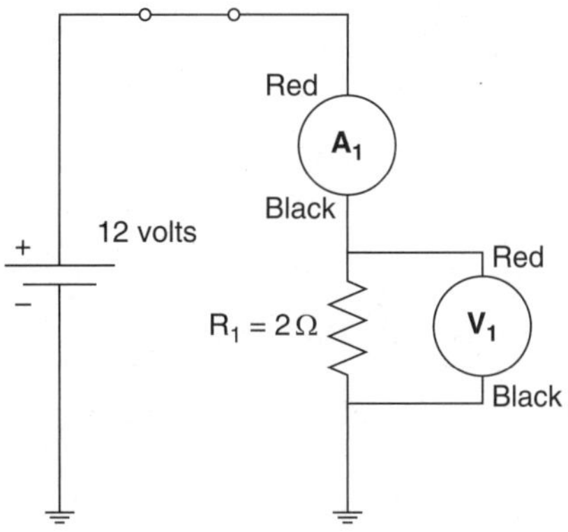

   A. Current flow at A_1: _____.

   B. Voltage at V_1: _____.

2. Measuring amperage in a circuit with two resistors in series with a 12-volt DC supply:

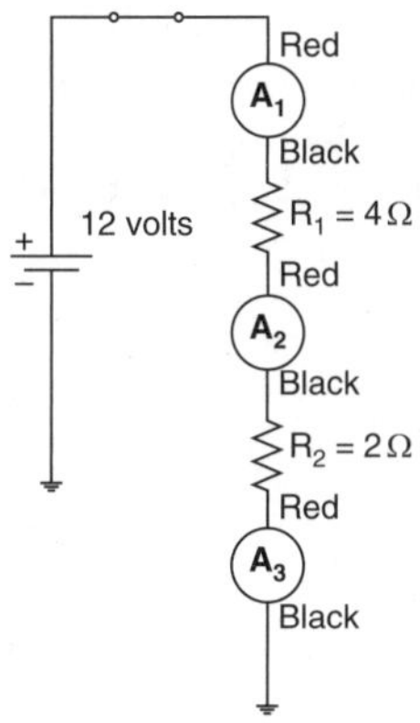

   A. Current flow at A_1: _____.

   B. Current flow at A_2: _____.

   C. Current flow at A_3: _____.

3. Measuring amperage in a circuit with a relay controlled by a switch and a single resistor with a 12-volt DC supply:

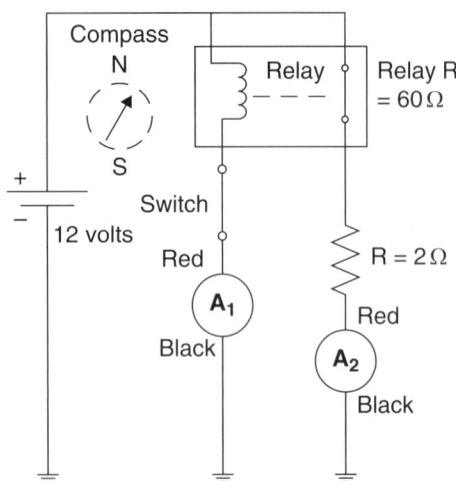

A. Current flow at A_1: _____.

B. Current flow at A_2: _____.

4. Measuring volts, amps, and ohms in a circuit with a 100 ohm resistor and a 12-volt DC supply:

A. Current flow at A_1: _____.

B. Voltage at V_1: _____.

C. Resistance at R_1: _____.

5. Measuring volts, amps, and ohms in a circuit with a 100 ohm resistor and a 12-volt DC supply in series with an LED:

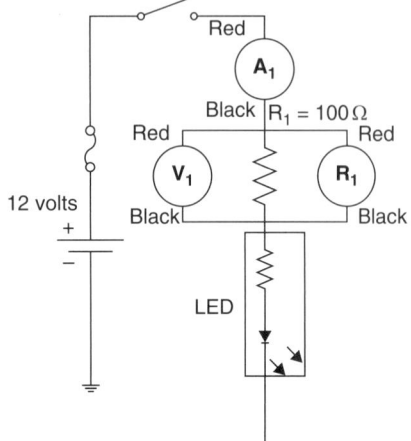

A. Current flow at A_1: _____.

B. Voltage at V_1: _____.

C. Resistance at R_1: _____.

6. Voltage measurements in a series circuit with unequal resistances:

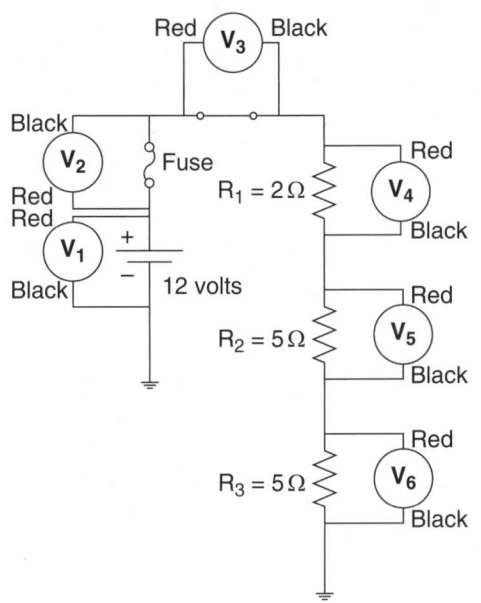

A. Voltage at V_1: _____.

B. Voltage at V_2: _____.

C. Voltage at V_3: _____.

D. Voltage at V_4: _____.

E. Voltage at V_5: _____.

F. Voltage at V_6: _____.

7. Current flow in a circuit with three unequal resistors in series:

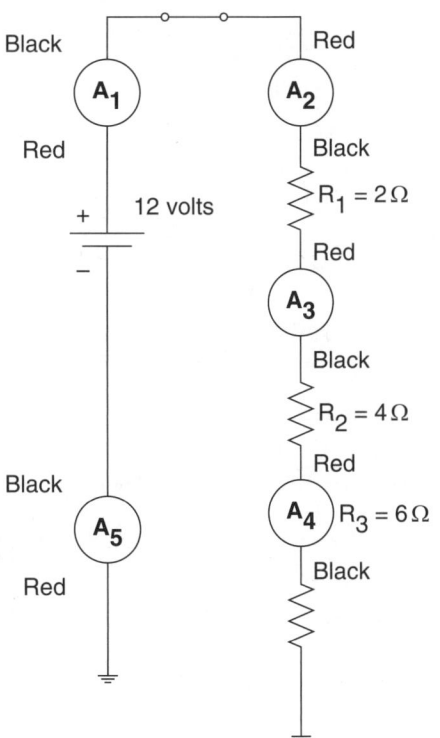

A. Current flow at A_1: _____.

B. Current flow at A_2: _____.

C. Current flow at A_3: _____.

D. Current flow at A_4: _____.

E. Current flow at A_5: _____.

8. Measuring voltage across three unequal resistors connected in parallel across a 12-volt supply:

A. Voltage at V_1: _____.

B. Voltage at V_2: _____.

C. Voltage at V_3: _____.

D. Voltage at V_4: _____.

9. Measuring amperage in a circuit with three resistors in parallel:

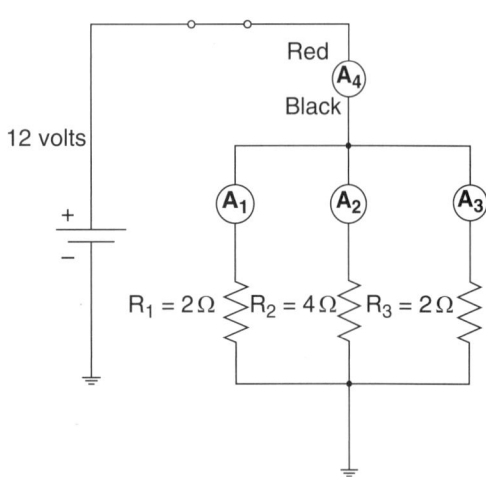

A. Current flow at A_1: _____.

B. Current flow at A_2: _____.

C. Current flow at A_3: _____.

D. Current flow at A_4: _____.

10. Series-parallel circuit formed by resistors R_1, R_2, R_3, and R_4:

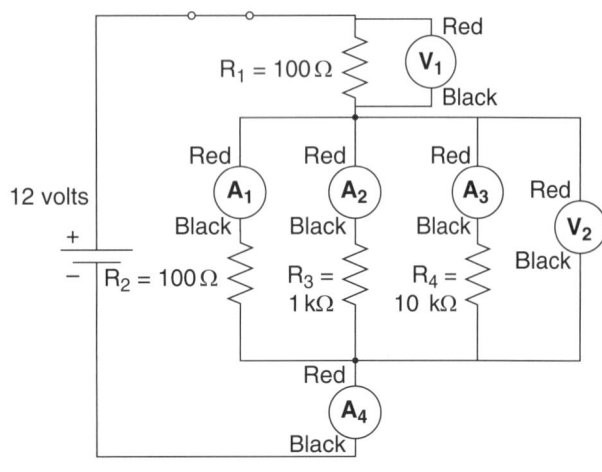

A. Voltage at V_1: _____.

B. Voltage at V_2: _____.

C. Current flow at A_1: _____.

D. Current flow at A_2: _____.

E. Current flow at A_3: _____.

F. Current flow at A_4: _____.

Skill Drills

Test your knowledge of skill drills by filling in the correct words in the photo captions.

1. Checking Circuit Waveforms:

Step 1: Determine the circuit to be tested and the likely _____ and frequency of the _____ to be measured. Set the voltage level and _____ _____.

Step 2: Connect the _____ to the _____ in the circuit to be measured.

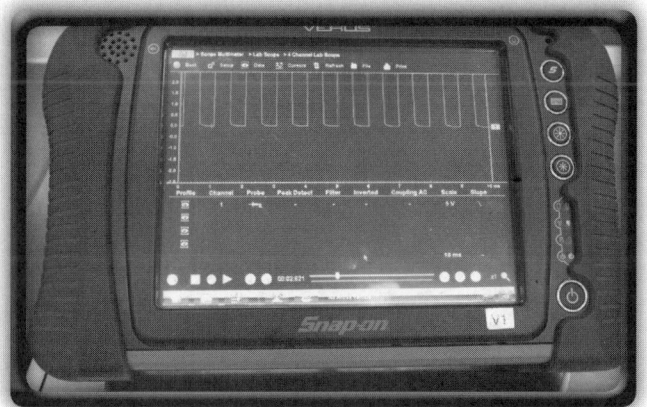

Step 3: _____ waveforms from the circuit being tested. _____ the waveform, comparing it to the manufacturer's specifications or _____ good waveforms.

2. Using a DVOM to Measure Voltage:

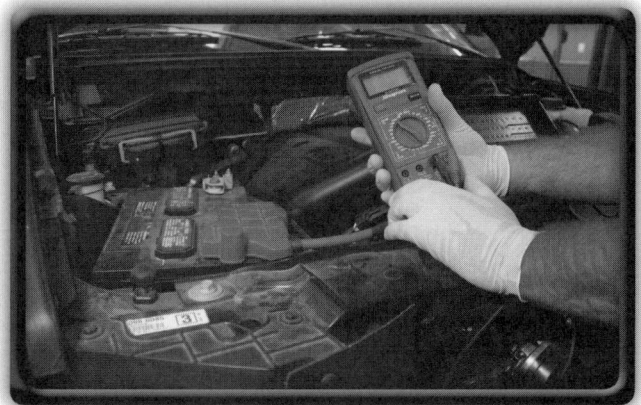 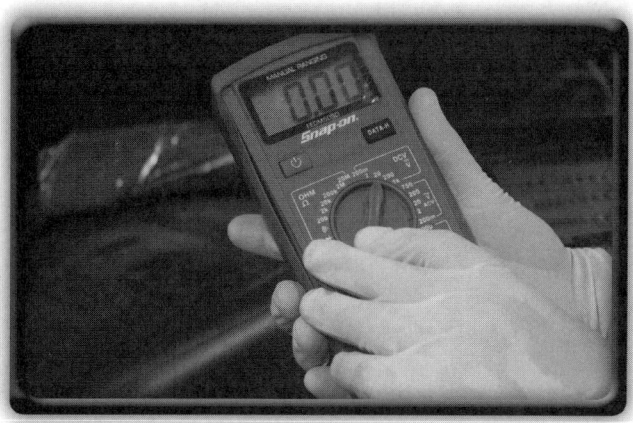

Step 1: Prepare the DVOM for testing _____ by connecting the _____ lead to the COM terminal and the _____ lead to the Volt/Ohms (V/_) terminal.

Step 2: Turn the _____ dial until you have selected the _____ for volts DC. The reading on the DVOM should now be at _____.

Step 3: Connect the black lead to the _____ battery post and the red lead to the _____ post. Measure the _____, and interpret the results.

3. Checking a Circuit with a Test Light:

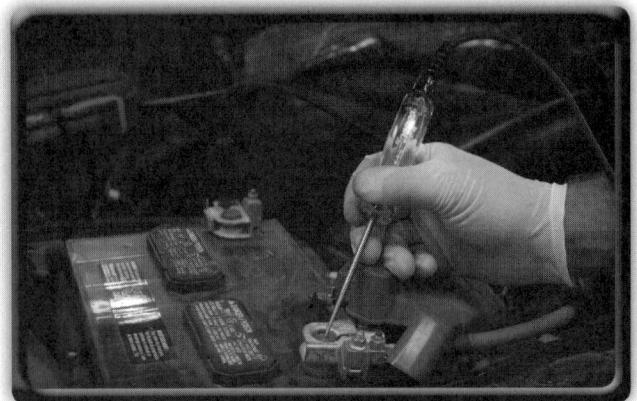

Step 1: Connect the end of the light with the _____ on it to the negative battery _____. Touch the _____ end of the test light to the positive battery terminal. The _____ should come on.

Step 2: Connect the clip to a known good _____. A typical known good ground is any unpainted _____ surface on the vehicle that is directly _____ to the battery ground return system.

Step 3: Place the _____ on the terminal to be _____. If _____ is present, the light will come on.

Crossword Puzzle

Use the clues to complete the puzzle.

Across

1. A test instrument that graphs voltage over time and displays the result on a screen.
3. Fault conditions in a circuit where the circuit is unintentionally contacting a grounded component or wire. This may result in a short, in the case of a power wire, or it could cause a circuit to stay live in the case of a switched ground circuit.
5. A device that reduces the power of a signal without distorting its waveform.
6. A condition in which the current flows along an unintended route.

Down

2. A circuit that has a break that prevents current from flowing.
4. A graphical plot of voltage over time that is displayed on an oscilloscope.

ASE-Type Questions

Read each item carefully, and then select the best response.

_____ 1. Tech A says that total resistance goes up as more parallel paths are added. Tech B says that total amperage goes up as more parallel paths are added. Who is correct?
   A. Tech A
   B. Tech B
   C. Both A and B
   D. Neither A nor B

_____ 2. Tech A says that when reading DC voltage on a meter, a "+" before the number means that there is a higher voltage at the red lead than the black lead. Tech B says that a "–" before the number means that there is a lower voltage at the red lead than the black lead. Who is correct?
   A. Tech A
   B. Tech B
   C. Both A and B
   D. Neither A nor B

3. Tech A says that to read amperage the meter needs to be hooked up in series in a circuit. Tech B says that to read amperage at a load, place one lead of the ammeter on the input side of the load and the other lead on the output. Who is correct?
   A. Tech A
   B. Tech B
   C. Both A and B
   D. Neither A nor B

4. Tech A says that when checking amperage on a battery, hook the red lead to positive and the black lead to ground. Tech B says that a resistance reading on a light bulb requires the DVOM to be hooked to each side of the bulb and the switch turned on. Who is correct?
   A. Tech A
   B. Tech B
   C. Both A and B
   D. Neither A nor B

5. Tech A says that when checking a voltage drop across an open switch, a measurement of 12 volts means the circuit is open. Tech B says that when checking voltage drop across an open switch, a measurement of 12 volts means the switch contacts are closed. Who is correct?
   A. Tech A
   B. Tech B
   C. Both A and B
   D. Neither A nor B

6. A customer complains of slow engine cranking. Tech A says that the starter is faulty and should be replaced. Tech B says that performing a voltage drop test on the high current side of the starter circuit is a valid test in this situation. Who is correct?
   A. Tech A
   B. Tech B
   C. Both A and B
   D. Neither A nor B

7. Tech A says that when resistance increases, current flow decreases. Tech B says that when voltage decreases, current increases. Who is correct?
   A. Tech A
   B. Tech B
   C. Both A and B
   D. Neither A nor B

8. Tech A says that an open circuit will typically cause higher current flow, which will blow the fuse. Tech B says that a short to ground can typically be found by checking voltage at different points in the circuit. Who is correct?
   A. Tech A
   B. Tech B
   C. Both A and B
   D. Neither A nor B

9. Tech A says that when a fuse has popped, it just needs to be replaced, since it performed its job. Tech B says that when a fuse has blown, the circuit needs to be diagnosed because the fuse was probably not the problem. Who is correct?
   A. Tech A
   B. Tech B
   C. Both A and B
   D. Neither A nor B

10. Two technicians are discussing scope testing. Tech A says that the vertical axis of the screen shows voltage. Tech B says that the horizontal axis shows time. Who is correct?
    A. Tech A
    B. Tech B
    C. Both A and B
    D. Neither A nor B

Batteries, Starting, and Charging Systems

CHAPTER 38

Tire Tread:
© AbleStock

Chapter Review

The following activities have been designed to help you refresh your knowledge of this chapter. Your instructor may require you to complete some or all of these activities as a regular part of your training program. You are encouraged to complete any activity that your instructor does not assign as a way to enhance your learning.

Matching

Match the following terms with the correct description or example.

- **A.** Absorbed glass mat
- **B.** Battery terminal configuration
- **C.** Cold cranking amps (CCA)
- **D.** Counter-electromotive force (CEMF)
- **E.** Cranking amps (CA)
- **F.** Hold-in winding
- **G.** Keep alive memory (KAM)
- **H.** Pull-in winding

_____ 1. A low-current winding found in starter solenoids that holds the plunger in the activated position.

_____ 2. A standard for rating the ability of a vehicle battery to supply high current under cold operating conditions.

_____ 3. The placement of positive and negative battery terminals.

_____ 4. A certain minimum amount of parasitic current draw that is used by the vehicle's circuits to maintain memory functions and monitor systems.

_____ 5. Voltage created in the field windings as the motor rotates, which opposes battery voltage and limits motor speed.

_____ 6. A high-current winding found in starter solenoids that pulls the solenoid plunger into the activated position.

_____ 7. A type of lead-acid battery.

_____ 8. A standard similar to CCA but that measures the battery's function at a higher temperature—32°F (0°C).

Multiple Choice

Read each item carefully, and then select the best response.

_____ 1. A standard 12-volt car battery consists of _____ cells connected in series.
   - **A.** 1
   - **B.** 2
   - **C.** 6
   - **D.** 12

_____ 2. All of the following are methods used to rate automotive battery capacity, *except*:
   - **A.** cold cranking amps
   - **B.** voltage output
   - **C.** cranking amps
   - **D.** reserve capacity

_____ 3. A(n) _____ battery typically has no removable cell covers, so you cannot adjust or test the fluid levels inside.
   - **A.** deep cycle
   - **B.** absorbed glass mat
   - **C.** low-maintenance
   - **D.** reserve capacity

_____ 4. Which of the following is a type of rechargeable cell battery?
   A. Lead-acid
   B. Nickel-cadmium
   C. Lithium ion
   D. All of the above

_____ 5. All of the following are advantages of lithium-ion batteries, *except*:
   A. high energy density
   B. shelf life
   C. low internal resistance
   D. low self-discharge

_____ 6. Which tool is commonly used to measure parasitic draw from the battery?
   A. Low-amp clamp
   B. Scan tool
   C. Ammeter
   D. Either A or C

_____ 7. Performing a battery state-of-charge test with a(n) _____ will give a good indication of whether the battery needs to be charged or not.
   A. refractometer
   B. hydrometer
   C. DVOM
   D. any of the above

_____ 8. The higher the _____, the higher the percentage of acid in the electrolyte, which corresponds to a high battery state of charge.
   A. specific gravity
   B. water level
   C. temperature
   D. amperage

_____ 9. The starter motor is mounted on the transmission or cylinder block in a position to engage a _____ around the outside edge of the engine flywheel, flex plate, or torque converter.
   A. magnet
   B. commutator
   C. ring gear
   D. sleeve

_____ 10. In the direct-drive system, the starter drive is mounted directly on one end of the _____.
   A. flywheel
   B. armature shaft
   C. transmission
   D. reduction gear

_____ 11. The commutator end frame carries copper-impregnated carbon _____, which conduct current through the armature when it is being rotated in operation.
   A. spur gears
   B. brushes
   C. coils
   D. strips

_____ 12. A(n) _____ consists of two semicircular segments that are connected to the two ends of the loop and are insulated from each other.
   A. helix
   B. armature
   C. commutator
   D. electromagnet

_____ 13. The _____ is typically a cylindrical device mounted on the starter motor that switches the high current flow required by the starter motor on and off and engages the starter drive with the ring gear.
   A. solenoid
   B. armature
   C. commutator
   D. pinion

Chapter 38   Batteries, Starting, and Charging Systems

_____ 14. For many years, manufacturers have placed switches in _____ with the starter solenoid windings, which prevents the starter from being activated.
   A. parallel
   B. series
   C. series/parallel
   D. either A or C

_____ 15. The _____ prevents the starter motor from being driven by the engine once the engine starts, which would spin the armature faster than it could handle.
   A. flywheel
   B. starter control circuit
   C. pole shoe
   D. overrunning clutch

_____ 16. The _____ converts mechanical energy into electrical energy by electromagnetic induction.
   A. DC generator
   B. inverter
   C. alternator
   D. rectifier

_____ 17. The voltage potential induced by an AC generator is called _____.
   A. electromotive force
   B. electromagnetic induction
   C. counter-electromotive force
   D. either A or B

_____ 18. To change AC to DC, automotive alternators use a rectifier assembly consisting of _____ in a specific configuration.
   A. transistors
   B. diodes
   C. resistors
   D. semiconductors

_____ 19. In the _____ method of connection, one end of each phase winding is taken to a central point where the ends are connected together.
   A. wye
   B. delta
   C. triangle
   D. square

_____ 20. The _____ is an electromagnet that rotates freely in the alternator and is supported on each end by ball bearings.
   A. stator
   B. slip ring
   C. voltage regulator
   D. rotor

True/False

If you believe the statement to be more true than false, write the letter "T" in the space provided. If you believe the statement to be more false than true, write the letter "F".

_____ 1. Batteries store electricity in chemical form.

_____ 2. The sulfuric acid contained in batteries is highly corrosive and can also be very harmful to metal, painted surfaces, and skin.

_____ 3. Sizing has to do with the electrical specifications of the battery, while ratings have to do with the battery's physical attributes.

_____ 4. Reserve capacity is the time in minutes that a new fully charged battery at 80°F (27°C) will supply a constant load of 25 amps without its voltage dropping below 10.5 volts for a 12-volt battery.

_____ 5. The typical cell voltage of a lithium-ion battery is 1.2 volts.

_____ 6. Installing a battery into a vehicle backward can instantly destroy on-board electronics.

_____ 7. When jump starting a vehicle the slave battery is connected in series to the host (discharged) battery in order to provide additional capacity to crank and start the vehicle.

_____ 8. Many municipalities require battery recycling and levy a "core charge" on every new automotive battery sold.

_____ 9. State-of-charge testing tells us how much capacity the battery has left.

_____ 10. In some cases, it may be possible to use a 9-volt memory minder or memory saver to maintain the vehicle's memory while the vehicle battery is disconnected.

_____ 11. All vehicles equipped with an automatic transmission use a neutral safety switch or a similar device.

_____ 12. The starter motor converts mechanical energy to electrical energy for the purpose of cranking the engine over.

_____ 13. A conductor loop that can freely rotate within the magnetic field is the most efficient motor design.

_____ 14. The hold-in winding draws a higher current and creates a stronger magnetic field than the pull-in winding.

_____ 15. Some Ford vehicles use a separate starter relay in the engine compartment, instead of a solenoid, to control the high current for the starter motor.

_____ 16. Some immobilizer systems now use keyless starting. The vehicle has a start button on the dash and does not require the key to be inserted into an ignition switch.

_____ 17. A diode bridge gets its name from two diodes in series bridged with a wire.

_____ 18. In the wye method, the windings are connected in the shape of a triangle.

_____ 19. It is not unusual for a typical modern light vehicle alternator to have an output of about 150 amps or more to ensure the electrical system is capable of powering the electrical systems.

_____ 20. The alternator's cooling fan is mounted on the rotor shaft and may be an integral part of the drive pulley or part of the rotor.

Fill in the Blank

Read each item carefully, and then complete the statement by filling in the missing word(s).

1. Each cell of a fully charged "12-volt" battery has a nominal _____ volts, for a total of _____ volts.

2. The more plate _____ _____ there is, the higher the electrical capacity of the battery.

3. Because the electrolyte in _____ _____ _____ batteries is a gel, which does not spill, this type of battery is especially handy for rough handling or tipping.

4. The charging process increases the amount of _____ in the electrolyte, making the electrolyte stronger.

5. The _____ the battery temperature, the higher the rate of charging.

6. Lithium-ion batteries may suffer from _____ _____ and cell rupture if overheated or overcharged.

7. Battery cables and terminals are designed to carry _____ _____ currents that are required during cranking of the automotive engine.

8. Battery terminals are usually _____ or _____ onto the battery cables to ensure strong, low-resistance connections.

9. Always remove the _____ or ground terminal first when disconnecting battery cables.

10. Some manufacturers say that their batteries should not be load tested, and instead should be _____ tested.

11. The _____ system provides a method of rotating the vehicle's internal combustion engine to begin the combustion cycle.

12. _____ _____ starters use an extra gear between the armature and the starter drive mechanism.

13. The _____ _____ _____ activates the solenoid winding to draw the plunger forward.

14. A(n) _____ _____ system is a computer-managed security system that disables the vehicle starter and engine systems by using an electronic system to uniquely identify each vehicle key by a security code system.
15. Most hybrid vehicles use a _____-_____ electric motor for engine start-up, auxiliary power, and regenerative braking functions.
16. A _____-_____ AC generator has only one stationary coil, which creates a single sine wave.
17. The _____ _____ monitors battery voltage and adjusts the current flow through the rotor appropriately.
18. Many alternators are designed with A-type circuits and a(n) _____ _____ on the regulator that provides a method of full-fielding the charging system with the engine running.
19. The _____ consists of a cylindrical, laminated iron core, which carries the three (or four) phase windings in slots on the inside.
20. The diodes for rectification are mounted on _____ _____ to assist in dissipating the heat generated in the diodes.

Labeling

Label the following diagrams with the correct terms.

1. Components of a simple battery:

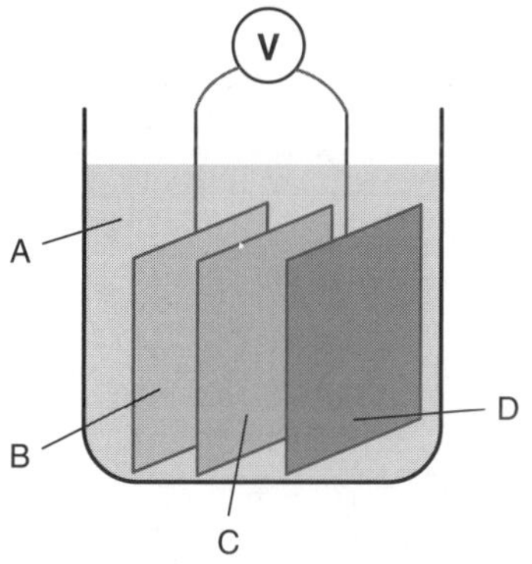

   A. _____
   B. _____
   C. _____
   D. _____

2. Typical plate arrangement in a wet cell battery:

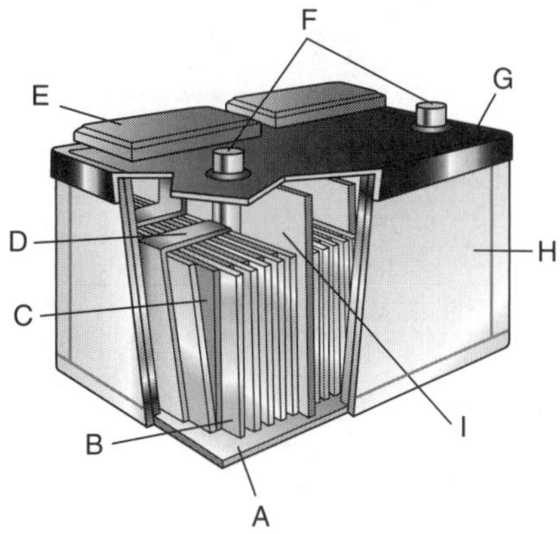

   A. _____
   B. _____
   C. _____
   D. _____
   E. _____
   F. _____
   G. _____
   H. _____
   I. _____

3. Starter motor:

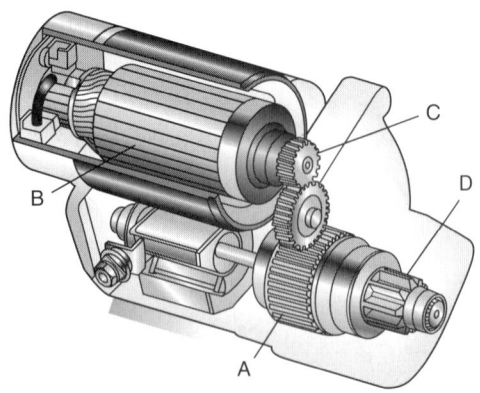

A. _____
B. _____
C. _____
D. _____

4. Series-wound starter:

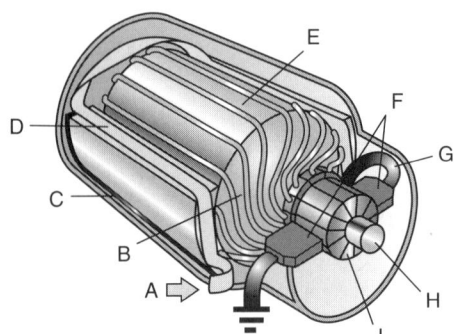

A. _____
B. _____
C. _____
D. _____
E. _____
F. _____
G. _____
H. _____
I. _____

5. Simple multiloop motor and electromagnetic fields—with commutator and brushes:

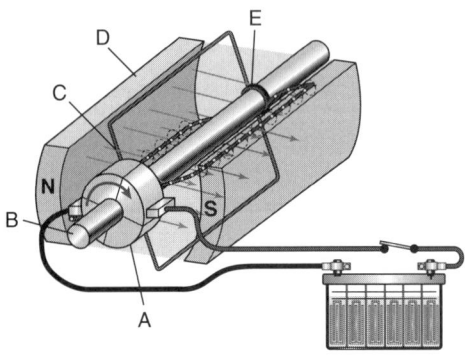

A. _____
B. _____
C. _____
D. _____
E. _____

6. Hold-in and pull-in solenoid windings:

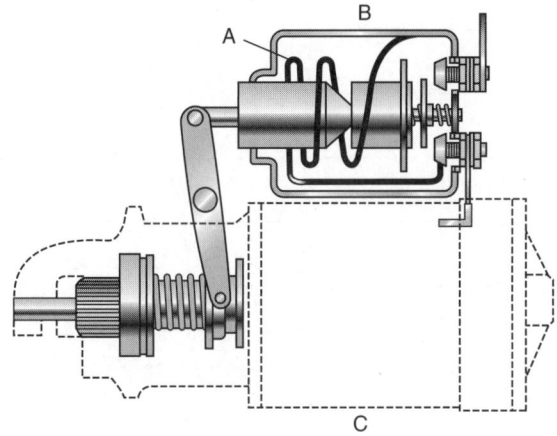

A. _____

B. _____

C. _____

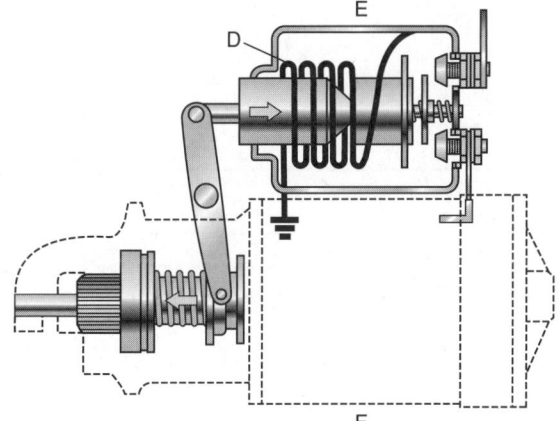

D. _____

E. _____

F. _____

7. Ford moveable pole shoe starter:

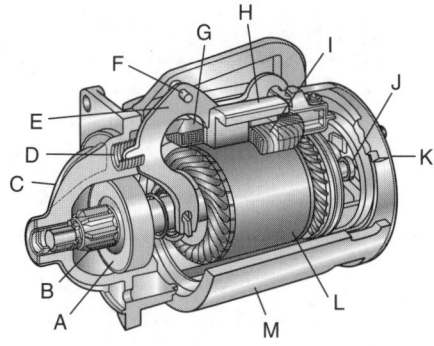

A. _____

B. _____

C. _____

D. _____

E. _____

F. _____

G. _____

H. _____

I. _____

J. _____

K. _____

L. _____

M. _____

8. Basic starter control circuit:

Automatic Transmission

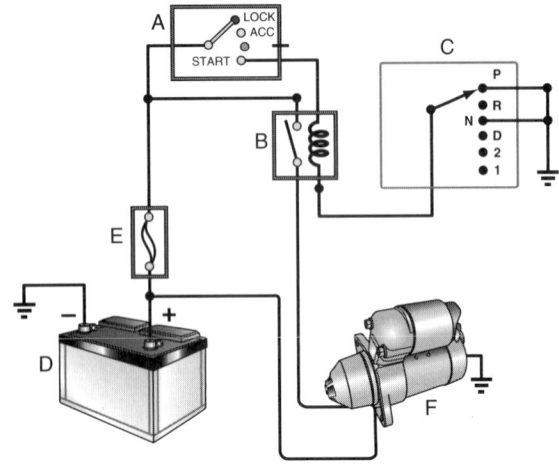

A. _____
B. _____
C. _____
D. _____
E. _____
F. _____

Manual Transmission

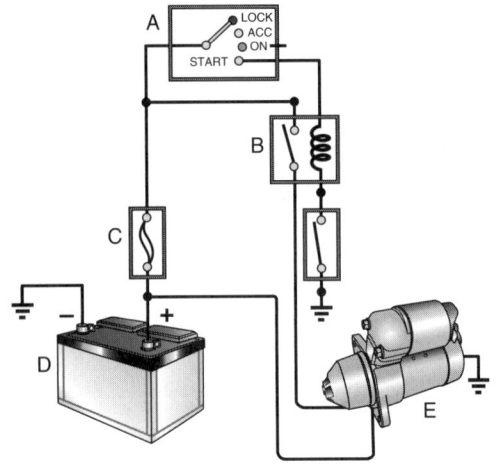

A. _____
B. _____
C. _____
D. _____
E. _____

9. Starter drive one-way clutch:

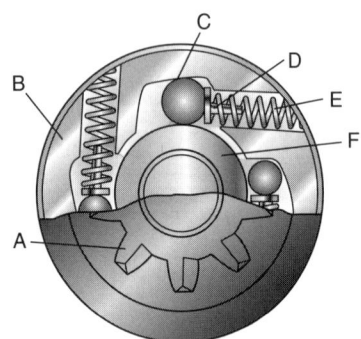

A. _____
B. _____
C. _____
D. _____
E. _____
F. _____

10. The alternator:

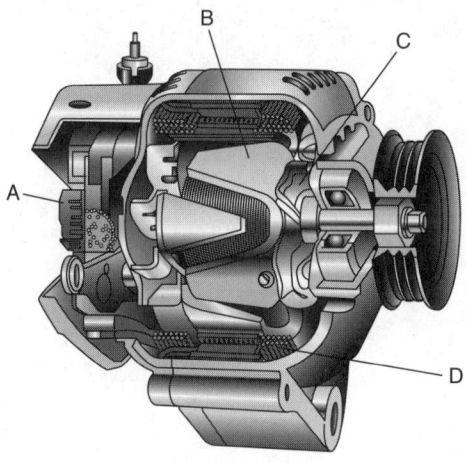

A. _____

B. _____

C. _____

D. _____

Skill Drills

Place the skill drill steps in the correct order.

1. Inspecting, Cleaning, Filling, and Replacing the Battery and Cables:

_____ **A.** Carefully clean the battery case and the battery tray.

_____ **B.** Reconnect the positive battery terminal and tighten it in place. Once the positive terminal is finished, reconnect the negative terminal and tighten it.

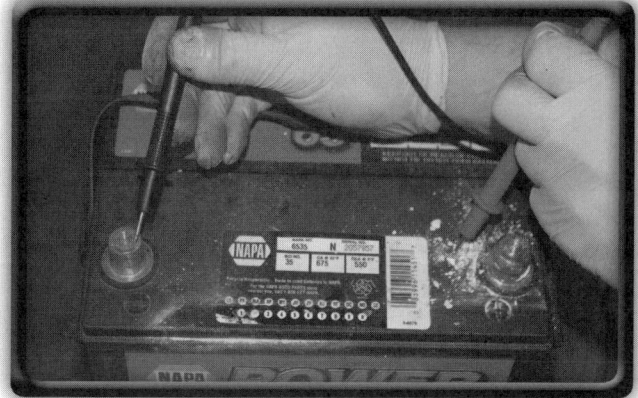

_____ **C.** Measure the voltage on the top of the battery with a DVOM. Place the black lead on the negative post, and move the red lead across the top of the battery until you find the highest reading.

_____ **D.** Coat the terminal connections with anticorrosive paste or spray to keep oxygen from the terminal connections. Test that you have a good electrical connection by starting the vehicle.

_____ **E.** Keeping it upright, remove the battery from its tray and place it on a clean work surface. Inspect the battery for damage.

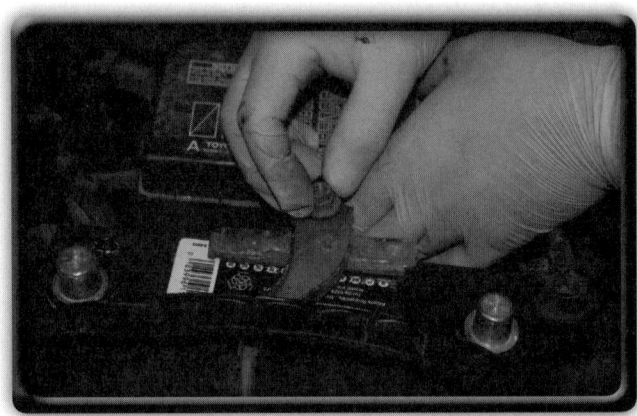

_____ **F.** Reinstall the cleaned and serviced battery. Replace the hold-downs, and make sure the battery is securely held in position. If installing a new battery, ensure that it meets the original manufacturer's specifications.

_____ **G.** Remove the battery hold-downs or other hardware securing the battery.

_____ **H.** Clean the battery posts with a battery terminal tool. Clean the cable terminals with the same battery terminal tool. Examine the battery cables for fraying or corrosion.

_____ **I.** Remove the cable clamp from the negative terminal first. Then remove the positive terminal. Bend the cables back out of the way so that they cannot fall back and touch the battery terminals accidentally.

2. Jump-Starting a Vehicle:

_____ **A.** Connect the black lead to the negative terminal of the charged battery.

_____ **B.** Position the vehicle with the charged battery close to the vehicle with the discharged battery, but not touching. Connect the red or orange lead to the positive terminal of the discharged battery.

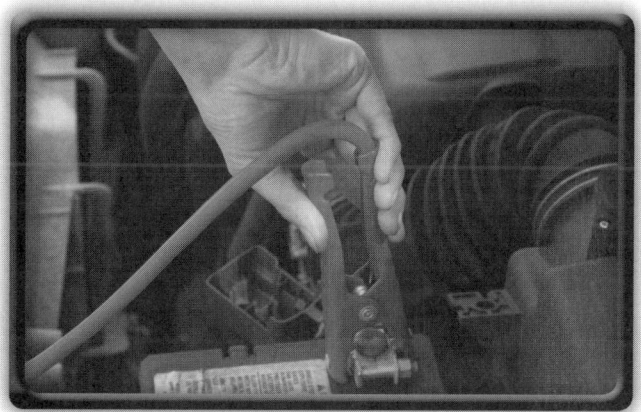

_____ **C.** Start the vehicle with the discharged battery. Turn the headlights on to prevent a possible voltage spike damaging the electronic equipment. Disconnect the jumper leads in the reverse order that you connected them. If the charging system is working correctly and the battery is in good condition, the battery will be recharged while the engine is running.

_____ **D.** Connect the other end of this red or orange lead to the positive terminal of the charged battery.

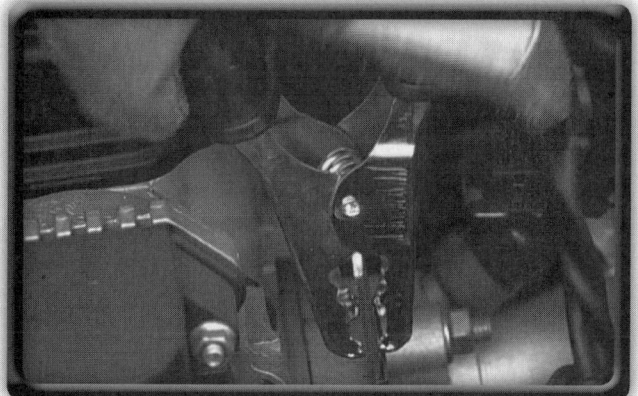

_____ **E.** Connect the other end of the black lead to a paint-free ground on the engine block of the vehicle with the discharged battery.

3. Inspecting and Testing Relays and Solenoids:

_____ A. If the solenoid does *not* click with the key in the crank position, remove the electrical connection for the control circuit at the solenoid.

_____ B. Measure the voltage across the contacts with the relay *not* activated. This should read near battery voltage if both sides of the switched circuit are OK. If not, perform voltage drop tests on each side of the switch circuit. Activate the relay while measuring the voltage drop across the contacts. If it is more than 0.5 volts, the relay will need to be replaced.

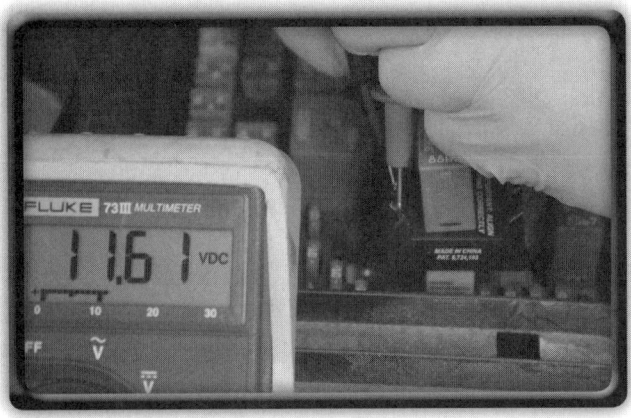

_____ C. Use a relay adapter to mount the relay on top of the relay socket so you can check the control circuit wiring and perform voltage drop tests on the contacts. Activate the relay while measuring the voltage across the relay winding. If it is near battery voltage, the control circuit wiring is OK.

_____ D. Use a jumper wire to apply battery voltage to the control circuit terminal on the solenoid and see if the solenoid clicks. Determine any necessary actions.

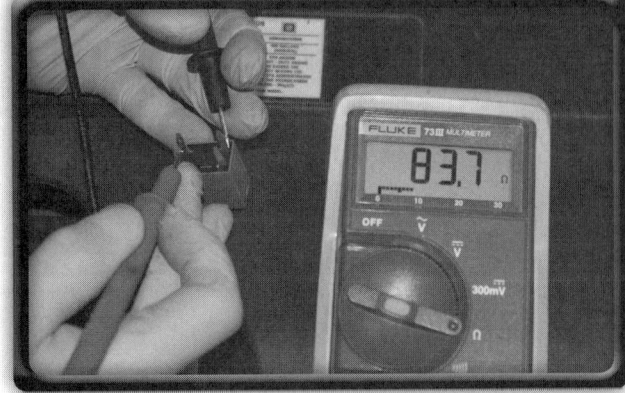

_____ E. To test a relay, measure the resistance of the relay winding and compare to specifications. If out of specifications, replace the relay.

_____ F. To test a starter solenoid, measure the voltage drop across the solenoid contact terminals with the key in the crank position. If more than 0.5 volts, replace the solenoid or starter assembly.

4. Removing and Installing a Starter:

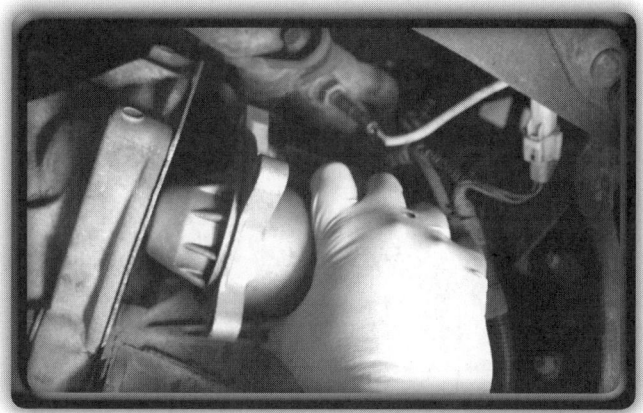

_____ **A.** Remove the starter motor, being careful to catch any shims that might be between the starter and the block. Reinstall the starter motor by reversing these steps, and verify its proper operation.

_____ **B.** Disconnect the negative terminal of the battery after determining whether a memory minder is required.

_____ **C.** Loosen the starter motor mounting bolts, and remove them while holding the starter so it does not fall.

_____ **D.** Remove any engine covers or components required to gain access to the starter. Remove starter motor electrical connections, noting how the wires were routed. In some vehicles, the wires cannot be accessed until the starter is removed.

5. Replacing an Alternator:

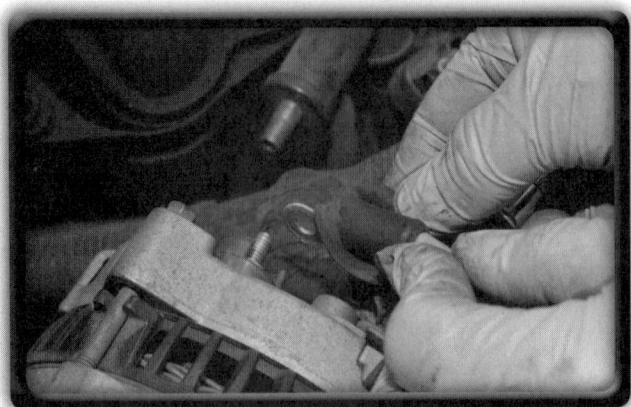

_____ **A.** Locate the electrical connections at the rear of the alternator, and note their positions. Loosen any securing fasteners or covers, and remove terminals one at a time.

_____ **B.** Turn the ignition to the on position, and make sure the charge light on the instrument panel illuminates. Start the engine to see if the charge light goes off. Measure the regulated voltage and maximum alternator amperage output. Remove the fender covers, and return any tools used to their correct place.

_____ **C.** Reinstall the electrical wires to their correct terminals, referring to the manufacturer's information. Check the security of any fastening devices.

_____ **D.** Reattach to the negative post of the battery. Make sure the fastener is tight, and replace any battery terminal covers.

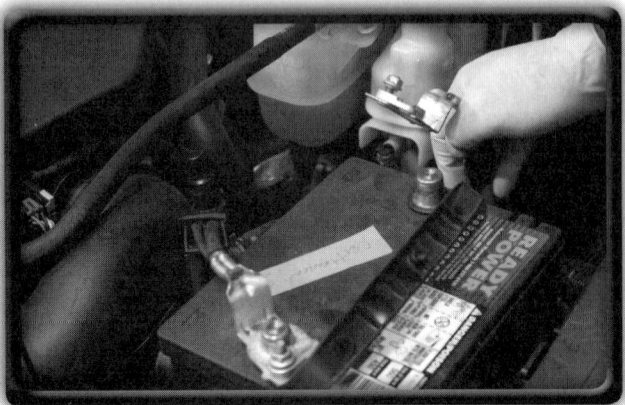

_____ **E.** Install fender covers. Verify any memory issues, and remove the negative terminal of the battery.

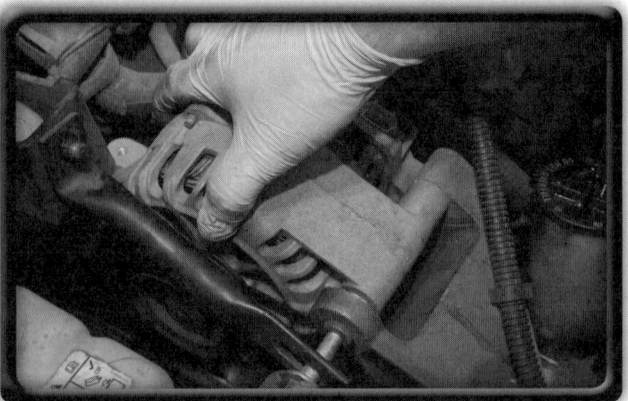

_____ **F.** Reinstall the alternator. Situate the alternator in the mounting bracket(s) and, while still supporting the alternator, loosely start the securing fasteners that hold the alternator to its mounting bracket(s).

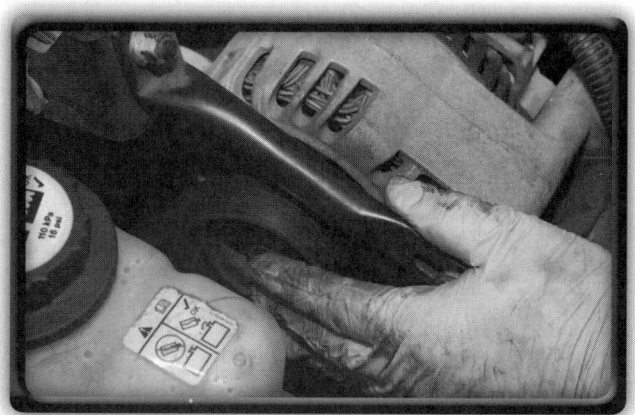

_____ G. Install the drive belt over the alternator drive pulley, and, using the correct tools, adjust the belt to the correct tension.

_____ H. Loosen the drive belt, and remove it from the alternator pulley. Check the condition of the belt to see if it is still serviceable.

_____ I. Loosen the securing fasteners that hold the alternator to its mounting bracket(s), making sure the alternator is supported. Remove the alternator.

Crossword Puzzle

Use the clues to complete the puzzle.

Across

5. The ability of a circuit or component to carry electrical loads.
7. A cycle during battery charging in which heating lowers resistance, which in turn increases current flow, which in turn further increases heat created. During this cycle, dangerous gases may build up, creating the potential for an explosion or damage through excessive current flow.
8. A process of converting AC into DC required by the battery and nearly all of the automobile systems.

Down

1. A standard used to specify the time in minutes that a battery will supply a load of 25 amps at 80°F (27°C) without its voltage dropping below 10.5 volts.
2. A device that clamps around a conductor to measure current flow. It is often used in conjunction with a DVOM.
3. Unwanted drain on the vehicle battery when the vehicle is off.
4. The escape of gas from the battery.
6. The liquid in lead-acid battery cells. It has a mixture of about 67% water and 33% sulfuric acid.

Chapter 38 Batteries, Starting, and Charging Systems   433

ASE-Type Questions

Read each item carefully, and then select the best response.

_____ 1. Tech A says that a battery stores electrical energy in chemical form. Tech B says that a battery creates direct current. Who is correct?
   A. Tech A
   B. Tech B
   C. Both A and B
   D. Neither A nor B

_____ 2. Tech A says that a 12-volt battery has six cells. Tech B says that the more plates a cell in a battery has, the more voltage it creates. Who is correct?
   A. Tech A
   B. Tech B
   C. Both A and B
   D. Neither A nor B

_____ 3. Tech A says that a parasitic draw is measured in volts. Tech B says that pulling fuses one at a time can help locate a parasitic draw. Who is correct?
   A. Tech A
   B. Tech B
   C. Both A and B
   D. Neither A nor B

_____ 4. Tech A says that when disconnecting the battery, the negative terminal should be disconnected first. Tech B says that baking soda and water will remove oxidation from battery terminals. Who is correct?
   A. Tech A
   B. Tech B
   C. Both A and B
   D. Neither A nor B

_____ 5. Tech A says that checking the specific gravity will indicate the battery's cold cranking amps. Tech B says that a battery load test should be performed when the battery is heavily discharged. Who is correct?
   A. Tech A
   B. Tech B
   C. Both A and B
   D. Neither A nor B

_____ 6. Tech A says that some starters use gear reduction to improve efficiency. Tech B says that a starter converts electrical energy to mechanical energy. Who is correct?
   A. Tech A
   B. Tech B
   C. Both A and B
   D. Neither A nor B

_____ 7. Tech A says that the pull-in winding is short-circuited when the solenoid is fully engaged. Tech B says that the starter drive has a built-in one-way clutch. Who is correct?
   A. Tech A
   B. Tech B
   C. Both A and B
   D. Neither A nor B

_____ 8. Tech A says that a voltage drop of 0.8 volts on the starter ground circuit is within specifications. Tech B says that high starter draw current could be caused by a spun main bearing in the engine. Who is correct?
   A. Tech A
   B. Tech B
   C. Both A and B
   D. Neither A nor B

_____ 9. Tech A says that the first item to check if an engine does not crank is the voltage to the S-terminal on the starter. Tech B says that the first item to check if an engine does not crank is battery voltage. Who is correct?
   A. Tech A
   B. Tech B
   C. Both A and B
   D. Neither A nor B

_____ 10. Tech A says that the voltage regulator controls the strength of the rotor's magnetic field. Tech B says that the voltage regulator is installed between the output terminal of the alternator and the positive terminal of the battery. Who is correct?
   A. Tech A
   B. Tech B
   C. Both A and B
   D. Neither A nor B

Lighting Systems

CHAPTER 39

Tire Tread:
© AbleStock

Chapter Review

The following activities have been designed to help you refresh your knowledge of this chapter. Your instructor may require you to complete some or all of these activities as a regular part of your training program. You are encouraged to complete any activity that your instructor does not assign as a way to enhance your learning.

Matching

Match the following terms with the correct description or example.

A. Controlled area network bus (CAN-bus)
B. High-intensity discharge (HID)
C. Incandescent lamp
D. Light-emitting diode (LED)
E. Vacuum tube fluorescent (VTF)

_____ 1. A type of lighting that produces light with an electric arc rather than a glowing filament.

_____ 2. A type of lighting used for instrumentation displays on vehicle instrument panel clusters. This type of lighting emits a very bright light with high contrast and can shine in various colors. Also called vacuum fluorescent display (VFD).

_____ 3. The data network used to connect various electronic systems on vehicles to share data.

_____ 4. A type of lighting used in various automotive applications, such as warning indicators and alphanumeric displays.

_____ 5. The traditional bulb that uses a heated filament to produce light.

Multiple Choice

Read each item carefully, and then select the best response.

_____ 1. Incandescent bulbs are inefficient, converting only about _____ of the electricity to visible light.
   A. 10%
   B. 15%
   C. 25%
   D. 30%

_____ 2. What kind of incandescent lamps are filled with bromine or iodine gas?
   A. Incandescent
   B. Vacuum tube fluorescent
   C. Halogen
   D. High-intensity discharge

_____ 3. LEDs are often required to give off a specified amount of light; to achieve this they are usually connected in groups called _____.
   A. clusters
   B. series strings
   C. parallel sets
   D. packs

_____ 4. Which type of bulb works well as a combination taillight and brake light because it has one filament that emits a small amount of light, and a second filament that emits more light?
   A. LED cluster
   B. Festoon-style bulb
   C. Tandem bulb
   D. Dual-filament bulb

_____ 5. Which type of lamp base gets its name from the two retaining pins on the side of the base?
   A. Bayonet-style
   B. Festoon-style
   C. Wedge-style
   D. Pin-type

_____ 6. Which type of lights usually light up as part of a self-test when the ignition initially comes on to show they are in working order?
   A. Marker lights
   B. Warning lamps
   C. Tail lights
   D. Courtesy lights

_____ 7. For safety reasons, _____ continue to operate when the light switch is moved to the headlight position.
   A. park lights
   B. license plate illumination lamps
   C. taillights
   D. both A and C

_____ 8. Many vehicles by law now have a higher additional third brake light mounted on top of the trunk lid or on the rear window called a _____ light.
   A. pedestal light
   B. courtesy light
   C. center high mount stop light
   D. festoon light

_____ 9. White lights mounted at the rear of a vehicle that provides the driver with vision behind the vehicle at night are called _____.
   A. back-up lights
   B. tail lights
   C. reverse lights,
   D. either A or C

_____ 10. Which type of lights simultaneously cause a pulsing in all exterior indicator lights and both indicator lights on the instrument panel?
   A. Hazard warning lights
   B. Turn signal lights
   C. Backup lights
   D. Driving lights

_____ 11. Existing lights that turn on when the vehicle is running and turn off when the engine stops in order to improve the vehicle's visibility to other drivers in all weather conditions are called _____.
   A. hazard warning lights
   B. dual-filament lights
   C. courtesy lights
   D. daytime running lights

_____ 12. A(n) _____ headlight has a highly polished aluminized glass reflector that is fused to the optically designed lens.
   A. dual-filament
   B. semi-sealed beam
   C. sealed-beam
   D. HID

_____ 13. Which type of headlight is filled with xenon gas and does not use a filament-style lamp?
   A. LED
   B. HID
   C. Sealed-beam
   D. Halogen

_____ **14.** Which type of lights are used with other vehicle lighting in poor weather such as thick fog, driving rain, or blowing snow?
  A. Driving lights
  B. Cornering lights
  C. Fog lights
  D. Smart lights

_____ **15.** In comprehensive adaptive lighting systems, otherwise known as smart lighting, all lighting system decisions are determined by the _____.
  A. Driver
  B. CAN-bus
  C. Power control module
  D. Body control module

True/False

If you believe the statement to be more true than false, write the letter "T" in the space provided. If you believe the statement to be more false than true, write the letter "F".

_____ **1.** Wattage is found by multiplying the voltage used by the lamp by the current flowing through it.
_____ **2.** High-intensity halogen lamp light comes from metallic salts that are vaporized within an arc chamber.
_____ **3.** Some countries mandate that HID headlamps may only be installed on vehicles with lens-cleaning systems and automatic self-leveling systems.
_____ **4.** Wattage is an indication of light output.
_____ **5.** Dome lights use festoon-style lights, which have a base on each end of a cylindrical light bulb.
_____ **6.** Government regulations control the height of taillights and their brightness.
_____ **7.** Taillights are the only white lights on the rear of the vehicle.
_____ **8.** Imported vehicles tend to have separate amber-colored turn signal lamps on both the front and the rear of the vehicle.
_____ **9.** The beam selector switch is a double-pole, single-throw switch, meaning it has two movable poles but makes contact in one position.
_____ **10.** Some vehicles use LED lights as headlights.
_____ **11.** Passive night vision systems use a heat-sensing camera to pick up thermal radiation emitted by objects.
_____ **12.** Cornering lights are white lights usually installed into the bumper or fender and are designed to provide side lighting when the vehicle is turning corners.
_____ **13.** Automatic headlight-leveling systems ensure the headlights are always correctly aligned, regardless of the load the vehicle is carrying.
_____ **14.** To reduce the amount of wiring required, a system is used that integrates sensors into a common wiring harness, called a BCM.
_____ **15.** Not all wiring diagrams use the same symbols or the same numbering system.

Fill in the Blank

Read each item carefully, and then complete the statement by filling in the missing word(s).

  1. Modern vehicles use many different kinds and sizes of _____, also known in some places as light bulbs or light globes.
  2. One of the advantages of a(n) _____ is that it turns on instantly.
  3. In a bulb marked 12V/21W, the _____ will consume 21 watts of power when 12 volts is applied across it.
  4. Many newer bulbs use a _____ base either made from the glass bulb itself or with a built-in plastic base.
  5. _____ lights are usually controlled by the vehicle body computer with inputs from the ignition and door switches either in the handle, latch, or door pillar.
  6. _____ lights are used to mark the sides of some vehicles and are often located down the sides of the vehicle or trailer.

438 Fundamentals of Automotive Technology Student Workbook

7. Today's computer-controlled brake lights are activated by the _____ _____ _____ when the computer senses an input from the brake pedal switch.

8. The cancelling mechanism of the _____ _____ lights operates to return the switch to its central or "off" position after a turn has been completed and the steering wheel is returned to the straight-ahead position.

9. Some domestic vehicles use the rear brake lamps as turn signals by flashing the _____ _____ on one side to indicate the turn.

10. Most vehicle _____ require two beams to provide for a high beam and low beam operation.

11. A(n) _____-_____ headlight system produces a high-intensity forward beam and uses a lens system to project the light forward, rather than the traditional reflector system.

12. Active _____ _____ systems use an infrared light generator that projects infrared light in front of and to the side of the roadway ahead.

13. _____ _____ are installed on the front of the vehicle and provide higher intensity illumination over longer distances than standard headlight systems.

14. _____ headlights have the ability to adjust the length of the headlight beam based on the distance of oncoming traffic.

15. Manufacturers design the circuits to operate in specific ways and then create schematics or _____ _____, which are a diagrammatic layout of the entire circuit.

Labeling

Label the following diagrams with the correct terms.

1. Projector bulb assembly:

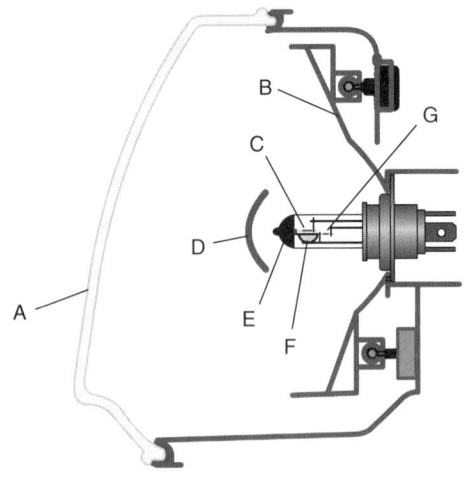

A. _____
B. _____
C. _____
D. _____
E. _____
F. _____
G. _____

2. Automatic headlight system:

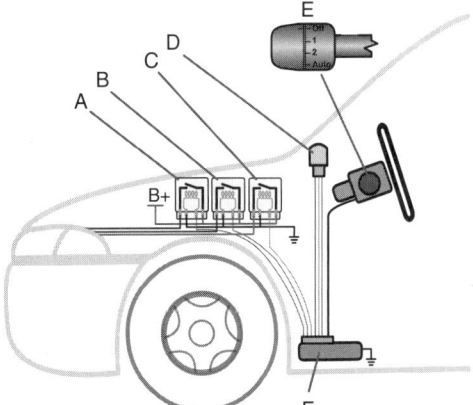

A. _____
B. _____
C. _____
D. _____
E. _____
F. _____

3. Battery and fuse symbols:

 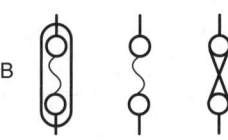

A. _____

B. _____

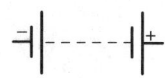 

C. _____

D. _____

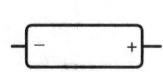 

4. Ground and connector symbols:

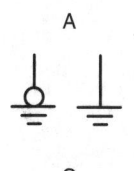 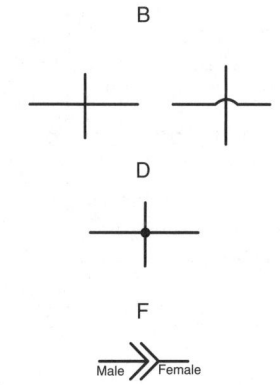

A. _____

B. _____

C. _____

D. _____

E. _____

F. _____

5. Switch symbols:

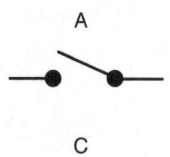 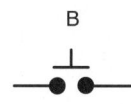

A. _____

B. _____

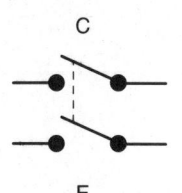 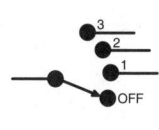

C. _____

D. _____

E. _____

F. _____

6. Resistors, coils, and relay symbols:

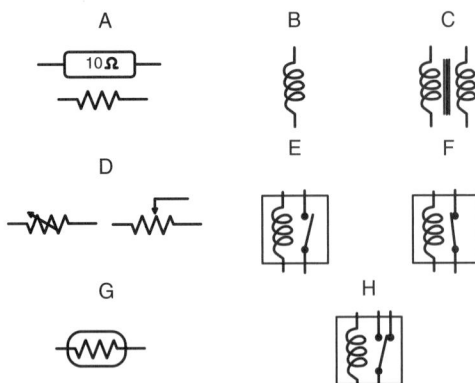

A. _____
B. _____
C. _____
D. _____
E. _____
F. _____
G. _____
H. _____

7. Semiconductor symbols:

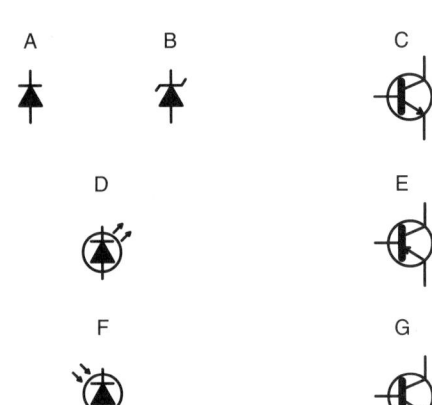

A. _____
B. _____
C. _____
D. _____
E. _____
F. _____
G. _____

8. Capacitor and device symbols:

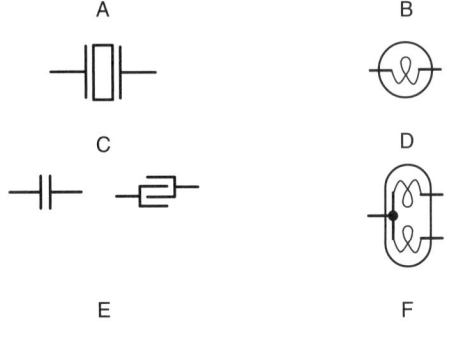

A. _____
B. _____
C. _____
D. _____
E. _____
F. _____

Chapter 39 Lighting Systems

9. Motors, generators, and solenoid symbols:

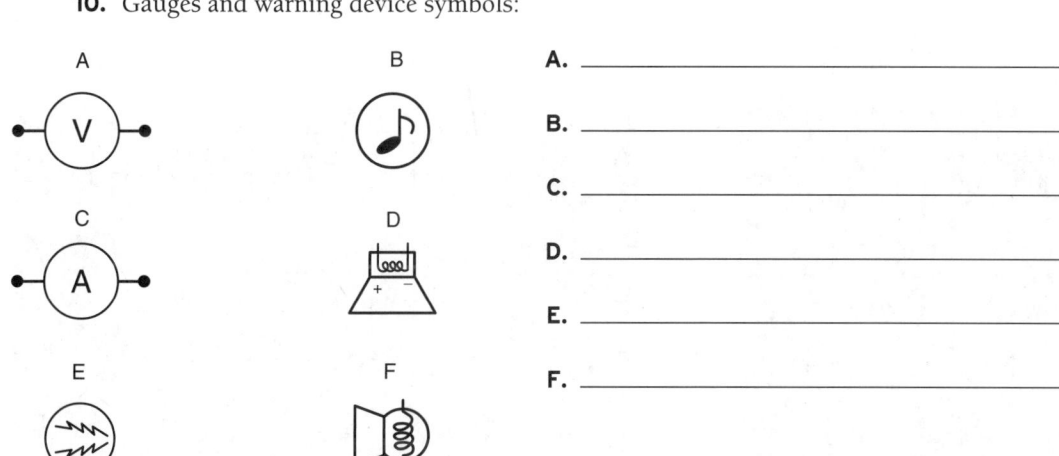

A. _____
B. _____
C. _____
D. _____
E. _____
F. _____

10. Gauges and warning device symbols:

A. _____
B. _____
C. _____
D. _____
E. _____
F. _____

Skill Drills

Test your knowledge of skill drills by filling in the correct words in the photo captions.

1. Checking and Changing an Exterior Light Bulb:

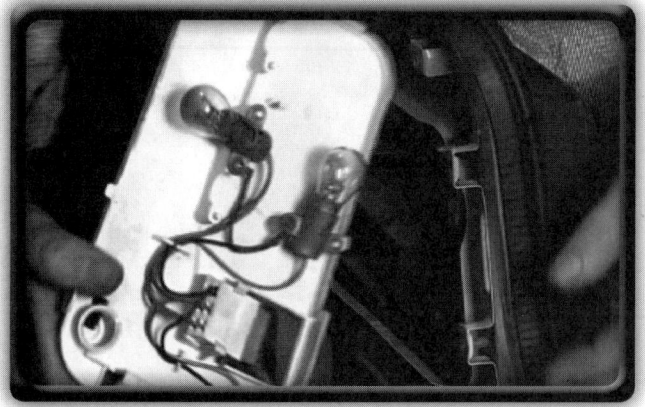

Step 1: Remove the _____ to expose the bulb. If the bulb is _____ mounted, gently grip the bulb and push it inward. Turn the bulb slightly _____, and remove it from the bulb holder.

Step 2: Inspect the bulb holder to make sure there is no _____. If there is, clean it with a bulb socket _____ _____ or _____ _____.

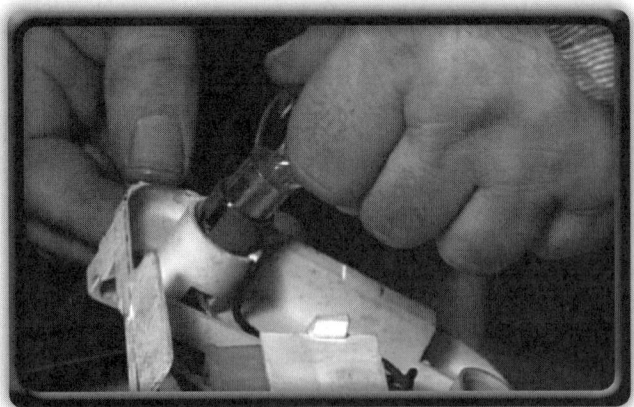

Step 3: Insert the new bulb into the bulb holder, _____ it fully, turn it slightly _____, and release it. Test it by _____ it on and off. Then replace the cover, and _____ it again.

2. Checking and Changing a Headlight Bulb:

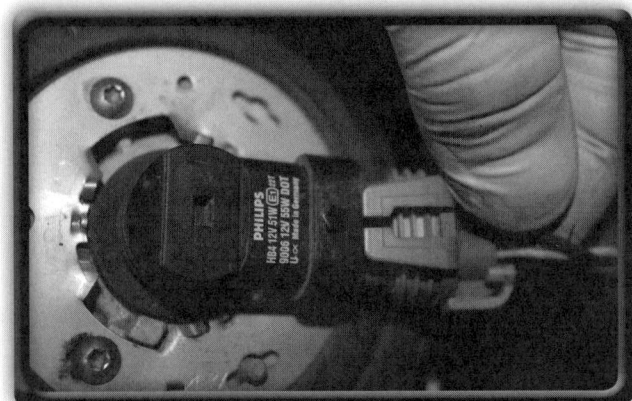

Step 1: Test the vehicle headlights. Obtain the replacement _____ for the vehicle. Unplug the _____ _____ at the _____ of the lamp unit.

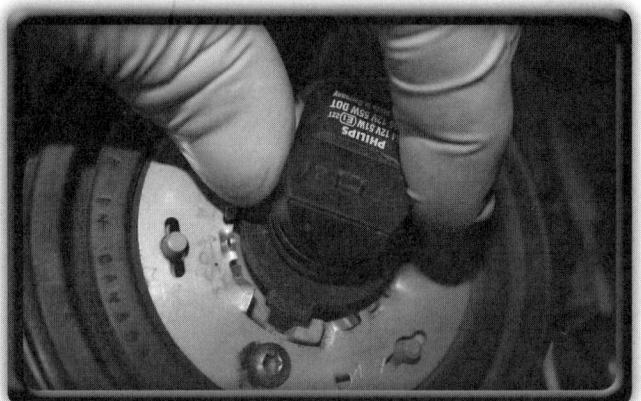

Step 2: _____ the old bulb, and _____ it with the new one. Handle the new bulb only by its _____ or, if supplied, by the _____ cover.

Step 3: Replace the unit and the _____ _____ or bulb assembly, and then plug in the _____. Switch on the lights again to _____ that they are both operating correctly.

3. Aiming Headlights:

Step 1: Make sure the tires are _____ properly, the _____ point straight ahead, and there is no extra _____ in the vehicle. Position the vehicle correctly in relation to the headlamp _____ unit following the equipment manufacturer's instructions. Calibrate the aligner for any _____ _____ and for the vehicle being tested.

Step 2: On the types of _____ that require the headlights to be on during alignment, turn the headlights on to a _____ _____ setting. The _____ of the illuminating beams should be in the _____ _____ quadrants of the chart or wall markings or as specified by the manufacturer.

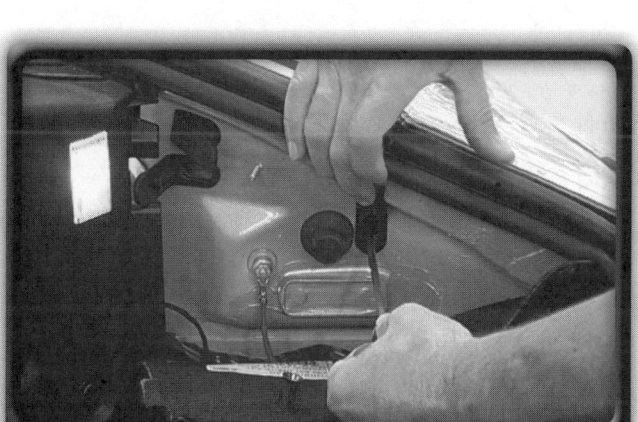

Step 3: The high beam should be _____, falling on the _____ of the horizontal and vertical marks or as specified by the manufacturer. If necessary, turn the _____ _____ on the headlight so the lights _____ to the correct places or _____ on the levels are centered, depending on the type of aligner equipment you are using.

Crossword Puzzle

Use the clues to complete the puzzle.

Across

3. The carrying of multiple signals on one wiring circuit. Digital signals are multiplexed on a dedicated CAN-bus or similar network. Also referred to by the abbreviation MUX.
4. A system of networking circuits in which various on-board modules are connected to one another, allowing the exchange of information between them.

Down

1. A type of bulb that produces a bright white light.
2. A device that increases lighting voltage substantially and controls the current to the bulb.

ASE-Type Questions

Read each item carefully, and then select the best response.

_____ 1. Tech A says that a voltage drop on the ground side of a bulb won't affect its brightness because the electricity has already been used up. Tech B says that a voltage drop is one way to determine if there is unwanted resistance in a circuit. Who is correct?
   A. Tech A
   B. Tech B
   C. Both A and B
   D. Neither A nor B

_____ 2. Tech A says that incandescent bulbs resist vibration well. Tech B says that HID headlamps require up to approximately 25,000 volts to start. Who is correct?
   A. Tech A
   B. Tech B
   C. Both A and B
   D. Neither A nor B

_____ 3. Tech A says that many automotive light bulbs have more than one filament inside. Tech B says that some lights use a filament made of quartz. Who is correct?
   A. Tech A
   B. Tech B
   C. Both A and B
   D. Neither A nor B

Chapter 39   Lighting Systems

_____ 4. Tech A says that LED brake lights illuminate faster than incandescent bulbs. Tech B says that LED brake lights have more visibility and last longer. Who is correct?
   A. Tech A
   B. Tech B
   C. Both A and B
   D. Neither A nor B

_____ 5. Tech A says that if the brake light switch is open, neither brake light will illuminate. Tech B says that the back-up lights are connected in parallel with the taillights. Who is correct?
   A. Tech A
   B. Tech B
   C. Both A and B
   D. Neither A nor B

_____ 6. Tech A says that some brake lights get power from the brake switch through the turn signal switch. Tech B says many turn signals use amber lights. Who is correct?
   A. Tech A
   B. Tech B
   C. Both A and B
   D. Neither A nor B

_____ 7. Tech A says that daytime running light systems illuminate the taillights to enhance visibility. Tech B says that the CHMSL is illuminated when the taillights are activated. Who is correct?
   A. Tech A
   B. Tech B
   C. Both A and B
   D. Neither A nor B

_____ 8. Tech A says that some turn signals are flashed by a flasher can. Tech B says that some turn signals are flashed by the BCM. Who is correct?
   A. Tech A
   B. Tech B
   C. Both A and B
   D. Neither A nor B

_____ 9. Tech A says that mutiplexed lights are controlled by an electronic control module. Tech B says that some headlights can adjust side to side when the vehicle is cornering. Who is correct?
   A. Tech A
   B. Tech B
   C. Both A and B
   D. Neither A nor B

_____ 10. Tech A says that if a bulb is dim when operated, the circuit likely has a short to ground. Tech B says that if a bulb is dim, you should perform a voltage drop test on the power and ground side of the bulb. Who is correct?
   A. Tech A
   B. Tech B
   C. Both A and B
   D. Neither A nor B

CHAPTER 40

Body Electrical System

Chapter Review

The following activities have been designed to help you refresh your knowledge of this chapter. Your instructor may require you to complete some or all of these activities as a regular part of your training program. You are encouraged to complete any activity that your instructor does not assign as a way to enhance your learning.

Matching

Match the following terms with the correct description or example.

- **A.** Blower motor
- **B.** Body electronic module (BEM)
- **C.** Brushless DC motor
- **D.** Controller area network (CAN)
- **E.** Data link connector (DLC)
- **F.** Passive keyless entry (PKE)
- **G.** Permanent magnet electric motor
- **H.** Remote keyless entry (RKE)
- **I.** Stepper motor
- **J.** Supplemental restraint system (SRS)

_____ 1. A type of brushless motor with a key difference: It is designed to rotate in fixed steps through a set number of degrees.

_____ 2. An active system that senses the proximity of a fob and locks or unlocks the vehicle.

_____ 3. The port to which a scan tool can be connected.

_____ 4. An electric motor that does not have any brushes and is sometimes called an "electronically commutated motor." In this type of motor, an electronic control module replaces the brushes and commutator.

_____ 5. A system that remotely unlocks and locks the vehicle without the use of a traditional key.

_____ 6. An electronic control module for body electrical systems.

_____ 7. The network that connects the vehicle's electronic control modules.

_____ 8. A passenger safety system, such as airbags and seat belt pretensioners.

_____ 9. An electric motor, usually the permanent magnet type that moves air over the air-conditioning evaporator and heater core.

_____ 10. An electric motor in which the magnetic field in the casing is produced by permanent magnets, while the armature has an electromagnetic field generated by passing electrical current through loops or windings, thereby producing the motor action.

Multiple Choice

Read each item carefully, and then select the best response.

_____ 1. Which of the following is a high-speed network used for critical data such as supplemental restraint systems, antilock brake systems, and engine controls?
- **A.** CAN-bus A
- **B.** CAN-bus B
- **C.** CAN-bus C
- **D.** CAN-bus D

_____ 2. A group of eight bits of binary data is called a _____.
- **A.** microbyte
- **B.** byte
- **C.** kilobyte
- **D.** megabyte

_____ 3. Which of the following is a standard CAN-bus network configuration?
   A. Bus parallel
   B. Star parallel
   C. Looped series
   D. All of the above

_____ 4. What type of motor can be used to rotate a short distance and then stop or go in the reverse direction for a set number of degrees by controlling each coil individually through the microcontroller?
   A. Brush-type motor
   B. Potentiometer
   C. Stepper motors
   D. Induction motor

_____ 5. The horn switch on vehicles with a driver's side airbag is usually mounted in the steering wheel and requires a _____ to maintain an electrical connection to the circuit as the steering wheel rotates.
   A. clock spring
   B. slot switch
   C. slip ring and brush assembly
   D. transmitter

_____ 6. What type of keyless entry system uses a fob transmitter but does not require any action by the user?
   A. Remote keyless entry
   B. Passive keyless entry
   C. Magnetic proximity entry
   D. Either A or C

_____ 7. What controls the intermittent operation of a windshield wiper timer circuit?
   A. Discrete electronic timer
   B. Time-delay cube relay
   C. Vehicle body computer
   D. Either A or C

_____ 8. Which of the following is an example of a secondary vehicle safety system?
   A. Seat belt
   B. Collapsible steering column
   C. SRS airbag
   D. Seat belt pretensioners

_____ 9. To prevent incorrect and unnecessary airbag deployment, systems include a(n) _____ mounted within the SRS control unit.
   A. mercury switch
   B. accelerometer
   C. safing sensor
   D. proximity sensor

_____ 10. The vehicle's onboard GPS system needs to know _____ to determine the location of the vehicle.
   A. signal travel time
   B. the accurate time
   C. satellite position
   D. all of the above

True/False

If you believe the statement to be more true than false, write the letter "T" in the space provided. If you believe the statement to be more false than true, write the letter "F".

_____ 1. CAN-bus-compliant diagnostic systems have been required on all vehicles sold in the United States since 2008.

_____ 2. Some networks transmit and receive signals over a single wire, but most have dual wires and are commonly called "CAN-bus high (H)" and "CAN-bus low (L)," with the same message sent on both lines.

_____ 3. Using the scan tool may require the vehicle's ignition system to be on for an extended period of time. Use a standard battery charger to support the battery when the engine is not running.

_____ 4. Two magnetic fields are required for motor action, one in the casing and the other in the rotating armature.

_____ 5. Blower motor speed can be controlled by using a number of resistors connected in series and a switch to select between combinations of resistors or a more complex electronic speed control module.

_____ 6. More sophisticated electric seat systems are controlled by an electronic control unit and can "remember" seat and mirror positions for each driver.

_____ 7. The clock spring is a device within the steering wheel with a flexible ribbon cable that can rotate endlessly in a single direction.

_____ 8. Keyless entry systems and the engine immobilizer or security systems are two parts of the same system.

_____ 9. For safety reasons, the cruise control has a minimum speed that must be met before it can be set and is deactivated when the driver touches the brake pedal or, in the case of manual transmissions, the clutch pedal.

_____ 10. The GPS receiver on the vehicle has to locate four or more satellites, determine the distance to each, and use this information to establish its own location.

Fill in the Blank

Read each item carefully, and then complete the statement by filling in the missing word(s).

1. In electrical and electronic systems, a _____ is a means of connecting many electrical or electronic components for either data or power sharing.

2. Regardless of how the network is physically connected, the CAN-bus-H and CAN-bus-L systems will have terminating resistors connected to the data lines to form a(n) _____ through the resistors.

3. Diagnosing and repairing vehicles may require software transfers, software updates, or _____ reprogramming of the vehicle's electronic modules with the manufacturer's latest update.

4. _____ _____ in modern vehicles are usually switched by relays, which are controlled by an electronic control unit based on information from the coolant temperature sensor.

5. The electric motors used for power mirrors and widows are _____ so they can provide the movements in either direction.

6. Electric pedal height adjustment is achieved by using a reversible motor driven adjustment _____ that moves the pedals in toward the firewall or out toward the driver.

7. The driver's side door is usually a _____ door lock, which means that when it is locked or unlocked with the key (or the remote fob) the other locks follow suit.

8. Heated glass and seats tend to draw a substantial amount of electrical _____.

9. Seat belt _____ are used to tighten the seat belt in a severe frontal accident.

10. To allow for safer parking, ultrasonic _____ _____ can be mounted in the front or rear bumpers.

11. The entertainment system _____ _____, mounted in the dash, is the main control unit for controlling the subsystems installed on the vehicle.

12. A(n) _____ system allows for vehicle tracking and security features, monitoring of onboard systems, messaging, delivery of travel information, entertainment functions, and the use of safety and fleet management systems.

13. _____ are small plastic particles that can have either the VIN or a unique number printed on them. They are practically invisible to the naked eye but can be seen by using an ultraviolet light and a magnifying glass.

14. _____ occurs when the theft-deterrent system prevents the vehicle's engine from starting or the transmission from operating without the properly authorized key.

15. Vehicles may be equipped with _____ systems designed to sound high-decibel sirens or horns if an intruder attempts to break into or move a vehicle.

Chapter 40   Body Electrical System

Labeling
Label the following diagrams with the correct terms.

1. CAN-bus configurations:

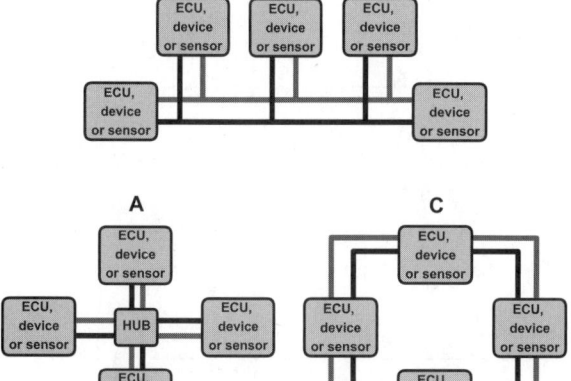

A. _____

B. _____

C. _____

2. Brush-type motor:

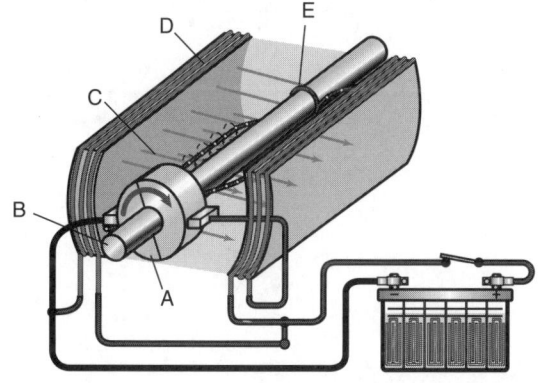

A. _____

B. _____

C. _____

D. _____

E. _____

3. Stepper motor:

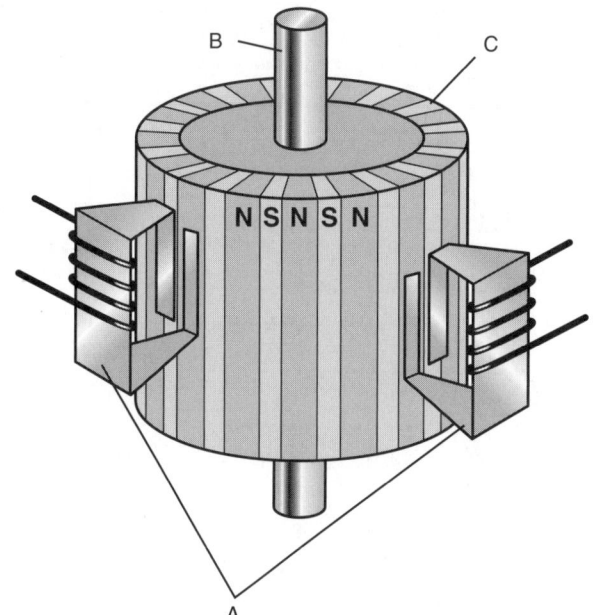

A. _____

B. _____

C. _____

4. Parts of an airbag assembly:

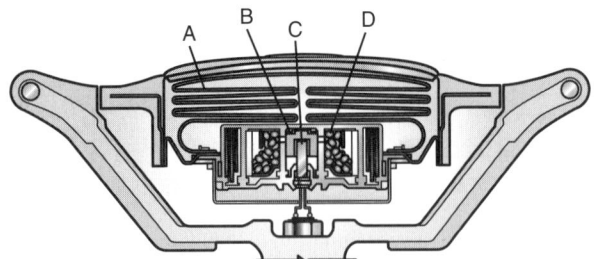

A. _____

B. _____

C. _____

D. _____

5. Seat belt pretensioner:

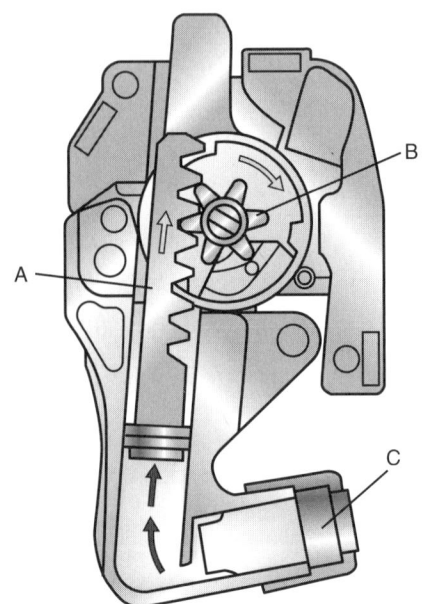

A. _____

B. _____

C. _____

Skill Drills

Place the skill drill steps in the correct order.

1. Checking for Module Communication Errors Using a Scan Tool:

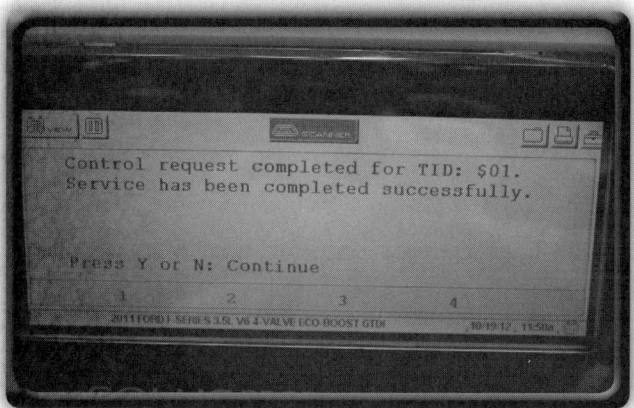

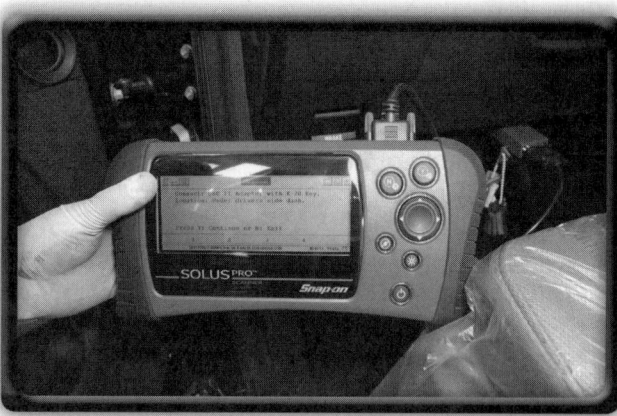

_____ **A.** Command the module to take the appropriate action, and observe the response. If you do not receive the proper response, refer to the service information for the diagnostic procedure.

_____ **B.** After researching the correct procedure in the manufacturer's information, locate the DLC and connect the scan tool. Power on the scan tool, and turn the ignition on. Establish scan tool communications with the vehicle.

_____ **C.** Check fault codes using the scan tool, and follow the diagnostic procedure in the service information. Check to see if any modules are inactive, or not listed as active, that should be.

2. Performing Software Transfers, Software Updates, or Flash Reprogramming on Modules:

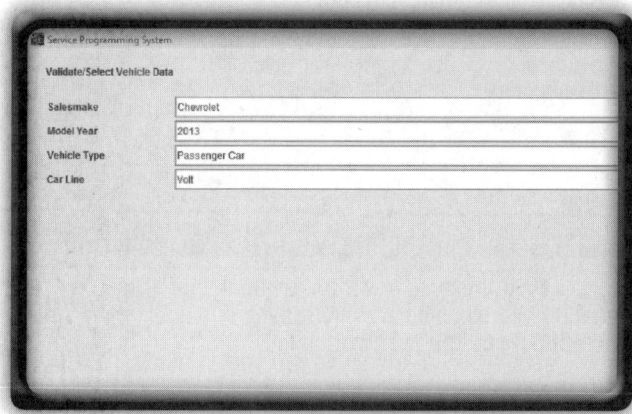

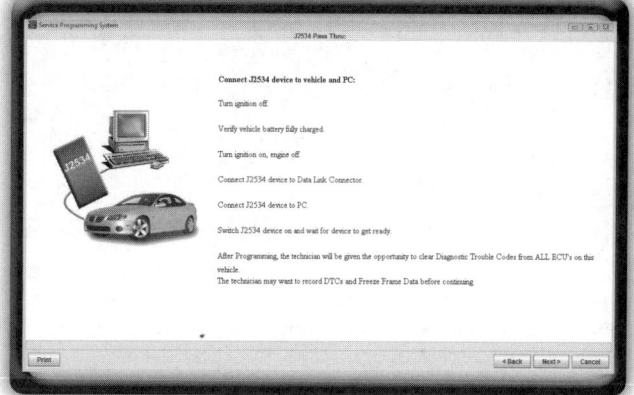

_____ **A.** Establish flash program tool communication with the vehicle.

_____ **B.** Locate the DLC. Connect and power on the flash program tool. Turn the ignition on.

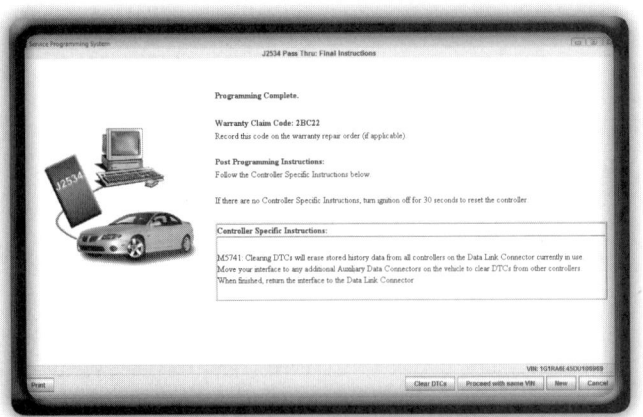

_____ **C.** Disconnect the flash programmer, and check the vehicle's functionality.

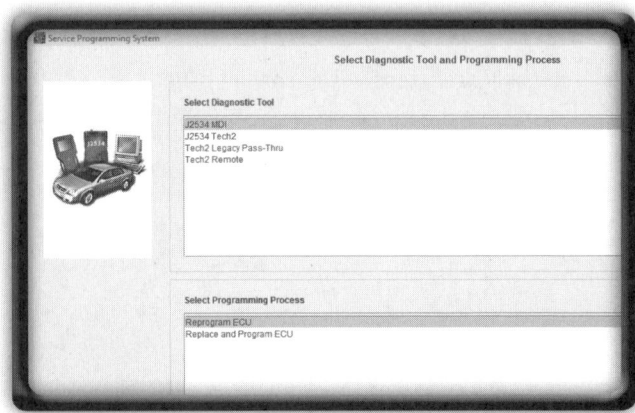

_____ **D.** Obtain the latest vehicle software program updates from the manufacturer, and load the updates into the flash programmer. Connect a battery support unit to the vehicle.

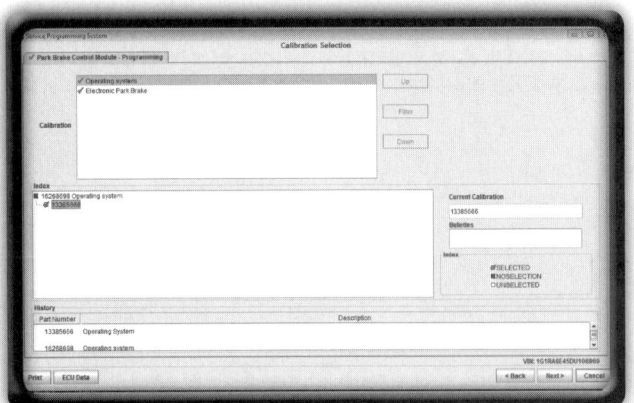

_____ **E.** Identify the vehicle modules to be updated, and use the flash program tool to perform software transfers, software updates, or flash reprogramming.

3. Testing the Horn System:

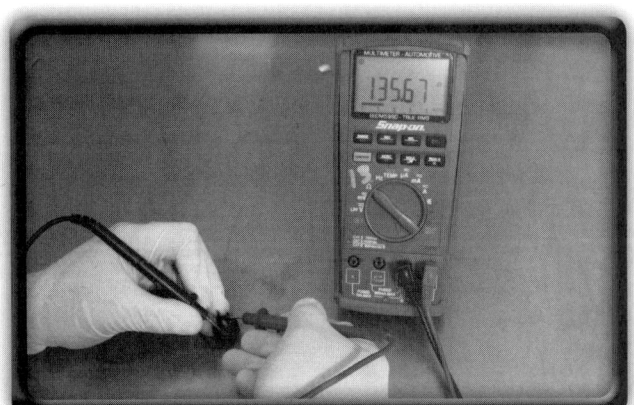

_____ **A.** If the switch leg is good, measure the resistance between terminals 85 and 86 on the horn relay itself. The resistance should be between about 40 and 100 ohms. If not, replace the relay.

_____ **B.** If the horn now honks, check that either terminal 85 or 86 has battery voltage. If one does, then the other should be the horn switch leg.

_____ **C.** Confirm that the horn does not operate. Research circuit operation and circuit diagrams for the horn from the service information. Check for power and ground at horn when the horn switch is operated. If power and ground are present, the horn is faulty and needs to be replaced.

_____ **D.** If battery voltage is present, jump terminal 30 (or 3) to terminal 87 (or 5). The horn should honk. If it does not, check for an open wire between terminal 87 and the horn with a DVOM.

_____ **E.** If there is no power to the horn, remove the relay and check for battery voltage at terminal 30 (or 3). If battery voltage is not present, check the horn fuse and the rest of the feed circuit.

_____ **F.** If the resistance is good, either measure the voltage drop across the switch contacts of the relay (30 and 87) or substitute a known good relay.

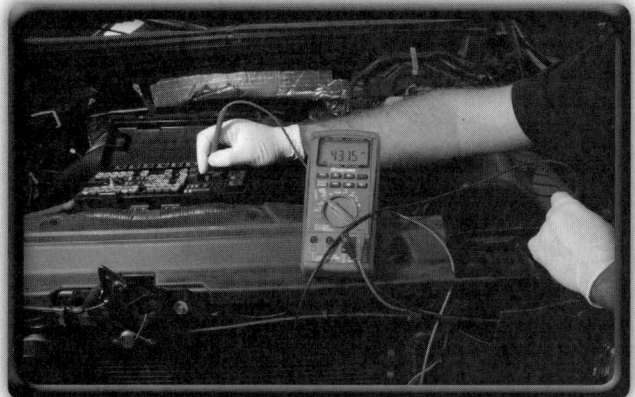

_____ **G.** Connect a DVOM between the switch leg terminal and ground. Operate the horn switch; resistance should decrease to less than an ohm. If it does not, search out the high resistance or open circuit back to the horn switch. Suspect either the clock spring or the horn switch itself.

4. Removing the Door Panel:

_____ **A.** Remove the inner liner from the frame, preserving it for reuse.

_____ **B.** Research removal and reinstallation of the door panel from the manufacturer's service information. Remove fixtures such as the arm rest.

_____ **C.** Remove the cables, and carefully pry the panel from the frame.

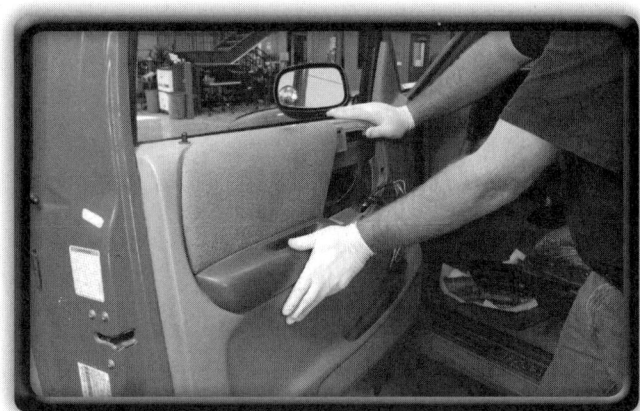

_____ **D.** Reinstall the door panel using the reverse procedure while ensuring all electrical connections are reinstalled.

_____ **E.** Remove the switch panel.

Chapter 40   Body Electrical System   455

5. Testing the Washer System:

_____ A. If an electrical fault is indicated by lack of a washer motor sound, check for battery voltage and ground at the pump, when it is activated.

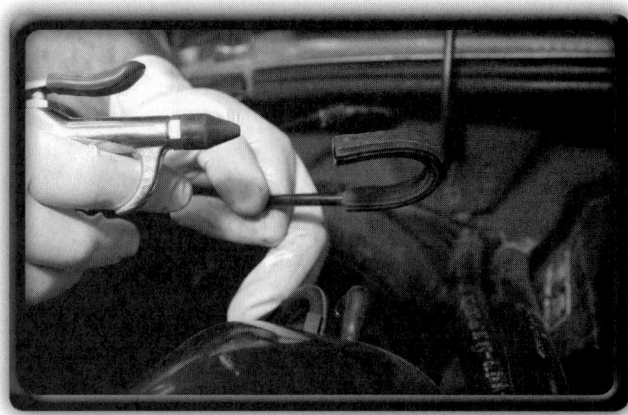

_____ B. If the pump spins when activated but does not pump washer fluid, check the hoses and nozzles for obstructions. Blow out with compressed air if necessary.

_____ C. Check for the correct solution level in tanks, and listen for washer pump operation.

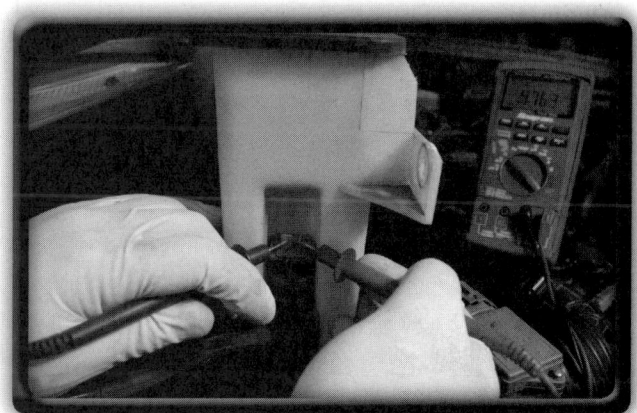

_____ D. If the power and ground circuits are good, measure the resistance of the pump motor and compare to specifications.

_____ E. If the power or ground circuit is faulty, perform voltage drop tests on the faulty side, until the fault is located.

Crossword Puzzle

Use the clues to complete the puzzle.

Across

3. A method of locating points by the measurement of distances using geometry.
4. The component inside the airbag inflator that triggers the airbag deployment.

Down

1. Satellite-based, two-way communication with the vehicle's electronic systems.
2. A method of locating an object using mathematics based on forming a triangle with two known points.
5. The data transport system for electronic control modules.

ASE-Type Questions

Read each item carefully, and then select the best response.

_____ 1. Tech A says that CAN A is a low-speed data network. Tech B says that CAN C is a high-speed network. Who is correct?
   A. Tech A
   B. Tech B
   C. Both A and B
   D. Neither A nor B

_____ 2. Tech A says that CAN C uses two 120-ohm resistors with a total circuit resistance of 60 ohms. Tech B says that CAN C uses two 120-ohm resistors with a total circuit resistance of 240 ohms. Who is correct?
   A. Tech A
   B. Tech B
   C. Both A and B
   D. Neither A nor B

_____ 3. Tech A says that a twisted pair of wires in a wire loom makes it easier to route wires to their destination. Tech B says that the twisted pair of wires transmits signals of opposite polarity on the two lines. Who is correct?
   A. Tech A
   B. Tech B
   C. Both A and B
   D. Neither A nor B

_____ 4. Tech A says that during troubleshooting, you should never disconnect an electronic control unit or module that is powered up. Tech B says that when flashing an electronic control unit or module, make sure no power loss or disconnect occurs, as the electronic control unit or module may be damaged. Who is correct?
   A. Tech A
   B. Tech B
   C. Both A and B
   D. Neither A nor B

_____ 5. Tech A says that stepper motors are used as blower motors to achieve various speeds controlled by the driver. Tech B says that stepper motors are used where precise movement must occur, such as in electronic throttle controls. Who is correct?
   A. Tech A
   B. Tech B
   C. Both A and B
   D. Neither A nor B

_____ 6. Tech A says that microdots are used as a theft deterrent in some vehicles. Tech B says that microdot is the name given to a single datum transmitted over a CAN-bus. Who is correct?
   A. Tech A
   B. Tech B
   C. Both A and B
   D. Neither A nor B

_____ 7. Tech A says that the clock spring is a device used in the steering wheel to transmit signals across the rotating electrical connection. Tech B says that the clock spring must be wound correctly during installation, as it only allows turning both ways a certain number of rotations. Who is correct?
   A. Tech A
   B. Tech B
   C. Both A and B
   D. Neither A nor B

_____ 8. Tech A says that most SRS systems use a safing sensor to reduce the possibility of accidental deployment of the airbags. Tech B says that airbags are designed to act like a nice soft pillow in an accident. Who is correct?
   A. Tech A
   B. Tech B
   C. Both A and B
   D. Neither A nor B

_____ 9. Tech A says that if airbags have been deployed, the affected seat belts and pre-tensioners need to be replaced. Tech B says that there are relatively large holes in the rear of an airbag that allow it to quickly deflate after deployment. Who is correct?
   A. Tech A
   B. Tech B
   C. Both A and B
   D. Neither A nor B

_____ 10. Tech A says that in some vehicles, airbag deployment can be based on the speed of the impact and the weight of the driver. Tech B says that the SRS seat belt pre-tensioners tighten the seat belt when you connect the seat belt. Who is correct?
   A. Tech A
   B. Tech B
   C. Both A and B
   D. Neither A nor B

CHAPTER

Principles of Heating and Air-Conditioning Systems

Chapter Review

The following activities have been designed to help you refresh your knowledge of this chapter. Your instructor may require you to complete some or all of these activities as a regular part of your training program. You are encouraged to complete any activity that your instructor does not assign as a way to enhance your learning.

Matching

Match the following terms with the correct description or example.

- A. Accumulator
- B. Axial piston compressor
- C. Barrier-type hose
- D. Chlorofluorocarbon (CFC)
- E. Closed-loop system
- F. Condensation
- G. Convection
- H. Double crimping
- I. Electric servo
- J. Evaporator
- K. Filter media
- L. Fixed-orifice tube system
- M. Heat transfer
- N. High-pressure switch
- O. Parallel flow condenser
- P. Refrigerant
- Q. Restriction
- R. Swash plate
- S. Tubes
- T. Vaporization

_____ 1. The name given to a chemical compound designed to meet the needs of the refrigeration system.

_____ 2. The process of transferring heat by the circulatory movement that occurs in a gas or fluid as areas of differing temperatures exchange places due to variations in density and the action of gravity.

_____ 3. A system with a fixed-orifice tube that uses an accumulator between the evaporator and the compressor.

_____ 4. An air-conditioning door actuator controlled by electricity.

_____ 5. Metal pipes running side to side or up and down that the coolant or refrigerant travels through.

_____ 6. A totally self-contained system with no materials entering or exiting.

_____ 7. A condenser with multiple parallel tubes flowing from one side tank to the other.

_____ 8. The changing of a gas into a liquid through cooling.

_____ 9. The changing of a liquid to a gas through boiling.

_____ 10. Overcrimping or recrimping a fitting to keep it from leaking. It normally results in a bigger leak and is not recommended.

_____ 11. Another name for an axial plate.

_____ 12. A design of compressor that uses an angled plate (swash plate) to create the piston movement.

_____ 13. A screen designed to keep debris from the TXV that is normally located in the receiver dryer.

_____ 14. The flow of heat from a hotter part to a cooler part; it can occur in solids, liquids, or gases.

_____ 15. A manufactured compound designed to be used as a refrigerant. It is now illegal due to the high chlorine content.

_____ 16. The air-conditioning component used on fixed-orifice systems to protect the compressor by storing liquid refrigerant so it does not reach the compressor.

_____ 17. The air-conditioning component normally located in the passenger compartment designed to allow low-pressure refrigerant liquid to change states to a gas.

_____ 18. A blockage that partially stops or slows the flow of a material such as refrigerant.

_____ 19. A switch designed to open at a predetermined high pressure to protect the compressor; it is normally located on the high-side line near the compressor.

_____ 20. A rubber hose made with a nylon bladder inside to contain substances with small molecular structures, such as R-134a.

Multiple Choice

Read each item carefully, and then select the best response.

_____ 1. The federal law that regulates air emissions from stationary and mobile sources and contains all of the motor vehicle air-conditioning requirements and laws is called the _____.
   A. National Air Quality Standards Act
   B. Clean Air Act
   C. Air Conditioning Act
   D. Environmental Protection Act

_____ 2. The amount of water vapor in the air expressed as a percentage is called _____.
   A. saturation level
   B. condensation point
   C. relative humidity
   D. evaporation point

_____ 3. The process of transferring heat through matter by the movement of heat energy through solids from one particle to another is called _____.
   A. convection
   B. conduction
   C. radiation
   D. evaporation

_____ 4. Heat energy is measured in _____.
   A. degrees Fahrenheit
   B. degrees Celsius
   C. British thermal units
   D. relative humidity

_____ 5. The heat required to change the water at its maximum temperature from a liquid to a gas is called _____.
   A. latent heat of evaporation
   B. latent heat of condensation
   C. latent heat of freezing
   D. latent heat of vaporization

_____ 6. The _____ changes a low-pressure refrigerant gas to a high-pressure gas and provides the needed refrigerant movement in the system.
   A. condenser
   B. compressor
   C. evaporator
   D. restrictor

_____ 7. Hot high-pressure gas enters the _____ from the compressor; the gas flows through a series of coils, ambient air removes the heat from the hot high-pressure gas, which begins to change into a liquid.
   A. condenser
   B. evaporator
   C. restriction
   D. orifice tube

_____ 8. Dichlorodifluoromethane, commonly known as _____, is a member of the CFC family of gases and was the first common refrigerant to be used in an automotive air conditioner.
   A. R-10
   B. R-12
   C. R-124b
   D. R-134a

_____ 9. All of the following are common oils used in refrigeration systems, *except*:
   A. POE oil
   B. mineral oil
   C. PAG oil
   D. copaiba oil

_____ 10. The _____ compressor works in much the same way as the axial piston compressor except that it has only a single row of pistons instead of two opposing rows of pistons.
   A. rotary vane
   B. scroll-type
   C. swash plate
   D. wobble plate

_____ 11. The air-conditioning system may use a _____ to dampen pulsations and noise from the compressor.
   A. muffler
   B. restrictor
   C. noise damper
   D. serpentine condenser

_____ 12. A _____ has a sensing device on the outlet side of the evaporator that senses changes in temperature and a valve that can vary the amount of opening to allow more or less gaseous refrigerant through.
   A. Bernoulli valve
   B. thermal expansion valve
   C. fixed-orifice tube
   D. velocity restrictor

_____ 13. According to the automotive industry numbering system a ½" air-conditioning hose would be referred to as a(n) _____ hose.
   A. #4
   B. #6
   C. #8
   D. #10

_____ 14. Nitrile butadiene rubber and hydrogenated nitrile butadiene rubber have both been used to make _____ for air-conditioning systems.
   A. gaskets
   B. barrier hose cores
   C. O-rings
   D. all of the above

_____ 15. Fixed-orifice tube systems use a(n) _____ to ensure that a pure gas is delivered to the compressor.
   A. receiver filter drier
   B. accumulator
   C. Schrader valve
   D. Bernoulli filter

_____ 16. The _____ is designed to ensure that the temperature of the refrigerant leaving the evaporator is between 5°F and 20°F higher than the boiling point of the refrigerant at the current operating pressure.
   A. superheat spring
   B. accumulator
   C. desiccant
   D. Schrader valve

_____ 17. The thermal expansion valve uses an _____ to open and close the port.
   A. internally equalized valve
   B. externally equalized valve
   C. H block valve
   D. any of the above

_____ 18. The _____ is located inside the cabin in the plastic box with the air-conditioning evaporator.
   A. defroster
   B. heater core
   C. condenser
   D. accumulator

_____ 19. Air doors are controlled by _____.
   A. cables
   B. vacuum servos
   C. electric servos
   D. any of the above

_____ 20. The most common type of blower motor control is a(n) _____ that uses resistors in series to regulate the speed of the fan.
   A. resistor pack
   B. potentiometer
   C. resistor block
   D. either A or C

True/False

If you believe the statement to be more true than false, write the letter "T" in the space provided. If you believe the statement to be more false than true, write the letter "F".

_____ 1. Automotive air-conditioning service and repair technicians need to have a special license.

_____ 2. When pressure is lowered on a gas, its temperature rises; when pressure is raised, its temperature drops.

_____ 3. The amount of heat given up by steam to turn back into a liquid is 970 Btu and is called latent heat of condensation.

_____ 4. With the use of refrigerants, maintaining vaporization and condensing points at normal ambient temperatures is simply done by raising or reducing the pressure of the refrigerant.

_____ 5. PAG oil mixed with mineral oil creates a hazardous gas that can corrode the air-conditioning system.

_____ 6. When the volume above a compressor piston reduces, the refrigerant is forced out of the cylinder through the suction reed valves.

_____ 7. Sanden scroll-type compressors are used as air-conditioning compressors as well as transmission oil pumps and power steering pumps.

_____ 8. The clutch electrical winding typically has a resistance value of 3–5 ohms on a 12-volt DC air-conditioning compressor.

_____ 9. The condenser fan is designed to pull or push air from in front of the vehicle through the condenser to remove heat and reduce the pressure on the high side.

_____ 10. Some vehicles, such as conversion vans, minivans, and SUVs, use a dual evaporator system composed of two different units to cool the vehicle.

_____ 11. The site glass is a glass portal used only with R-134a refrigerant that allows you to look into the air-conditioning system and see the liquid refrigerant flow.

_____ 12. A fixed-orifice tube with a diameter of 0.057" (1.4 mm) would be blue in color.

_____ 13. Desiccant is a chemical compound that can absorb moisture and help keep the air-conditioning system dry.

_____ 14. The receiver filter drier is located between the condenser outlet and the TXV inlet.

_____ 15. The heater core is a small radiator consisting of tubes, fins, and tanks.

Fundamentals of Automotive Technology Student Workbook

Fill in the Blank

Read each item carefully, and then complete the statement by filling in the missing word(s).

1. Automotive air conditioning is federally regulated by the _____ _____ _____.
2. Heat always transfers from _____ objects to _____ objects.
3. The transfer of heat through the emission of energy in the form of invisible waves is called _____.
4. One pound of water at _____ _____ and at a temperature of 32°F (0°C) would require 1 Btu of energy added to increase the temperature 1 degree Fahrenheit.
5. At 32°F, the latent heat of _____ begins.
6. If the velocity of a liquid rises, the pressure of the liquid must drop according to the _____ principle.
7. Tetrafluoroethane, commonly known as _____, was the replacement for R-12 refrigerant.
8. When the piston is drawn away from the compressor head, the pumping chamber volume increases and the _____ _____ _____ allows low-pressure refrigerant to enter the pumping chamber.
9. The air-conditioning _____ _____ is an electromagnetic device that uses a metal plate with no friction material and spring steel to engage and disengage the compressor to the drive pulley.
10. The condenser is normally positioned in front of the radiator where it is exposed to maximum _____ _____ when the vehicle is in motion.
11. A(n) _____ condenser is constructed from a single tube that snakes back and forth across the condenser.
12. The job of the _____ is to take a low-pressure liquid and transform it to a low-pressure gas.
13. Barrier-type hose is constructed with a nylon inner core to prevent moisture _____ and permeation of refrigerant from the air-conditioning system.
14. _____ _____ are placed on both the low and the high side of an air-conditioning system to allow an entry point for checking pressures.
15. If the air-conditioning system loses all refrigerant, the _____-_____ cycling switch will turn off the compressor clutch and possibly save the compressor from failure due to no refrigerant and oil.
16. The _____ _____ valve is located at the entry to the evaporator and provides a throttling or restricting function to control the quantity of refrigerant entering the evaporator.
17. The _____ _____ switch is a normally closed, temperature-controlled switch that controls the compressor clutch.
18. Attached to the spinning shaft of the blower motor is a plastic cage with fins to pull or push air called a(n) _____ _____.
19. Some blower motor systems use _____ _____ _____ to control fan speed, this is done by controlling "on" time compared to "off" time on the ground side of the circuit.
20. Some manufacturers use an activated charcoal cabin _____ _____, which helps trap odors and airborne pollutants such as carbon monoxide and oxides of nitrogen.

Labeling

Label the following diagrams with the correct terms.

1. Components of an air-conditioning unit:

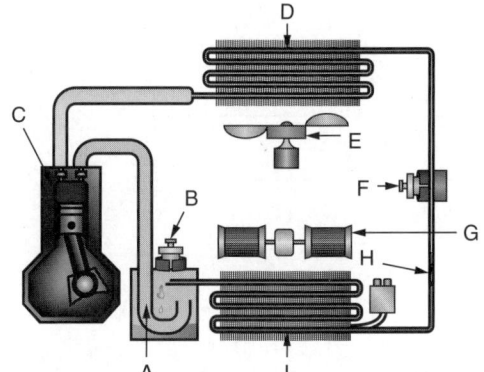

A. _____
B. _____
C. _____
D. _____
E. _____
F. _____
G. _____
H. _____
I. _____

2. Axial piston compressor:

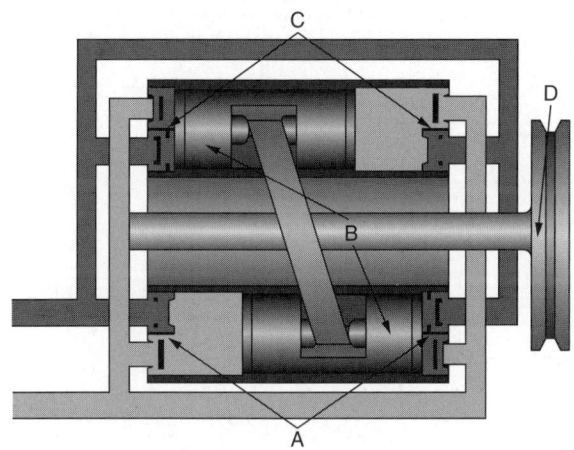

A. _____
B. _____
C. _____
D. _____

3. Rotary vane compressor:

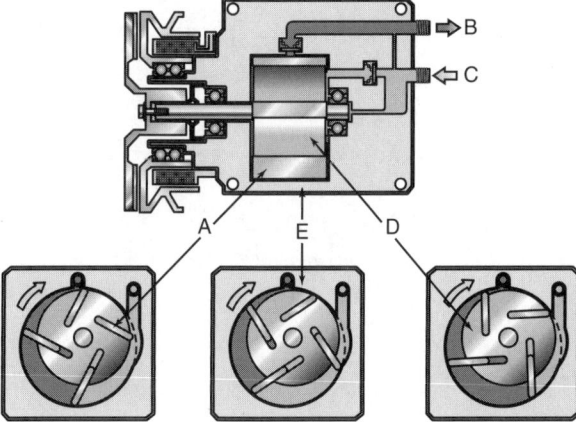

A. _____
B. _____
C. _____
D. _____
E. _____

4. Sanden scroll-type air-conditioning compressor and clutch:

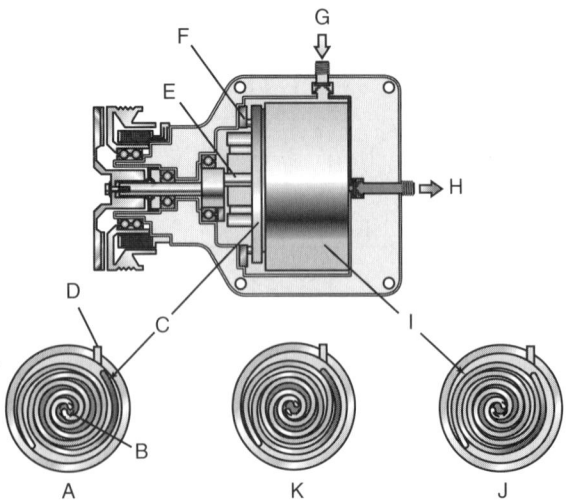

A. _____
B. _____
C. _____
D. _____
E. _____
F. _____
G. _____
H. _____
I. _____
J. _____
K. _____

5. Pressure drop and speed increase through a restriction:

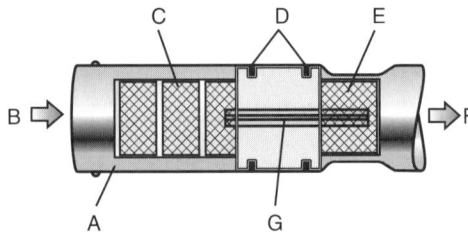

A. _____
B. _____
C. _____
D. _____
E. _____
F. _____
G. _____

Chapter 41 Principles of Heating and Air-Conditioning Systems

6. Accumulator:

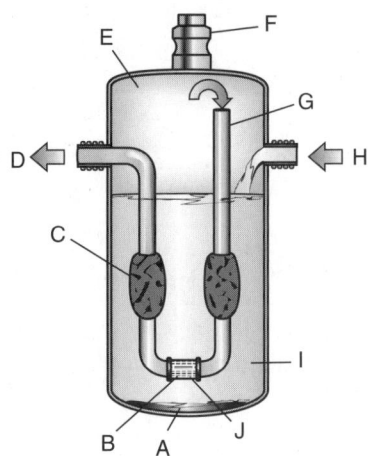

A. _____
B. _____
C. _____
D. _____
E. _____
F. _____
G. _____
H. _____
I. _____
J. _____

7. Thermal expansion valve:

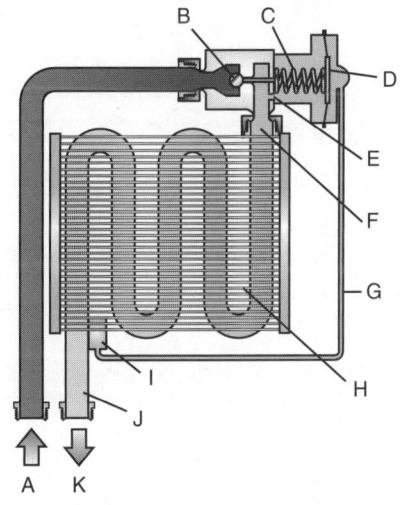

A. _____
B. _____
C. _____
D. _____
E. _____
F. _____
G. _____
H. _____
I. _____
J. _____
K. _____

8. TXV block:

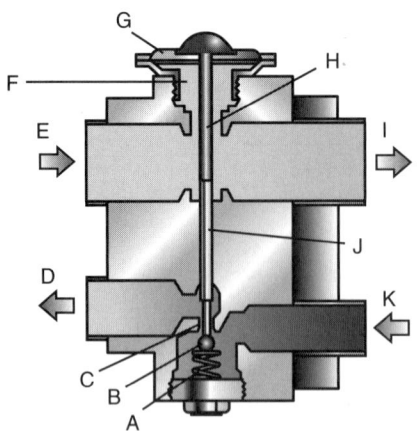

A. _____
B. _____
C. _____
D. _____
E. _____
F. _____
G. _____
H. _____
I. _____
J. _____
K. _____

9. Receiver filter drier:

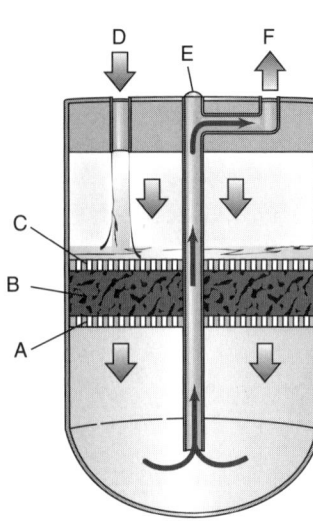

A. _____
B. _____
C. _____
D. _____
E. _____
F. _____

Chapter 41   Principles of Heating and Air-Conditioning Systems

10. Heater controls:

Heater control valve:

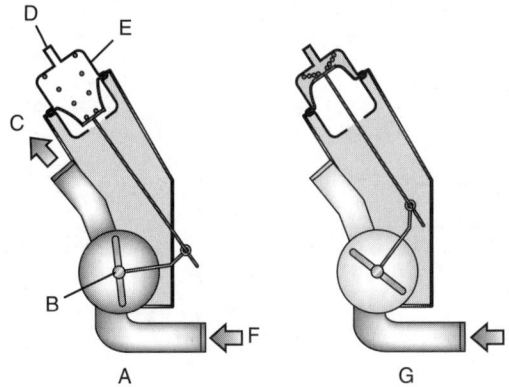

Blend door in the "Cold Air" position:

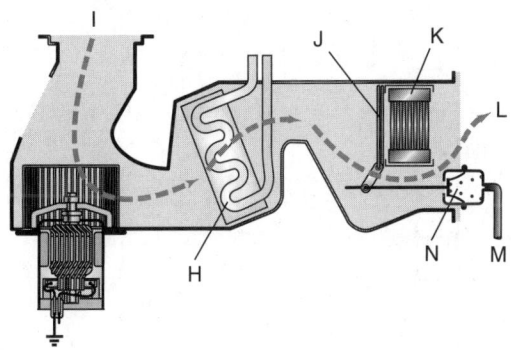

A. _____

B. _____

C. _____

D. _____

E. _____

F. _____

G. _____

H. _____

I. _____

J. _____

K. _____

L. _____

M. _____

N. _____

Crossword Puzzle

Use the clues to complete the puzzle.

Across

4. A flat spring-loaded valve that allows gaseous refrigerant to enter the compressor.
7. A measure of heat energy. It takes 1 Btu to raise the temperature of 1 pound to 1 degree Fahrenheit.
8. A heat-exchanging device that transfers heat converted by the fuel burning in the engine to the passenger compartment.
11. The passing of foreign bodies, such as moist warm air, through the air-conditioning lines or fittings.
12. The transfer of heat through the emission of energy in the form of invisible waves.
14. The disc shaped device that is driven by the accessory drive belt and used to power the compressor.
15. A small, flat piece of metal placed between the tubes to help with the transfer of heat from the coolant to the air, refrigerant to air, or air to refrigerant.

Down

1. A drying agent used in air-conditioning systems to absorb moisture.
2. An electromagnetic control device that creates a magnetic field that pulls in the clutch to make contact with the pulley.
3. A policy signed into law in 1990 that sets standards for air pollution to eliminate ozone-depleting elements.
5. The pressure of the air surrounding everything caused by gravity and the weight of air. The higher from sea level, the lower the atmospheric pressure.

Chapter 41 Principles of Heating and Air-Conditioning Systems

6. A belt- or electrically driven device designed to increase refrigerant pressure and cause refrigerant to travel through the air-conditioning system.
9. The air-conditioning component located in the front of the vehicle designed to allow high-pressure refrigerant to change states from gas to liquid.
10. The flow of air created by a vehicle moving down the road.
13. Metal or plastic pieces that line the pipes used to connect the tubes together to allow the coolant to continue to flow.

ASE-Type Questions

Read each item carefully, and then select the best response.

_____ 1. Tech A says that one advantage of an air conditioner is that the system removes water from the air. Tech B says that air conditioners are needed to remove moisture from the inside of the windshield to meet the mandated defrost time. Who is correct?
   A. Tech A
   B. Tech B
   C. Both A and B
   D. Neither A nor B

_____ 2. Tech A says that the EPA regulates automotive air-conditioning systems. Tech B says that EPA tests air-conditioning systems every 5 years. Who is correct?
   A. Tech A
   B. Tech B
   C. Both A and B
   D. Neither A nor B

_____ 3. Tech A says that an accumulator creates the pressure needed for the air-conditioning system to operate. Tech B says that the high side refers to refrigerant entering the compressor. Who is correct?
   A. Tech A
   B. Tech B
   C. Both A and B
   D. Neither A nor B

_____ 4. Tech A says that heat is removed from the passenger compartment by the heater core. Tech B says that heat in the cab is removed through by the refrigerant and is released to the outside air at the condenser. Who is correct?
   A. Tech A
   B. Tech B
   C. Both A and B
   D. Neither A nor B

_____ 5. Tech A says that conduction is when heat transfers through solids. Tech B says that radiation is when heat travels through space. Who is correct?
   A. Tech A
   B. Tech B
   C. Both A and B
   D. Neither A nor B

_____ 6. Tech A says that the air-conditioning compressor creates high-pressure liquid. Tech B says that the metering device creates a physical change in refrigerant from a liquid to a gas. Who is correct?
   A. Tech A
   B. Tech B
   C. Both A and B
   D. Neither A nor B

_____ 7. Tech A says that when an air-conditioning system freezes up, the refrigerant freezes and stops the cooling process. Tech B says that when an air-conditioning unit freezes up, the evaporator coils are covered with frozen water molecules that stop airflow across the coils, thus preventing cooling. Who is correct?
   A. Tech A
   B. Tech B
   C. Both A and B
   D. Neither A nor B

_____ 8. Tech A says that the evaporator is on the high side of the system. Tech B says that the condenser transfers heat to the atmosphere. Who is correct?
   A. Tech A
   B. Tech B
   C. Both A and B
   D. Neither A nor B

_____ 9. Tech A says that R-134a replaced R-12 as an approved refrigerant. Tech B says that HFO-1234yf refrigerant is flammable. Who is correct?
   A. Tech A
   B. Tech B
   C. Both A and B
   D. Neither A nor B

_____ 10. Tech A says that the engine thermostat controls the available coolant temperature for the heater system. Tech B says that the heater core can have engine coolant or refrigerant flowing through it depending on whether heat or cool is commanded. Who is correct?
   A. Tech A
   B. Tech B
   C. Both A and B
   D. Neither A nor B

Heating and Air-Conditioning Systems and Service

CHAPTER 42

Chapter Review

The following activities have been designed to help you refresh your knowledge of this chapter. Your instructor may require you to complete some or all of these activities as a regular part of your training program. You are encouraged to complete any activity that your instructor does not assign as a way to enhance your learning.

Matching

Match the following terms with the correct description or example.

- A. Accumulator
- B. Air-conditioning compressor clutch
- C. Air-conditioning machine
- D. Anemometer
- E. Charge
- F. Heated diode
- G. Microleak detector
- H. Micron gauge
- I. Muffler
- J. Oiler
- K. Overcharging
- L. Performance testing
- M. Pressure transients
- N. Reclaim/recycle machine
- O. Reclaiming
- P. Refrigerant identifiers
- Q. Sniffer
- R. State of charge
- S. Ultrasonic
- T. Vacuum pump

_____ 1. A device to quiet the pipes of the air-conditioning system with baffles placed inside to deaden the sound of refrigerant moving.

_____ 2. Minor fluctuations on the gauges that may indicate a problem.

_____ 3. The process of recreating a driving situation to check air-conditioning performance and vent temperature.

_____ 4. A pump used to evacuate the air-conditioning system and put it into a deep vacuum or low pressure to remove moisture.

_____ 5. A solution used to detect refrigeration leaks.

_____ 6. An electronic device used to determine the source of leaks.

_____ 7. A method of leak detection that uses a sensitive microphone to hear small refrigerant leaks by amplifying the hissing noise.

_____ 8. A machine designed to recover, recycle, evacuate, leak test, and recharge (R/R/R) the air-conditioning system.

_____ 9. Overfilling of the air-conditioning system; may result in poor cooling or mechanical failure of the system.

_____ 10. A device placed between the evaporator and the compressor to collect liquid refrigerant and prevent it from entering the compressor.

_____ 11. A device designed to measure vacuum very precisely.

_____ 12. The process of removing refrigerant from the air-conditioning system by using an air-conditioning machine; also called recovering.

_____ 13. A device that measures airflow in feet per minute (fpm).

_____ 14. Devices used to check for impurities in the air-conditioning system.

_____ 15. The amount of refrigerant present in the system or the process of installing refrigerant in the system.

_____ 16. An engagement device connected to the compressor crankshaft to engage the crankshaft with a belt-driven pulley.

_____ 17. The amount of refrigerant in a system compared to how much should be in it.
_____ 18. An air-conditioning machine designed to remove and recycle refrigerant for reuse.
_____ 19. A sniffer that uses electricity to determine if there is a refrigerant leak; considered the best sniffer for R-134a.
_____ 20. A device used to add oil to the air-conditioning system.

Multiple Choice

Read each item carefully, and then select the best response.

_____ 1. The only way to determine the type of refrigerant in a system is with the aid of a(n) _____.
   A. refrigerant identifier
   B. performance test
   C. pressure gauge set
   D. anemometer

_____ 2. The _____ test is used to check refrigerant for acidic contaminants and water intrusion.
   A. purity
   B. acid
   C. refrigerant identification
   D. contaminant

_____ 3. When testing the air-conditioning system, an engine's revolutions per minute should be at _____, which is the ideal and average rpm for the majority of air-conditioning systems on vehicles.
   A. 600
   B. 1200
   C. 1800
   D. 2400

_____ 4. Generally, if a(n) _____ is triggered, the cause is overcurrent, meaning the current flow has become elevated.
   A. circuit breaker
   B. thermal unit
   C. fuse
   D. any of the above

_____ 5. A(n) _____ comes in both mechanical and digital styles and has a low-side and a high-side gauge connected to a common manifold that can be used for manual servicing of the air-conditioning system.
   A. air-conditioning machine
   B. reclaim/recycle machine
   C. sealant detector
   D. pressure gauge set

_____ 6. A(n) _____ is used to determine what the high- and low-side pressures should be at a given outside temperature and humidity in a properly functioning system.
   A. vacuum pump
   B. anemometer
   C. PT chart
   D. relative humidity chart

_____ 7. What type of gauge is designed to read negative pressure?
   A. Pressure gauge
   B. Vacuum gauge
   C. Oil-filled gauge
   D. Transient gauge

_____ 8. Supersensitive microphones that can pick up the hiss of a leak that is too small to be heard by human ears are called _____ sniffers.
   A. corona-suppression
   B. heated diode
   C. ultrasonic
   D. supersonic

Chapter 42   Heating and Air-Conditioning Systems and Service

_____ 9. What type of sniffer is considered the best for working with R-134a refrigerant?
   A. Heated diode
   B. Ultrasonic
   C. Corona-suppression
   D. Ultraviolet

_____ 10. While performing a dye test to find a leak, ultraviolet light will cause the refrigerant dye to glow with a(n) _____ color.
   A. orange
   B. blue
   C. green
   D. either A or C

_____ 11. What type of testing works great for testing discharged systems or for prechecking for a leak after repairs have been made and before recharging the system?
   A. Ultraviolet
   B. Nitrogen
   C. Ultrasonic
   D. Heated diode

_____ 12. The reclaiming or recovering process uses a(n) _____ to remove refrigerant from the system.
   A. vacuum pump
   B. air-conditioning machine
   C. oiler
   D. drier

_____ 13. While thoroughly flushing a system usually removes all debris and contaminants, installing a(n) _____ provides an extra level of protection for the compressor and expansion valve or orifice tube.
   A. desiccant bag
   B. drier
   C. additional filter
   D. Schrader valve

_____ 14. The _____ needs to be removed and inspected whenever there are leaks or the desiccant is failing to dry the refrigerant.
   A. receiver/drier
   B. condenser
   C. accumulator
   D. either A or C

_____ 15. With a micron gauge, the pressure is measured in microns, or _____ of 1 millimeter.
   A. one-tenth
   B. one-hundredth
   C. one-thousandth
   D. one-hundred thousandth

True/False

If you believe the statement to be more true than false, write the letter "T" in the space provided. If you believe the statement to be more false than true, write the letter "F".

_____ 1. The proper charge for the air conditioner can be found under the hood on the air-conditioning identification sticker.

_____ 2. Overcharging the air-conditioning system will affect the orifice tube and TXV systems in the same manner.

_____ 3. An undercharged air-conditioning system will cause an unusual smell throughout the cabin when running the air conditioner.

_____ 4. A high-pressure switch may have a cutoff pressure of 400 psi; at that pressure the switch opens because the pressure is high enough to cause the hoses and lines to rupture.

_____ 5. Allowing sealant to be drawn into a refrigerant identifier or air-conditioning machine can ruin them.

_____ 6. Oil-filled pressure gauges do not allow the needle to move fast enough for you to see pressure transients.

_____ 7. The air-conditioning technician should perform maintenance and repair of the air-conditioning machine after every use.

_____ 8. An air-conditioning machine allows you to remove and store the old refrigerant by refilling the same tank that the new refrigerant came in.

_____ 9. The air-conditioning system should be flushed if the compressor comes apart or if the desiccant bag breaks open.

_____ 10. Any time lines are loosened, the O-rings in the fittings must be replaced or there is potential for leaks.

_____ 11. Typically, the TXV is mounted on the front of the evaporator.

_____ 12. The refrigerant R-12 used a mineral oil, while the refrigerant R-134a uses a polyalkylene glycol (PAG) oil.

_____ 13. Charging an air-conditioning system is performed after the air-conditioning system has been evacuated and before any oil is replaced.

_____ 14. Changing the vacuum pump oil regularly will extend the life of the pump and make it reach maximum vacuum quicker.

_____ 15. Leaks from the heater core can result in coolant misting out of the vents or coolant dripping onto the passenger-side carpet, or both.

Fill in the Blank

Read each item carefully, and then complete the statement by filling in the missing word(s).

1. Carrying out a (n) _____ _____ of the air-conditioning system is the first step in the diagnostic process.

2. The _____ _____ _____ is best tested by removing all of the refrigerant from the air-conditioning system and comparing that amount to the manufacturer's specifications.

3. If the _____ _____ becomes clogged with leaves or debris over time, the water will not be able to drain, it will stagnate and begin growing bacteria, creating an unpleasant odor.

4. A(n) _____ _____ set comes in both mechanical and digital styles and has a low-side and a high-side gauge connected to a common manifold that can be used for manual servicing of the air-conditioning system.

5. An aftermarket _____ _____ has the fittings and oil to change an R-12 unit over to an R-134a unit.

6. With a(n) _____-_____ sniffer, refrigerant gas is used as an insulator to slow current between electrodes.

7. To inspect the air gap of the compressor clutch use feeler gauges to measure the air gap between the _____ _____ and the clutch drive plate with the clutch disengaged.

8. Often, the entire dash assembly must be removed before the air-conditioning _____ box can be accessed.

9. After all repairs are made, the air-conditioning system needs to be _____ to remove all of the moisture from the air introduced into the lines from opening up the air-conditioning system.

10. The oiler uses _____ _____ to push oil into the air-conditioning system.

Chapter 42　Heating and Air-Conditioning Systems and Service

Labeling

Label the following diagrams with the correct terms.

Identify the following tools used in the service of heating and air-conditioning systems:

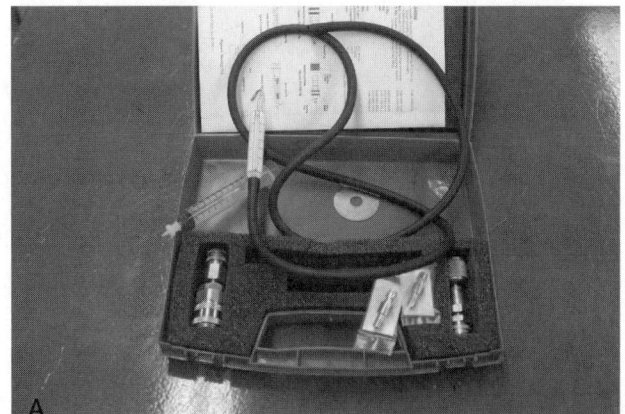

A. _____

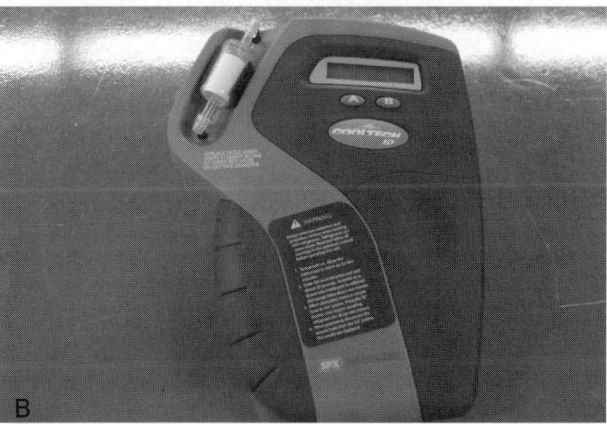

B. _____

C. _____

Fundamentals of Automotive Technology Student Workbook

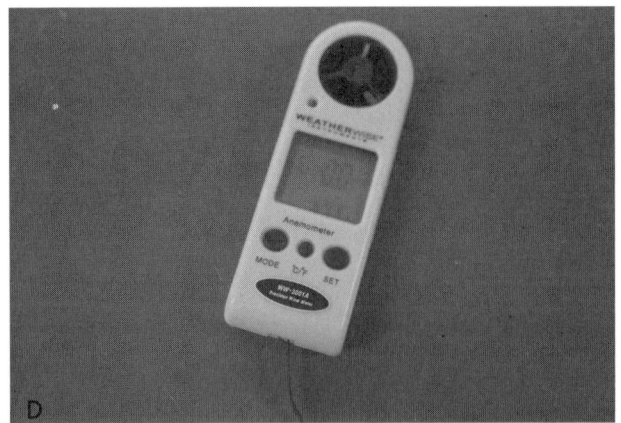

D. _____

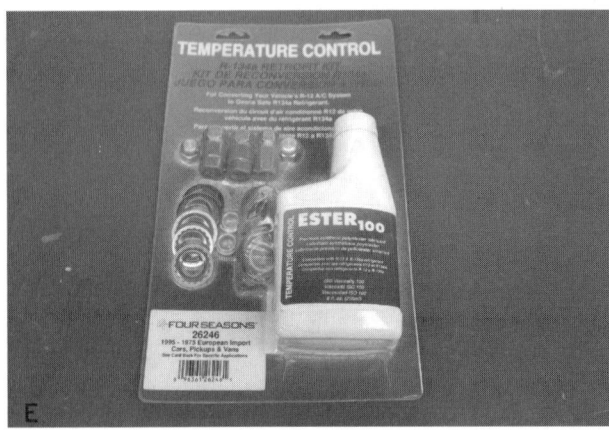

E. _____

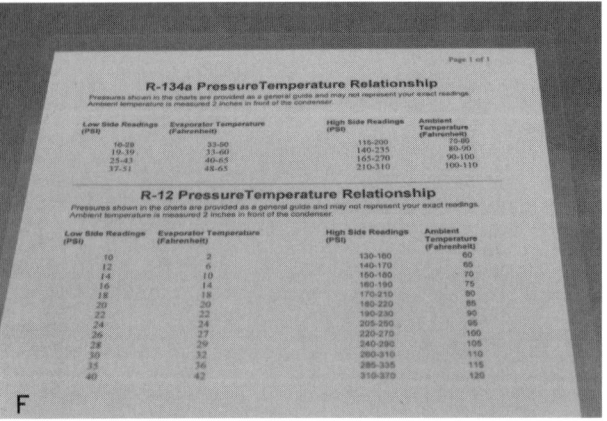

F. _____

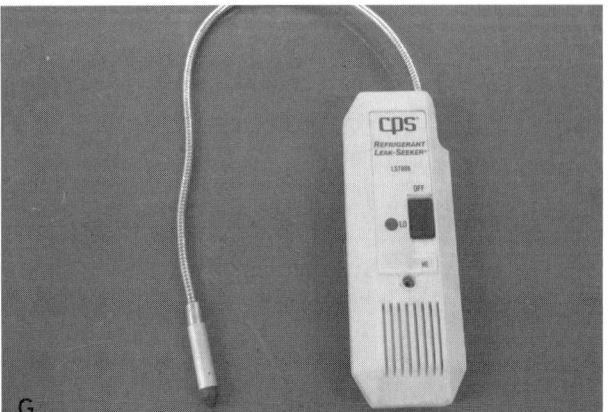

G. _____

Chapter 42   Heating and Air-Conditioning Systems and Service

H. _____

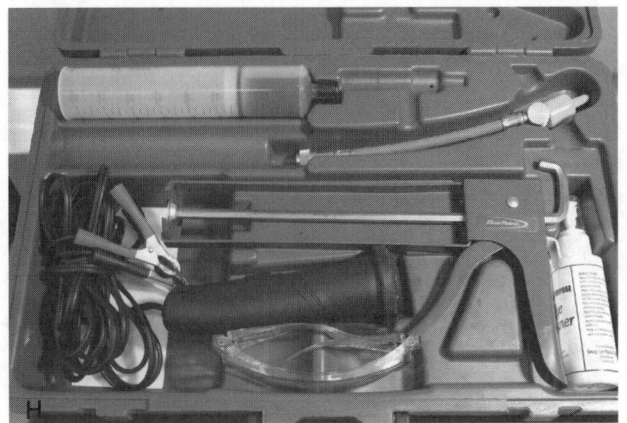

I. _____

J. _____

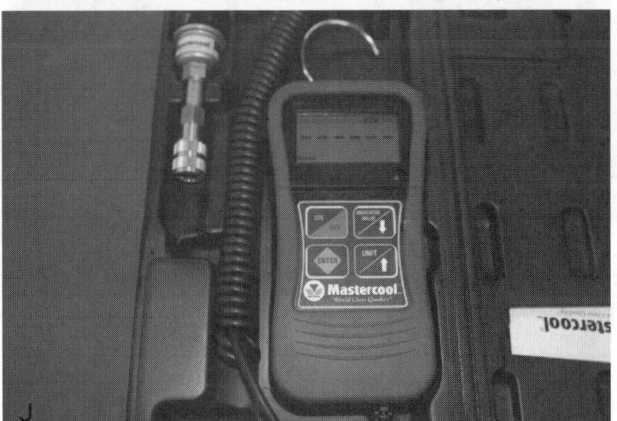

Skill Drills

Place the skill drill steps in the correct order.

1. Identifying the Refrigerant Type:

_____ **A.** After selecting the pressure gauge set, make sure the service valves on the gauges are in the off or shut position.

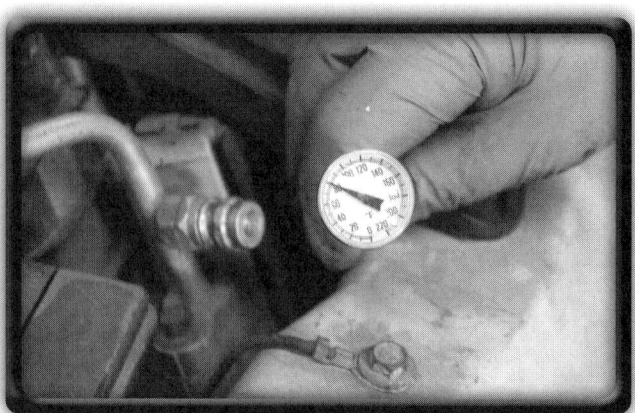

_____ **B.** Measure the ambient temperature (6" to 8" [15 to 20 cm] in front of the condenser) and the pressure on the pressure gauge set. Compare the pressure to the PT chart for that type of refrigerant.

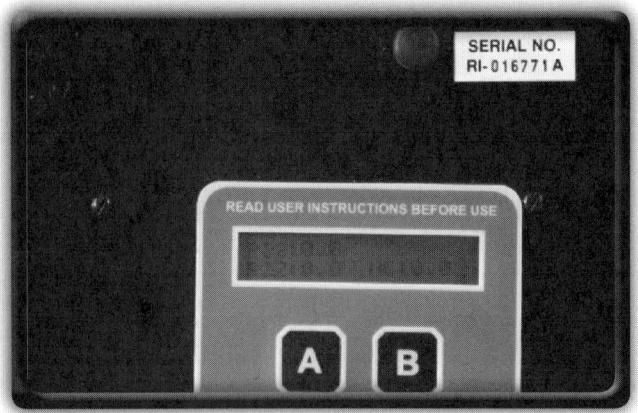

_____ **C.** Turn off the valve, and disconnect the refrigerant identifier. If the refrigerant does not match the under-the-hood sticker, reclaim the contaminated refrigerant into a contaminated tank and label the refrigerant for disposal. If the refrigerant is 100% pure and matches the under-the-hood sticker, continue to the next step.

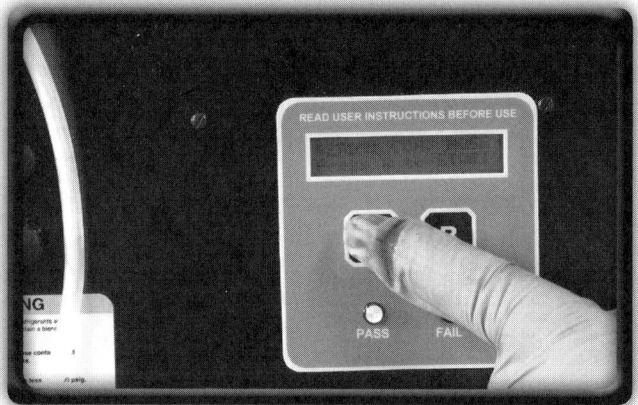

_____ **D.** Use a sealant identifier to check whether there is any sealant in the system. Turn on the refrigerant identifier and allow the machine to warm up.

Chapter 42 Heating and Air-Conditioning Systems and Service

_____ **E.** Connect the service chuck of the pressure gauge to the air-conditioning system. Open the valves, watch the high- and low-pressure gauges, and record the pressure. If the high- and low-side gauges are the same pressure and you could not identify the refrigerant, use the PT chart to determine whether there are noncondensable gases in the air-conditioning system.

_____ **F.** Connect the refrigerant identifier to the low side of the air-conditioning system, and open the service valve. Follow the prompts on the refrigerant identifier, and record the refrigerant type and amount of air, if any.

   2. Removing, Inspecting, and Reinstalling the Compressor:

_____ **A.** Unplug the compressor clutch.

_____ **B.** Plug in the compressor clutch, and install the drive belt. Proceed to the evacuation procedure.

_____ **C.** Check the resistance of the clutch coil, and check the pulley for wobble that may be caused by a faulty bearing. Check the plugs for corrosion. Check the tag on the new compressor for oil.

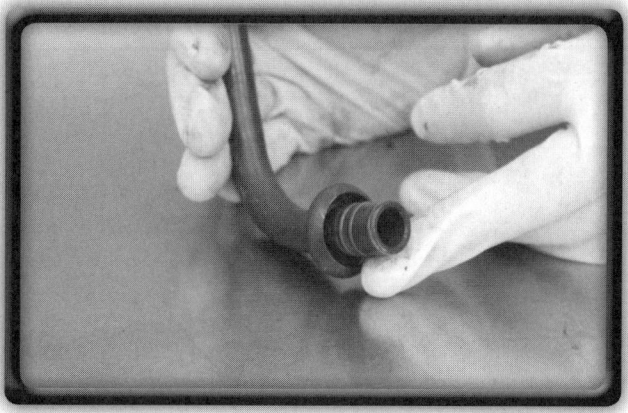

_____ **D.** Install the air-conditioning lines with new sealing rings.

_____ **E.** Remove the hoses from the air-conditioning system. Cap the lines.

_____ **F.** Install the new compressor; tighten the mounting hardware to specifications.

_____ **G.** Reclaim the refrigerant, making note of the amount removed. Compare this amount to the factory specifications. Remove the belt using a serpentine belt tool or the proper wrench. Remove tension from the tensioner, and slide the belt off, noting routing for reinstallation.

_____ **H.** Remove the mounting bolts following the manufacturer's specifications. Pour oil from the compressor into a graduated cylinder, and check the oil for acid using acid test strips.

3. Removing, Inspecting, and Reinstalling the Condenser:

_____ **A.** Remove the inlet and outlet lines on the condenser.

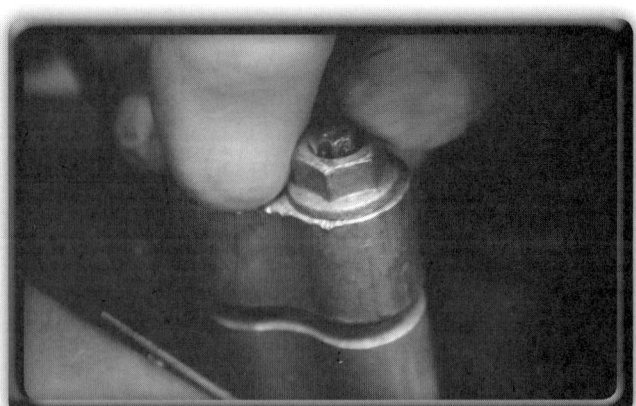

_____ **B.** Install the inlet and outlet lines. Replace all of the components in reverse order from removal. Evacuate the air-conditioning system if all other air-conditioning repairs are complete.

Chapter 42 Heating and Air-Conditioning Systems and Service

_____ **C.** Install the new condenser. Fasten the condenser with the mounting hardware.

_____ **D.** After reclaiming the air-conditioning system, remove all necessary components to access the condenser. Refer to the manufacturer's specific procedures on component removal.

_____ **E.** Install new O-rings on the air-conditioning lines.

_____ **F.** Remove the hold-down bolts for the condenser.

4. Removing, Inspecting, and Installing a TXV:

_____ **A.** Install the new expansion valve and torque the bolts to specification.

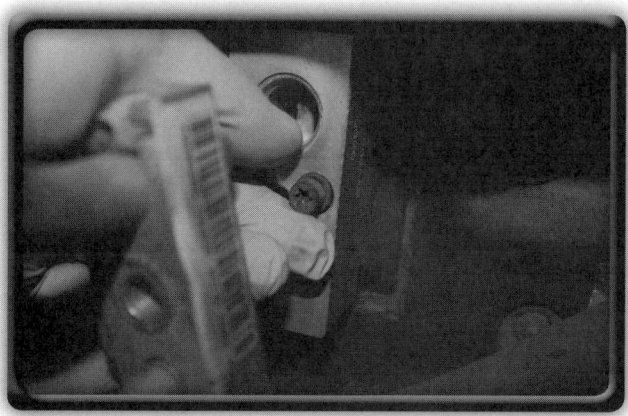

_____ **B.** Remove the bolts holding the expansion valve to the evaporator inlet.

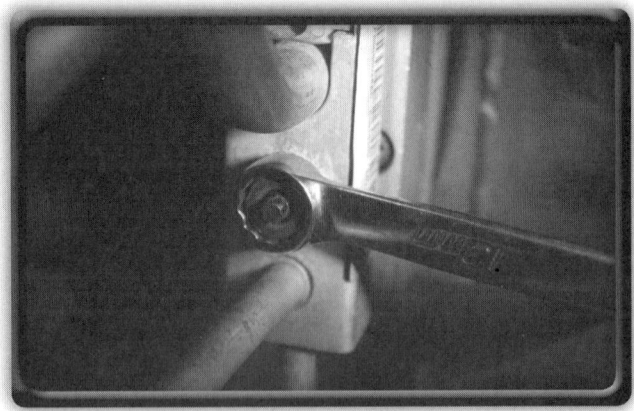

_____ **C.** Install a new mounting stud. Reinstall the lines to the expansion valve, and tighten retaining nut. Evacuate the system if all other repairs were made.

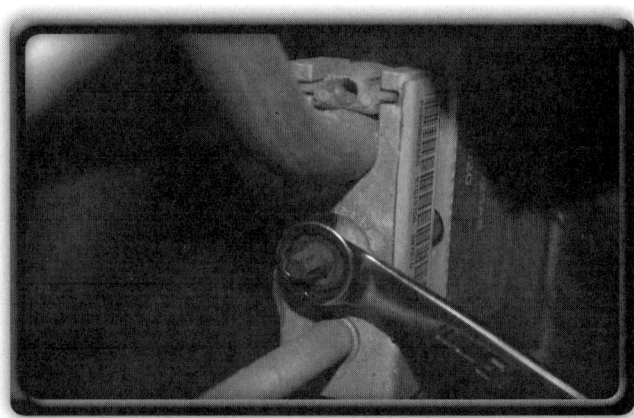

_____ **D.** After reclaiming the air-conditioning system, use a wrench to loosen the bolt holding the liquid and suction line block to the expansion valve.

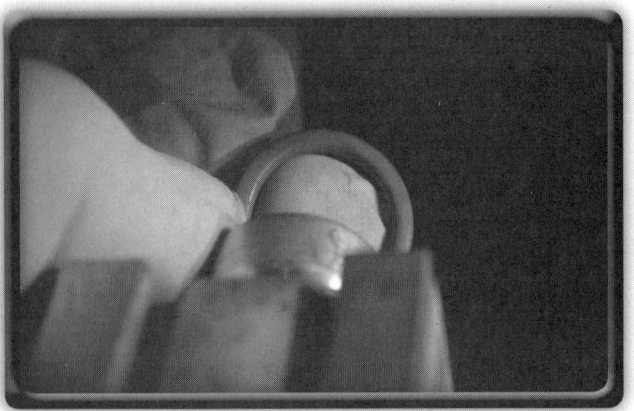

_____ **E.** Install new gaskets or O-rings onto the new expansion valve and the suction and liquid lines.

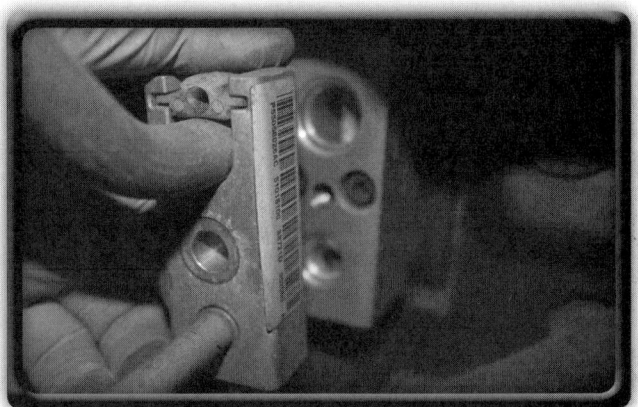

_____ **F.** Remove the center stud going through the expansion valve. Gently remove the lines from the expansion valve.

5. Removing, Inspecting, and Installing an Orifice Tube:

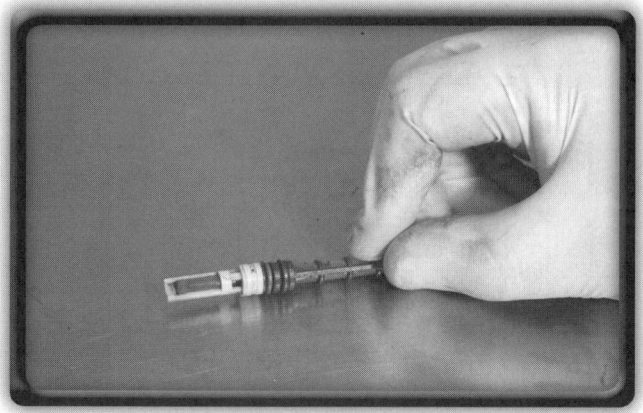

_____ **A.** Select the new orifice tube, making sure to replace with the properly colored tube. Lubricate the orifice tube O-rings, and install the new one with the direction arrow pointing in the direction of refrigerant flow.

_____ **B.** Locate the orifice tube, and determine whether it is serviceable. After locating the serviceable orifice tube, remove the line where the orifice tube is housed.

_____ **C.** Lubricate and install new O-rings on the line, and reinstall the line. Proceed to the evacuation if all other repairs were made.

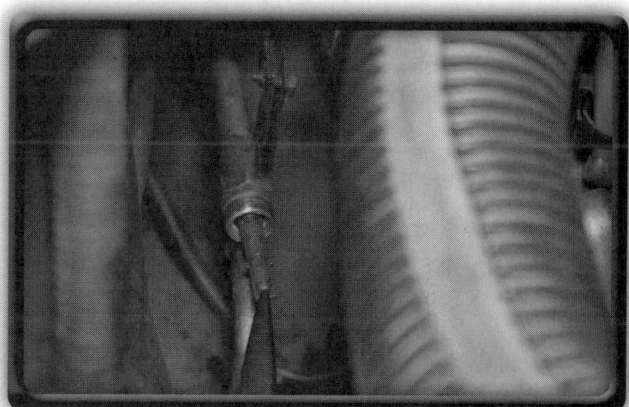

_____ **D.** Using small needle-nose pliers or the orifice tube removal tool, pull out the old orifice tube from the housing. Note the color of the old orifice tube.

Crossword Puzzle

Use the clues to complete the puzzle.

Across

2. A pressure-temperature chart that shows the relationship between air-conditioning pressures and evaporator temperature.
5. Rubber-type ring available in different sizes used to seal pipe fittings.
7. An electronic sniffer used for detecting refrigerant leaks.
8. Any belt-driven pulley used to power an accessory such as power steering or the air-conditioning compressor.
9. Term used interchangeably with pipes or tubes.

Down

1. An aftermarket kit that has the fittings and oil to change an R-12 unit to an R-134a unit.
3. Flexible lines used to direct liquids or gases.
4. Another word for reclaiming; the process of removing refrigerant from the air-conditioning system by using an air-conditioning machine.
6. A gauge designed to read negative pressure or vacuum.

ASE-Type Questions

Read each item carefully, and then select the best response.

_____ 1. Tech A states that the wider the gap on an air-conditioning clutch, the greater the ohm reading when checking the windings. Tech B states that the air-conditioning clutch is electromagnetically operated. Who is correct?
   A. Tech A
   B. Tech B
   C. Both A and B
   D. Neither A nor B

_____ 2. Tech A states that an air-conditioning performance test usually requires that an auxiliary condenser fan be used during the test. Tech B states that a performance test will show if the air-conditioning system is contaminated with sealer. Who is correct?
   A. Tech A
   B. Tech B
   C. Both A and B
   D. Neither A nor B

_____ 3. Tech A states that refrigerant in a vehicle should be identified before recovering the refrigerant. Tech B states that refrigerant doesn't need to be identified if you are only topping up a system with refrigerant. Who is correct?
   A. Tech A
   B. Tech B
   C. Both A and B
   D. Neither A nor B

_____ 4. Tech A states that when evacuating an air-conditioning system, the vacuum should be maintained for approximately 10 minutes after the system reaches the boiling point pressure of water. Tech B states that the primary purpose of evacuating an air-conditioning system is to remove any moisture from the system. Who is correct?
   A. Tech A
   B. Tech B
   C. Both A and B
   D. Neither A nor B

_____ 5. Tech A states that the system should be flushed if the compressor came apart. Tech B states that the system should be flushed if the oil is contaminated. Who is correct?
   A. Tech A
   B. Tech B
   C. Both A and B
   D. Neither A nor B

_____ 6. Tech A states that when using pressurized nitrogen to locate a leak, an electronic sniffer should be used. Tech B states that electronic sniffers are used when the system has at least a minimal refrigerant charge. Who is correct?
   A. Tech A
   B. Tech B
   C. Both A and B
   D. Neither A nor B

_____ 7. Tech A states that microns are a much more accurate unit of measuring vacuum than inches of mercury (Hg). Tech B states that microns are a much more accurate measure of time than seconds. Who is correct?
   A. Tech A
   B. Tech B
   C. Both A and B
   D. Neither A nor B

8. Tech A states that to determine how much Freon is needed in a system, you must refer to identifying labels on the vehicle or the service manual. Tech B states that to determine the amount of Freon needed, you just charge the system until the pressures look correct. Who is correct?
   A. Tech A
   B. Tech B
   C. Both A and B
   D. Neither A nor B

9. Tech A states that when removing any part of an air-conditioning system, the oil should be drained from it and measured so that the same amount can be reinstalled. Tech B states that oil should only be in the compressor, and if any oil is found in any other components, it means that the receiver drier is faulty. Who is correct?
   A. Tech A
   B. Tech B
   C. Both A and B
   D. Neither A nor B

10. Tech A states that if moisture enters the air-conditioning system, acid will be created. Tech B states that evacuating an air-conditioning system will boil moisture, which will be removed from the system as a gas. Who is correct?
    A. Tech A
    B. Tech B
    C. Both A and B
    D. Neither A nor B

Electronic Climate Control

CHAPTER 43

Tire Tread:
© AbleStock

Chapter Review

The following activities have been designed to help you refresh your knowledge of this chapter. Your instructor may require you to complete some or all of these activities as a regular part of your training program. You are encouraged to complete any activity that your instructor does not assign as a way to enhance your learning.

Matching

Match the following terms with the correct description or example.

- A. Actuator
- B. Air-conditioning compressor clutch
- C. Air-conditioning pressure sensor
- D. Ambient air temperature sensor
- E. Aspirator
- F. Cabin air temperature sensor
- G. Dual-drive air-conditioning compressor
- H. Electric servo motor
- I. Engine coolant temperature sensor
- J. Evaporator temperature sensor
- K. Feedback signal
- L. Heater control cables
- M. Limit switches
- N. Manual climate control system
- O. Negative temperature coefficient (NTC) thermistor
- P. Photodiode
- Q. Printed circuitry
- R. Semiautomatic climate control system
- S. Sun load (solar) sensor
- T. Vacuum servo

_____ 1. Circuitry that forms the framework for electronic module construction. A printed circuit board holds electronic components that are soldered into place.

_____ 2. A voltage signal sent back to an electronic control unit. The feedback signal is how the control module is able to interpret temperature or other information from its sensors.

_____ 3. A thermistor that measures air temperature inside the vehicle.

_____ 4. A climate control system fully controlled by the operator.

_____ 5. A photodiode that varies voltage based on light. It is used to determine the radiant heat coming from the sun into the passenger cabin and gives an input signal of sunlight load to the ECU.

_____ 6. A thermistor that is used to measure the air temperature outside the vehicle.

_____ 7. A vacuum-controlled device that moves the air doors of the air box. It is controlled by the vacuum that comes from the vacuum-type climate control panel.

_____ 8. An air compressor drive used on some hybrid vehicles. The compressor can be driven by the accessory belt or by an electric motor.

_____ 9. Switches that turn off power flow to an electric motor when a particular limit is reached.

_____ 10. A tube that is used to direct airflow across the cabin air temperature sensor.

_____ 11. A thermistor that measures the temperature of the engine coolant.

_____ 12. A sensor that gives an input signal of refrigerant pressure in the air-conditioning system to the ECU.

_____ 13. A system that provides automatic function of the heater or cooling only, leaving fan speed and mode selection to the operator.

_____ 14. A device that moves the air doors of the air box. It is controlled by input from the climate control panel.

_____ 15. Also referred to as an electric actuator, a motor that provides movement to operate the air doors in an air box to control air temperature and air movement.

_____ 16. Cables that control the air doors in an air box as part of the air distribution system.

_____ 17. The mechanical coupler that is electromagnetically engaged and provides a way of uncoupling the compressor from the accessory drive belt.

_____ 18. A thermistor that reads the temperature of the evaporator, used to ensure that the evaporator does not freeze.

_____ 19. An electronic component that creates a varying voltage or current output based on the amount of light striking it.

_____ 20. A thermistor that gains resistance as temperature goes down and loses resistance as temperature goes up.

Multiple Choice

Read each item carefully, and then select the best response.

_____ 1. The climate control system manages all of the following subsystems, *except*:
   A. heating
   B. cooling
   C. ventilation
   D. air conditioning

_____ 2. The electronic control unit compares the _____ from the temperature sensors with values stored in its memory, programmed in when the module is installed into the vehicle.
   A. signal voltage values
   B. feedback signals
   C. reference voltages
   D. pulse-width

_____ 3. The ECU can control a blower motor's speed by sending a(n) _____ signal that works by repeatedly turning on and off the current flow to the component.
   A. analog
   B. high frequency
   C. pulse-width–modulated
   D. low frequency

_____ 4. Which sensor is usually located in front of the air-conditioning condenser where the forward motion of the vehicle and the action of the condenser fans force air over the sensor?
   A. Ambient air temperature sensor
   B. Engine coolant temperature sensor
   C. Air-conditioning pressure sensor
   D. Cabin air temperature sensor

_____ 5. The _____ sensor is a photodiode.
   A. evaporator
   B. sun load
   C. engine coolant
   D. cabin air

_____ 6. Either the low-pressure sensor or the high-pressure sensor can send a _____ signal to the ECU based on the pressure in the system.
   A. pulse-width–modulated
   B. resistance
   C. analog
   D. variable voltage

_____ 7. The _____ sensor is located in the airstream where the air leaves the evaporator and signals to the ECU the temperature of the air.
   A. ambient air temperature
   B. evaporator temperature
   C. air-conditioning pressure
   D. cabin air temperature sensor

_____ 8. The operator can tell the ECU where to deliver air and at what temperature, using switches that produce a(n) _____ signal.
   A. on/off–type
   B. pulse-width–modulated
   C. variable voltage
   D. resistance

_____ 9. The _____ temperature control panel is designed to allow the driver and passenger to control the temperature delivered to their side of the vehicle.
   A. manual
   B. semiautomatic
   C. fully automatic
   D. dual-zone

_____ 10. The vacuum _____ is installed on the vacuum system to ensure vacuum is present for the control panel to use even when the engine is at wide open throttle.
   A. switch
   B. canister
   C. servo
   D. relay

_____ 11. In a manual air conditioner, the fan control switch connects _____ with the blower motor to control its speed.
   A. resistors in series
   B. resistors in parallel
   C. resistors in series/parallel
   D. none of the above

_____ 12. The _____ is a plastic housing that holds other components of the climate control system, such as the heater core and the evaporator.
   A. blend box
   B. plenum
   C. air box
   D. air duct

_____ 13. The _____ is a sealed container with a spring and diaphragm inside.
   A. vacuum servo
   B. electric actuator
   C. solenoid actuator
   D. recirculation door

_____ 14. Some electric servo motors use a _____ to signal to the ECU the position of the blend door and, if it is moving, the direction in which it is traveling.
   A. thermistor
   B. variable voltage resistor
   C. potentiometer
   D. limit switch

_____ 15. Most hybrids now use a _____ motor, which directly drives the compressor.
   A. single-phase
   B. high-voltage electric
   C. low-voltage
   D. servo

True/False

If you believe the statement to be more true than false, write the letter "T" in the space provided. If you believe the statement to be more false than true, write the letter "F".

_____ 1. The manual climate control system gives the operator full control of the climate control system and does not use a computer or electronic control unit.

_____ 2. The resistance value of a positive temperature coefficient thermistor goes down as it is heated.

_____ 3. Negative temperature coefficient thermistors are usually used in automotive applications.

_____ 4. The actuator is a tube that directs airflow to the in-vehicle temperature sensor.

_____ 5. Typically, air-conditioning systems will use two air-conditioning pressure sensors to monitor the air-conditioning system pressure.

_____ 6. Transducers are electronic devices that convert one form of energy to another.

_____ 7. All of the buttons used with the manual control panel are present on the automatic panel as well.

_____ 8. Most manual air-conditioning systems use four resistors to give three operator-selectable speeds of the blower motor.

_____ 9. The mode door directs air through or around the heater core to deliver hot or cold air into the passenger cabin, and the blend door changes the location of air delivery to floor, defrost, or vent.

_____ 10. Geared motor actuators are used in HVAC systems, while solenoid actuators are used in systems such as door locks and fuel door latches.

Fill in the Blank

Read each item carefully, and then complete the statement by filling in the missing word(s).

1. The _____ _____ _____ automatically maintains the operator-selected climate (temperature) preference within the passenger cabin.

2. The _____ loaded into the electronic control unit contains all the data to make the system operate properly.

3. The electronic control unit receives signals in the form of _____ from each of the sensors.

4. In most instances the temperature sensors used in the climate control air-conditioning system are _____.

5. The _____ _____ is a safety device built into the compressor by manufacturers to ensure pressures cannot get too high in the system and rupture a line.

6. Most of the switches located on a modern control panel are _____ to the electronic control unit.

7. Relay windings are normally equipped with a _____ _____ mechanism in the form of either a diode or a resistor wired in parallel with the winding.

8. The _____ _____ is an electric motor that is attached to a circular fan sometimes referred to by technicians as a squirrel cage fan.

9. The higher the resistance connected in line with the blower motor, the _____ the blower motor will operate.

10. The _____ _____ directs air through or around the heater core to deliver hot or cold air into the passenger cabin.

11. The _____ door is designed to work with the maximum air-conditioning button or switch and will shut airflow off from outside the vehicle and instead recirculate air from inside the passenger compartment back through the evaporator.

12. The electric actuator may incorporate _____ _____ to stop current to the electric actuator when the maximum travel of the door has been reached.

13. Most hybrids now use a high-voltage electric motor, which directly drives the _____.

14. Some hybrids use _____-_____ air-conditioning compressors that can be belt driven by the ICE or by a high-voltage motor.

15. The compressor in the hybrid air-conditioning system may be driven by the internal combustion engine through a standard belt and _____ clutch arrangement, just like a standard air-conditioning compressor clutch arrangement.

Chapter 43 Electronic Climate Control

Labeling

Label the following diagrams with the correct terms.

1. Blend door:

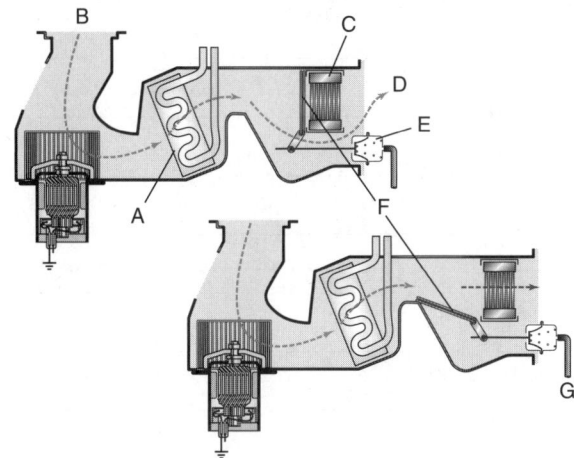

A. ___
B. ___
C. ___
D. ___
E. ___
F. ___
G. ___

2. Recirculation door:

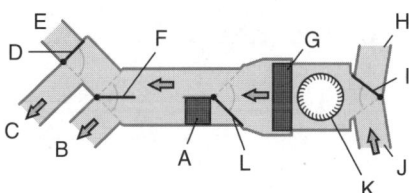

A. ___
B. ___
C. ___
D. ___
E. ___
F. ___
G. ___
H. ___
I. ___
J. ___
K. ___
L. ___

3. Vacuum solenoid:

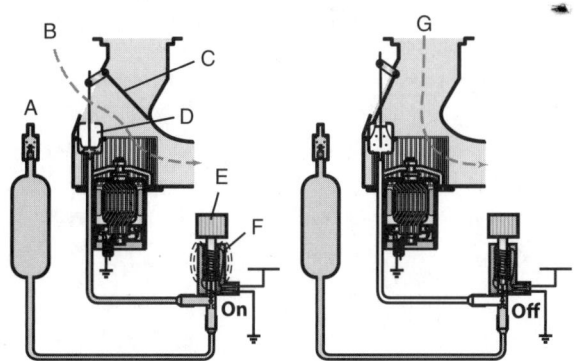

A. ___
B. ___
C. ___
D. ___
E. ___
F. ___
G. ___

4. Electrically driven air-conditioning compressor:

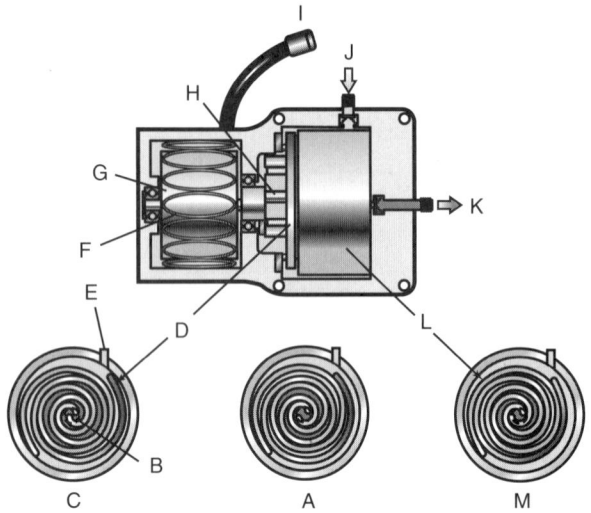

A. _____
B. _____
C. _____
D. _____
E. _____
F. _____
G. _____
H. _____
I. _____
J. _____
K. _____
L. _____
M. _____

5. Dual-drive air-conditioning compressor:

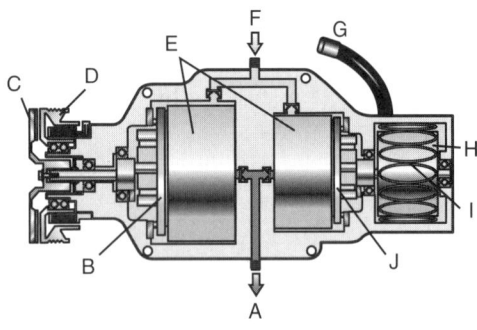

A. _____
B. _____
C. _____
D. _____
E. _____
F. _____
G. _____
H. _____
I. _____
J. _____

Skill Drills

Test your knowledge of skill drills by filling in the correct words in the photo captions.

1. Inspecting and Testing the Electric Cooling Fan:

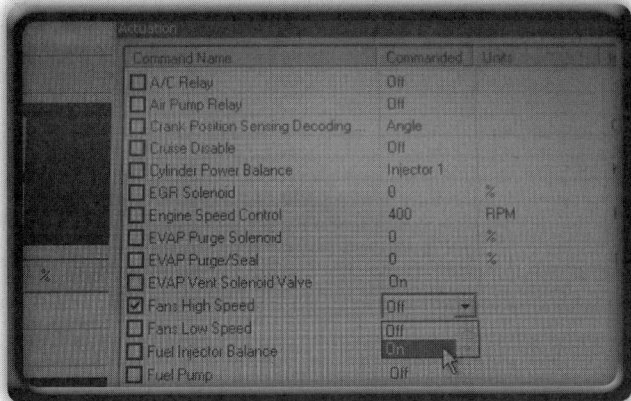

Step 1: If the cooling fan does not come on when the system is turned on, install the _____ _____ and find the _____ controls for the cooling fan; then turn the cooling fan on.

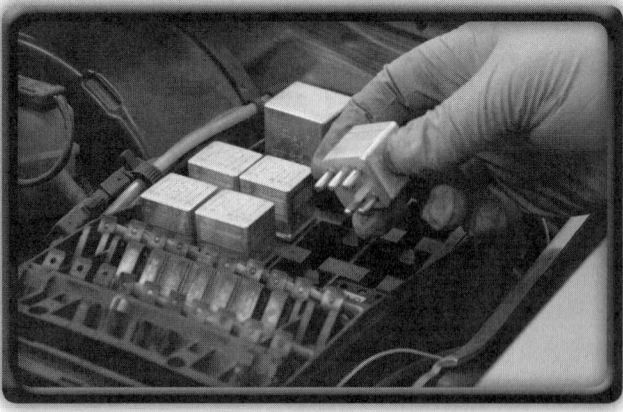

Step 2: If the cooling fan still does not run, find the coolant _____ relay, _____ the relay, and find the two _____ that provide power to the cooling fan (this is the _____ side of the fan relay).

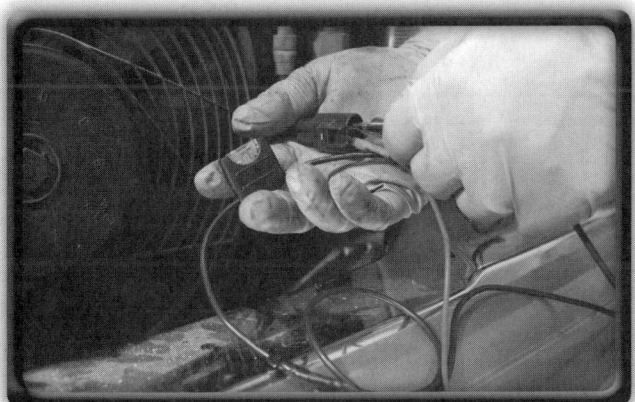

Step 3: Install the _____ _____ wire to the connections to provide power to the cooling fan. CAUTION: Make sure you _____ the _____ _____ and not the control circuit. Jumping the _____ _____ can burn up the ECU, even with fused jumper wires!

Step 4: Inspect the cooling fan for operation. If the fan is still not operating, _____-_____ the connector at the cooling fan with the jumper wire _____ _____ from the previous step. Connect the _____ leads to both back-probed wires (power and ground); the meter should read _____ voltage.

Step 5: If battery voltage is present, _____ the faulty _____ _____. If battery voltage is not present, use the _____ to locate the _____ or high _____ in the circuit.

2. Inspecting and Testing a Blower Motor:

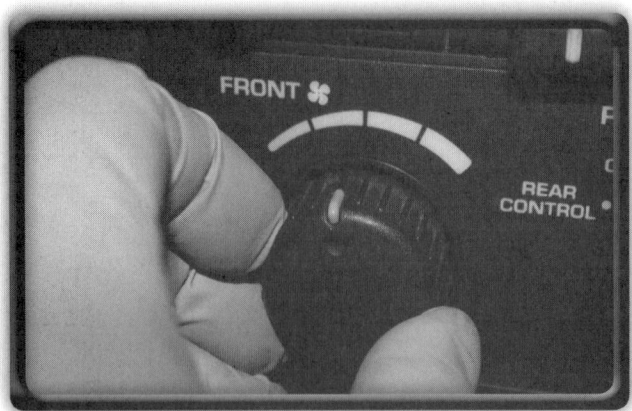

Step 1: Turn on the blower motor, and check whether the motor is _____ on each _____. Research the _____ _____ for the condition found.

Step 2: Use a _____ to test for _____ and _____ by back-probing the wires at the _____ of the blower motor. Turn the _____ switch to "run" and the _____ switch to the speed that does not operate, and _____ the meter. If the meter reads more than about _____ _____, the fan should be operating. If it is not, the blower motor needs to be _____.

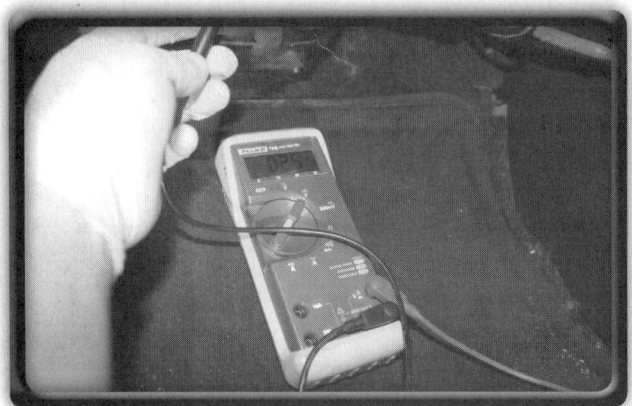

Step 3: If the reading is less than 3 volts, then perform a _____ _____ test on the power and ground sides of the _____. If the voltage drop is excessive, use a wiring _____ and voltmeter to track down the high _____ or opening in the circuit, and _____ as necessary.

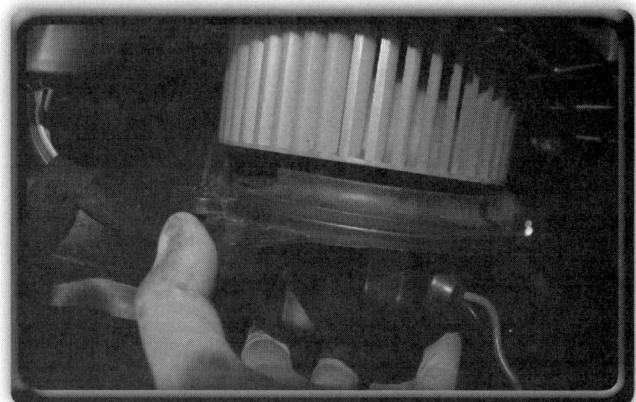

Step 4: If the control circuit tests _____, but the fan is making unusual _____ or is not operating correctly, remove the blower motor by _____ the electrical connector and removing any _____, nuts, or _____ retaining the blower motor to the _____ _____.

Step 5: Inspect the fan _____ and fins for _____ objects or _____.

Step 6: If the fan is in good operating condition, measure the _____ _____ with the blower motor _____ from the vehicle, either (1) with a low-amp _____ (inductive clamp) around the _____ to or from the motor, or (2) with an _____ set to 20 amps and wired in _____ with the motor. Run the motor while _____ the amperage. If necessary, _____ the blower motor and retest.

3. Testing Air-Conditioning Compressor Clutch Control Systems:

Step 1: _____ the testing procedure and specifications in the appropriate service information. Unplug the compressor _____ electrical two-wire _____ located at the front of the air-conditioning _____. Check the _____ of the clutch windings. If the resistance does not match the manufacturer's specifications, replace the _____ _____.

Step 2: Check supply _____ and ground to the compressor clutch _____. Place the _____ lead on the _____ side of the clutch connector, and place the _____ lead on the _____ side of the clutch connector. Ensure that the air-conditioning _____ is turned on. Your meter should display near battery _____.

Step 3: Perform a _____ _____ test of power and ground. Connect the red lead of your voltmeter to the positive terminal of the _____ _____ _____. Connect the black lead to _____ positive. Ensure the air-conditioning clutch is turned on with the _____ panel. Look for a _____ or loss of voltage on the _____ wire of no more than 0.4 volts. If more than 0.4 volts is found, the power wire has _____ resistance in it. Repeat the step for the _____ side by connecting the _____ lead to the negative terminal and the _____ lead to the battery _____.

4. Inspecting and Testing the HVAC Control Panel Assembly:

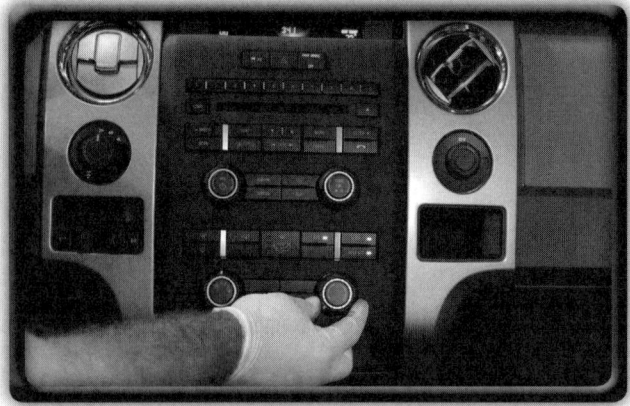

Step 1: After researching the operation and _____ procedure for the control panel _____, start by _____ all of the heater and air-conditioning _____. Note any _____. Remove clips and bolts holding _____ _____ under the _____.

Step 2: Move the _____ control switch from maximum to _____.

Step 3: Observe the blend door _____ for movement. If no movement is detected, test the _____ system according to service information. If movement is present, _____ the actuator by undoing the _____ securing it to the air box.

Step 4: Move the _____ door by hand, and check whether it is _____ at the _____ or is _____. If the blend door is broken, the entire _____ _____ must be removed. Refer to service information for removal and repair.

Step 5: Move each _____, and watch the appropriate _____ for proper _____.

5. Diagnosing Temperature Control Problems in the HVAC System:

 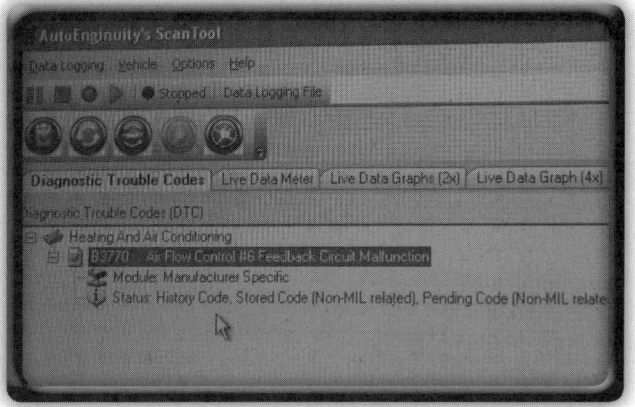

Step 1: Research the operation and diagnostic procedure for the control panel assembly, and _____ the customer _____. Check for full operating temperature of the _____ by pointing an infrared _____ at the thermostat housing. If the temperature is not hot enough, the _____ most likely will need to be _____.

Step 2: Test for _____ with the _____ tool. If a code is set, follow the _____ chart for the code pulled. If no code is set, follow the _____-_____ troubleshooting chart.

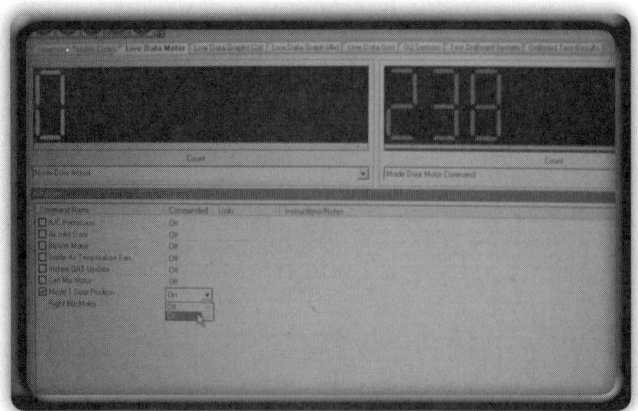

Step 3: If a code is found for the electric _____ for the blend door, pull up _____ controls in the _____ _____ and _____ the blend door to operate while _____ inspecting the movement of the actuator.

Step 4: If _____ is found at the actuator, inspect for a _____ or binding door by turning the _____ by hand. If no movement is found at the actuator, perform the test for _____ (power and ground) at the actuator while _____ it.

Step 5: If no power and ground are found, _____-_____ the wires for the actuator at the connector of the _____ voltage (power and ground) while commanding the _____. If no power and ground are coming from the ECU, the ECU will need to be _____.

Crossword Puzzle

Use the clues to complete the puzzle.

Across

3. The electric motors and fan that move air through the passenger compartment.
4. Measured voltage in a signal return circuit that is compared to a specified voltage value published by the manufacturer.
7. A resistor added to the circuit before or after the load to drop voltage to the load.
8. The pressure aspect of electricity, measured to show the potential of a circuit to do work.

Down

1. A signal that changes based on what the sensor is reading; for example, as temperature varies, so does the voltage signal to the ECU.
2. A system that provides the heating and cooling of air inside the passenger compartment for passenger comfort.
5. A sensor that acts as a variable resistor when a wiper arm moves across its resistive material.
6. A temperature-controlled variable resistor. As temperature changes, so does the resistance.

ASE-Type Questions

Read each item carefully, and then select the best response.

_____ 1. Tech A says that a temperature blend air door controls how much air bypasses the heater core. Tech B says that the regulation blend air door controls fresh air. Who is correct?
   A. Tech A
   B. Tech B
   C. Both A and B
   D. Neither A nor B

_____ 2. Tech A says that in a temperature blend air system, the air conditioner runs continuously. Tech B says that in a blend air system, there is always heated coolant in the heater core. Who is correct?
   A. Tech A
   B. Tech B
   C. Both A and B
   D. Neither A nor B

_____ 3. Tech A says that airflow in the vents is controlled by adding or removing a second or third motor to increase or decrease airflow. Tech B says that blower motor speed is controlled by controlling current to make the blower motor run faster or slower. Who is correct?
   A. Tech A
   B. Tech B
   C. Both A and B
   D. Neither A nor B

_____ 4. Tech A says that the "fresh air" setting pulls air from the cabin. Tech B says that when it is hot outside, the "recirculate" setting will cool the cabin better. Who is correct?
   A. Tech A
   B. Tech B
   C. Both A and B
   D. Neither A nor B

_____ 5. Tech A says that when an electric actuator fails, replace it with a new one. Tech B says that when an electric or vacuum actuator fails, always verify the door moves freely and there are no vacuum leaks. Who is correct?
   A. Tech A
   B. Tech B
   C. Both A and B
   D. Neither A nor B

_____ 6. Blend air systems use actuators to move doors to mix airflow. Tech A says that actuators are positioned by an A/C ECU, by checking airflow with sensors. Tech B says that actuators are positioned by an A/C ECU and the actuator has a potentiometer to tell the A/C ECU the position of the door. Who is correct?
   A. Tech A
   B. Tech B
   C. Both A and B
   D. Neither A nor B

_____ 7. Tech A says some newer vehicles use pulse width modulation to control blower motor speeds, which leads to smooth changes in speeds. Tech B says that the evaporator temperature sensor is used to control cab temperature in the passenger compartment. Who is correct?
   A. Tech A
   B. Tech B
   C. Both A and B
   D. Neither A nor B

_____ 8. Tech A says that the A/C high pressure switch controls the condenser cooling fan. Tech B says that the A/C ECU controls the condenser cooling fan. Who is correct?
   A. Tech A
   B. Tech B
   C. Both A and B
   D. Neither A nor B

_____ 9. Tech A says that the air gap between the clutch and the compressor directly relates to how much voltage is required to engage the clutch. Tech B says that if system voltage is low or the gap between the A/C clutch and compressor is too large, the clutch will not engage. Who is correct?
   A. Tech A
   B. Tech B
   C. Both A and B
   D. Neither A nor B

_____ 10. Tech A says that when troubleshooting an automatic temperature controlled A/C or heating problem one of the first steps should be to check codes for any related faults. Tech B says that active codes make troubleshooting more effective. Who is correct?
   A. Tech A
   B. Tech B
   C. Both A and B
   D. Neither A nor B

CHAPTER 44
Motive Power Types—Spark-Ignition (SI) Engines

Chapter Review

The following activities have been designed to help you refresh your knowledge of this chapter. Your instructor may require you to complete some or all of these activities as a regular part of your training program. You are encouraged to complete any activity that your instructor does not assign as a way to enhance your learning.

Matching

Match the following terms with the correct description or example.

- A. Base circle
- B. Bottom dead center (BDC)
- C. Cam
- D. Camshaft
- E. Column inertia
- F. Cylinder bore
- G. Duration
- H. Exhaust stroke
- I. Fulcrum
- J. Horsepower
- K. Ignition
- L. Lobe
- M. Piston clearance
- N. Piston skirt
- O. Reed valve
- P. Spark ignition (SI) engine
- Q. Torque
- R. Valve margin
- S. Vane-type phaser
- T. Work

_____ 1. The hole in the engine block that the piston fits into.

_____ 2. An engine that relies on an electrical spark to ignite the air and fuel mixture.

_____ 3. A cam phaser that uses vanes inside to allow oil pressure to push against and change cam timing as it is turning.

_____ 4. The stroke of piston during which the exhaust valve is open and the piston is moving from bottom dead center to top dead center to push exhaust gas out of the cylinder.

_____ 5. The clearance between the piston and the cylinder wall that allows for lubricating oil to reduce friction.

_____ 6. The flat surface on the outer edge of the valve head between the valve head and the valve face.

_____ 7. The principle that as a column of air flows, it creates inertia, which keeps air flowing until its inertia energy is spent; sometimes referred to as a "ram effect" when using tuned intake or exhaust systems.

_____ 8. The amount of twisting force applied in a turning application, usually measured in foot-pounds.

_____ 9. The egg-shaped lobe machined to a shaft used to cause opening and closing of the valves of a four-stroke cycle engine.

_____ 10. The amount of time the valve stays open, given in degrees of rotation of the crankshaft.

_____ 11. A half-round bearing that the rocker moves on as a bearing surface.

_____ 12. The position of the piston at the end of its stroke when it is closest to the crankshaft.

_____ 13. The part of the engine that activates the valve train by using lobes riding against lifters.

_____ 14. The lighting of the fuel and air mixture in the combustion chamber.

_____ 15. The result of force creating movement.

_____ 16. The area below the ring groove area of the piston that prevents the piston from cocking and becoming jammed in the cylinder bore.

Chapter 44   Motive Power Types—Spark-Ignition (SI) Engines

_____ 17. The rounded bottom part of the camshaft (off the lobe) where the valves remain closed or at rest.
_____ 18. A small flexible metal plate that covers the inlet port of a two-stroke engine and opens and closes to let air and fuel into the crankcase.
_____ 19. An amount of work performed in a given time.
_____ 20. The raised portion on a camshaft; used to lift the lifter and open the valve.

Multiple Choice

Read each item carefully, and then select the best response.

_____ 1. The steam engine and the Stirling engine are examples of what kind of engine?
   A. Internal combustion engine
   B. External combustion engine
   C. Rotary engine
   D. Reciprocating piston engine

_____ 2. Which of the following types of fuel contains latent heat energy?
   A. Diesel
   B. Ethanol
   C. Hydrogen
   D. All of the above

_____ 3. The effort to produce a push or pull action is referred to as _____.
   A. power
   B. work
   C. force
   D. torque

_____ 4. The unit of measurement for torque in the metric system is _____.
   A. ft-lb
   B. Newton meters
   C. horsepower
   D. rpms

_____ 5. How many radians are there in the circumference of a circle?
   A. 1
   B. 3.14
   C. 6.28
   D. Depends on the radius of the circle

_____ 6. The _____ occurs as extreme force moves the piston from TDC to BDC with both valves remaining closed.
   A. power stroke
   B. compression stroke
   C. intake stroke
   D. exhaust stroke

_____ 7. The distance the piston travels from TDC to BDC, or from BDC to TDC, is called the _____.
   A. throw
   B. piston stroke
   C. displacement
   D. compression ratio

_____ 8. The displacement of an engine can be altered by changing the _____.
   A. cylinder bore
   B. piston stroke
   C. number of cylinders
   D. all of the above

_____ 9. The Miller cycle engine and the Atkinson cycle engine are both variations on the traditional _____ engine.
   A. two-stroke SI
   B. four-stroke SI
   C. external combustion engine
   D. rotary engine

_____ 10. The _____ are the areas between the ring grooves that support the rings as the piston moves.
   A. offsets
   B. journals
   C. ring lands
   D. skirts

_____ 11. The piston _____ prevents the piston from rocking and jamming in the cylinder bore.
   A. skirt
   B. pin boss
   C. connecting rod
   D. head

_____ 12. What type of engines use pushrods to transfer the camshaft's lifting motion to the valves by way of rocker arms on top of the cylinder head?
   A. Flathead engines
   B. Cam-in-block engines
   C. Overhead cam
   D. Dual overhead cam (DOHC) engines

_____ 13. Which part rides on the camshaft lobes to actuate the pushrods, rocker arms, and valves?
   A. Tappets
   B. B Ramps
   C. Lifters
   D. Either A or C

_____ 14. The number of degrees between the centerline of the intake lobe and the centerline of the exhaust lobe is called _____.
   A. cam lobe centerline
   B. cam lobe separation
   C. cam lobe duration
   D. cam lobe margin

_____ 15. A machined surface on the back of the valve head is known as the valve _____.
   A. face
   B. margin
   C. keeper
   D. stem

_____ 16. The valve train operates off of the camshaft, and the part that rides against the cam lobe is the _____.
   A. valve stem
   B. valve lifter
   C. rocker arm
   D. fulcrum

_____ 17. The _____ typically takes the place of the standard cam gear or pulley and uses oil pressure from the engine oil pump to move the camshafts when commanded by the power train control module.
   A. lobe
   B. fulcrum
   C. freewheel
   D. phaser

_____ 18. The _____ engine has enough clearance between the pistons and the valves so in the event the timing belt breaks any valve that is hanging all the way open will not contact the piston, thus preventing engine damage.
   A. interference
   B. rotary
   C. freewheeling
   D. stirling

Chapter 44   Motive Power Types—Spark-Ignition (SI) Engines   505

_____ 19. The intake system of a two-stroke engine is called a _____ system.
  A. multiport
  B. vane-type
  C. piston port
  D. phaser

_____ 20. The _____ engine is also called the Wankle engine because it was improved upon by Felix Wankle for automotive use in the 1940s.
  A. two-stroke
  B. four-stroke
  C. interference
  D. rotary

True/False

If you believe the statement to be more true than false, write the letter "T" in the space provided. If you believe the statement to be more false than true, write the letter "F".

_____ 1. The internal combustion engine has almost completely replaced the external combustion engine and has been around for well over a century.

_____ 2. Compression-ignition engines do not use spark plugs.

_____ 3. Pressure and volume are inversely related; as one rises, the other falls.

_____ 4. Force is measured in ft-lb per second or ft-lb per minute.

_____ 5. Movement must occur to produce torque.

_____ 6. A turbocharger or supercharger will increase an engine's volumetric efficiency well above 100%.

_____ 7. When the piston in the cylinder is at a position closest to the crankshaft, it is said to be at top dead center.

_____ 8. The power stroke starts with the exhaust valve closed, the intake valve(s) opening, and the piston moving from TDC to BDC.

_____ 9. The block deck is the machined surface of the block farthest from the crankshaft.

_____ 10. Increasing the diameter of the bore or increasing the length of the stroke will produce a larger piston displacement.

_____ 11. The time that both the intake and exhaust valves are open is called valve overlap.

_____ 12. A long block replacement includes the engine block from below the head gasket to above the oil pan.

_____ 13. The piston body has a piston pinhole machined through a reinforced area called the piston pin boss.

_____ 14. On many engines, the oil pan houses the oil pump, which is the heart of the engine in that it supplies critical lubrication and cooling for the internal moving parts of the engine.

_____ 15. Intake and exhaust valves may all be actuated by a single camshaft, or there may be two camshafts per head, called dual overhead cam engines.

_____ 16. The cam lobe ramp is where the rise of the cam lobe starts from the base circle to the top of the lobe.

_____ 17. L-head engines have not been used in automobiles for many years due to inefficiency and exhaust emissions.

_____ 18. If valve clearance is too large, the valve can be held open longer than it should be.

_____ 19. Valve timing is a mechanical setting, and once set, it cannot be changed without physically changing the timing components in some way.

_____ 20. The rotary engine is not an internal combustion engine.

Fill in the Blank

Read each item carefully, and then complete the statement by filling in the missing word(s).

1. The branch of physical science that deals with heat and its relation to other forms of energy such as mechanical energy is generally defined as _____.

2. The gasoline _____ engine uses a crankshaft to convert the reciprocating movement of the pistons in their cylinder bores into rotary motion at the crankshaft.

3. Heating a gas in a sealed container will _____ the pressure in the container.

4. One _____ equals 550 ft-lb per second, or 33,000 ft-lb per minute.
5. A(n) _____ describes how many radius distances there are in the circumference of a circle.
6. A(n) _____ _____ is the operating range of an engine from start-up until shutdown when finished performing its job.
7. When the piston in a cylinder is at the position farthest away from the crankshaft, it is at _____ _____.
8. Internal combustion engines are designated by the amount of space their pistons displace as they move from TDC to BDC, which is called _____ _____.
9. The process of using a column of moving air to create a low-pressure area behind it to assist in removing any remaining burned gases from the combustion chamber and replacing these gases with a new charge is called _____.
10. The _____ _____ is the single largest part of the engine.
11. A(n) _____ is a weighted assembly that stores kinetic energy from each power stroke and helps keep the crankshaft turning through nonpower strokes.
12. The _____ _____ connects the piston to the crankshaft and transfers piston movement and combustion pressure to the crankshaft rod journals.
13. The upper two piston rings are compression rings, which prevent combustion pressure, called _____ _____, from leaking past the pistons into the crankcase.
14. A(n) _____ is a combination of materials that has properties that are different from the original materials.
15. Up until the 1950s, many engines had their valves installed in the engine block. Such engines are called _____ engines.
16. Coil _____ occurs when the coils of the spring touch each other.
17. The opposite end of the valve stem has grooves machined into its end to receive locking pieces, sometimes referred to as valve _____, that retain a valve spring retainer and spring on the valve.
18. Air/fuel has a fairly constant burn time, so as engine rpm increases, the spark will have to occur earlier _____ top dead center to make sure maximum pressure is developed shortly _____ top dead center.
19. In a _____-_____ engine, the inlet and exhaust ports are opened and closed by the movement of the piston.
20. A(n) _____ _____ is the circular movement around the perimeter of another circle.

Labeling

Label the following diagrams with the correct terms.

1. External combustion engines:

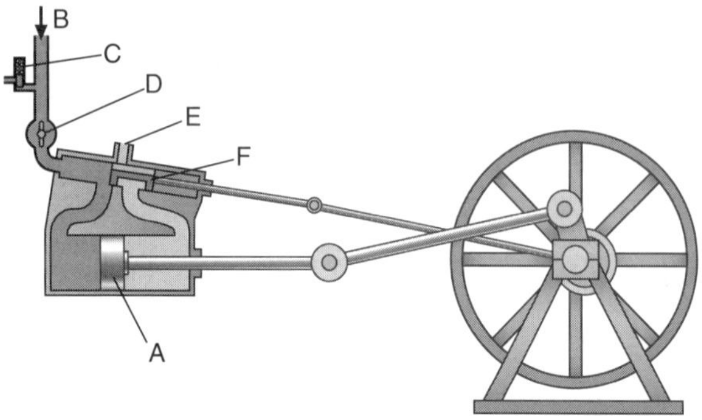

Steam engine

A. _____
B. _____
C. _____
D. _____
E. _____
F. _____

Chapter 44 Motive Power Types—Spark-Ignition (SI) Engines 507

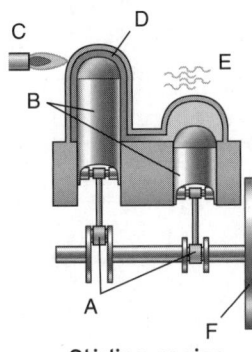

Stirling engine

A. _____
B. _____
C. _____
D. _____
E. _____
F. _____

2. Temperature changes pressure:

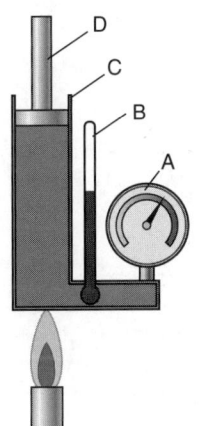 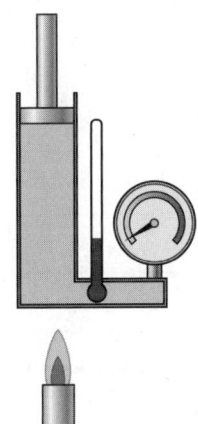

A: As temperature increases, the pressure increases

B: As temperature decreases, the pressure decreases

A. _____
B. _____
C. _____
D. _____

3. Volume affects pressure:

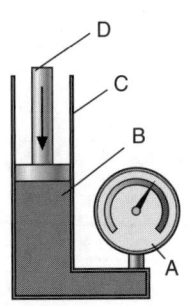 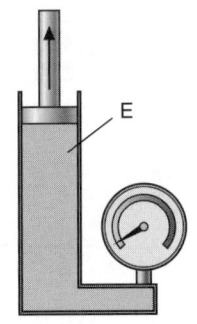

A: As the volume decreases, the pressure increases

B: As the volume increases, the pressure decreases

A. _____
B. _____
C. _____
D. _____
E. _____

4. The basic four-stroke cycle:

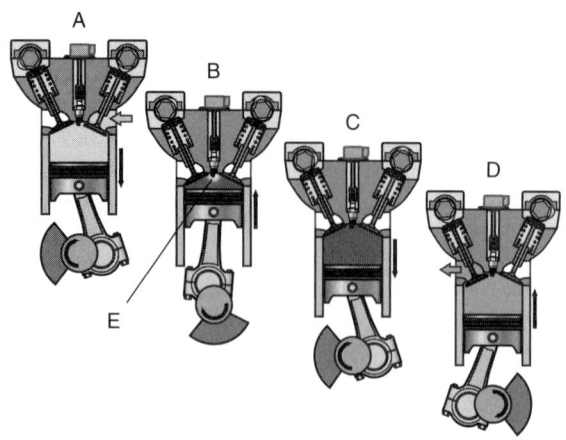

A. _____

B. _____

C. _____

D. _____

E. _____

5. Engine block:

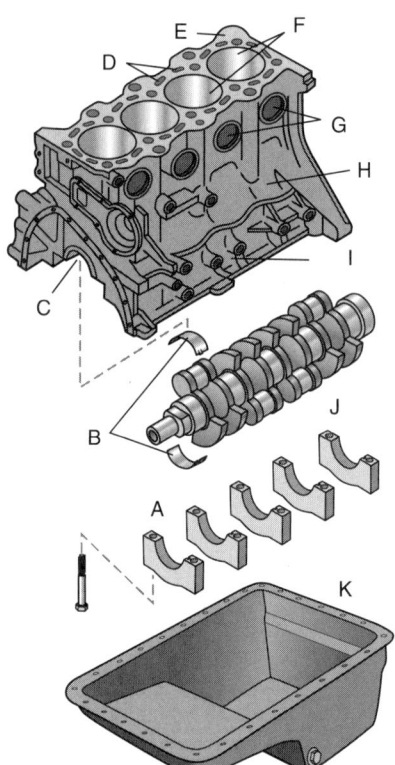

A. _____

B. _____

C. _____

D. _____

E. _____

F. _____

G. _____

H. _____

I. _____

J. _____

K. _____

Chapter 44 Motive Power Types—Spark-Ignition (SI) Engines 509

6. Basic parts of the crankshaft:

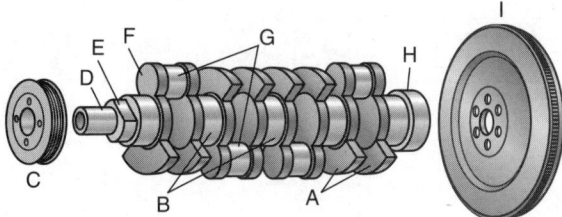

A. _____
B. _____
C. _____
D. _____
E. _____
F. _____
G. _____
H. _____
I. _____

7. Piston and piston rings:

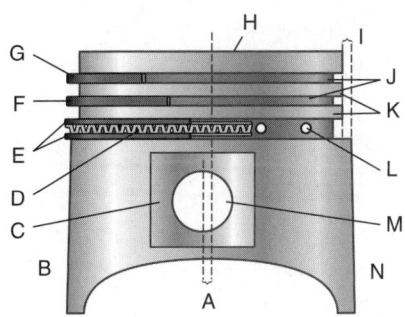

A. _____
B. _____
C. _____
D. _____
E. _____
F. _____
G. _____
H. _____
I. _____
J. _____
K. _____
L. _____
M. _____
N. _____

8. Camshaft lobe:

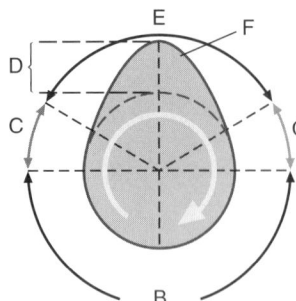
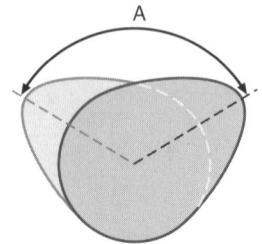

A. _____
B. _____
C. _____
D. _____
E. _____
F. _____
G. _____

9. Two-stroke reed valve operation:

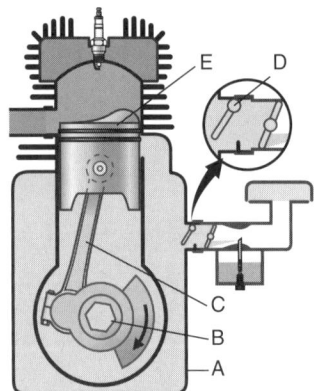

A. _____
B. _____
C. _____
D. _____
E. _____

10. Rotary engine:

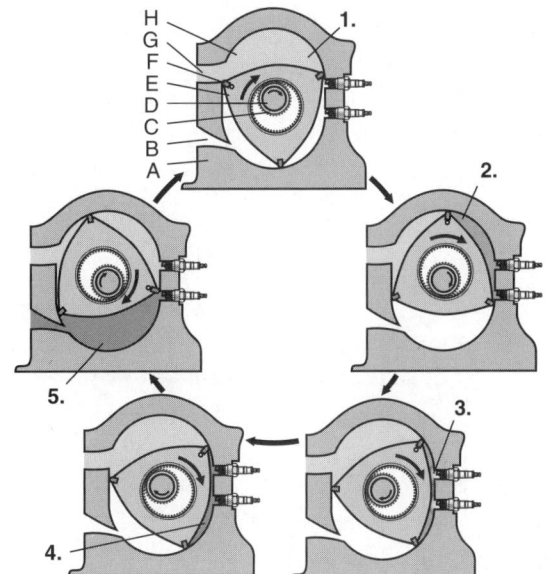

1. _____
2. _____
3. _____
4. _____
5. _____

A. _____
B. _____
C. _____
D. _____
E. _____
F. _____
G. _____
H. _____

Crossword Puzzle

Use the clues to complete the puzzle.

Across

3. A gas that will not react chemically.
7. The result of combustion gases leaking past the compression rings and getting into the crankcase.
11. The operating range of an engine from start-up until shutdown when finished performing its job.
13. The size of the engine given in cubic inches, cubic centimeters, and liters. It is found by multiplying the piston displacement by the number of cylinders the engine has. Sometimes called "swept volume."
14. A loss of engine efficiency caused by internal friction, inefficient breathing, etc.
15. A device that is able to shift camshaft timing while the camshaft is turning. This device takes the place of the standard timing belt pulley or timing chain gear.

Down

1. An engine cycle that uses a longer effective exhaust stroke than intake stroke to reduce exhaust emissions. This type of engine is widely used in hybrid-electric vehicles.
2. An L-head engine with valves in the block.
4. The stroke of the piston during which the air and fuel is being compressed into a small area prior to ignition.
5. The mixture of materials to make a substance that has properties different from the original materials.
6. The heavy, circular flat plate that keeps the engine rotating when power is not produced, such as on the exhaust, intake, and compression strokes.
8. The effort to produce a push or pull action.

9. The "top" of the engine block and cylinder bore where the cylinder head is bolted on.
10. The amount the valve will open. The more the valve lifts off its seat, the more air can get into and out of the engine.
12. Locking devices that keep the valve retained by the valve spring seat.

ASE-Type Questions

Read each item carefully, and then select the best response.

_____ 1. Tech A says that engines using compression ignition control timing by regulating when fuel is injected into the cylinder. Tech B says that engines using spark ignition control timing by regulating when fuel is injected into the cylinder. Who is correct?
   A. Tech A
   B. Tech B
   C. Both A and B
   D. Neither A nor B

_____ 2. Tech B says that horsepower is a measurement of the amount of work being performed. Tech B says that horsepower can be calculated by multiplying torque by rpm and dividing by 5252. Who is correct?
   A. Tech A
   B. Tech B
   C. Both A and B
   D. Neither A nor B

_____ 3. Tech B says that in a four-stroke engine, the piston is at TDC four times to complete the cycle. Tech B says that the air/fuel mixture is ignited once every two strokes. Who is correct?
   A. Tech A
   B. Tech B
   C. Both A and B
   D. Neither A nor B

_____ 4. Tech A says that spark ignition typically occurs before TDC. Tech B says that spark ignition typically occurs after TDC. Who is correct?
   A. Tech A
   B. Tech B
   C. Both A and B
   D. Neither A nor B

_____ 5. Tech A says that valve overlap occurs between the exhaust stroke and the intake stroke. Tech B says that valve overlap occurs to assist in scavenging the cylinder. Who is correct?
   A. Tech A
   B. Tech B
   C. Both A and B
   D. Neither A nor B

_____ 6. Tech A says that compression ratio is the comparison of the volume above the piston at BDC to the volume above the piston at TDC. Tech B says that scavenging of the exhaust gases occurs once the exhaust valve closes. Who is correct?
   A. Tech A
   B. Tech B
   C. Both A and B
   D. Neither A nor B

_____ 7. Tech A says that the weight of the flywheel smooths out the engine's power pulses. Tech B says that the flex plate and torque converter perform the same function as the flywheel. Who is correct?
   A. Tech A
   B. Tech B
   C. Both A and B
   D. Neither A nor B

_____ **8.** Tech A says that an interference engine is designed so that the pistons can hit the valves if the timing belt breaks. Tech B says that the shape of the cam lobe determines how long and far the valves are held open. Who is correct?
   A. Tech A
   B. Tech B
   C. Both A and B
   D. Neither A nor B

_____ **9.** Tech A says that blowby gases occur when compression and combustion gases leak past the piston rings. Tech B says that a rotary engine uses reed valves to control air flow into the cylinder on intake. Who is correct?
   A. Tech A
   B. Tech B
   C. Both A and B
   D. Neither A nor B

_____ **10.** Tech A says that the principle of thermal expansion is what pushes the piston down the cylinder on the power stroke. Tech B says that the piston is pulled down the cylinder on the intake stroke. Who is correct?
   A. Tech A
   B. Tech B
   C. Both A and B
   D. Neither A nor B

Engine Lubrication

CHAPTER 45

Tire Tread:
© AbleStock

Chapter Review

The following activities have been designed to help you refresh your knowledge of this chapter. Your instructor may require you to complete some or all of these activities as a regular part of your training program. You are encouraged to complete any activity that your instructor does not assign as a way to enhance your learning.

Matching

Match the following terms with the correct description or example.

- A. Antifoaming agents
- B. Bypass filter
- C. Corrosion inhibitors
- D. Crescent pump
- E. Detergents
- F. Dispersants
- G. Extreme loading
- H. Gelling
- I. Hydro-cracking
- J. Hydrogenating
- K. Lubrication system
- L. Oil cooler
- M. Oil slinger
- N. Oxidation inhibitor
- O. Polyalphaolefin (PAO)
- P. Rotor-type oil pump
- Q. Scavenge pump
- R. Synthetic blend
- S. Viscosity index improver
- T. Windage tray

_____ 1. A thickening effect of oil in cold weather, wax content in base stock mineral oil makes it worse.

_____ 2. A device used on small engines, located on the crankshaft or driven by the camshaft. It works to fling oil up onto moving engine parts.

_____ 3. Oil additives that keep contaminants held in suspension in the oil, to be removed by the filter or when the oil is changed.

_____ 4. A plate that bolts onto the bottom of the engine between the crankshaft and the oil pan, helping to keep the crankshaft from contacting the engine oil.

_____ 5. A system of parts that work together to deliver lubricating oil to the various moving parts of the engine.

_____ 6. A pump used with a dry sump oiling system to pull oil from the dry sump pan and move it to an oil tank outside the engine.

_____ 7. Oil additives that help to keep carbon from sticking to engine components.

_____ 8. A blend of conventional engine oil and pure synthetic oil.

_____ 9. An oil filter system that only filters some of the oil.

_____ 10. A process in which group 2 and group 3 oils are refined with hydrogen at much higher temperatures and pressures.

_____ 11. An oil pump that uses a crescent-shaped part to separate the oil pump gears from each other, allowing oil to be moved from one side of the pump to the other.

_____ 12. An oil additive that resists a change in viscosity over a range of temperatures.

_____ 13. An oil additive that helps keep hot oil from combining with oxygen to produce sludge or tar.

_____ 14. Oil additives that keep acid from forming in the oil.

_____ 15. An oil pump that uses rounded gears to squeeze oil through.

_____ 16. Oil additives that keep oil from foaming as it moves through the engine.

_____ 17. A man-made base stock (synthetic) used in place of mineral oil. Oil molecules are more consistent in size, and no impurities are found in this oil since it is made in a lab.

_____ 18. Large amount of pressure placed on two bearing surfaces to press oil from between them.

_____ 19. A process used during refining of crude oil. Hydrogen is added to crude oil to create a chemical reaction to take out impurities such as sulfur.

_____ 20. A device that takes heat away from engine oil by passing it near either engine coolant or outside air.

Multiple Choice

Read each item carefully, and then select the best response.

_____ 1. What kind of oil varies in color from a dirty yellow to dark brown to black and can be thin like gasoline or a thick oil- or tarlike substance?
   A. Lubricating oil
   B. Motor oil
   C. Grease
   D. Crude oil

_____ 2. _____ is a measure of how easily a liquid flows.
   A. Pour point
   B. Viscosity
   C. Solidity
   D. Reluctance

_____ 3. What kind of additive coats parts with a protective layer so that the oil resists being forced out under heavy load?
   A. Extreme-pressure additives
   B. Viscosity index improver
   C. Pour point depressants
   D. Dispersants

_____ 4. The _____ is the certifying body for engine oil.
   A. American Petroleum Institute
   B. American Society of Automotive Engineers
   C. International Lubricant Standardization and Approval Committee
   D. all of the above

_____ 5. The letter W in an SAE viscosity rating stands for _____.
   A. water content
   B. winter viscosity
   C. wax content
   D. weight

_____ 6. Which of the following organizations sets classification standards for motorcycle engines, both two-stroke and four-stroke?
   A. American Society of Automotive Engineers
   B. Association des Constructeurs Européens d'Automobiles
   C. Japanese Automotive Standards Organization
   D. All of the above

_____ 7. The American Petroleum Institute classifies oils into _____ groups.
   A. three
   B. four
   C. five
   D. six

_____ 8. What type of oil can be man-made or highly processed petroleum?
   A. Group 1
   B. Synthetic
   C. Conventional
   D. Blended

Chapter 45   Engine Lubrication   517

_____ 9. Lubrication oil is stored in the _____.
   A. oil slinger
   B. pickup tube
   C. oil gallery
   D. oil sump

_____ 10. Flat pieces of steel placed around the oil pump pickup in the oil pan to prevent oil from surging away from the pickup during cornering, braking, and accelerating are called _____.
   A. windage trays
   B. baffles
   C. crescents
   D. barriers

_____ 11. In a _____ oil pump, the driving gear meshes with a second gear; as both gears turn, their teeth separate, creating a low-pressure area.
   A. geared
   B. crescent
   C. rotor-type
   D. spline-type

_____ 12. The spin-on type oil filter has a(n) _____ that fits into a groove in the base of the filter.
   A. garter spring
   B. O-ring
   C. plastic cap
   D. compression ring

_____ 13. Passageways called _____ allow oil to be fed to the crankshaft bearings first, then through holes drilled in the crankshaft to the connecting rods.
   A. pickup tubes
   B. sumps
   C. oil jackets
   D. galleries

_____ 14. On horizontal-crankshaft engines, a _____ on the bottom of the connecting rod scoops up oil from the crankcase for the bearings.
   A. crescent pump
   B. dipper
   C. rotor
   D. strainer

_____ 15. When the crankshaft hits the oil in the oil pan and it tries to stop the crankshaft from turning it is called _____.
   A. drag
   B. windage
   C. reluctance
   D. hydro-cracking

True/False

If you believe the statement to be more true than false, write the letter "T" in the space provided. If you believe the statement to be more false than true, write the letter "F".

_____ 1. Clearances, such as those between the crankshaft journal and crankshaft bearing, fill with lubricating oil so engine parts move or float on layers of oil instead of directly on each other.

_____ 2. A power stroke can put as much as 2 tons of force on the main bearings.

_____ 3. Oxidation allows air bubbles to form in the engine oil, reducing the lubrication quality of oil and contributing to the breakdown of the oil.

_____ 4. The International Lubricant Standardization and Approval Committee requires that the oil provide increased fuel economy over a base lubricant.

_____ 5. If you are servicing a European vehicle, it is advised that you do not go by any API recommendations; instead, make sure the oil meets the recommended JASO rating specified by the manufacturer.

_____ 6. One of the impurities found in all crude oil is wax, which is removed during refining and is used for candle wax.

_____ 7. Type 4 synthetic lubricating oil is not a true synthetic.

_____ 8. On a wet sump lubricating system, the oil pan holds the entire volume of the oil required to lubricate the engine.

_____ 9. The bypass filtering system is more common on diesel engines and is used in conjunction with a full-flow filtering system.

_____ 10. The oil pressure sensor is also commonly called a sending unit since it sends a signal to the light, gauge, or message center in the dash.

_____ 11. Oil analysis is typically used in heavy vehicle applications that may use 3 or more gallons of engine oil.

_____ 12. Oil would be good for thousands of miles longer in a vehicle driven in stop-and-go traffic with long periods of idling than if it were driven in moderate temperatures for long distances.

_____ 13. Because there is no oil storage sump under the engine, the engine can be mounted much lower in a dry sump system.

_____ 14. Most two-stroke gasoline engines use a specified gasoline–oil mixture for lubrication.

_____ 15. Some two-stroke gasoline engines use an oil injection system that does not require the oil and gasoline be mixed manually.

Fill in the Blank

Read each item carefully, and then complete the statement by filling in the missing word(s).

1. _____ oil is distilled from crude oil and used as a base stock.
2. _____ occurs between all surfaces that come into contact with each other.
3. Base stock derived from crude oil will not retain its viscosity if the temperature gets cold enough, so viscosity _____ improvers are added to the stock.
4. The American _____ _____ symbol is the donut symbol located on the back of the oil bottle.
5. _____ oils flow easily during cold engine start-up but do not thin out as much as the engine and oil come up to operating temperature.
6. Using the wrong _____ can void the customer's warranty, leaving the customer, or your shop, responsible for repairs.
7. Group 2 and group 3 API classified oils are refined with _____ at much higher temperatures and pressures, in a process known as hydro-cracking.
8. Crude oil is broken down into _____ oil, which is then combined with additives to enhance the lubricating qualities.
9. True synthetic oils are based on man-made _____, commonly polyalphaolefin oil, which is a man-made oil base stock.
10. Oil is drawn through the oil pump _____ from the oil sump by an oil pump.
11. In a(n) _____ _____ system, the oil is not stored under the engine in an oil pan.
12. Between the oil sump and oil pump is a(n) _____ _____ with a flat cup and a wire mesh strainer immersed in the oil.
13. In a rotor-type oil pump, outside atmospheric pressure forces oil into the pump, and the oil fills the spaces between the _____ _____.
14. An oil pressure _____ _____ stops excess pressure from developing.
15. The location of the _____-_____ filter right after the oil pump ensures that all of the oil is filtered before it is sent to the lubricated components.
16. Some connecting rods have oil _____ holes that are positioned to receive oil from similar holes in the crankshaft.
17. Oil _____ systems are used to inform the driver when the oil needs to be changed.
18. Modern vehicle engines use a pressure, or _____-_____, lubrication system where the oil is forced throughout the engine under pressure.

19. Diesel fuel has more _____ thermal units of heat energy than gasoline, so it produces more heat when it is ignited, placing more stress on the engine's moving parts.
20. A two-stroke engine may require a 40:1 mixture, which is 40 parts _____ to 1 part oil.

Labeling

Label the following diagrams with the correct terms.

1. The lubrication system:

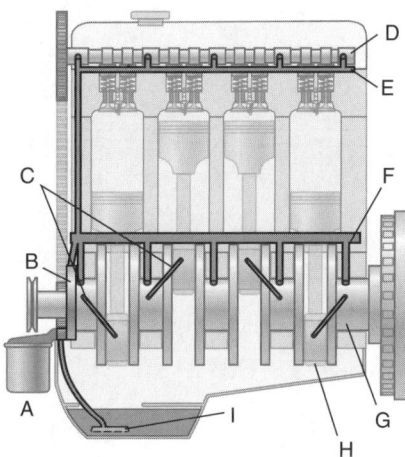

A. _____
B. _____
C. _____
D. _____
E. _____
F. _____
G. _____
H. _____
I. _____

2. Oil pump types:

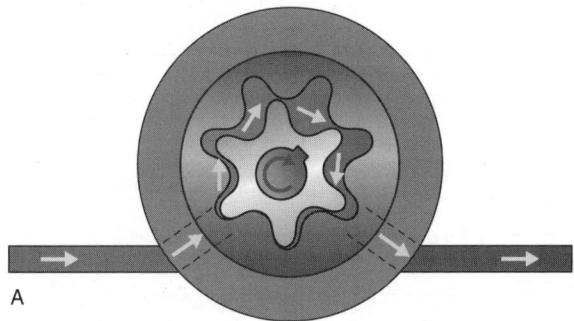

A. _____

B.

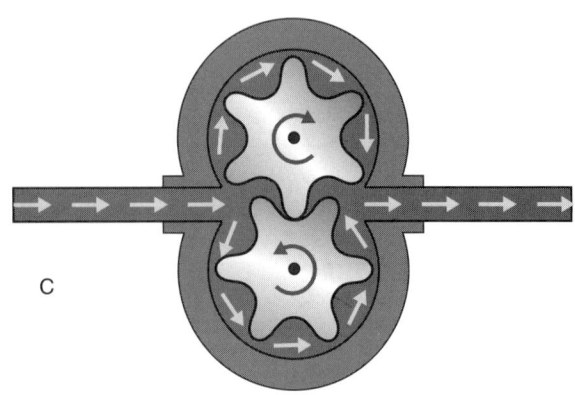

C.

B. _____

C. _____

3. Full-flow filtering system:

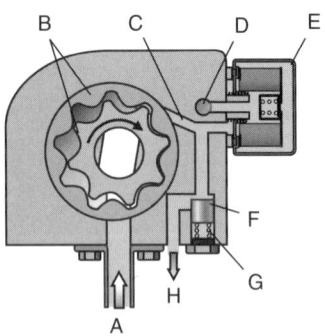

A. _____
B. _____
C. _____
D. _____
E. _____
F. _____
G. _____
H. _____

Chapter 45 Engine Lubrication 521

4. Bypass filtering system:

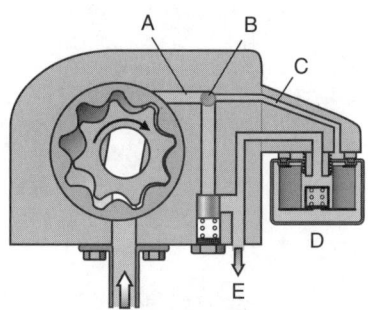

A. _____
B. _____
C. _____
D. _____
E. _____

5. Oil injection system in a two-stroke engine:

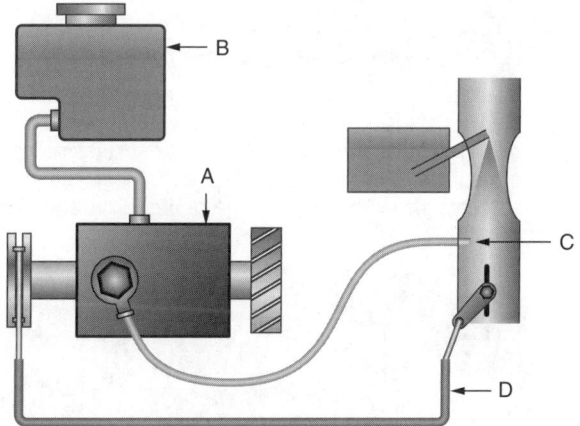

A. _____
B. _____
C. _____
D. _____

6. Tools for lubrication repair:

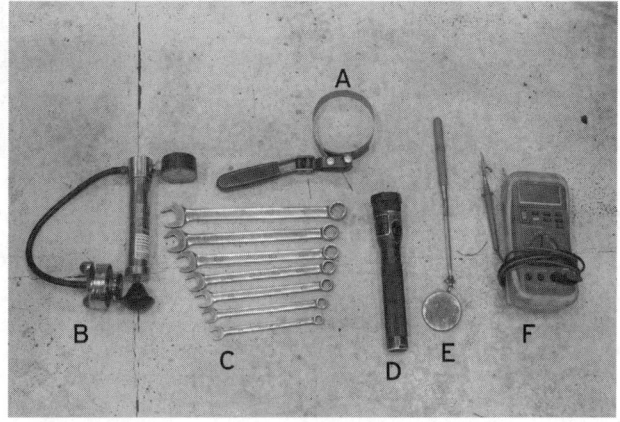

A. _____
B. _____
C. _____
D. _____
E. _____
F. _____

Skill Drills

Place the skill drill steps in the correct order.

1. Checking the Engine Oil:

_____ **A.** Replace the oil filler cap, and check the dipstick again to make sure the oil level is now correct.

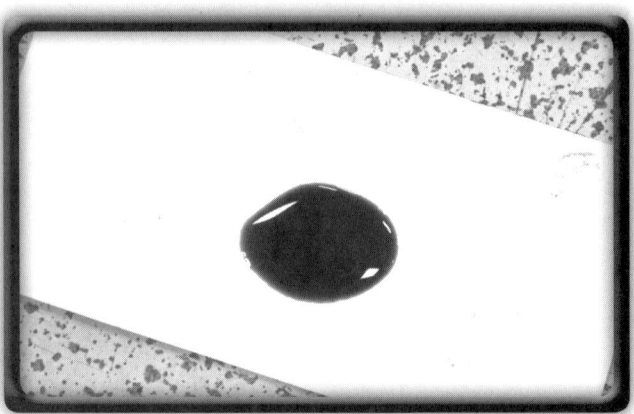

_____ **B.** Check the oil for any conditions such as unusual color or texture. Report these to your supervisor. Check the oil monitoring system, oil sticker, or service record to determine if the oil needs to be changed. (Some oil monitoring systems show the percentage of life left in the oil.)

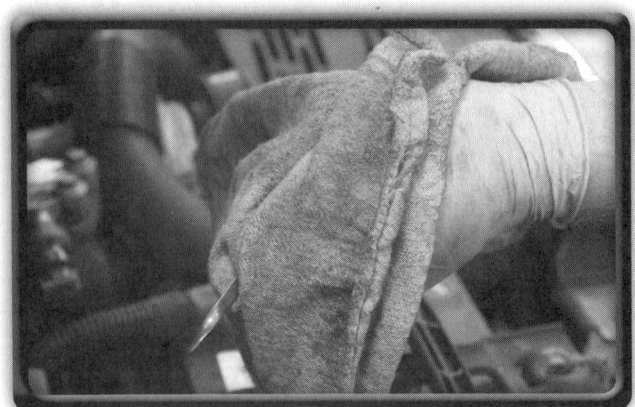

_____ **C.** Locate the dipstick. With the engine off, remove the dipstick, catching any drops of oil on a rag, and wipe it clean. Observe the markings on the lower end of the stick, which indicate the "full" and "add" marks or specify the "safe" zone.

_____ **D.** If additional oil is needed, estimate the amount by checking the service manual guide to the dipstick markings. Unscrew the filler cap at the top of the engine, and using a funnel to avoid spillage, turn the oil bottle so the spout is on the high side of the bottle and gently pour the oil into the engine. Recheck the oil level.

Chapter 45 Engine Lubrication

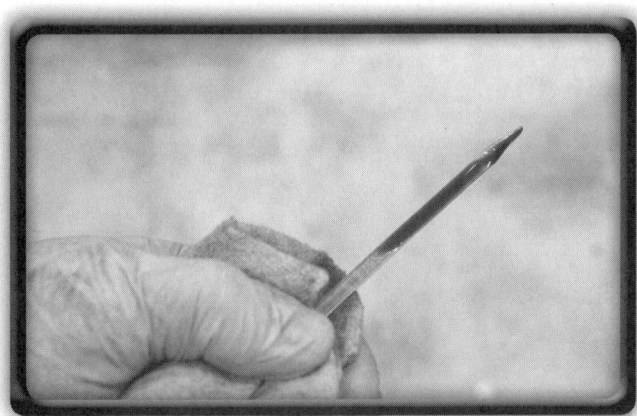

_____ **E.** Replace the dipstick, and push it back down into the sump as far as it will go. Remove it again, and hold it level while checking the level indicated on the bottom of the stick. If the level is near or below the "add" mark, then you will need to determine if the engine just needs to be topped up to the full level with fresh oil or replaced with new oil and oil filter.

2. Draining the Engine Oil:

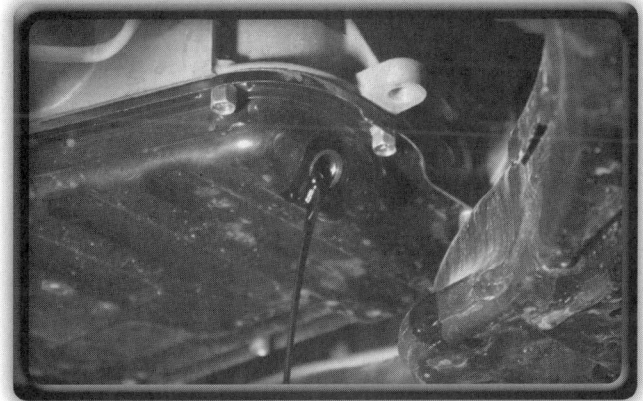

_____ **A.** Allow the oil to drain while you are dealing with the drain plug, gasket, and oil filter.

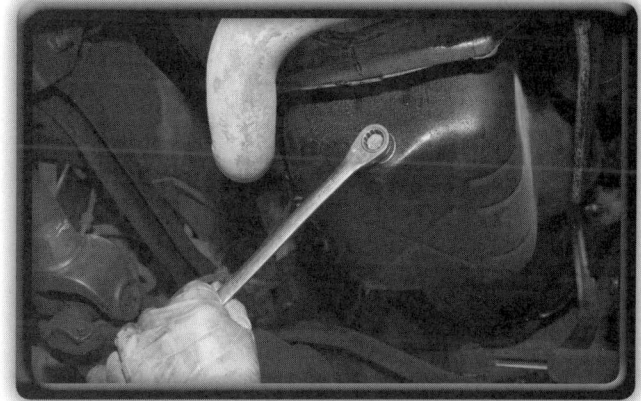

_____ **B.** Identify the location of the oil drain plug. Some vehicles have two drain plugs, draining separate sump areas. If the drain plug is leaking, damaged, or does not look right, inform your supervisor. Use a box wrench or socket to remove and replace the drain bolt. Be careful that you do not remove the transmission drain plug by mistake.

_____ **C.** Safely dispose of the drained oil according to all local regulations.

_____ **D.** Position the drain pan so it will catch the oil. Remove and inspect the drain plug and gasket; replace as necessary.

_____ **E.** Before you begin, clean up any oil spills, obtain the oil drain container (and make sure it has enough room for the oil to be drained), have enough new oil of the correct type to refill the engine, have the correct oil filter, and ensure that the engine oil is up to operating temperature before starting the oil change.

_____ **F.** Screw in the drain plug all the way by hand and then tighten it to the torque specified by the manufacturer. Wipe any drips from the underside of the engine.

3. Refilling the Engine Oil:

_____ **A.** If the oil pressure is good, turn the engine off and check underneath the vehicle to make sure no oil is leaking from the oil filter or drain plug.

_____ **B.** Reset the maintenance reminder system to remind the owner when the next oil change is due.

_____ **C.** Start the engine, and check the oil pressure indicator on the dash. If the oil pressure is inadequate, stop. Do not continue to run the engine.

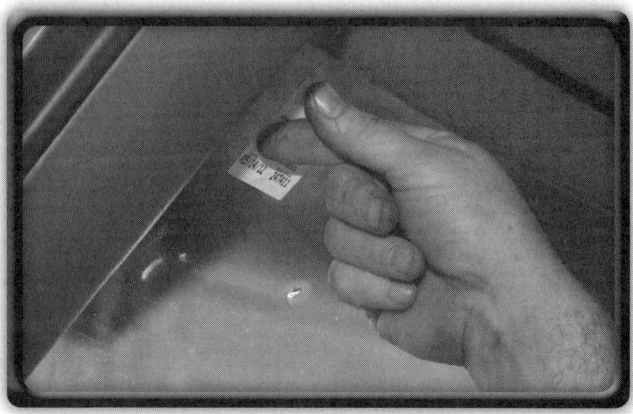

_____ **D.** Refer to the owner's manual or the service information, and install a static sticker.

Chapter 45 Engine Lubrication   525

_____ **E.** Using the service information, research the correct grade and the quantity of oil you will need to fill the engine. Turn the container of oil so that the spout is on the high side of the bottle. Pour the oil into the funnel carefully so no oil is spilled onto the outside of the engine, and pour slowly enough to avoid the risk of blowback or overflow. Fill the engine only to the level indicated on the engine dipstick. Replace the filler cap.

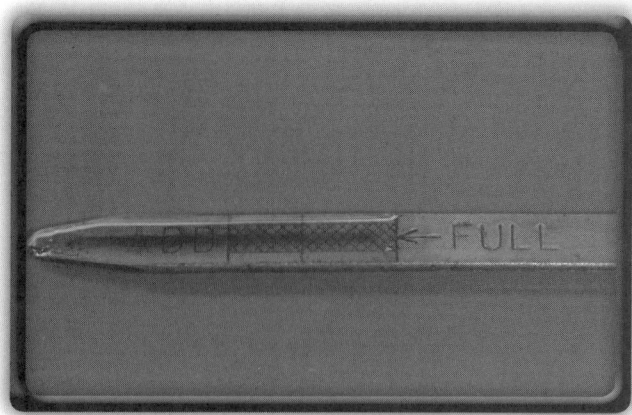

_____ **F.** With a level vehicle, check the oil level again with the dipstick. It may be necessary to top off the engine by adding a small quantity of oil to compensate for the amount absorbed by the new filter. Do not overfill.

4. Testing the Oil Pressure:

_____ **A.** Connect the oil pressure gauge, then start and completely warm up the engine. Compare the readings to the specifications in the service information.

_____ **B.** Match the adapter thread to the thread on the pressure sensor to ensure the correct adapter will be screwed into the engine.

_____ **C.** Remove the manual pressure gauge and adaptor. Using a thread sealant, reinstall the oil sending unit and torque to specifications.

_____ **D.** Locate the oil pressure sensor by using the component locator in the service information.

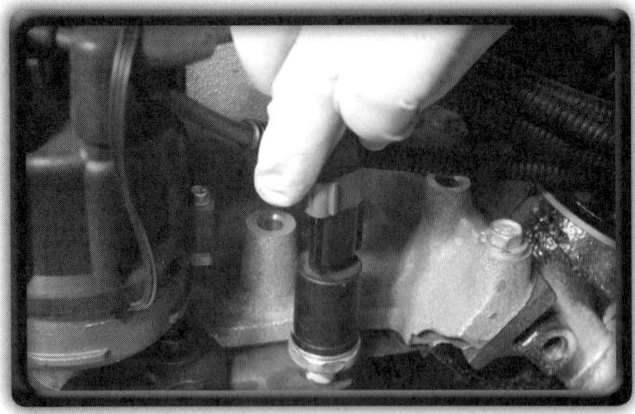

_____ **E.** Install the wire harness connector, and start the engine. Inspect for oil leaks and proper pressure gauge operation.

_____ **F.** Place an oil drain pan under the engine to catch any oil when you remove the oil pressure sensor.

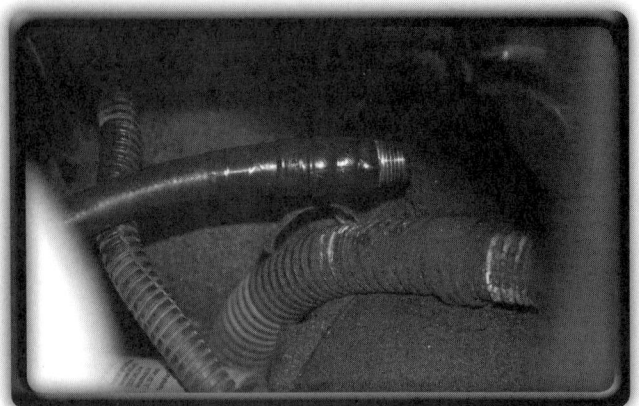

_____ **G.** Carefully thread the adapter into the engine, ensuring that you do not cross-thread the adapter.

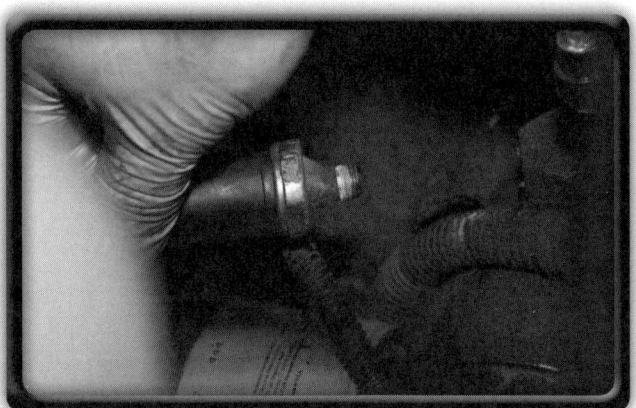

_____ **H.** Disconnect the wire harness connector, and remove the oil pressure sensor from the engine using the recommended tool in the service information.

5. Inspecting, Testing, and Replacing Oil Temperature and Pressure Switches:

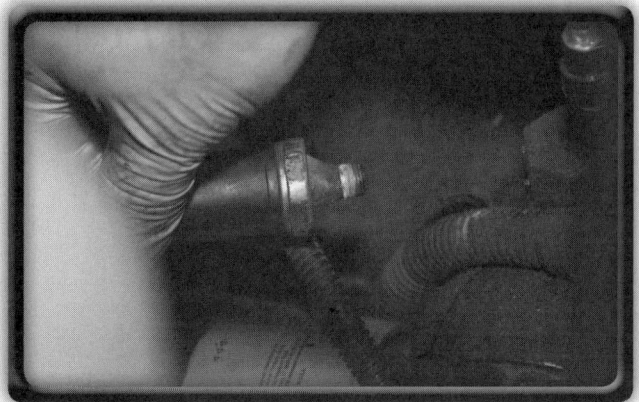

_____ **A.** Replace the oil pressure sensor or switch. Use the proper socket to remove the sensor or switch. If the new sensor does not come with thread sealant, use thread sealant on the threads, then install and torque the new sensor to the manufacturer's specifications. Double-check the oil level after replacement, as the oil level may have dropped.

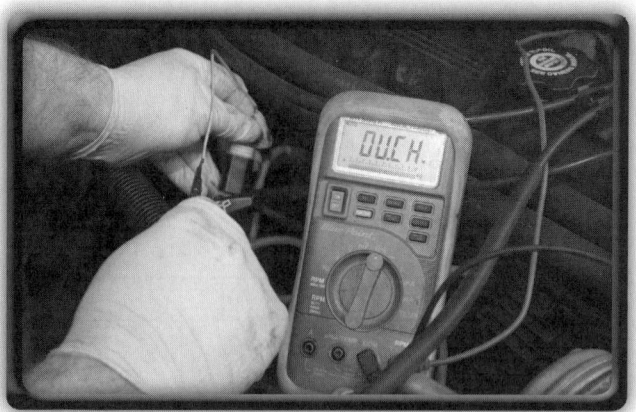

_____ **B.** If the gauge or light does not respond, then use an ohmmeter to test the wiring coming back to the gauge or light for an open or high resistance. If the gauge or light does respond, then the sensor or switch is bad.

_____ **C.** Locate the oil pressure switch or sensor. Disconnect the connector to the switch or sensor. If it is a sensor, determine which wire is the feedback to the gauge or warning light. You may need to use a connector end view to ensure the correct wire is found.

_____ **D.** Use a nonpowered test light to ground the feedback wire to see if the gauge or light responds. The gauge should move one direction or the warning light should glow.

Crossword Puzzle

Use the clues to complete the puzzle.

Across

1. Lobes or rounded edges on rotors that squeeze oil and create pressure.
3. A device that pumps lubricating oil through the engine.
7. Oil that is processed from crude oil; about 20% of oil is additives.
9. The slowing down of the crankshaft when it strikes the oil in the crankcase.
10. A type of splash lubricating system used in small engines. It works like a spoon scooping up oil and throwing it upward onto the crankshaft and other wear surfaces.
11. Oil additives that keep wax crystals from forming and causing the oil to gel during cold operation.
14. The ability of a liquid to flow.
15. Material pulled from the earth, originating from organic compounds broken down over time and formed into petroleum. This material is processed in a refinery to break down into various hydrocarbon substances such as diesel, gasoline, and mineral oil, among others.

Down

2. Oil passages that are drilled into the engine block and cylinder head(s). These passageways carry oil from the oil pump to critical moving parts.
4. Base stock processed from crude oil in a refinery, used as the base material of all conventional oil.
5. Passageways drilled or cast into the engine block or head(s), which carry pressurized lubricating oil to various moving parts in the engine, such as camshaft bearings.
6. Metal plates that are welded into the oil sump. These plates help to keep the oil at the oil pickup screen when the vehicle is cornering, braking, or accelerating hard.
8. The metal pan that covers the bottom of the engine contains oil sump where engine oil is held.
12. This oil, in its pure form, uses man-made base stocks and is not derived from crude oil. This oil lasts longer and performs better than normal oil.
13. A tube connected to the oil pump that acts like a straw for the oil pump to pull oil from the sump of the oil pan.

ASE-Type Questions

Read each item carefully, and then select the best response.

_____ 1. Tech A says that one function of oil is to clean. Tech B says that one function of oil is to cushion. Who is correct?
   A. Tech A
   B. Tech B
   C. Both A and B
   D. Neither A nor B

_____ 2. Two techs are discussing 5W20 oil. Tech A says the W stands for "weight." Tech B says the W stands for "winter." Who is correct?
   A. Tech A
   B. Tech B
   C. Both A and B
   D. Neither A nor B

_____ 3. Tech A says that the higher the viscosity number, the thicker the oil. Tech B says that most modern vehicles use single weight oil. Who is correct?
   A. Tech A
   B. Tech B
   C. Both A and B
   D. Neither A nor B

_____ 4. Tech A says that spin-on oil filters need RTV gasket sealer to seal the gasket. Tech B says that some oil filters use a replaceable paper filter cartridge. Who is correct?
   A. Tech A
   B. Tech B
   C. Both A and B
   D. Neither A nor B

_____ 5. Tech A says that the oil pressure will typically be low if the oil level is at the "add" line on the dipstick. Tech B says that a cracked pickup tube could cause low oil pressure. Who is correct?
   A. Tech A
   B. Tech B
   C. Both A and B
   D. Neither A nor B

_____ 6. Tech A says that most oil pumps are of the positive displacement style. Tech B says that oil pumps are designed to deliver more oil than is needed for an engine. Who is correct?
   A. Tech A
   B. Tech B
   C. Both A and B
   D. Neither A nor B

_____ 7. Tech A says that it takes about 1 pint of oil to raise the oil level from "add" to "full." Tech B says that it takes about 1 quart to raise it that much. Who is correct?
   A. Tech A
   B. Tech B
   C. Both A and B
   D. Neither A nor B

_____ 8. Tech A says that oil pressure is reduced when bearing clearances increase. Tech B says that oil pressure is regulated by the pressure relief valve. Who is correct?
   A. Tech A
   B. Tech B
   C. Both A and B
   D. Neither A nor B

_____ 9. Tech A says that a full-flow oil filter filters all of the oil going to the bearings. Tech B says that a bypass filter bypasses the pump, so that any particles in the oil won't damage the pump. Who is correct?
   A. Tech A
   B. Tech B
   C. Both A and B
   D. Neither A nor B

_____ 10. Tech A says that if the oil pressure reads low, the oil pressure sending unit should be replaced. Tech B says that if the oil pressure is low, the oil pressure should be measured using a separate pressure tester. Who is correct?
   A. Tech A
   B. Tech B
   C. Both A and B
   D. Neither A nor B

Engine Cooling

CHAPTER 46

Chapter Review

The following activities have been designed to help you refresh your knowledge of this chapter. Your instructor may require you to complete some or all of these activities as a regular part of your training program. You are encouraged to complete any activity that your instructor does not assign as a way to enhance your learning.

Matching

Match the following terms with the correct description or example.

- **A.** Coolant
- **B.** Coolant control valve
- **C.** Cooling hoses
- **D.** Cross-flow radiator
- **E.** Down-flow radiator
- **F.** Ethylene glycol
- **G.** Heat dissipation
- **H.** Overflow tank
- **I.** Propylene glycol
- **J.** Radiator
- **K.** Radiator hoses
- **L.** Rotor housing
- **M.** Surge tank
- **N.** Thermo-control switch
- **O.** Thermostat
- **P.** Water jackets

_____ 1. A device that takes hot coolant and cools it by passing heat energy to the surrounding air.

_____ 2. Flexible hoses that connect the stationary components of the cooling system such as the heater core and radiator to the engine, which is mounted on flexible mounts.

_____ 3. Regulates the flow of coolant, allowing coolant to flow from the engine to the radiator when the engine is running at its operating temperature.

_____ 4. The engine block of the rotary engine. The rotor moves within it.

_____ 5. The passages in the engine block and cylinder head that surround the cylinders, valves, and ports.

_____ 6. Rubber hoses that connect the radiator to the engine. Because they are subject to pressure, they are reinforced with a layer of fabric, typically nylon.

_____ 7. A fluid that contains anti-freeze mixed with water.

_____ 8. A temperature-sensitive switch that is mounted into the radiator or into a coolant passage on the engine to control electric fan operation.

_____ 9. The spreading of heat over a large area to increase heat transfer.

_____ 10. A valve that blocks off coolant flow to keep hot water from entering the heater core when less heat is requested by the operator.

_____ 11. A chemical used as anti-freeze. It is labeled as a nontoxic anti-freeze.

_____ 12. A radiator that uses cooling tubes that run horizontal with tanks on each end. This design allows lower hood profile for better vehicle aerodynamics.

_____ 13. A tank used to catch any coolant that is released from the radiator cap (works like a catch can).

_____ 14. A chemical used as anti-freeze that provides the lower freezing point of coolant and raises the boiling point. It is a toxic anti-freeze.

_____ 15. A sealed tank that captures coolant coming from the head that has turned to steam and changes the steam back to coolant to be reused by the cooling system.

_____ 16. A radiator that uses cooling tubes that run vertically. This design requires a higher hood profile.

Multiple Choice

Read each item carefully, and then select the best response.

_____ 1. Heat transfers through liquids and gases by a process called _____.
   A. conduction
   B. convection
   C. radiation
   D. transference

_____ 2. Engines have an ideal operating temperature somewhere around _____, give or take 20°F, depending on the vintage of the vehicle.
   A. 200°F
   B. 212°F
   C. 250°F
   D. 325°F

_____ 3. The circulation of coolant is controlled by the _____.
   A. water pump
   B. radiator temperature sensor
   C. air temperature
   D. thermostat

_____ 4. The best coolant is a _____ balance of water and anti-freeze, making it an ideal coolant for both hot and cold climates and providing adequate corrosion protection.
   A. 60/40
   B. 40/60
   C. 50/50
   D. 70/30

_____ 5. A force pulling outward on a rotating body is known as _____.
   A. roll
   B. convection
   C. centrifugal force
   D. spin

_____ 6. Coolant absorbs heat by _____ from the engine and becomes hotter.
   A. conduction
   B. convection
   C. radiation
   D. all of the above

_____ 7. As coolant moves through the hottest part of an engine with a reverse-flow design, steam tends to form and get stuck at the head cooling passages; this problem was fixed by adding a(n) _____.
   A. overflow tank
   B. cooling fan
   C. surge tank
   D. recirculation pump

_____ 8. What type of belt is used to drive the water pump and other accessories on the front of the engine?
   A. V-belt
   B. Serpentine belt
   C. Toothed belt
   D. Any of the above

_____ 9. _____ are used around the radiator fan to help draw air through the entire radiator core and not just in front of the fan blades.
   A. Fins
   B. Baffles
   C. Shrouds
   D. Ducts

_____ 10. The _____ is called a fan clutch since it can engage and disengage the cooling fan from the pulley.
   A. torque connector
   B. viscous coupler
   C. thermo-control switch
   D. electric solenoid

_____ 11. The _____ is typically located on the water pump and connects to the intake manifold on many V-configured engines, such as a V6 or V8.
   A. bypass hose
   B. upper hose
   C. lower hose
   D. throttle body coolant line

_____ 12. A _____ belt has a flat profile with a number of small V-shaped grooves running lengthwise along the inside of the belt.
   A. V-type
   B. serpentine
   C. stretch
   D. toothed

_____ 13. The _____ operates by using a signal sent from a coolant temperature sensor located on the engine in a coolant passage.
   A. temperature gauge
   B. thermostat
   C. warning light
   D. both A and C

_____ 14. The aluminum, brass, or steel plugs designed to seal the openings left from the casting process where the casting sand was removed are called _____.
   A. core plugs
   B. soft plugs
   C. expansion plugs
   D. all of the above

_____ 15. What type of coolant contains a mixture of inorganic and organic additives?
   A. Organic acid technology
   B. Inorganic acid technology
   C. Hybrid organic acid technology
   D. Poly organic acid technology

True/False

If you believe the statement to be more true than false, write the letter "T" in the space provided. If you believe the statement to be more false than true, write the letter "F".

_____ 1. Heat is transferred from one solid to another by a process called radiation.

_____ 2. No matter how efficiently fuel burning occurs, and no matter the size of the engine, the heat energy generated never completely transforms into kinetic energy.

_____ 3. Some newer vehicles have a coolant heat storage system that uses a vacuum-insulated container similar to a Thermos bottle.

_____ 4. Water alone is by far the best coolant there is, since it can absorb a larger amount of heat than most other liquids.

_____ 5. Ethylene glycol is a chemical that resists freezing but is not toxic and is used in nontoxic anti-freezes.

_____ 6. Because atmospheric pressure is lower at higher elevations, the boiling temperature of a liquid in an unsealed system is higher.

_____ 7. In an automotive cooling system, electrolysis is possible when the coolant breaks down and becomes more acidic.

_____ 8. Cooling systems, such as those used for turbocharger intercoolers/aftercoolers or hybrid high-voltage battery packs, may use air-to-air systems to cool the engine's intake air or high-voltage battery.

Fundamentals of Automotive Technology Student Workbook

_____ 9. All automobiles are now water cooled, although small engines can still be found in older automobiles that are air cooled.

_____ 10. All engines operate best when they are at their full operating temperature.

_____ 11. The hottest part of any engine is the cylinder head, since this is where the combustion chamber is located.

_____ 12. Radiator hoses connect the water pump and engine to the heater core. They carry heated coolant to the heater core to be used to heat the passenger compartment.

_____ 13. In both the cross-flow and the down-flow radiator designs, the core is built of the same components.

_____ 14. The thermostat is a spring-loaded valve that is controlled by a wax pellet located inside the valve.

_____ 15. Thermo-control switches often use a bimetallic strip that consists of two different metals or alloys laminated back to back that expand and contract at different rates.

_____ 16. Some radiator hoses, especially lower hoses, have a spiral wire inside to keep the hose from collapsing during heavy acceleration when the water pump is drawing a lot of water from the radiator.

_____ 17. The bottom radiator hose is typically attached to the thermostat housing, which allows the heated coolant to enter the inlet side of the radiator.

_____ 18. The heater control valve, if used, controls the flow of coolant to the heater core to control the temperature of the air desired by the operator.

_____ 19. Coolant can be easily identified according to its color, which may be anything from green or purple to yellow/gold, orange, blue, or pink.

_____ 20. Poly organic acid technology coolant contains a proprietary blend of corrosion inhibitors and provides up to 7 years or 250,000 miles of protection.

Fill in the Blank

Read each item carefully, and then complete the statement by filling in the missing word(s).

1. Heat is _____ energy. It cannot be destroyed; it can only be transferred.
2. Heat is transferred through space by a process called _____.
3. The principle that heat always moves from hot to cold is known as the second law of _____.
4. Coolant is a mixture of water and _____-_____, which is used to remove heat from the engine.
5. _____ glycol is a chemical that resists freezing but is very toxic to humans and animals.
6. A liquid under pressure higher than atmospheric pressure has a _____ boiling point than when that liquid is at atmospheric pressure.
7. _____ engine cooling uses heat-dissipating fins on the engine cylinders and heads to allow the movement of air to absorb heat and carry it away from the engine through special ducting.
8. In a typical _____-_____ design, coolant starts from the radiator and flows through the radiator outlet hose to the thermostat and then to the water pump.
9. The _____ is usually made of copper, brass, or aluminum tubes with copper, brass, aluminum, or plastic tanks on the sides or top for coolant to collect in.
10. _____ coolers are used to cool automatic transmission fluid, power steering fluid, EGR gasses, and compressed intake air.
11. One way to prevent coolant from boiling is to use a radiator _____ cap.
12. The coolant _____ system consists of an overflow bottle, a sealed radiator pressure cap, and a small hose connecting the bottle to the radiator neck.
13. The _____ _____ is usually belt driven from a pulley on the front of the crankshaft.
14. Engine-driven _____ _____ may be located on the water pump shaft, or in a few cases it may be attached directly to the engine crankshaft.
15. The bottom or lower _____ _____ is connected between the outlet of the radiator and the inlet of the water pump.
16. A(n) _____ belt looks like an ordinary serpentine belt but is found on vehicles without a tensioner.

17. _____ are used to keep the drive belt tight around the pulleys to ensure the least amount of slippage without causing damage to component bearings.
18. Temperature _____ can come in two forms: a temperature gauge or a temperature light located in the instrument cluster.
19. The _____ _____ is simply a small radiator that is mounted inside the heater box in the passenger compartment.
20. Air doors are moved by one of three methods: cable, vacuum actuator, or an electric _____, called a stepper motor.

Labeling

Label the following diagrams with the correct terms.

1. Heat transfer:

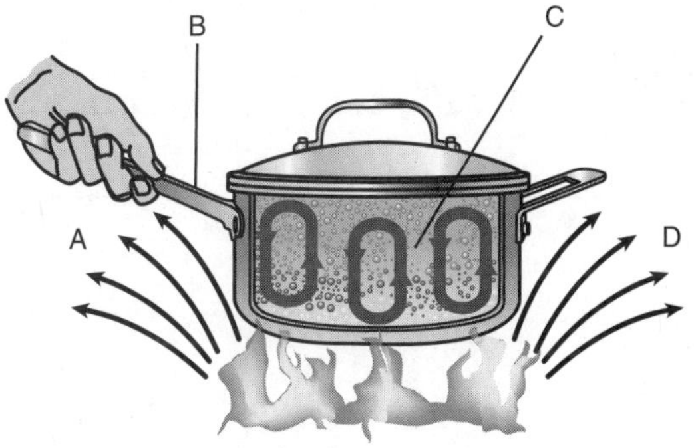

    A. _____
    B. _____
    C. _____
    D. _____

2. Percent of heat loss in a gasoline engine:

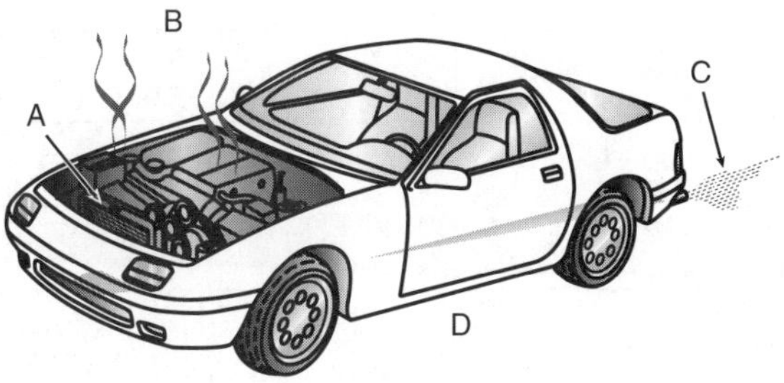

    A. _____
    B. _____
    C. _____
    D. _____

3. Types of radiators:

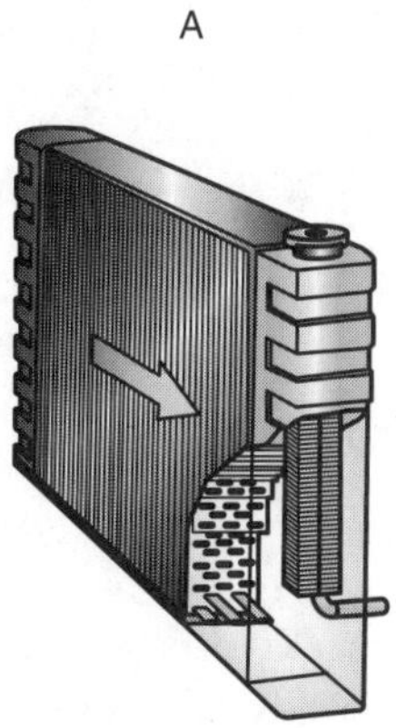

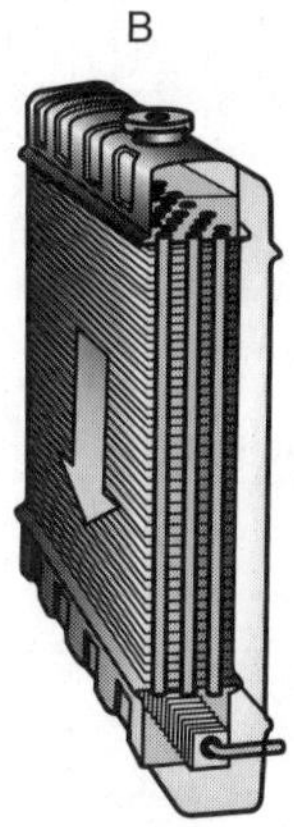

    A. _____
    B. _____

Fundamentals of Automotive Technology Student Workbook

4. Coolant recovery system:

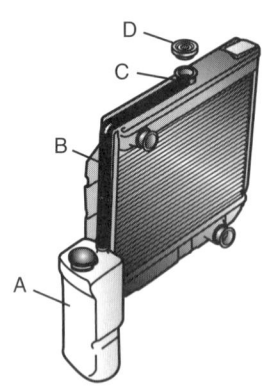

A. _____

B. _____

C. _____

D. _____

5. Thermostat valve:

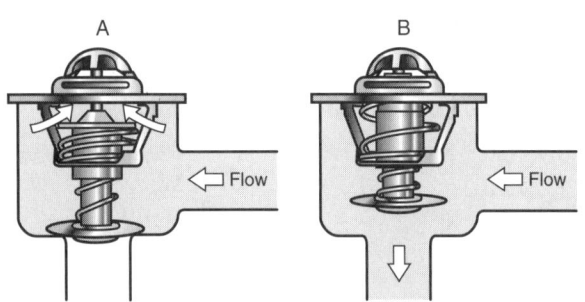

A. _____

B. _____

6. Water pump:

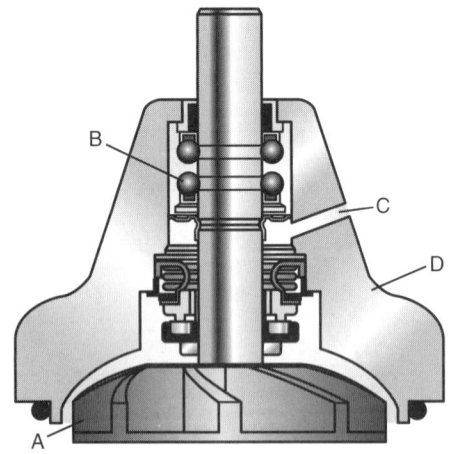

A. _____

B. _____

C. _____

D. _____

Chapter 46 Engine Cooling 537

7. Hydraulically operated cooling fan:

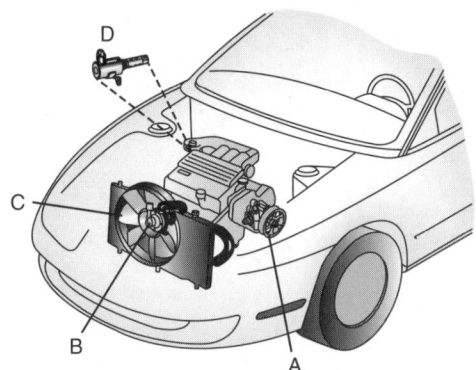

A. _____
B. _____
C. _____
D. _____

8. Cooling system:

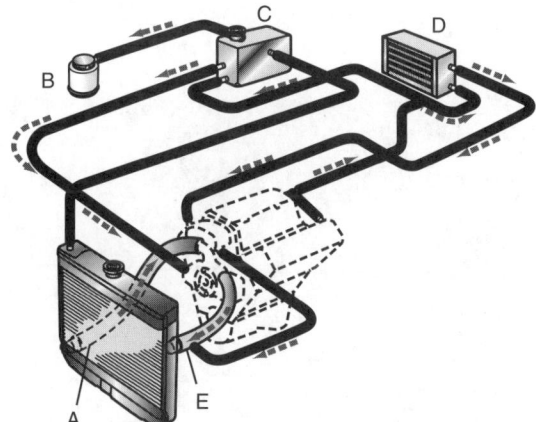

A. _____
B. _____
C. _____
D. _____
E. _____

Skill Drills

Test your knowledge of skill drills by filling in the correct words in the photo captions.

1. Testing the Cooling System Pressure:

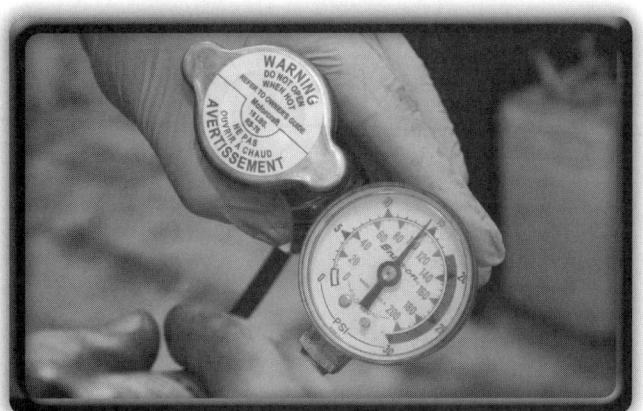 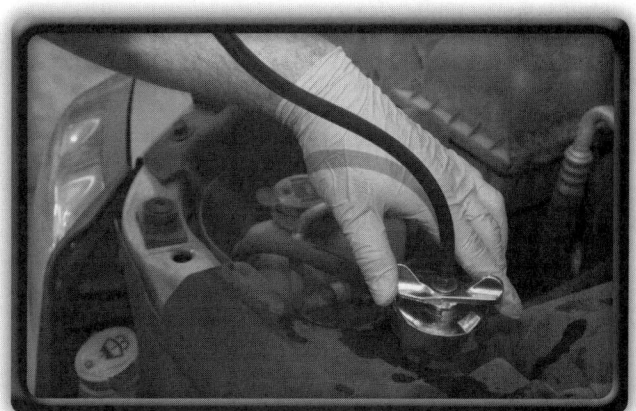

Step 1: Inspect for _____. Verify specified cooling system _____, install the radiator _____ on the _____ tester, and pressurize the cap to the correct pressure. It should hold pressure at approximately the _____ pressure and _____ at slightly above the rated pressure.

Step 2: Top off the radiator with _____ or _____, and install the tester. _____ the system to the _____ cap pressure.

Step 3: Watch the pressure reading for a _____ while performing a _____ check for any leaks. Check heater _____, soft _____, and any heater _____; determine necessary action.

2. Using a Hydrometer to Test the Freeze Point of the Coolant:

Step 1: Remove the pressure cap. Place the _____ tube in the _____, and _____ the ball on top.

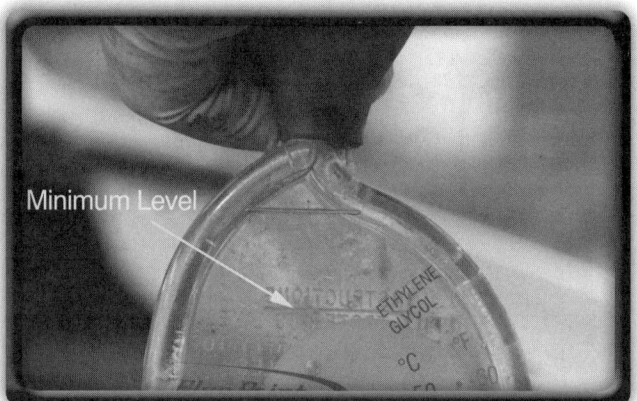

Step 2: _____ the ball to pull a coolant _____ into the hydrometer. _____ it is above the _____ level in the tester.

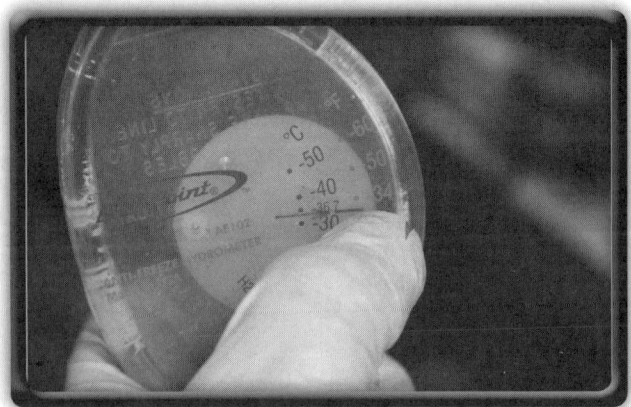

Step 3: Read the _____ on the tool to verify the _____ _____ of the coolant. Return coolant sample to the _____ or _____ tank.

3. Draining and Refilling Coolant:

Step 1: Locate the radiator _____ _____, if equipped, and place a catch _____ marked for coolant underneath the _____ _____. Drain the radiator into the catch pan.

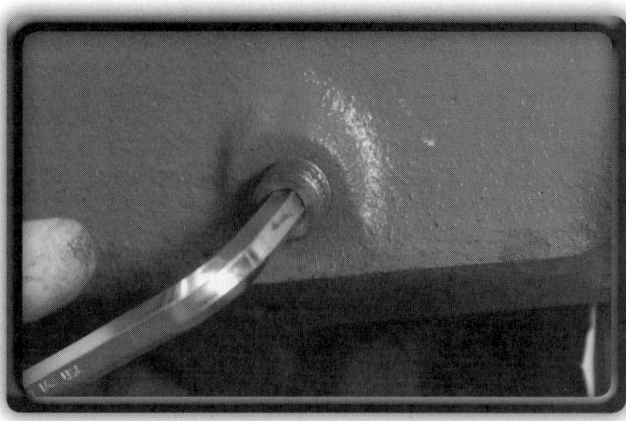

Step 2: Remove the _____ drain plugs, and allow the _____ to drain into the pan.

Step 3: Refill the cooling _____ with the proper coolant _____ after _____ the drain plugs. _____ the engine, and _____ the proper level. _____ of coolant in an _____ way.

4. Checking and Replacing a Coolant Hose:

Step 1: Inspect the hoses by _____ them, and _____ inspect the _____. Remove the _____ _____. Carefully pull or _____ the hose off the fittings.

Step 2: Verify that the _____ hose is correct. Cut to _____ if necessary. Reinstall the new _____ all the way into position. Install new _____.

Step 3: Refill _____, verify correct _____, and _____ check for leaks.

5. Removing and Replacing a Radiator:

Step 1: Place a _____ _____ below the radiator, and _____ the drain plug. _____ the coolant from the system. Replace the drain plug, and _____ of the drained _____.

Step 2: Unscrew any _____ or covers from the radiator. Remove the _____ or _____ that hold the _____ in position in the engine bay, and _____ the radiator from its location.

Step 3: Inspect the radiator, then place it into position and replace the _____ bolts or screws. Refit the shrouds or _____. Rotate the _____ and _____ by _____ to check that the covers do not _____ movement. Attach the coolant _____ to the radiator. Refill the system, and _____ the correct level.

Crossword Puzzle

Use the clues to complete the puzzle.

Across

3. The process of pulling metals apart by using electricity or by creating electricity through the use of chemicals and dissimilar metals.
5. The temperature at which a substance begins to change from a liquid to a gas.
6. Called a fan clutch, a hub that connects the water pump drive to the cooling fan uses a temperature-sensitive viscous fluid to cause the fan to turn faster as the temperature of the air pulled through the radiator increases.
7. A device that is electrically or vacuum controlled and is used to physically move doors within the heater box to control airflow.
9. A force pulling outward on a rotating body.
10. The angle of a fan blade. A steeper pitch draws more air, while a shallower pitch draws less air.

Down

1. Movement of heat energy through gases or liquids.
2. A temperature-sensitive switch that is mounted into the radiator or into a coolant passage on the engine to control fan operation.
4. Movement of heat energy through solids.
8. The movement of energy through space such as the movement of energy from the sun to the earth.

ASE-Type Questions

Read each item carefully, and then select the best response.

_____ 1. Tech A says that the cooling system is designed to keep the engine as cool as possible. Tech B says that the heater core can remove heat from the cooling system. Who is correct?
   A. Tech A
   B. Tech B
   C. Both A and B
   D. Neither A nor B

_____ 2. Tech A says that the thermostat is open until the engine warms up, and then it closes. Tech B says that a faulty radiator cap can be the cause of boiling coolant. Who is correct?
   A. Tech A
   B. Tech B
   C. Both A and B
   D. Neither A nor B

_____ 3. Tech A says that when pure anti-freeze is used in the cooling system, the protection level is –70°F. Tech B says that pure anti-freeze will cool the engine better than water. Who is correct?
   A. Tech A
   B. Tech B
   C. Both A and B
   D. Neither A nor B

_____ 4. Tech A says that some drive belts are a stretch fit design. Tech B says that the thermostat may have a bleed valve that should be accurately positioned when the thermostat is replaced. Who is correct?
   A. Tech A
   B. Tech B
   C. Both A and B
   D. Neither A nor B

_____ 5. Tech A says that the design of a surge tank system helps to purge air from the cooling system. Tech B says that overflow tanks are pressurized. Who is correct?
   A. Tech A
   B. Tech B
   C. Both A and B
   D. Neither A nor B

_____ 6. Tech A says that when you find coolant hoses collapsed, the radiator pressure cap has failed. Tech B says that overflow tanks are reservoirs designed to hold coolant and allow flow back to the radiator. Who is correct?
   A. Tech A
   B. Tech B
   C. Both A and B
   D. Neither A nor B

_____ 7. Tech A says that electric cooling fans are used to cause a large airflow over the radiator at high vehicle speeds. Tech B says that electric cooling fans are used to cause a large airflow over the radiator at low vehicle speeds. Who is correct?
   A. Tech A
   B. Tech B
   C. Both A and B
   D. Neither A nor B

_____ 8. Tech A says that there are a number of coolant types that each has its own life span. Tech B says that mixing of coolants is generally ok, since all coolants use the same chemical base. Who is correct?
   A. Tech A
   B. Tech B
   C. Both A and B
   D. Neither A nor B

_____ **9.** Tech A says that coolant leaking out of the water pump weep hole is normal. Tech B says that a refractometer measures the specific gravity of a liquid. Who is correct?
   A. Tech A
   B. Tech B
   C. Both A and B
   D. Neither A nor B

_____ **10.** Tech A says that a cooling system pressure tester can be used to find leaks in the cooling system. Tech B says that the cooling system pressure tester can be used to test the pressure at which the radiator cap vents. Who is correct?
   A. Tech A
   B. Tech B
   C. Both A and B
   D. Neither A nor B

Ignition Systems Overview

CHAPTER 47

Chapter Review

The following activities have been designed to help you refresh your knowledge of this chapter. Your instructor may require you to complete some or all of these activities as a regular part of your training program. You are encouraged to complete any activity that your instructor does not assign as a way to enhance your learning.

Matching

Match the following terms with the correct description or example.

- A. Ballast resistor
- B. Breaker plate
- C. Center electrode
- D. Cranking
- E. Direct ignition system
- F. Dwell angle
- G. Electronic ignition system
- H. Firing line
- I. Heat range
- J. Ignition coil
- K. Ignition switch
- L. Inductive current
- M. Oscilloscope
- N. Primary winding
- O. Rotor
- P. Secondary winding
- Q. Spark plug
- R. Spark timing
- S. Throttle body
- T. Waste cylinder

___N___ 1. The low-voltage coiled copper wiring found in an ignition coil.

___T___ 2. The cylinder in a waste spark ignition system that receives a spark near the top of its exhaust stroke.

___O___ 3. A high-voltage rotating switch that transfers voltage from the distributor cap's center terminal to the outer terminals.

___R___ 4. The point at which a spark occurs at the spark plug relative to the position of the piston.

___J___ 5. A device used to amplify an input voltage into the much higher voltage needed to jump the electrodes of a spark plug.

___M___ 6. A tool that shows graphically what is happening to voltage over a period of time; it is used to diagnose electrical faults.

___S___ 7. The housing on an intake manifold that is used to control the amount of filtered air that enters the cylinders.

___I___ 8. The rating of a spark plug's operating temperature.

___A___ 9. Used to limit the amount of current flowing in the ignition primary circuit.

___G___ 10. An ignition system that uses a nonmechanical method of triggering the ignition coil's primary circuit.

___P___ 11. The high-voltage copper wiring found in an ignition coil.

___B___ 12. The movable plate the breaker points are mounted on that pivots as the vacuum advance pulls on it.

___E___ 13. May refer to a waste spark ignition or a coil-on-plug ignition system, in which the coils are directly attached to the spark plugs.

___L___ 14. The current that has been created across a conductor by moving it through a magnetic field.

___C___ 15. The electrode located in the center of a spark plug. It is the hottest part of the spark plug.

___F___ 16. The amount of time that the primary circuit is energized, measured in degrees of distributor rotation.

___D___ 17. Rotating the engine by turning the ignition key to the start position.

__K__ 18. A switch operated by a key or start/stop button and used to turn on or off a vehicle's electrical and ignition system.

__H__ 19. The tall lines on a parade pattern that indicate the voltage required to initially jump the spark plug gap.

__Q__ 20. A device that provides a gap for the high-voltage spark to occur in each cylinder.

Multiple Choice

Read each item carefully, and then select the best response.

__B__ 1. For an engine to run smoothly and efficiently, the high-voltage spark must cross the spark plug electrode as the piston approaches top dead center of the _____.
   A. intake stroke
   B. compression stroke
   C. power stroke
   D. exhaust stroke

__A__ 2. What type of ignition system eliminates the distributor by using dedicated ignition coils—one coil for each pair of cylinders?
   A. Contact breaker point ignition system
   B. Electronic ignition system
   C. Waste spark ignition system
   D. Direct ignition system

__B__ 3. The _____ replaced the contact breaker points with an electronic switching device but still used a distributor to dispense the spark to the various cylinders.
   A. electronic ignition system—distributor type
   B. waste spark ignition system
   C. direct ignition system
   D. coil-on-plug ignition system

__B__ 4. The ignition system uses a(n) _____ to convert relatively low-voltage and high-current flow into very high-voltage and very low-current flow.
   A. spark plug
   B. ignition module
   C. induction coil
   D. reluctor

__B__ 5. Faraday's law states that relative movement between a conductor and a magnetic field allows _____ ways by which voltage can be induced in a conductor.
   A. three
   B. four
   C. five
   D. six

__D__ 6. The correct spark timing varies according to _____.
   A. transmission gear selected
   B. detected knock
   C. engine temperature
   D. all of the above

__D__ 7. The power train control module may also be called a(n) _____.
   A. engine control module
   B. body control module
   C. electronic control unit
   D. either A or C

__A__ 8. When the ignition switch is turned to the _____ position, most warning lamps on the instrument panel should illuminate.
   A. accessory
   B. on/run
   C. start/crank
   D. lock and off

Chapter 47 Ignition Systems Overview

___A___ 9. When battery voltage pushes current through the _____ of an ignition coil, a magnetic field is created.
  A. primary winding
  B. secondary winding
  C. iron core
  D. high-tension terminal

___D___ 10. Spark plugs are identified by _____.
  A. thread size or diameter
  B. reach or length of the thread
  C. heat range or operating temperature
  D. all of the above

___C___ 11. Contact breaker points are opened by _____ on the distributor shaft, and closed by a spring.
  A. rubbing blocks
  B. sensors
  C. cam lobes
  D. gears

___C___ 12. The _____ is made up of two plates constructed from narrow strips of aluminum foil that are insulated from each other by a special waxed paper, called a dielectric.
  A. rotor
  B. condenser
  C. contact breaker
  D. ballast resistor

_____ 13. It is the main function of the _____ to distribute the spark to the spark plugs in the correct sequence and at the correct time in the engine cycle.
  A. condenser
  B. ballast resistor
  C. distributor
  D. high-tension terminal

___D/B___ 14. The vacuum advance mechanism is designed to operate on _____ vacuum.
  A. ported
  B. manifold
  C. venturi
  D. any of the above

___A___ 15. A _____ is an example of an inductive pickup.
  A. Hall-effect switch
  B. throttle position sensor
  C. mass airflow sensor
  D. manifold absolute pressure sensor

___D___ 16. Ignition coils employed with electronic systems are referred to as _____.
  A. low-inductance coils
  B. high-inductance coils
  C. dual-inductance coils
  D. stators

___D___ 17. The _____ is shaped like a very shallow cup with slits, or windows, cut into it at evenly spaced intervals and made of a ferrous metal.
  A. optical-type sensor
  B. interrupter ring
  C. reluctor
  D. stator

___A___ 18. What type of sensor, located inside the distributor, can be used to sense the position of the crankshaft and send an appropriate voltage signal to the ignition module or PCM?
  A. Hall-effect
  B. Induction-type
  C. Optical-type
  D. Variable-resistance

_D_ 19. A distributorless ignition system uses a _____ to calculate engine speed and determine engine position.
   A. crankshaft position sensor
   B. cam lobe
   C. camshaft position sensor
   D. both A and C

_C_ 20. A _____ on an oscilloscope or lab scope shows all the cylinders firing in sequence.
   A. parade pattern
   B. firing line
   C. raster
   D. spark line

True/False

If you believe the statement to be more true than false, write the letter "T" in the space provided. If you believe the statement to be more false than true, write the letter "F".

_T_ 1. The ignition coil's function is to amplify the battery's low voltage and high current into very high voltage and low current.

_F_ 2. The low-voltage side of the induction coil is called the secondary circuit, and the high-voltage side is called the primary circuit.

_F_ 3. Available voltage is the maximum amount of voltage available to try to push current to jump the spark plug gap if the gap were infinite.

_T_ 4. In early vehicles that used a distributor, base spark timing was set at idle speeds by positioning the distributor body in relation to its rotating cam.

_T_ 5. Ignition coils are basically step-up transformers.

_T_ 6. The secondary ignition coil winding has 15,000 to 30,000 turns of very thin insulated aluminum wire wound around its core.

_T_ 7. High-tension leads conduct the high-output voltage generated in the secondary ignition circuit between the high-tension terminal(s) of the ignition coil, the distributor cap, and the spark plugs.

_F_ 8. The same spark plug can sometimes be used in different engines with different gap settings.

_F_ 9. The condenser absorbs the surge of inductive current that occurs when the contact breaker points begin to separate.

_T_ 10. The most common automotive use for a ballast resistor is to reduce the voltage and current to the ignition coil by being inserted in series in the primary circuit between the ignition switch and the positive terminal of the ignition coil.

_T_ 11. The distributor cap covers the end of the distributor to protect the components and provide a connection point between the rotor and the spark plug leads.

_F_ 12. The centrifugal advance mechanism controls ignition advance in relation to engine load.

_T_ 13. In electronic ignition systems, the contact breaker points are eliminated and the primary circuit is switched or triggered electronically with a power transistor located in the ignition module.

_T_ 14. The stator has one tooth for each cylinder; as it spins the teeth interact with the reluctor to trigger the ignition module.

_T_ 15. The dwell control section of the ignition module determines when the primary circuit will be switched on and for how long current will flow in the primary winding.

_T_ 16. The switching transistor acts as a relay; when activated, it allows current to flow through the ignition coil primary winding and through the collector/emitter to ground.

_T_ 17. In direct ignition systems, there is one ignition coil for each pair of companion cylinders.

_T_ 18. In a waste spark system the cylinder on the compression stroke is said to be the event cylinder and the cylinder on the exhaust stroke is the waste cylinder.

_T_ 19. The spark timing on a distributorless ignition system is controlled by the powertrain control module.

_T_ 20. Raster allows a technician to compare each spark line to each other, which tells how much voltage it takes to keep the spark burning and for how long.

Fill in the Blank

Read each item carefully, and then complete the statement by filling in the missing word(s).

1. The purpose of the ignition system is to create the __High__-__voltage__ __spark__ and deliver it at the right time to each cylinder.
2. The original ignition system was called the __contacts__ __breaker__ __points__ ignition system, a mechanical system with a switch that was opened and closed as the engine was running.
3. The amount of voltage required to initially get current to jump the spark plug gap is called __required__ voltage.
4. Spark timing is almost always indexed to the __number__ __one__ cylinder.
5. The __distributor__ __base__ plate, also called a breaker plate, is a moveable metal plate located in the distributor, beneath the distributor cap on which the contact breaker points are mounted.
6. The __battery__ supplies the electrical energy to the ignition circuit during start-up.
7. The immobilizer system uses a randomly selected __rolling__ __code__, a constantly changing, numeric code that is communicated to the engine immobilizer and security system by a radio frequency signal.
8. __High__-__tension__ __leads__ connect the secondary ignition components together, such as the coil to the distributor cap and the distributor cap to the spark plugs.
9. If the insulation around the cable is insufficient and the leads are not spaced far enough apart and run parallel to each other, then a(n) __spark__ __plug__ can be generated in the other wires.
10. Spark plug __reach__ is the distance from the seat of the spark plug to the end of the spark plug threads.
11. The amount of time, in degrees of __distributor__ __rotation__, that the contacts are closed is called the dwell angle.
12. The __centrifugal__ advance mechanism controls ignition timing in relation to engine speed.
13. Ignition __advances__ are used in virtually every type of ignition system except the contact breaker point system.
14. __Inductive__-__type__ systems are electronic systems that use a magnetic pulse generator, also called a variable reluctor sensor, to generate an AC signal.
15. A(n) __stationary__ winding is an extended length of wire wrapped into a circle.
16. The induction winding is connected to the ignition module and forms part of an internal module __control__ __circuit__.
17. In a distributor, the Hall-effect generator and its __integrated__ __circuit__ are located on one leg of a U-shaped assembly mounted on the distributor base plate.
18. In __distributorless__ ignition systems, also known as electronic ignition systems, the distributor is eliminated and replaced by multiple ignition coils.
19. A(n) __oscilloscope__ or lab scope can be used to observe the electrical patterns created by the triggering devices and ignition coil.
20. Many oscilloscopes have the ability to stack each spark plug's ignition pattern vertically, which is called __raster__.

Labeling

Label the following diagrams with the correct terms.

1. Contact breaker point ignition system:

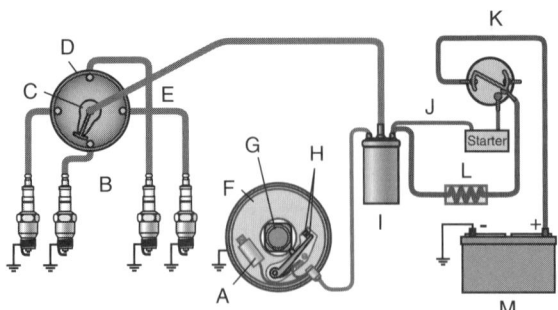

A. Condenser
B. Spark plugs
C. Rotor button
D. Distributor cap
E. High tension leads
F. Breaker plate
G. Distributor camshaft
H. Ignition points
I. Ignition
J. Ignition start bypass
K. Ballast resistor / Ignition switch
L.
M. Battery

2. Electronic ignition system—distributor type:

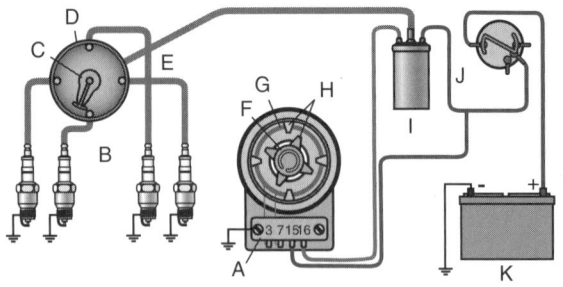

A. Electronic module
B. Spark plugs
C. Rotor button
D. Distributor cap
E. High-tension leads
F. Reluctor
G. Pickup coil
H. Poles
I. Ignition coil
J. Ignition switch
K. Battery

3. Waste spark ignition system:

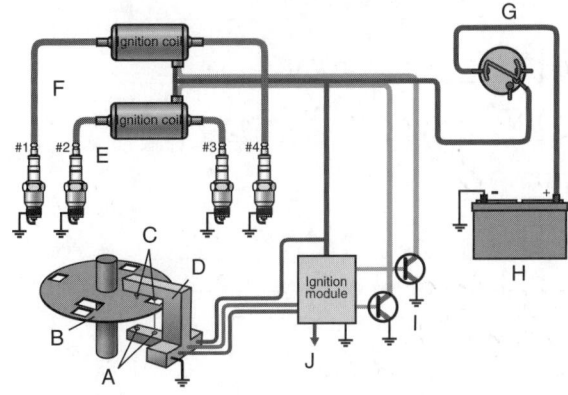

A. Infrored Recievers
B. Interropter plate
C. Infrared transmiters
D. Optical sensor
E. Spark plugs
F. High tensor leads
G. Ignition switch
H. Battery
I. To ECU
J. Power transwero

4. Coil-on-plug ignition system:

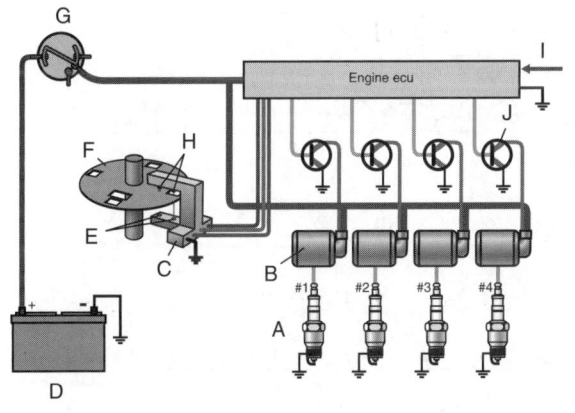

A. Spark plugs
B. Ignition coil
C. Optical sensor
D. Batter
E. Intared recievers
F. Interruptor plator
G. Ignition switch
H. Infared transmitted
I. Engine sensor
J. Power transistor

5. Vacuum advance mechanism:

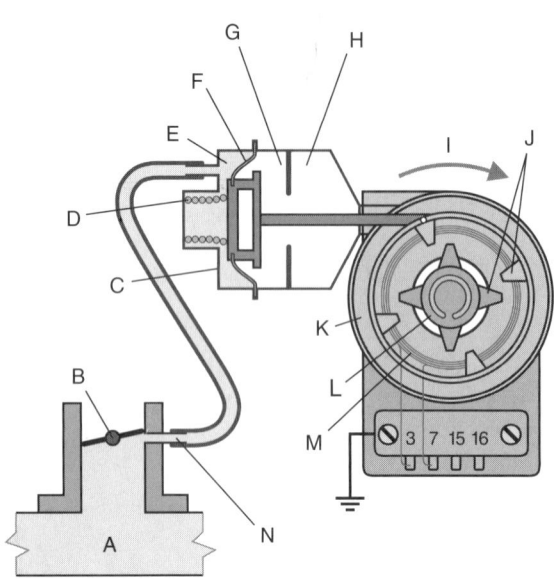

A. Intake manifold
B. Throttle
C. Vacuum advance body
D. Spring
E. Vacuum chamber
F. Diaphragm
G. Advance leg
H. Atmospheric chamber
I. Direction of rotation
J. Poles
K. Distributor body
L. Reluctor
M. Pickup coil
N. Vacuum port

6. Ignition coil:

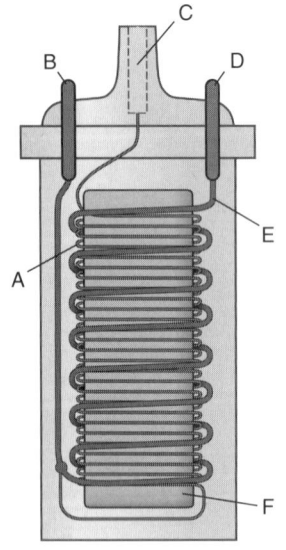

A. Secondary winding
B. Terminal
C. HT tower
D. Terminal
E. Primary winding
F. Laminated soft

7. Parts of a high-tension lead:

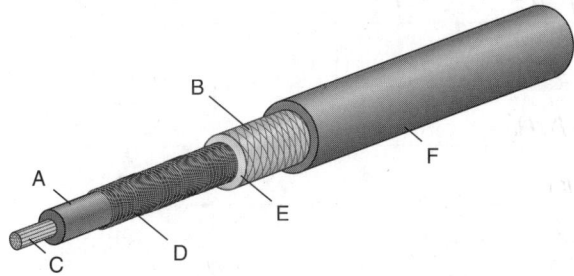

A. Ferrite or EPDM core
B. Braided glass
C. Glass fiber core
D. Variable pitch coil
E. High dielectric strength in stator
F. Silicon insulator layer

8. Spark plug:

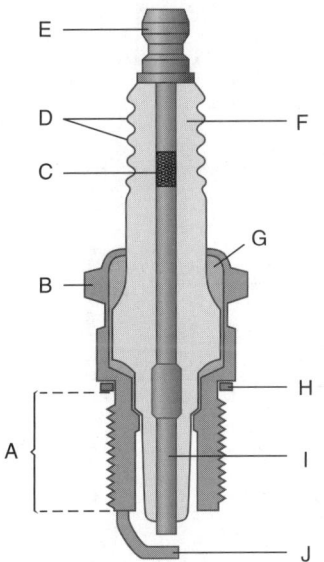

A. Thread reach
B. Metal shell
C. Resistor
D. Flashover ribs
E. Terminal nut
F. Insulator
G. Powder filling
H. Gasket
I. Center electrode
J. Ground electrodes

9. Contact breaker point operation:

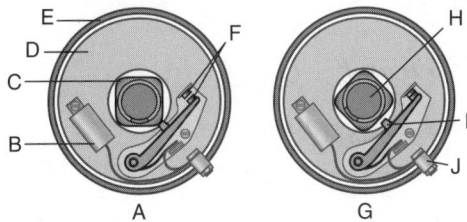

A. Points open
B. Condenser
C. Cam lob
D. Breaker plate
E. Distributor body
F. Breaker points
G. Point closed
H. Distributor camshaft
I. Rubbing block
J. Insulated terminal

10. Distributor:

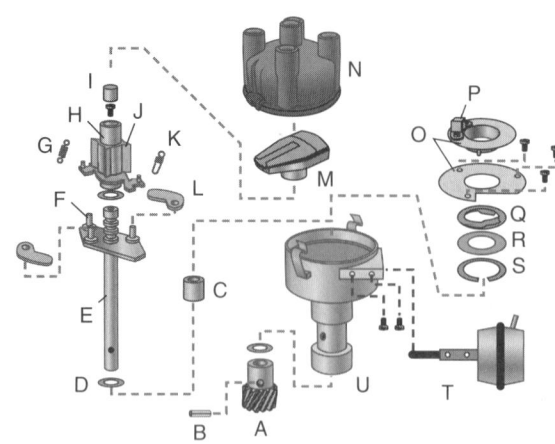

A. Drive gear
B. Roll pin
C. Bushing
D. Shim
E. lower shaft
F. Maxiumum advance stop
G. Primary spring
H. Upper shaft
I. Felt Picking
J. Relucter
K. lower shaft
L. Advance weights
M. rotor
N. Distrubtor cap
O. Breaker plates
P. Inductive sensor
Q. Spring
R. Shim
S. Snap ring
T. Vacuum advance unit
U. Distributor housing

11. Vacuum sources:

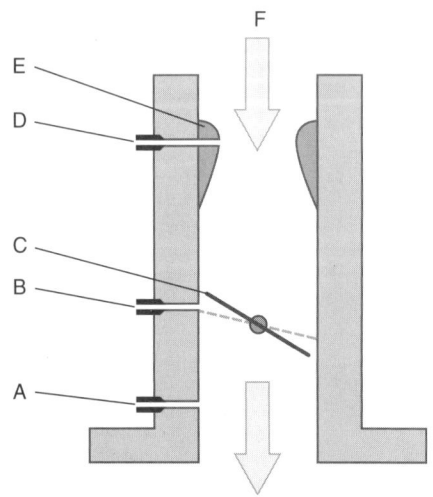

A. Manifold vacuum
B. Ported vacuum
C. Throttle valve
D. Venturi vacuum
E. Venturi
F. Air flow

Skill Drills

Test your knowledge of skill drills by filling in the correct words in the photo captions.

1. Performing a Spark Test:

Step 1: Remove the high-tension _lead_ or _coil_ from the end of the _spark plug_.

Step 2: Connect a _spark tester_ to the _boot_ of the coil or high-tension lead, and attach the _clamp_ to a good _ground_.

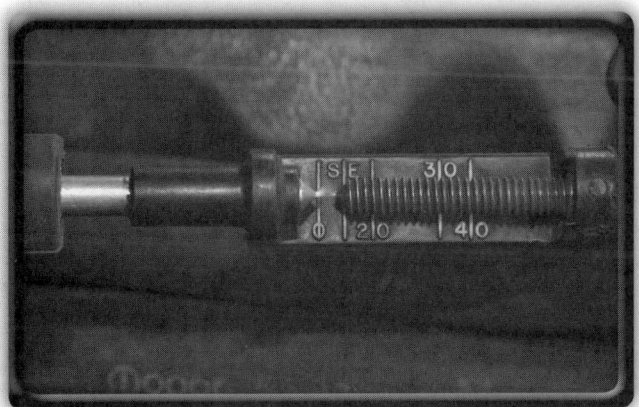

Step 3: Crank the engine, and watch for a _spark_ across the tester _electrodes_. A properly functioning _ignition_ system will produce a _blue_ spark. If there is no spark present, test the spark plug _wire_ or coil, distributor cap, and _rotor_ on that plug. If there is still no spark present, inspect the primary and secondary circuits.

2. Testing the Ignition Coil:

Step 1: Obtain the recommended testing procedure for the particular vehicle you are inspecting. Inspect and test ignition **primary** and **secondary** windings of the ignition coil(s). All the coil **terminals** should be clean, secure, and free of **corrosion**.

Step 2: To test the **resistance** of the primary windings, place the **ohmmeter** leads on each of the two primary winding **terminals**. If the reading is not within specifications, the coil is **faulty** and must be **replaced**.

Step 3: To test the **secondary** windings, place one ohmmeter **probe** on the secondary **tower** terminal and the other to one of the **primary** terminals, or the coil secondary **ground** if equipped. If the reading is not within specifications, the **coil** is faulty and must be replaced.

Step 4: If the coil readings are OK, perform a **spark test** to verify that the coil can produce a **spark** that will **jump** at least 1/2" (13 mm).

3. Testing a Spark Plug Wire:

Step 1: Disconnect the spark plug _wire_ at both ends by grasping the _boots_ on the ends and _twisting_ while pulling the wire off. _Inspect_ each wire for _cracks_, brittleness, or _burnt_ spots.

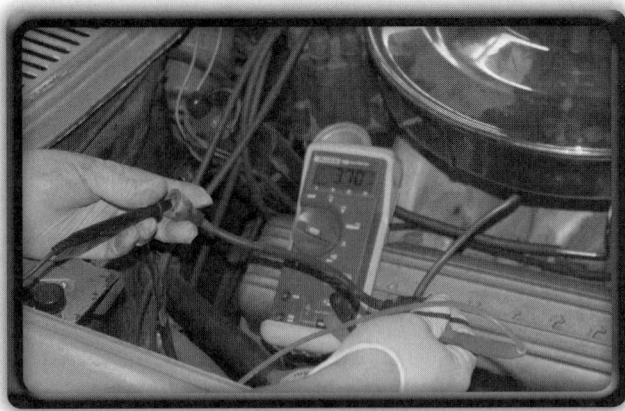

Step 2: Place one lead of the _ohmmeter_ on each _end_ of the spark plug wire. _Flex_ the wire while _reading_ the ohmmeter.

Step 3: Replace the _spark plug wire_, start the engine, and lightly _mist_ water on the spark plug wires or run a grounded _test lead_ along each wire. If the readings are not within specifications or if the wires _arc_, replace the spark plug wires, making sure to _route_ them in their factory positions to prevent damage.

4. Testing the Crankshaft and Camshaft Position Sensors:

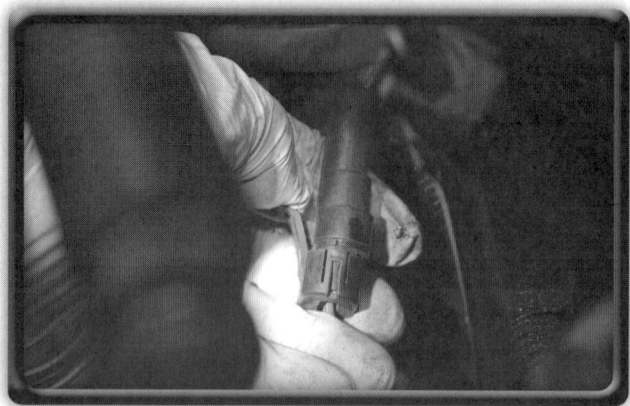

Step 1: Remove the connector from the _sensor_ by removing any _lock_ _clips_ that may be present, and _pull_ the connector off.

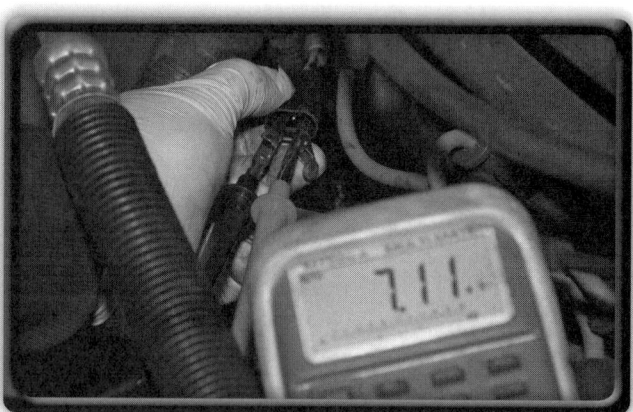

Step 2: If it is an _inductive type_ sensor, use an ohmmeter to read the _resistance_, and compare the readings to the _specification_. Replace the sensor if its resistance is too _high_ or too low, indicating that it is _defective_.

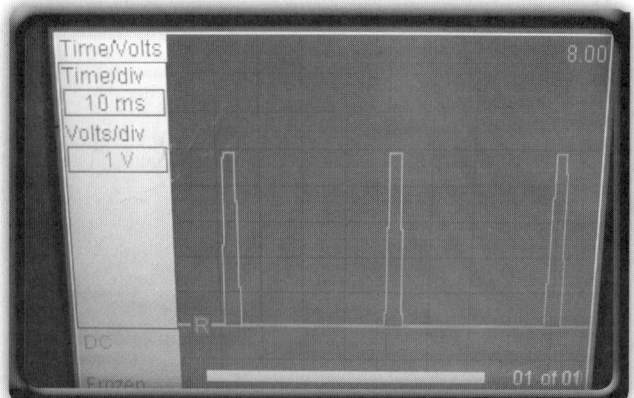

Step 3: If the sensor is a _Hall-effect_ or _optical_ sensor, or the previous reading is normal, use a _lab_ _scope_ to observe the sensor signal and compare to a known good _pattern_. If faulty, replace the sensor.

5. Testing an Ignition Module:

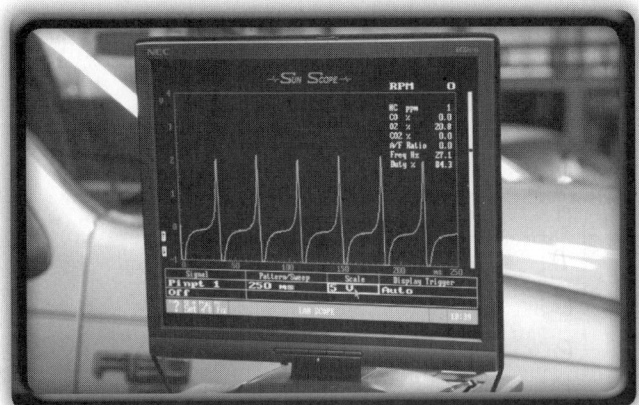

Step 1: Test all the _input_ wires leading to the _ignition_ module for proper _voltage_ and _ground_ signals.

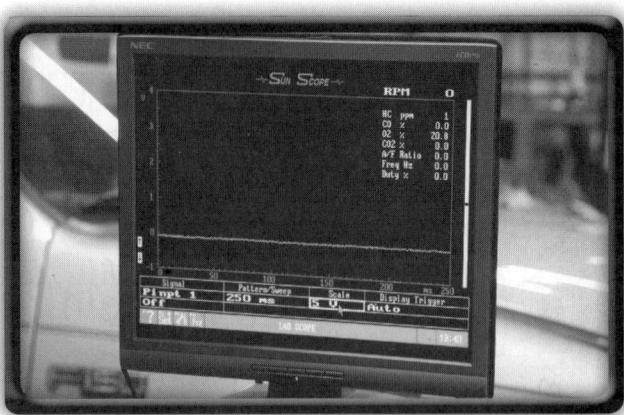

Step 2: Check for proper _output_ signals from the ignition module with the _key_ set to _run_ and while _cranking_ according to the manufacturer's specifications. If the specified _inputs_ are present but the ignition module will not produce the proper output _signals_ the ignition module should be replaced.

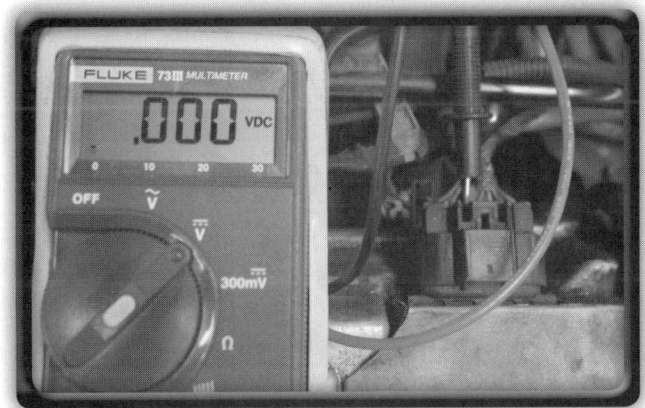

Step 3: With a multimeter set to read volts, check the _voltage drop_ between the ignition module and _ground_ on the engine with a multimeter set to read _volts_. If the voltage drop is more than 0.1 volt, check for _corrosion_ or rust between the ignition module and the _surface_ it is mounted to. Clean the corrosion or _rust_, and _reinstall_ the ignition module.

Crossword Puzzle

Use the clues to complete the puzzle.

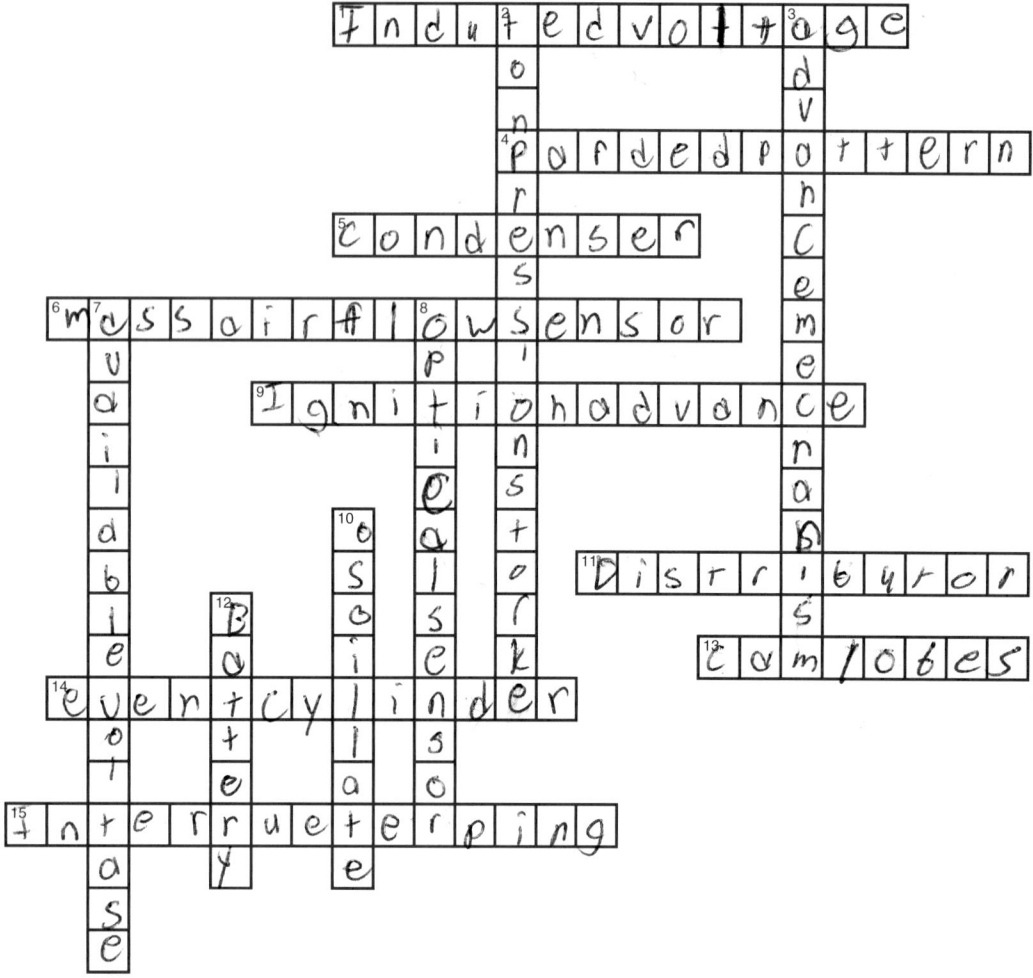

Across

1. The creation of voltage in a conductor by movement of a magnetic field that is near that conductor.
4. The display of all the cylinders firing in sequence on an oscilloscope.
5. Also known as a capacitor, a self-contained unit that is connected electrically in parallel with the contact breaker points and used to assist the rapid collapse of the magnetic field in the coil by preventing the contact breaker points from arcing. Also used on some coils to reduce radio frequency interference.
6. PCM input that measures the mass of the air entering the engine.
9. The means of causing the spark to occur earlier within the compression stroke for better performance and fuel economy during changing engine conditions.
11. The part of an ignition system that distributes the spark to the spark plugs in the correct sequence and at the correct time. It includes a distributor cap, rotor shaft, and usually a switching device.
13. Raised areas or protrusions on an otherwise round shaft.
14. The cylinder that uses the spark to ignite the air/fuel mixture on a waste spark ignition system.
15. A ferrous metal ring, shaped like a very shallow cup with slits or windows cut into it at evenly spaced intervals. The purpose of this ring is to systematically block, and expose, the magnetic field in a Hall-effect sensor in order to turn the primary ignition circuit on and off.

Down

2. The stroke in the four-stroke cycle in which the air/fuel mixture is compressed.
3. A device used to trigger an earlier spark based on engine conditions.
7. The maximum amount of voltage that the induction coil secondary is capable of putting out.
8. A sensor that generates a voltage when excited by a beam of light.
10. To cycle above and below a given value.
12. An electrochemical device used to supply voltage to a vehicle's electrical systems.

ASE-Type Questions

Read each item carefully, and then select the best response.

___C___ 1. Tech A says that contact breaker points are a mechanical switch that opens and closes once for every ignition spark that is created. Tech B says that contact breaker points send high voltage directly from the points to the spark plugs. Who is correct?
   A. Tech A
   B. Tech B
   C. Both A and B
   D. Neither A nor B

___B___ 2. Tech A says that as engines gain miles, the spark plug gap increases, which raises the ignition system's available voltage. Tech B says that misfire occurs when required voltage is higher than available voltage. Who is correct?
   A. Tech A
   B. Tech B
   C. Both A and B
   D. Neither A nor B

___A___ 3. Tech A says that as engine rpm increases, spark timing generally increases. Tech B says that as engine load increases, spark timing generally decreases. Who is correct?
   A. Tech A
   B. Tech B
   C. Both A and B
   D. Neither A nor B

___C___ 4. Tech A says that a Hall-effect switch uses light to turn a circuit on and off. Tech B says that waste spark systems don't need distributors. Who is correct?
   A. Tech A
   B. Tech B
   C. Both A and B
   D. Neither A nor B

___A___ 5. Tech A says that coil-on-plug ignition systems use one coil to fire two cylinders. Tech B says that you should twist spark plug boots before removing them. Who is correct?
   A. Tech A
   B. Tech B
   C. Both A and B
   D. Neither A nor B

___A___ 6. Tech A says that the ignition system will maintain spark at the spark plug for approximately 23 degrees of crankshaft rotation. Tech B says that the duration of spark in the spark plug only lasts 2 to 3 degrees of crankshaft rotation. Who is correct?
   A. Tech A
   B. Tech B
   C. Both A and B
   D. Neither A nor B

Fundamentals of Automotive Technology Student Workbook

___C___ 7. Tech A says that the positive side of the coil primary circuit is switched. Tech B says that the negative side of the coil primary circuit is switched. Who is correct?
   A. Tech A
   B. Tech B
   C. Both A and B
   D. Neither A nor B

___C___ 8. Tech A says that the advantage of distributorless ignition systems is no moving parts to maintain. Tech B says that the coil primary winding can best be tested with a voltmeter. Who is correct?
   A. Tech A
   B. Tech B
   C. Both A and B
   D. Neither A nor B

___B___ 9. Tech A says that an inductive pick-up coil can be tested for shorts and grounds with an ohmmeter. Tech B says that an inductive pick-up coil pattern can be tested with a lab scope. Who is correct?
   A. Tech A
   B. Tech B
   C. Both A and B
   D. Neither A nor B

___D___ 10. Tech A says that on a secondary ignition pattern, the firing line shows how much voltage it takes to jump the spark plug gap. Tech B says that the firing line tells how long the spark lasted. Who is correct?
   A. Tech A
   B. Tech B
   C. Both A and B
   D. Neither A nor B

Gaosline Fuel Systems

CHAPTER 48

Chapter Review

The following activities have been designed to help you refresh your knowledge of this chapter. Your instructor may require you to complete some or all of these activities as a regular part of your training program. You are encouraged to complete any activity that your instructor does not assign as a way to enhance your learning.

Matching

Match the following terms with the correct description or example.

- A. Accelerator pump
- B. Accelerometer
- C. Air supply system
- D. Choke
- E. Dieseling
- F. Element
- G. Float chamber
- H. Frequency
- I. Fuel metering system
- J. Fuel system
- K. Idle
- L. Injector
- M. Knocking
- N. Misfire
- O. Output signal
- P. Pollutant
- Q. Relay
- R. Throttle
- S. Vacuum
- T. Vapor lock

__E__ 1. A condition in which the engine continues to run after the ignition key is turned off. Also referred to as run-on.

__P__ 2. A potential threat to human health or the environment resulting from excessive amounts of chemicals and waste.

__K__ 3. The speed at which an engine runs without any throttle applied.

__ized__ 4. Equipment in a motor vehicle that delivers the proper amount of fuel to each cylinder.

__R__ 5. A device used to produce acceleration by controlling the air/fuel mixture.

__T__ 6. A situation in which vapor forms in the fuel line, and the bubbles of vapor block the flow of fuel and stop the engine.

__B__ 7. An electronic instrument that measures the amount of acceleration in a specific direction.

__O__ 8. The voltage sent to solenoids and motors by the PCM to do work.

__H__ 9. The rate of change in direction, oscillation, or cycles in a given time.

__Q__ 10. A magnetically operated switch used to make or break current flow in a circuit.

__A__ 11. A small pump usually located inside the carburetor that sprays an extra amount of fuel into the carburetor air horn during acceleration.

__J__ 12. Equipment in a motor vehicle that delivers fuel to the engine.

__F__ 13. The replaceable portion of a filter, such as an air filter or oil filter.

__S__ 14. A pressure in an enclosed area that is lower than atmospheric pressure.

__D__ 15. A device that provides a rich air/fuel mixture until the engine warms up by restricting the flow of air at the entrance to the carburetor, before the venturi.

M 16. A noise heard when the air/fuel mixture spontaneously ignites before the spark plug is fired at the optimum ignition moment.

C 17. Equipment in a motor vehicle that delivers air to the engine.

N 18. Failure of one or more cylinders to fire or complete combustion.

G 19. A chamber that holds a quantity of fuel at atmospheric pressure ready for use.

W 20. A valve that is controlled by a solenoid or spring pressure to inject fuel into the engine.

Multiple Choice

Read each item carefully, and then select the best response.

D 1. If liquid gasoline vaporizes in the fuel pump, bubbles of vapor can block the flow of fuel and stop the engine causing _____.
   A. engine knock
   B. dieseling
   C. choking
   D. vapor lock

B 2. A violent collision of flame fronts in the cylinder, caused by uncontrolled combustion is called _____.
   A. dieseling
   B. detonation
   C. misfire
   D. throttling

D 3. The stoichiometric air/fuel ratio is represented by the Greek letter _____.
   A. alpha
   B. delta
   C. omega
   D. lambda

D 4. All of the following are key components of internal combustion, *except*: _____.
   A. air
   B. fuel
   C. heat
   D. pressure/vacuum differential

D 5. Atmospheric pressure at sea level is calculated as _____.
   A. 14.7 psi
   B. 1 lambda
   C. 101 kilopascals
   D. either A or C

C 6. In a carburetor power circuit the size of the main _____ is selected to provide the best mixture for economy under cruising conditions.
   A. venturi
   B. metering jet
   C. barrel
   D. float bowl

7. The _____ should operate as briefly as possible; overuse produces rich mixtures that cause exhaust pollution and increases fuel consumption.
   A. choke
   B. accelerator pump
   C. metering jet
   D. fuel pump

D 8. Extra carburetor _____ (as many as four) improve performance, particularly at high speeds, by letting more air and fuel enter the cylinders.
   A. injectors
   B. metering jets
   C. float bowls
   D. barrels

Chapter 48   Gasoline Fuel Systems

__B__ 9. Vapor from the fuel tank is vented through a _____ where fuel vapors are stored until they are burned in the engine.
   A. drier
   B. charcoal canister
   C. desiccant
   D. filter sock

__C__ 10. The _____ is a variable resistor that is attached to a float mechanism in the fuel tank.
   A. sending unit
   B. fuel pump relay
   C. pressure regulator
   D. accelerometer

__A__ 11. A _____ is incorporated into the end of the fuel pickup tube and is the first line of defense against fuel contamination.
   A. fuel filter
   B. fuel rail
   C. filter sock
   D. fuel separator

__A__ 12. The _____ system uses one or two fuel injectors located centrally on the intake manifold, right above the throttle plates.
   A. throttle body injection
   B. fuel rail injection
   C. diesel fuel direct injection
   D. gasoline direct injection

__B__ 13. The _____ system uses a fuel injector for each cylinder located in the intake manifold near each intake valve that sprays fuel toward the valve.
   A. single-point injection
   B. multipoint fuel injection
   C. gasoline direct injection
   D. throttle body injection

__C__ 14. A vortex sensor is a(n) _____ type sensor.
   A. optical
   B. ultrasonic
   C. pressure
   D. all of the above

__A__ 15. A _____ is a mechanically variable resistor that, in electronic fuel injection applications, is normally a film type.
   A. potentiometer
   B. thermistor
   C. position sensor
   D. knock sensor

__B__ 16. The _____ sends constant data to the computer to let it know which cylinder is on its power stroke.
   A. throttle position sensor
   B. camshaft position sensor
   C. crankshaft position sensor
   D. either A or C

__C__ 17. If the air temperature sensor is installed in the airflow sensor, it is positioned in the airstream and is called a(n) _____.
   A. oxygen sensor
   B. mass airflow sensor
   C. intake air temperature sensor
   D. ambient air temperature sensor

566 Fundamentals of Automotive Technology Student Workbook

__A__ 18. A(n) _____ is used to monitor the noise that is created by an unwanted spike in pressure caused by preignition or detonation.
  A. knock sensor
  B. preignition sensor
  C. BARO sensor
  D. absolute pressure sensor

__C__ 19. Engine _____ is a condition that can cause catalytic converters to overheat in a very short time.
  A. detonation
  B. misfire
  C. knock
  D. vapor lock

__D__ 20. _____ signals are what the computer sends to motors and actuators to do work.
  A. High-frequency
  B. Reference
  C. Analog
  D. Output

True/False

If you believe the statement to be more true than false, write the letter "T" in the space provided. If you believe the statement to be more false than true, write the letter "F".

__T__ 1. The more effectively liquid gasoline is changed into vapor, the more efficiently it burns in the engine.

__T__ 2. There are two different methods used to measure the octane rating of a fuel—the Research Octane Number and the Motor Octane Number.

__F__ 3. A lean air/fuel mixture has less air in proportion to the amount of fuel.

__T__ 4. Dieseling may cause an engine to run backward for a brief time when it comes to a stop.

__T__ 5. A vacuum gauge can be calibrated in inches of mercury in a scale reading from 0" to 30", or if using millimeters of mercury, the scale reads from 0 to 760 mm.

__T__ 6. The off idle circuit enables the engine to keep running when there is not enough air speed through the venturi to create a vacuum.

__T__ 7. The fuel pump on a carbureted system can be either electric or mechanical.

__F__ 8. Modern vehicles are required by the Environmental Protection Agency to have a vented gas cap.

__T__ 9. The fuel filter typically consists of a pleated paper filter housed in a sealed container, and its primary function is to prevent contaminants from reaching the injectors.

__T__ 10. Some vehicles use a pressure regulator on the fuel rail, some are in the fuel tank, and others control fuel pressure by controlling the speed of the fuel pump.

__T__ 11. Gasoline direct injection engines can run fuel mixtures as lean as 65 to 1, much leaner than the stoichiometric ratio.

__F__ 12. Sequential injection means that the injectors operate twice per cycle, once each crankshaft revolution, each time delivering half the fuel for the cycle.

__T__ 13. In the event of the vehicle suffering a collision or rollover accident, the vehicle will shut off the electric fuel pump in an effort to avoid possible fire and/or leakage.

__T__ 14. During engine operation, there are three main pollutants created: nitrogen oxides, carbon monoxide, and hydrocarbons.

__T__ 15. Closed loop can mean the control unit receives feedback from the oxygen sensor and acts on it to alter the injection setting.

__F__ 16. A digital signal is smooth and gradually changing in strength, whereas an analog signal is a direct on/off with no in-between transition.

_T_ 17. The crankshaft position sensor may be mounted externally on the crankcase housing or it may be inside the housing of the ignition distributor.

_T_ 18. A manifold absolute pressure sensor can be used to determine if the exhaust gas recirculation valve is opened or not.

_T_ 19. In vehicles equipped with the circuitry to do so, the clutch electrical circuit signals the PCM that the air conditioning has been turned on and that there will therefore be greater load on the engine.

_F_ 20. In a system that uses inertia sensors, the fuel pump will only operate when the safety circuit containing the inertia sensor is complete.

Fill in the Blank
Read each item carefully, and then complete the statement by filling in the missing word(s).
1. The __Fuel__ __supply__ system draws in gasoline from the gas tank (fuel cell) and delivers it under pressure to a fuel metering device.
2. The less easily the fuel ignites, the higher the __oxtain__ __ratio__.
3. The term __stroimetric__ __ratio__ describes the chemically correct air/fuel ratio necessary to achieve complete combustion of the fuel and air.
4. The more effective, efficient, and reliable electronic __?__ _____ system can be either a circulation system or a returnless system.
5. When an extra squirt of fuel is needed for a burst of speed, the __excel eroter__ circuit comes into play.
6. A(n) __Computer__-__control__ carburetor normally uses an electronically controlled solenoid valve called a mixture control solenoid to respond to the PCM commands.
7. The fuel __fill__ __nock__ can incorporate the use of a blowback ball valve to prevent fuel from leaking from the vehicle during fill-ups and to deter gas theft.
8. The primary job of the __sending unit__ is to send constant electrical signals to the gas gauge located in the driver information center or to the BCM, which then controls the gas gauge.
9. A(n) __sensor injector unit__ is a special manifold designed to provide a reservoir of pressurized fuel for the fuel injectors.
10. The modern __fuel__ __? injector__ is simply a spring-loaded, electric-solenoid spray nozzle.
11. There are two types of __Direct__ fuel injection systems. One uses a pressure regulator in the fuel tank, and the other controls the speed of the fuel pump to modify pressure.
12. Modern vehicles use various __sensors__ to monitor key indicators of an accident requiring fuel shutoff.
13. The power train control module monitors the air/fuel ratio by using a(n) __oxygen__ sensor, also known as a lambda sensor.
14. A resistor that changes its resistance with changes in temperature is called a(n) __thermistor__.
15. __Inductive__-type sensors sense the movement of the ring gear teeth on the flywheel, or a toothed disc on the crank pulley.
16. The __throttle__ __position__ sensor is a potentiometer-type sensor, located on the throttle body and operated by rotation of the throttle shaft that monitors throttle position over its full range.
17. The __coolant__ __temperatur__ sensor is a hollow threaded pin with a resistor sealed inside it that is immersed in coolant in the cylinder head, block, or intake manifold.
18. A(n) __Manifold__ __absolute__ __pressure__ sensor measures pressure changes in engine speed and load and converts them into an electrical signal.
19. A(n) __barometric__ __pressure__ sensor is used to help calibrate the fuel injection system.
20. Commonly referred to as drive-by-wire or throttle-by-wire, a(n) __electronicly controlled__ throttle is the replacement device for the long-standing standard of a throttle cable.

Labeling

Label the following diagrams with the correct terms.

1. Fractional distillation tower:

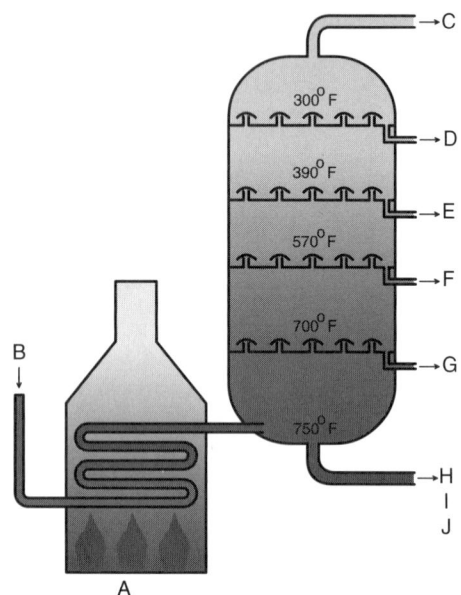

A. _____
B. _____
C. _____
D. _____
E. _____
F. _____
G. _____
H. _____
I. _____
J. _____

2. Types of carburetors:

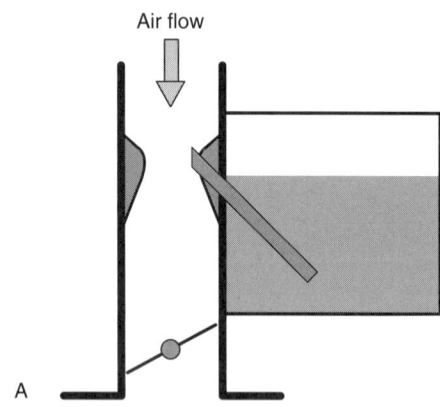

A. _____

Chapter 48   Gasoline Fuel Systems

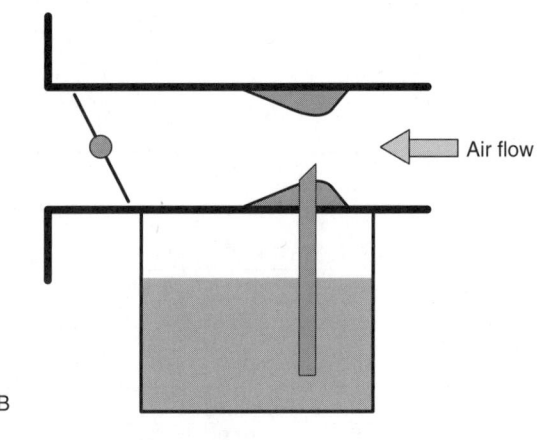

B

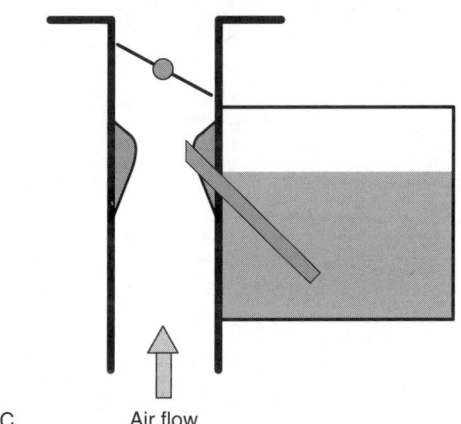

C   Air flow

B. _____

C. _____

3. Float bowl, float, and needle and seat:

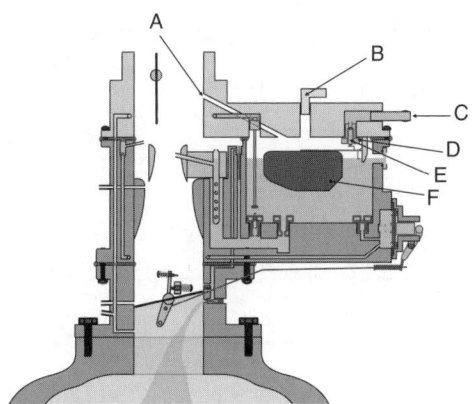

A. _____
B. _____
C. _____
D. _____
E. _____
F. _____

570 Fundamentals of Automotive Technology Student Workbook

4. Main metering jet and discharge nozzle:

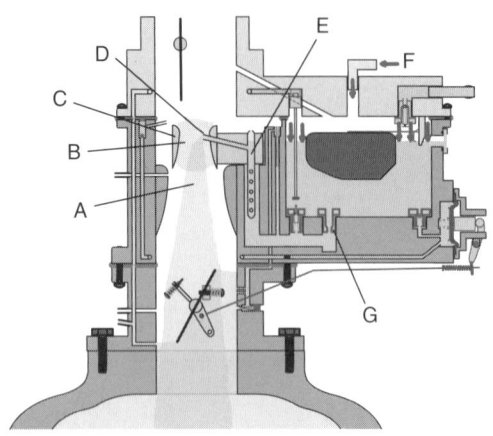

A. _____
B. _____
C. _____
D. _____
E. _____
F. _____
G. _____

5. The choke:

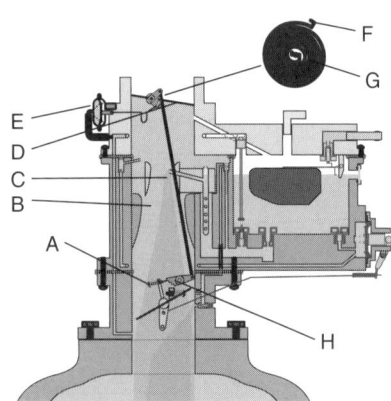

A. _____
B. _____
C. _____
D. _____
E. _____
F. _____
G. _____
H. _____

6. Mechanical fuel pump:

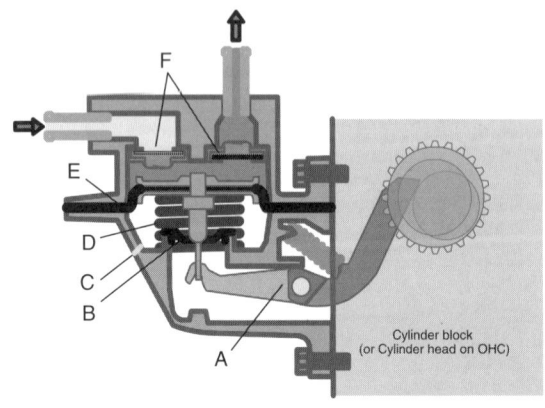

A. _____
B. _____
C. _____
D. _____
E. _____
F. _____

7. Diaphragm-type electric fuel pump:

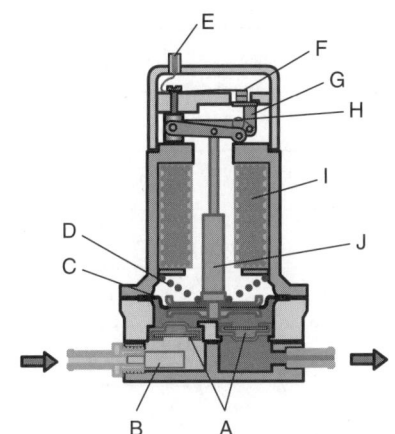

A. _____
B. _____
C. _____
D. _____
E. _____
F. _____
G. _____
H. _____
I. _____
J. _____

8. Fuel tank:

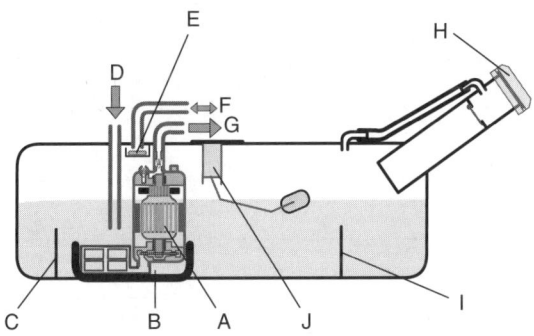

A. _____
B. _____
C. _____
D. _____
E. _____
F. _____
G. _____
H. _____
I. _____
J. _____

9. Vortex generator and sensor:

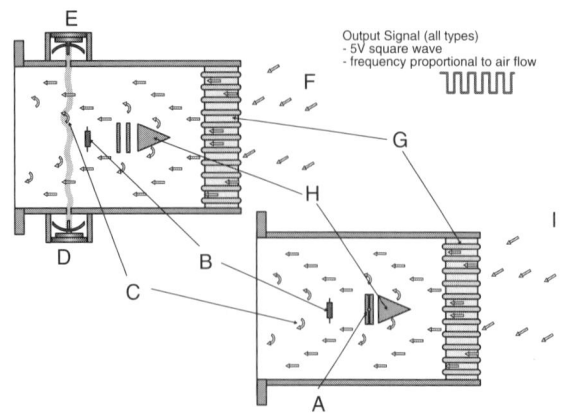

A. _____
B. _____
C. _____
D. _____
E. _____
F. _____
G. _____
H. _____
I. _____

10. Oxygen sensor:

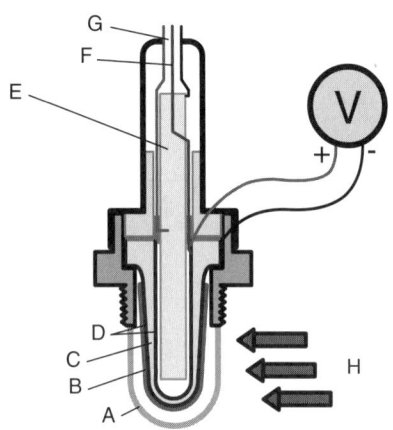

A. _____
B. _____
C. _____
D. _____
E. _____
F. _____
G. _____
H. _____

11. PCM sensor:

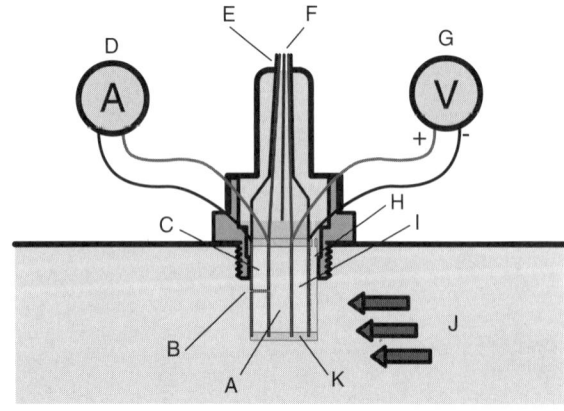

A. _____
B. _____
C. _____
D. _____
E. _____
F. _____
G. _____
H. _____
I. _____
J. _____
K. _____

Skill Drills

Place the skill drill steps in the correct order.

1. Replacing a Fuel Filter:

_____ **A.** Reinstall the filter, with the flow indicator arrow pointing toward the engine, and fully engage the lines, making sure they are secure.

_____ **B.** Using the correct tool, release the quick disconnect connectors from the outlet end of the filter, catching any leaking fuel in a fuel-proof container.

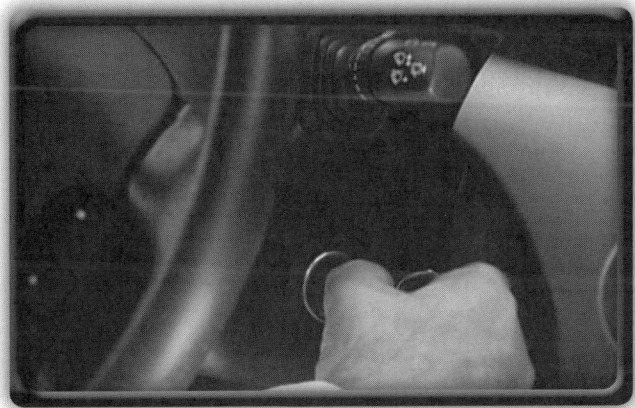

_____ **C.** Turn the key to the on position for 5 seconds, but do not start the engine, and then turn it back to off. Repeat the process two more times, checking the filter connections for leaks. If no leaks are found, start the vehicle, letting it run for 2 to 3 minutes before shutting it off. Recheck the filter connections for leaks.

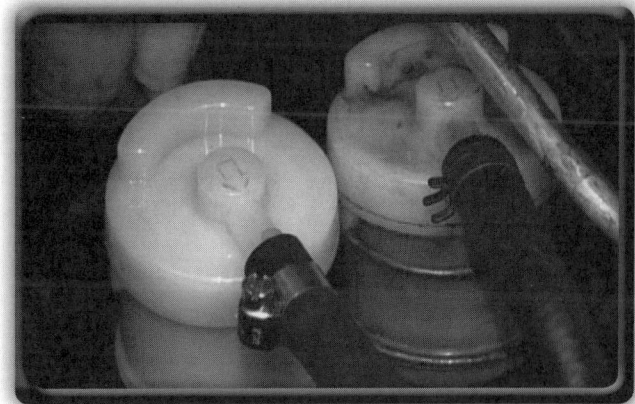

_____ **D.** Reinstall the filter, making sure you have the filter facing in the right direction, with the flow indicator arrow pointing toward the engine. Tighten the fittings on both ends using the double wrench method.

_____ **E.** Obtain the correct replacement filter and components. Loosen the bracket holding the filter in place, if equipped. Follow the steps below according to the type of filter you are replacing.

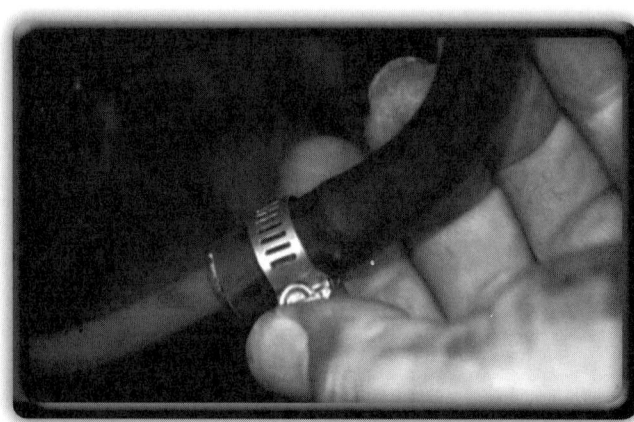

_____ **F.** Disconnect the fuel line on the inlet end of the fuel filter, and remove it from the lines.

_____ **G.** Refer to the service information to identify the location and type of fuel filter and the correct procedure for removing and replacing it. If the engine is equipped with an electric fuel pump, release the pressure according to the service information.

_____ **H.** Wipe any residual fuel off with a clean shop rag, and write the date and mileage on the filter. Remember to replace the fuel pump fuse, if removed.

_____ **I.** Disconnect the fuel line on the engine side of the filter using the double wrench method. If necessary, drain any excess fuel into a fuel-proof container.

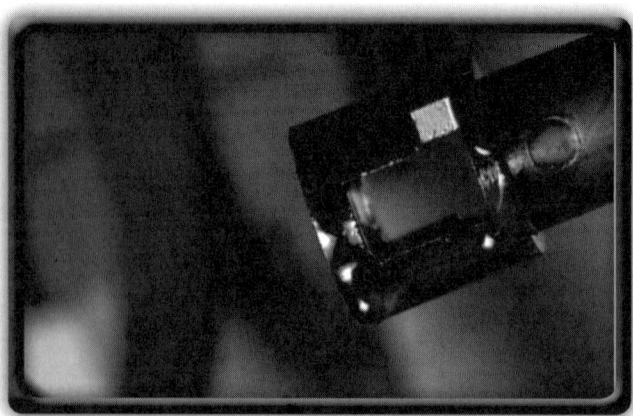

_____ **J.** Release the quick disconnect connectors from the inlet end of the filter, and remove the filter from the lines.

2. Inspecting and Testing Fuel Pumps:

_____ **A.** Start the engine, measure the pressure, and compare to specifications. With the engine running and the end of the fuel line from the fuel pressure gauge in a 1-quart plastic bottle, open the valve on the gauge and stop the fuel flow after 15 seconds.

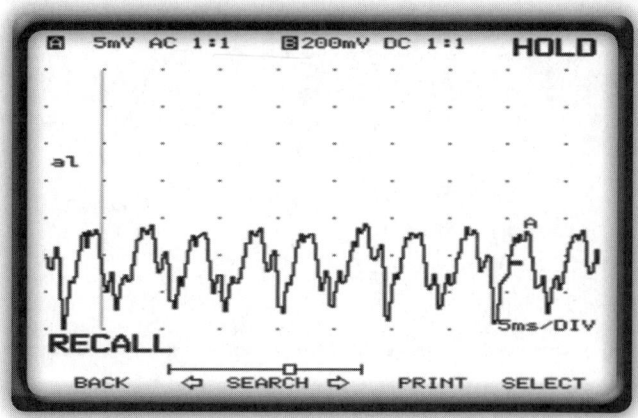

_____ **B.** Start the engine, look at the lab scope pattern, and compare the pattern to the service information.

_____ **C.** Turn the key to the run position and measure the pressure. If none, test the fuel pump's electrical circuit.

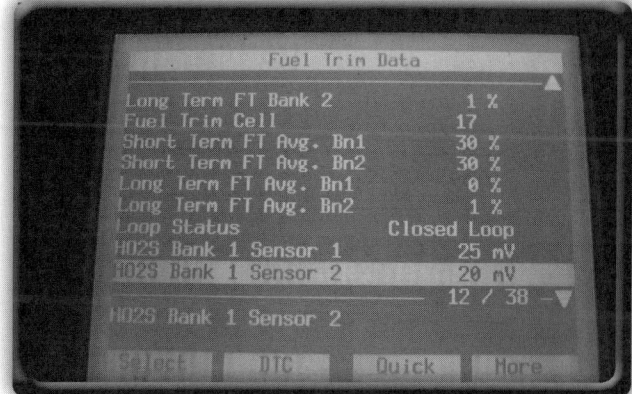

_____ **D.** Record rpm, VSS, front oxygen sensors, short-term fuel trim, and long-term fuel trim under heavy load and at higher rpm. After returning to the shop, look at your recording; if you see a lean oxygen signal with a large increase in fuel trim, the vehicle has a fuel delivery problem.

_____ E. If the pressure test or volume test was less than specified, you can use a lab scope and low-amps clamp to graph the current flow going to the fuel pump. Remove the fuel pump fuse, and connect a jumper lead between the terminals in the fuse box. Place the amps clamp around the jumper lead.

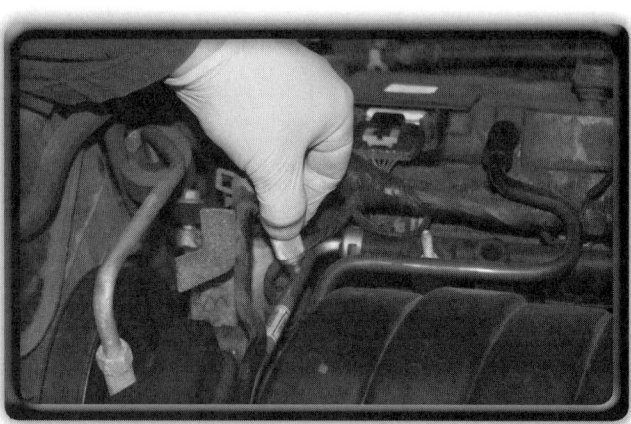

_____ F. After researching the procedure for testing the fuel pump in the service information, install a fuel pump gauge on the fuel rail test port.

_____ G. Measure the amount of fuel delivered in that time, and compare to specifications.

_____ H. Start the engine, measure the pressure, and compare to specifications. If fuel pressure is low, you may want to momentarily pinch off the return line, if equipped, which deadheads the pump. If the pressure was low before pinching off the return line and rises after pinching it off, the pressure regulator is faulty. If the pressure does not rise, either the fuel filter is restricted or the pump is likely to be faulty. **(There is no image associated with this step.)**

3. Checking Fuel for Contaminants and Quality:

_____ A. Allow the mixture to settle. Observe the level of the water in the bottom of the test tube. Anything higher than the initial 10 mL is the amount of alcohol in the fuel. List your observations and determine any necessary actions.

_____ B. Add 90 mL of gasoline, bringing the total volume to 100 mL.

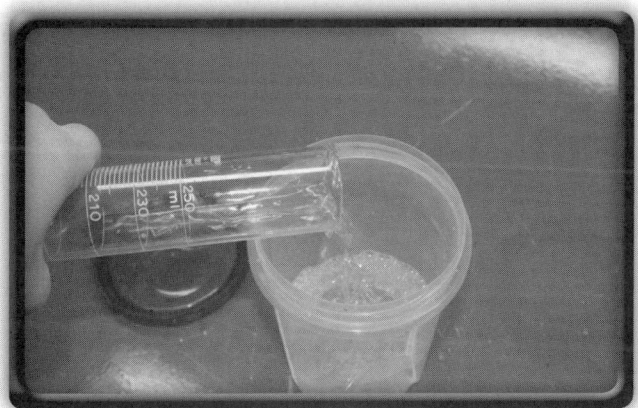

_____ C. Carefully pour off the fuel in the test tube back into the fuel container. Make sure no water leaves the test tube. Properly dispose of the remaining water-fuel mixture.

_____ D. Pour 10 mL of water into the 100 mL graduated test tube.

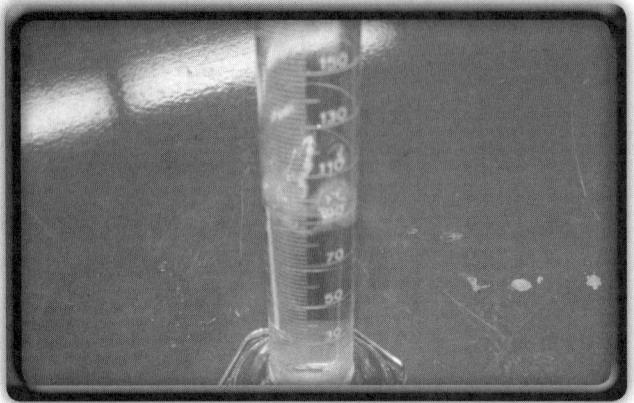

_____ **E.** Cap the test tube tightly. Slowly agitate the fuel-water mixture for 30 seconds. If there is any alcohol in the fuel, this motion will allow the water to be absorbed by the alcohol.

_____ **F.** Collect a quantity of fuel from the vehicle's fuel rail in a clear plastic fuel container. Let it settle and check for contaminants, cloudiness, or improper odor.

4. Inspecting and Testing Fuel Injectors:

_____ **A.** Activate the injector pulsing tool for the appropriate amount of time. Watch the pressure gauge, and record the pressure after the injector has been pulsed.

_____ **B.** Pressurize the fuel rail by turning on the ignition switch for a few seconds. Then turn the ignition switch off. Record the fuel pressure.

_____ **C.** Repeat this test on each fuel injector, pressurizing the fuel rail each time before activating the injector pulsing tool. Record the pressure readings before and after each test. Compare all readings to specifications and to each other. Determine any necessary actions.

_____ **D.** Install the fuel pressure gauge on the fuel rail. Connect the injector pulsing tool to one fuel injector, according to the tool maker's instructions, and set it for the appropriate number and time of pulses.

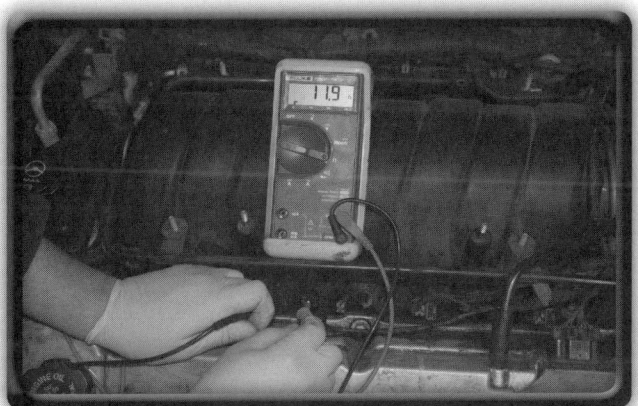

_____ **E.** Measure the resistance of the suspect fuel injector, and compare to specifications. If the resistance is not in specifications, replace the injector.

Crossword Puzzle

Use the clues to complete the puzzle.

Across

9. The upper end of the fuel filler tube leading down to the fuel tank, which accepts the fuel hose nozzle at the gas station pump.

10. Abbreviated as BARO, this sensor measures atmospheric pressure.

11. Tubing that connects several injectors to the main fuel line.

12. The condition in which the remaining fuel charge fires or burns too rapidly after the initial combustion of the air/fuel mixtures. It is audible through the combustion chamber walls as a knocking noise.

13. A valve that is controlled by a solenoid or spring pressure to inject fuel into the engine.

Down

1. A calibrated orifice in a carburetor for fuel to flow through. Often it is replaceable for performance or economy desires.

2. Also known as CTS, this is a thermistor which is usually screwed into a cylinder head water jacket.

3. Abbreviated as CMP, this detection device signals to the PCM the rotational position of the camshaft.

4. An engine sensor that detects preignition, detonation, and knocking.

5. A device used for transmitting a signal to control a fuel gauge.
6. The first line of defense in the fuel supply system. This typically consists of a fine mesh, which prevents most small particles from being drawn in to the fuel pump and sent through the rest of the fuel system.
7. A volatile, flammable liquid mixture of hydrocarbons, obtained from crude oil and used as fuel for internal combustion engines.
8. A device used to trap the fuel vapors.

ASE-Type Questions

Read each item carefully, and then select the best response.

_____ 1. Tech A says that most electric fuel pumps are now mounted inside of the fuel tank. Tech B says that fuel flows through the center of the electric pump and is used to cool the pump. Who is correct?
   A. Tech A
   B. Tech B
   C. Both A and B
   D. Neither A nor B

_____ 2. Tech A says that the stoichiometric ratio is 14.7 gallons of air to 1 gallon of fuel. Tech B says it is 14.7 lb of air to 1 lb of fuel. Who is correct?
   A. Tech A
   B. Tech B
   C. Both A and B
   D. Neither A nor B

_____ 3. Tech A says that some throttle plates are not mechanically connected to the gas pedal, but are operated by an electric motor. Tech B says that too much alcohol in gasoline can cause the engine to not run properly. Who is correct?
   A. Tech A
   B. Tech B
   C. Both A and B
   D. Neither A nor B

_____ 4. Tech A says that high octane fuel is harder to ignite than lower octane. Tech B says that using high octane fuel in an engine designed for lower octane will produce better fuel economy and power. Who is correct?
   A. Tech A
   B. Tech B
   C. Both A and B
   D. Neither A nor B

_____ 5. Tech A says that dieseling in a car today indicates a possible leaking injector. Tech B says that a tripped inertia switch means that there is a possible short in the fuel pump circuit. Who is correct?
   A. Tech A
   B. Tech B
   C. Both A and B
   D. Neither A nor B

_____ 6. Tech A says that a lambda greater than 1 means the engine is running rich. Tech B says that the amount of oxygen in the exhaust indicates how rich or lean the mixture is. Who is correct?
   A. Tech A
   B. Tech B
   C. Both A and B
   D. Neither A nor B

_____ 7. Tech A says that gauge pressure indicates pressure above atmospheric pressure. Tech B says that atmospheric pressure increases as elevation increases. Who is correct?
   A. Tech A
   B. Tech B
   C. Both A and B
   D. Neither A nor B

_____ 8. Tech A says that fuel pressure is regulated by a fuel pressure regulator on some vehicles. Tech B says that fuel pressure is regulated by controlling the speed of the fuel pump on some vehicles. Who is correct?
   A. Tech A
   B. Tech B
   C. Both A and B
   D. Neither A nor B

_____ 9. Tech A says that measuring the alcohol content in gasoline involves using water. Tech B says that as long as the fuel pressure is correct, you don't have to worry about fuel pump volume. Who is correct?
   A. Tech A
   B. Tech B
   C. Both A and B
   D. Neither A nor B

_____ 10. Tech A says that low battery voltage will require a longer injector pulse width time. Tech B says that engine temperature will affect the injector pulse width time. Who is correct?
   A. Tech A
   B. Tech B
   C. Both A and B
   D. Neither A nor B

On-Board Diagnostics

CHAPTER 49

Chapter Review

The following activities have been designed to help you refresh your knowledge of this chapter. Your instructor may require you to complete some or all of these activities as a regular part of your training program. You are encouraged to complete any activity that your instructor does not assign as a way to enhance your learning.

Matching

Match the following terms with the correct description or example.

- A. Aftermarket
- B. Bidirectional scanners
- C. Body control module (BCM)
- D. Clean Air Act
- E. Continuous monitoring
- F. Controller area network (CAN)
- G. Data link connector (DLC)
- H. Diagnostic trouble code (DTC)
- I. Drive cycle
- J. Electronic brake control module (EBCM)
- K. Emissions
- L. Emission analyzer
- M. Freeze-frame
- N. History code
- O. Malfunction indicator lamp (MIL)
- P. Module
- Q. Monitor
- R. Power train control module (PCM)
- S. Scan tool
- T. Transmission control module (TCM)

_____ 1. Tailpipe and volatile organic compound pollutants emitted by the automobile.

_____ 2. An OBDII test run to ensure that a specific component or system is working properly.

_____ 3. A series of prescribed automobile operating conditions during which emissions testing is performed.

_____ 4. A plug-in electronic device for extracting and interpreting fault codes (DTCs) in the automobile, plus much more.

_____ 5. A localized (on-board) vehicle network that enables computers and components to send and receive signals across a shielded twisted pair of wires.

_____ 6. An electronic computer that controls transmission function; it may include adaptive learning capabilities for driver preferences.

_____ 7. A computer-driven on-board report regarding component or system faults following the running of monitors.

_____ 8. An electronic computer or circuit board that controls specific functions.

_____ 9. The module that controls and monitors the anti-lock braking system.

_____ 10. Federal legislation enacted in the United States to reduce pollution, in part from automobiles.

_____ 11. A feature of OBDII that records events before, during, and after a fault occurs.

_____ 12. Scanners used to monitor engine compression, vacuum, internal engine anomalies, and so forth by causing various components and systems to operate for test purposes.

_____ 13. The module that controls and monitors the engine ignition, fuel, and emission system functions.

_____ 14. A term that describes OBDII monitors that run continuously throughout the drive cycle.

_____ 15. That segment of the trade that supplies parts, services, and repair for vehicles outside of the original equipment manufacturer (OEM) or the dealer network.

_____ 16. A fault code that has occurred but is not current and is saved in the PCM's memory for 40 drive cycles.

_____ 17. A device that enables a scan tool to access data stored in the vehicle's various computers.

_____ 18. A service bay or lab device used for detecting/measuring vehicle emissions.

_____ 19. An electronic unit that monitors and regulates electronic devices in the vehicle.

_____ 20. An indicator located in the instrument cluster that illuminates when the power train control module (PCM) detects a fault in one of the vehicle systems. Formerly called a check engine or service engine soon light.

Multiple Choice

Read each item carefully, and then select the best response.

_____ 1. When a vehicle's _____ is illuminated it means the vehicle is not complying with clean air regulations.
   A. malfunction indicator lamp
   B. VOC indicator light
   C. check engine light
   D. emission indicator lamp

_____ 2. Which of the following pollutants are monitored and controlled by on-board systems using state-of-the-art electronics?
   A. Hydrocarbons
   B. Carbon dioxide
   C. Oxides of nitrogen
   D. All of the above

_____ 3. Oxidizing catalytic converters were introduced in _____.
   A. 1968
   B. 1971
   C. 1975
   D. 1981

_____ 4. The second generation of on-board diagnostic (OBDII) systems operates under standards set by _____.
   A. Association des Constructeurs Européens d'Automobiles
   B. Society of Automotive Engineers
   C. Japanese Automotive Standards Organization
   D. all of the above

_____ 5. The complete listing of OBDI and OBDII standardized nomenclature for parts and systems used by engineers and technicians is known as _____.
   A. SAE J2012
   B. SAE J1930
   C. SAE J1962
   D. SAE J1968

_____ 6. To simplify wiring, _____ have become common place in today's vehicles.
   A. controller area networks
   B. data link connectors
   C. OBDI systems
   D. closed-loop systems

_____ 7. All diagnostic trouble codes have been standardized by the Society of Automotive Engineers and are listed in _____.
   A. SAE J1972
   B. SAE J1930
   C. SAE J2012
   D. SAE J1962

_____ 8. If the first character of an OBDII diagnostic code is the letter *U* then the fault is located in the _____.
   A. body
   B. communication system
   C. chassis controller
   D. emission system

_____ 9. Snapshots that are automatically recorded in the vehicle's power train control module when a vehicle fault occurs are referred to as _____ data.
   A. freeze-frame
   B. continuously monitored
   C. keep alive
   D. OBDI

_____ 10. A _____ includes the following events: A vehicle starts, warms up, is accelerated, cruises, slows down, accelerates once more, decelerates, stops, and cools down.
   A. warm-up cycle
   B. scan cycle
   C. drive cycle
   D. fault cycle

True/False

If you believe the statement to be more true than false, write the letter "T" in the space provided. If you believe the statement to be more false than true, write the letter "F".

_____ 1. Carbon monoxide and sulfur dioxide react together to create ground-level ozone, which is considered a health hazard.

_____ 2. Evaporative emission control systems were first introduced in 1971.

_____ 3. OBDI monitored mainly for parts and wiring malfunctions. OBDII is an enhanced diagnostic system that identifies faults in anything that may affect the vehicle's emission system.

_____ 4. Emission-related diagnostic trouble codes are the same across all vehicle makes and models, as are the SAE-recommended names used to describe components and systems.

_____ 5. If the catalytic converter is at risk, such as from overfueling or a continuous misfire, the malfunction indicator lamp will flash.

_____ 6. Diagnostic trouble codes are saved even if the vehicle's battery is disconnected.

_____ 7. OBDI systems monitor all emission-related components and circuits for opens, shorts, and abnormal operation.

_____ 8. There are hundreds of possible diagnostic trouble codes, and the list grows constantly as on-board systems become more sophisticated and unique.

_____ 9. An evaporative emission monitor will not run if the fuel tank is nearly full or nearly empty.

_____ 10. The bidirectional scan tool is like a "magic bullet." The fault code points directly to the problem.

Fill in the Blank

Read each item carefully, and then complete the statement by filling in the missing word(s).

1. _____ _____ compounds may be fuel or oil vapors emitted from the fuel tank, fuel lines, engine crankcase, or elsewhere.

2. OBDII faults and background data can be accessed and read by original equipment manufacturer (OEM) test equipment or by _____ test equipment made by other companies.

3. The first generation of _____-_____ diagnostic systems operated under manufacturer standards, starting with California vehicles and becoming nationwide in 1981.

4. The _____ _____ is a 16-pin connector with a common (SAE J1962) size and shape.

5. Diagnostic trouble _____ are "set" in one or more of the vehicle's on-board computers once a fault is detected.

6. Possible faults, including engine misfiring and an incorrect air/fuel mixture, are monitored on a continuous basis by the _____ _____ control module.

7. Lower-priority faults are monitored on a _____ basis, meaning they are checked only once during each engine warm-up cycle, or even less often, depending on certain circumstances such as ambient temperatures or fuel level.

8. A diagnostic trouble code will remain logged in the PCM's memory as a(n) _____ _____ for a set period.
9. A(n) _____ _____ is a device able to electronically communicate with and extract data from the vehicle's one or more on-board computers.
10. More sophisticated than simple scan tools, _____ scanners are used to cause various components and systems to operate for test purposes.

Skill Drills

Test your knowledge of skill drills by filling in the correct words in the photo captions.

1. Retrieving and Recording DTCs, OBD Monitor Status, and Freeze-Frame Data:

Step 1: Select the _____ _____ to provide the best coverage for the _____ and _____ of vehicle. Locate the _____, and connect the scan tool.

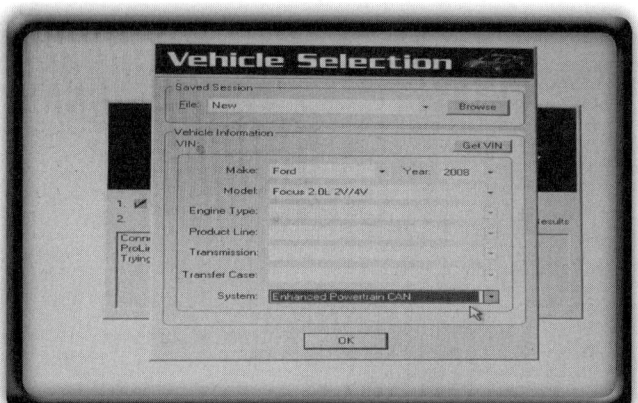

Step 2: Power on the scan tool, and turn the _____ on. Establish scan tool _____ with the _____.

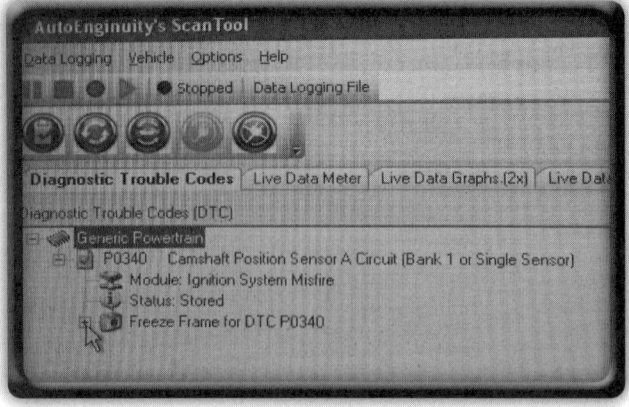

Step 3: _____ and _____ the DTCs.

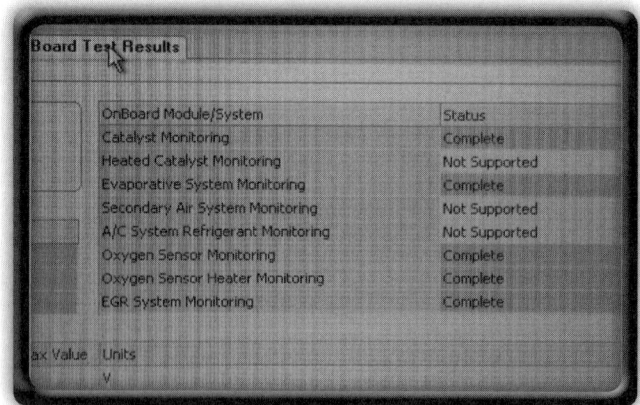

Step 4: Retrieve and record _____ _____ status.

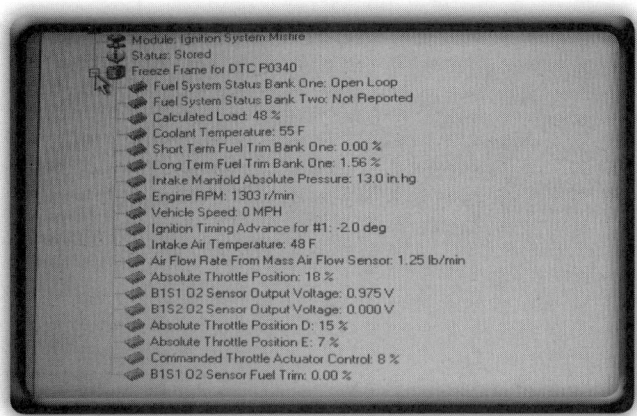

Step 5: Retrieve and record _____-_____ data applicable to DTCs and _____.

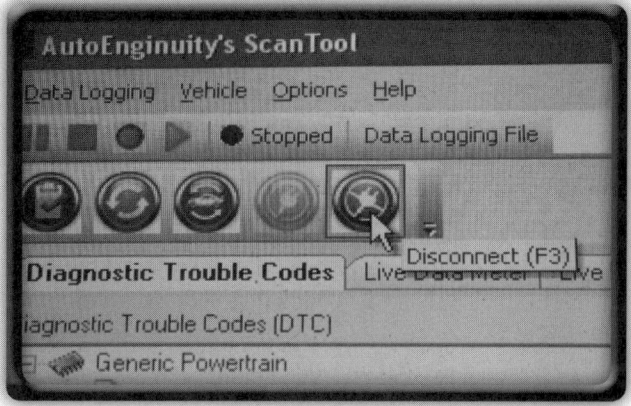

Step 6: Power _____ the scan tool, turn the _____ off, and _____ the scan tool.

2. Testing Engine Control System Sensors, the PCM, Actuators, and Circuits:

Step 1: Determine the _____ to be tested and the likely _____ and frequency of the _____ to be measured. Set up the _____ multimeter or _____, setting the voltage level and time base to commence measurements.

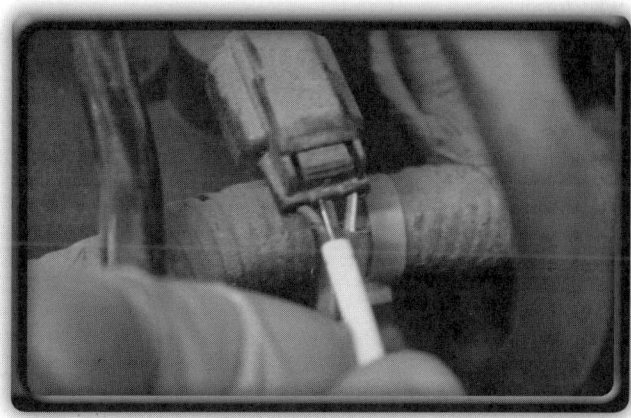

Step 2: Connect the _____ to the _____ to be measured.

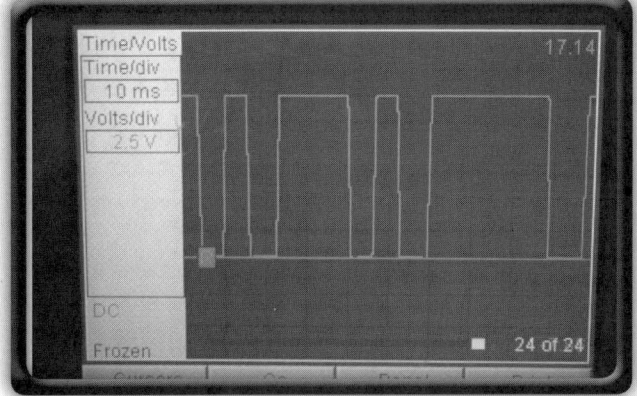

Step 3: _____ the waveforms from the circuit under test. _____ the waveform, _____ it to the manufacturer's specifications or _____ _____ waveforms. Determine the necessary actions.

Crossword Puzzle

Use the clues to complete the puzzle.

Across

3. A microscopic unburned fuel particle that contributes to photochemical smog and helps form ground-level ozone.

5. The first generation of on-board diagnostic systems that originated for California vehicles.

7. A pollutant resulting from sulfur in motor fuel and contributing to acid rain.

8. A vehicle emission that contributes to ground-level ozone. Oxides of nitrogen are produced when nitrogen and oxygen react during combustion, given sufficient temperatures and pressures.

9. Another name for a fault code.

Down

1. Unseen microscopic particles of carbon consisting of soot (especially prominent with diesel exhaust). Also known as PM, this clogs the lungs and is carcinogenic.

2. A monitor that runs only once per drive cycle.

4. A vehicle emission that is considered a primary greenhouse gas, though not toxic, and not yet regulated.

5. The second generation of on-board diagnostic systems, which have been in effect for all US vehicles since 1996.

6. Another word for diagnostic trouble, or a computer-driven on-board report regarding component or system faults following the running of monitors.

ASE-Type Questions

Read each item carefully, and then select the best response.

_____ 1. Tech A says that on OBDII vehicles, it is a good idea to clear the codes before diagnosis, and see if they reset. Tech B says that the DTC will tell you what part needs to be changed. Who is correct?
   A. Tech A
   B. Tech B
   C. Both A and B
   D. Neither A nor B

_____ 2. Tech A says that OBDI and OBDII use different DLC connectors. Tech B says that OBDII standardizes the designations for diagnostic trouble codes (DTCs). Who is correct?
   A. Tech A
   B. Tech B
   C. Both A and B
   D. Neither A nor B

_____ 3. Tech A says that monitors are designed to test if emission systems are working properly. Tech B says that monitors are designed to store sensor data about a fault if a DTC is set. Who is correct?
   A. Tech A
   B. Tech B
   C. Both A and B
   D. Neither A nor B

_____ 4. Tech A says that control modules communicate back and forth today using a CAN, which is a bundle of many individual wires, specially designed for fast communication speed. Tech B says that OBDII mandates that a DTC must be set if the emissions are more than 1.5 times the EPA FTP. Who is correct?
   A. Tech A
   B. Tech B
   C. Both A and B
   D. Neither A nor B

_____ 5. Tech A says that OBDII will allow a technician to hook up a generic scan tool to read DTCs and clear DTCs. Tech B says that when the MIL is on, the vehicle should not be driven. Who is correct?
   A. Tech A
   B. Tech B
   C. Both A and B
   D. Neither A nor B

_____ 6. Tech A says that something as simple as a loose gas tank fill cap will turn on the MIL. Tech B says that every digit of a DTC has identifying characteristics. Who is correct?
   A. Tech A
   B. Tech B
   C. Both A and B
   D. Neither A nor B

_____ 7. Tech A says that a "P" in the DTC stands for a powertrain code. Tech B says a "U" stands for an undetermined code. Who is correct?
   A. Tech A
   B. Tech B
   C. Both A and B
   D. Neither A nor B

_____ 8. Tech A says that on OBDII systems, the KAM stores DTCs forever. Tech B says that on OBDII systems, the KAM is erased when the battery is disconnected. Who is correct?
   A. Tech A
   B. Tech B
   C. Both A and B
   D. Neither A nor B

_____ 9. Tech A says that the MIL can turn off if the fault doesn't reappear in a certain number of tests. Tech B says that when an MIL is activated, the PCM stores the DTC until a predetermined number of drive cycles have been performed or cleared by a scan tool. Who is correct?
   A. Tech A
   B. Tech B
   C. Both A and B
   D. Neither A nor B

_____ 10. Tech A says some mechanical engine problems can cause OBDII DTCs to be set. Tech B says that OBDII codes only monitor the emission control system components. Who is correct?
   A. Tech A
   B. Tech B
   C. Both A and B
   D. Neither A nor B

Induction and Exhaust

CHAPTER 50

Tire Tread:
© AbleStock

Chapter Review

The following activities have been designed to help you refresh your knowledge of this chapter. Your instructor may require you to complete some or all of these activities as a regular part of your training program. You are encouraged to complete any activity that your instructor does not assign as a way to enhance your learning.

Matching

Match the following terms with the correct description or example.

- **A.** Blow-off valve
- **B.** Compressor surge
- **C.** Forced induction
- **D.** Helmholtz resonator
- **E.** Mandrel forming
- **F.** Mass airflow (MAF) sensor
- **G.** Plenum chamber

_____ **1.** The pressurization of airflow going into the cylinder through the use of a turbocharger or supercharger.

_____ **2.** The special bending of pipe to ensure the pipe does not collapse. The use of this pipe bender allows for very tight bends without creating kinks or reducing the size of the pipe.

_____ **3.** The PCM input sensor that tells the computer the amount of air coming into the engine. The sensor allows the correct calculated amount of fuel delivery to the engine. It is usually located in the ducting of the air cleaner system.

_____ **4.** The backup of air against the throttle plate as it is closed. The turbocharger is still spinning, pressurizing air, when the throttle plate is closed. Air will stack up, creating a rapid slowing of the turbocharger compressor wheel. This can damage the compressor wheel.

_____ **5.** A valve that allows the release of excessive boost pressure from the turbocharger when the throttle plate is quickly closed.

_____ **6.** A device that uses the principle of noise cancellation through the collision of sound waves. When necessary, it is used in addition to the muffler to cancel additional sounds. This device may also be used on the induction system to muffle noise of airflow through the induction system.

_____ **7.** A large portion of the intake manifold after the throttle plate and before the intake runner tubes. The manifold provides a reservoir of air and helps prevent interference with the flow of air between individual branches.

Multiple Choice

Read each item carefully, and then select the best response.

_____ **1.** The air cleaner element or filter is manufactured using _____.
- **A.** pleated paper
- **B.** oil-impregnated cloth or felt
- **C.** an oil bath configuration
- **D.** any of the above

_____ **2.** A(n) _____ is a container that is sealed and specially shaped to cancel noise created by pressure waves.
- **A.** air cleaner
- **B.** baffle
- **C.** Helmholtz resonator
- **D.** intake manifold

_____ 3. Cylinder heads that have intake and exhaust manifolds on opposite sides of the engine are known as _____ heads.
   A. parallel
   B. cross-flow
   C. in-line
   D. variable-intake

_____ 4. Manifolds that respond to changes in engine load and speed by changing their effective length in two or three stages are called _____ systems.
   A. variable inertia
   B. cross-flow
   C. intake charging
   D. either A or C

_____ 5. The method of warming the incoming air by passing it through a shroud around the exhaust manifold is called a(n) _____ system.
   A. early fuel evaporation
   B. manifold preheating
   C. mass airflow heating
   D. heated induction

_____ 6. The volume of air entering a cylinder during intake in relation to the internal volume of the cylinder when the piston is at bottom dead center is called _____ and is usually expressed as a percentage.
   A. cylinder volume
   B. induction volume
   C. volumetric efficiency
   D. compression efficiency

_____ 7. _____ increases air pressure in the intake manifold above atmospheric pressure.
   A. Back-pressure
   B. Forced induction
   C. Intake air heating
   D. Both A and C

_____ 8. In a turbocharged engine when air pressure in the intake manifold reaches a preset level, the _____ automatically directs the exhaust gases so they bypass the turbine.
   A. wastegate
   B. bypass valve
   C. blow-by valve
   D. aspirator

_____ 9. When manifold pressure rises above normal and works against the spinning exhaust turbine wheel, slowing it considerably, it is called _____.
   A. back-pressure
   B. blow-back
   C. forced induction
   D. compressor surge

_____ 10. When the outgoing pulse from one cylinder is timed to arrive at the header junction at exactly the right time to help draw out the pulse from another cylinder, it is known as a _____ exhaust.
   A. balanced
   B. cross-flow
   C. tuned
   D. mandrel-formed

_____ 11. The _____ is attached to the exhaust manifold and connects to the catalytic converter.
   A. engine pipe
   B. flexible connector
   C. down pipe
   D. either A or C

_____ 12. The _____ helps with the alignment of the exhaust pipes as the engine moves under load.
   A. engine pipe
   B. flexible connector
   C. exhaust manifold
   D. intermediate pipe

_____ 13. A _____ catalytic converter converts oxides of nitrogen back into nitrogen and oxygen, and the hydrocarbons and carbon monoxide to water and carbon dioxide.
   A. two-way
   B. three-way
   C. four-way
   D. all of the above

_____ 14. The _____ connects the catalytic converter to the muffler.
   A. flexible connector
   B. exhaust bracket
   C. intermediate pipe
   D. engine pipe

_____ 15. A(n) _____ is sometimes used between the muffler and the exhaust outlet to reduce any resonance levels that the muffler could not adequately suppress.
   A. baffle chamber
   B. intermediate pipe
   C. resonator
   D. secondary muffler

True/False

If you believe the statement to be more true than false, write the letter "T" in the space provided. If you believe the statement to be more false than true, write the letter "F".

_____ 1. Vaporization starts when the fuel system atomizes the fuel by breaking it up into very small particles by spraying it into the charge of air.

_____ 2. The throttle body controls airflow with a butterfly valve or valves, also called throttle valves.

_____ 3. On most heavy-duty diesel engines and a few gasoline engines, the air cleaner assembly uses an air filter indicator to tell the operator whether the filter needs to be cleaned or replaced.

_____ 4. The two main components of the four-stroke intake system are the air cleaner and the throttle body.

_____ 5. Diesel engines may have more than one air cleaner.

_____ 6. The air intake manifold for an electronic fuel injection multipoint engine normally has short branches of variable length.

_____ 7. Port fuel-injected or gasoline direct-injected intake manifolds do not normally need to be heated, since the manifold does not carry fuel.

_____ 8. Electric heaters can be placed between the intake manifold and the throttle body and used to preheat the air.

_____ 9. In a stock naturally aspirated engine, one without forced induction, volumetric efficiency can almost never be 100%.

_____ 10. In a turbocharged engine a bypass valve may be installed so if the wastegate should fail, it can prevent an abnormal rise in manifold pressure.

_____ 11. On many current vehicles, the exhaust manifold is often replaced with a header.

_____ 12. Leaded fuel must not be used in an engine with a catalytic converter, because lead will coat the catalyst and prevent it from doing its job.

_____ 13. Noise cancellation refers to putting a sound material around a perforated pipe that the exhaust gases flow through.

_____ 14. Adding a variable-flow exhaust to the baffle or chamber system reduces emission noise.

_____ 15. Exhaust gaskets can be found between the engine cylinder head and the exhaust manifold, the engine pipe and the catalytic converter, and possibly the catalytic converter and the muffler.

Fill in the Blank

Read each item carefully, and then complete the statement by filling in the missing word(s).

1. The _____ system ensures that clean, dry air is supplied to the engine.
2. The _____ system provides a path for the burned exhaust gases to safely exit the engine and travel out the rear of the vehicle.
3. The _____ _____ filters the incoming air.
4. A(n) _____ _____ sensor can measure air entering the engine down to tenths of a gram per second.
5. The _____ precleaner system uses angled vanes that give the incoming air a swirling motion.
6. The _____ _____ has several tubular branches and carries air and/or air/fuel mixture from the air cleaner to the cylinder head.
7. On vehicles with emission controls, air cleaners use a(n) _____ valve to control how much hot air enters the air cleaner.
8. A buildup of pressure in the system that interferes with the outward flow of exhaust gases is known as _____-_____.
9. A(n) _____ compresses the air in the intake system to above atmospheric pressure, which increases the density of the air entering the engine.
10. A(n) _____ is a forced induction system that uses wasted kinetic energy from the exhaust gases to increase the intake pressure.
11. The purpose of a(n) _____ is to reduce the intake air temperature up to a few hundred degrees Fahrenheit before it enters the intake manifold.
12. Headers provide a(n) _____ effect to help remove exhaust gases from the cylinders.
13. There may be a flexible connection between the _____ _____ and an intermediate pipe.
14. The exhaust components are supported by _____-mounted exhaust brackets that prevent vibrations from being felt by the driver.
15. The function of a vehicle's _____ is to minimize the sounds coming from the exhaust system.
16. Noise _____ is a system that prevents the sound waves from leaving the exhaust system by canceling them out inside the muffler.
17. The _____-type muffler uses a perforated tube or baffle that is wrapped in fiberglass material, and this will absorb the noise of combustion as gases flow past.
18. A(n) _____ muffler is designed to produce anti-noise without restricting exhaust flow.
19. The _____ takes the exhaust gases away from the vehicle and must not allow any of the exhaust gases to enter the vehicle.
20. Applying a(n) _____ _____ to the cylinder head's combustion chamber and exhaust ports has the effect of creating a thermal barrier from the hot gases to the cylinder head itself.

Labeling

Label the following diagrams with the correct terms.

1. Air induction system:

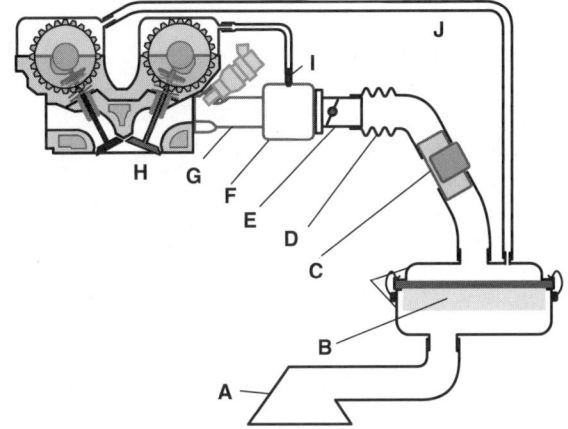

A. _____
B. _____
C. _____
D. _____
E. _____
F. _____
G. _____
H. _____
I. _____
J. _____

2. Using hot engine coolant to heat the manifold:

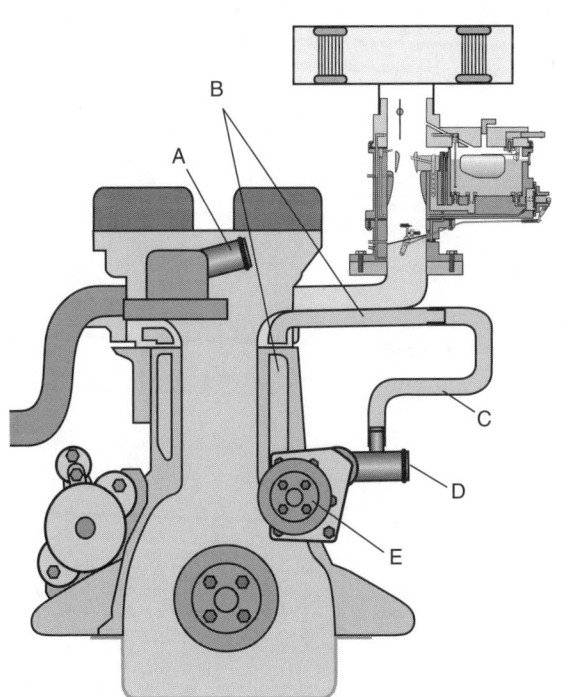

A. _____
B. _____
C. _____
D. _____
E. _____

3. Volumetric efficiency of a normally aspirated engine:

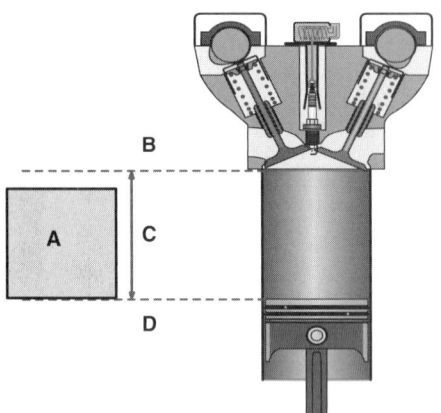

A. _____

B. _____

C. _____

D. _____

4. Volumetric efficiency of a forced-induction engine:

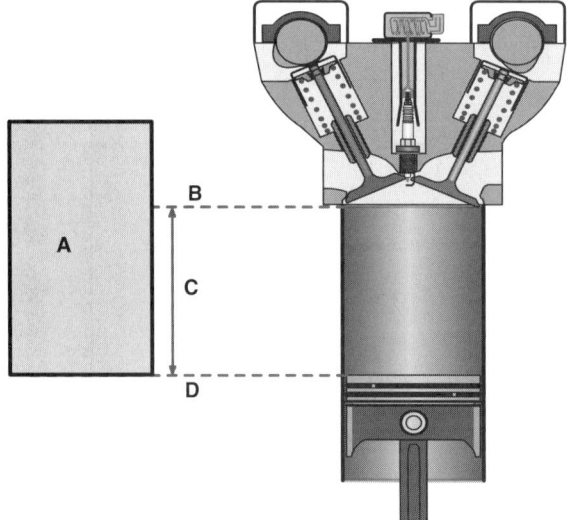

A. _____

B. _____

C. _____

D. _____

5. Identify the following exhaust system components:

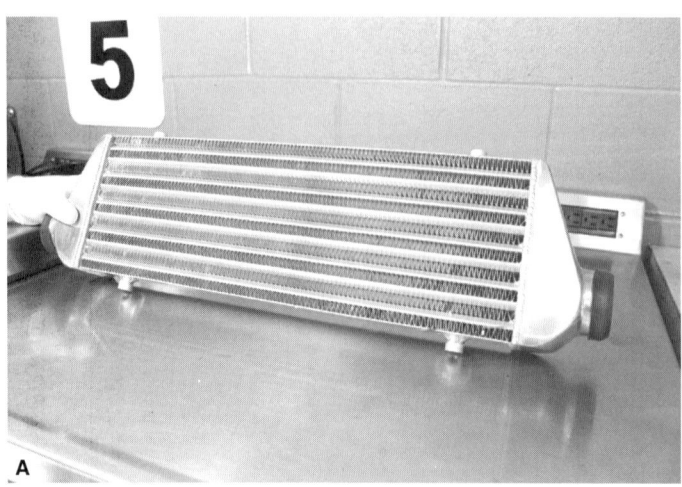

A. _____

Chapter 50 Induction and Exhaust

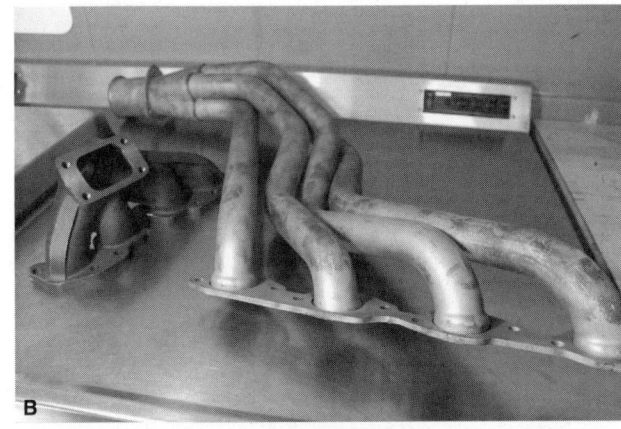

B. _____

C. _____

D. _____

E. _____

F. _____

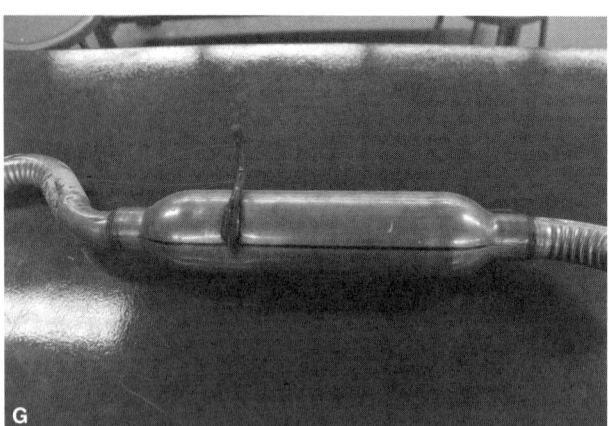

G. _____

Skill Drills

Test your knowledge of skill drills by filling in the correct words in the photo captions.

1. Inspecting the Throttle Body, Air Induction System, and Intake Manifold:

Step 1: Start the _____, and let the _____ stabilize. Using an _____ or _____ testing tool, place the hose near the suspected _____ area.

Step 2: Open the _____ _____ to slowly _____ the acetylene or propane.

Step 3: On vehicles _____ an idle control system, observe the rpm and smoothness of the engine. If the engine speed _____ or smooths out, then a _____ leak is present. Determine any necessary repairs. On vehicles with an _____ _____ system, connect a scan tool and select _____ _____. Observe the oxygen, short-term fuel trims, and _____ _____ width as you are _____ the acetylene or propane around.

Step 4: Shut off the _____, and connect a _____ _____ to the _____ manifold. Start the smoke machine, and _____ smoke into the intake manifold.

Step 5: Using a bright _____, look around the engine compartment for _____ of _____. Any smoke coming from the _____ _____ inlet is normal. Determine any necessary action(s).

2. Inspecting the Exhaust System for Leaks:

Step 1: _____ the vehicle on a _____, and secure it in place. Use a _____ _____ to inspect for leaks or _____ holes in the exhaust system. Check the exhaust _____ for evidence of exhaust leaks.

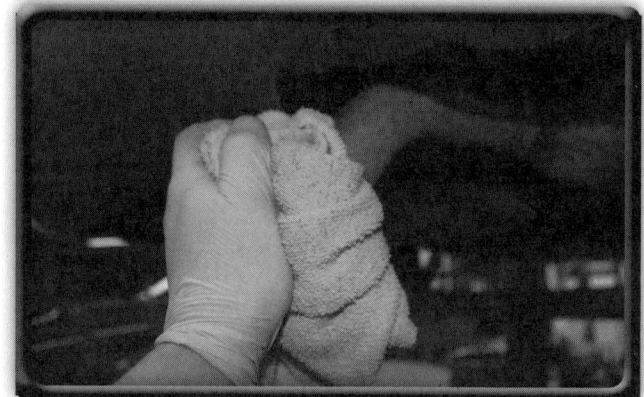

Step 2: Have a helper hold a _____ against the exhaust pipe(s) while the engine is _____ to increase the _____ in the system. _____ and _____ for any leaks.

Step 3: Use large adjustable _____ to test the _____ of the pipes by moderately _____ the pipes. Determine necessary repairs.

3. Testing the Exhaust System for Back-Pressure:

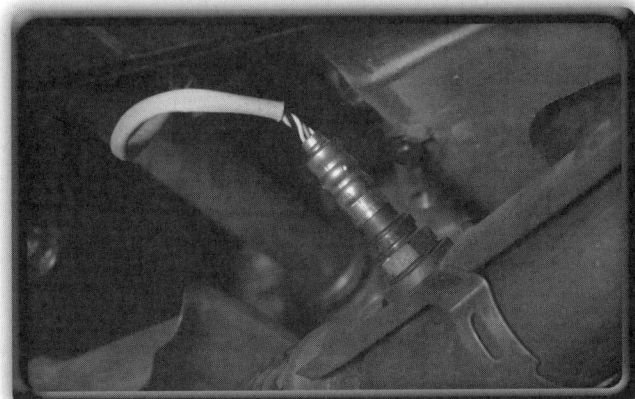

Step 1: Remove the _____ _____. Install an _____ to the oxygen sensor _____, and connect a _____-_____ gauge.

Step 2: Start the engine, and _____ any exhaust _____-_____, which should be no more than 1.5 psi (10.3 kPa) at _____ and no more than 3 psi (20.7 kPa) at 2000 rpm. Increase the engine _____ to 2000 rpm, and observe the _____. Determine any necessary actions.

Step 3: Inspect the oxygen sensor _____, put _____-_____ on the sensor _____, and _____ to the proper specifications.

Crossword Puzzle

Use the clues to complete the puzzle.

Across

2. A specially tuned exhaust manifold typically made of exhaust pipes. These pipes are usually made equal length to ensure equal flow between cylinders.

4. The pressure control valve of the supercharger. When this valve opens, it lets air move around the supercharger compressor section, reducing the boost pressure.

6. A device that pressurizes airflow into the engine. This device works similar to a supercharger, but is driven by the exhaust gases.

Down

1. A pressure regulator device that allows control of the pressure produced by the turbocharger. The device moves to allow exhaust gases to bypass the turbine wheel of the turbocharger and flow down the exhaust pipe.

3. The process of exhaust flow pulling fresh air into the cylinder as it creates a low-pressure area in the cylinder as it moves out of the cylinder.

5. A device that pressurizes the airflow into the engine, working similar to a turbocharger. This device is driven by the crankshaft through a belt or gears and does not require power from the engine.

ASE-Type Questions

Read each item carefully, and then select the best response.

_____ **1.** Tech A says that coolant circulates through some intake manifolds to help warm them up. Tech B says that some intake manifolds use an electric heater grid to warm up the intake air. Who is correct?
  A. Tech A
  B. Tech B
  C. Both A and B
  D. Neither A nor B

_____ **2.** Tech A says that in some gas engines, the intake manifold runner length can be changed. Tech B says that some intake systems use resonators to quiet intake noise. Who is correct?
  A. Tech A
  B. Tech B
  C. Both A and B
  D. Neither A nor B

_____ 3. Tech A says that diesel engine rpm is controlled by a throttle plate. Tech B says that in a diesel engine, air and fuel are mixed in the plenum. Who is correct?
   A. Tech A
   B. Tech B
   C. Both A and B
   D. Neither A nor B

_____ 4. Tech A says that scavenging creates a low pressure in the cylinder by means of the exhaust gases flowing through the exhaust pipe. Tech B says that back-pressure in the exhaust system increases the scavenging effect. Who is correct?
   A. Tech A
   B. Tech B
   C. Both A and B
   D. Neither A nor B

_____ 5. Tech A says that shorter intake manifolds produce higher torque at lower rpm. Tech B says that longer intake manifolds produce higher torque at high engine speeds. Who is correct?
   A. Tech A
   B. Tech B
   C. Both A and B
   D. Neither A nor B

_____ 6. Tech A says that air becomes heated as a supercharger compresses it. Tech B says that a turbocharger pulls the exhaust gases out of the engine, thereby increasing the scavenging effect. Who is correct?
   A. Tech A
   B. Tech B
   C. Both A and B
   D. Neither A nor B

_____ 7. Tech A says that a wastegate is designed to direct waste gases to the intake manifold. Tech B says that a blow-off valve vents excess boost pressure from the turbocharger when the throttle is closed quickly. Who is correct?
   A. Tech A
   B. Tech B
   C. Both A and B
   D. Neither A nor B

_____ 8. Tech A says that most engines with a turbocharger have turbo lag. Tech B says that a variable-geometry turbocharger will overcome most turbo lag. Who is correct?
   A. Tech A
   B. Tech B
   C. Both A and B
   D. Neither A nor B

_____ 9. Tech A says that a catalytic converter operates best at cold temperatures. Tech B says that the catalytic converter is typically located after the muffler. Who is correct?
   A. Tech A
   B. Tech B
   C. Both A and B
   D. Neither A nor B

_____ 10. Tech A says that small vacuum leaks can be found with a vacuum gauge. Tech B says that the catalytic converter is typically located after the muffler. Who is correct?
   A. Tech A
   B. Tech B
   C. Both A and B
   D. Neither A nor B

CHAPTER 51

Emission Control

Tire Tread:
© AbleStock

Chapter Review

The following activities have been designed to help you refresh your knowledge of this chapter. Your instructor may require you to complete some or all of these activities as a regular part of your training program. You are encouraged to complete any activity that your instructor does not assign as a way to enhance your learning.

Matching

Match the following terms with the correct description or example.

- A. Activated charcoal
- B. Blow-by pressure
- C. Carbon monoxide
- D. Compression-ignition engine
- E. Emission
- F. Exhaust gas recirculation (EGR) system
- G. Flame front
- H. Hydrocarbon
- I. Lambda
- J. Particulate matter
- K. PCV valve
- L. Pintle
- M. Pressure relief valve
- N. Pulse air system
- O. Purging
- P. Reed valve
- Q. Stroke
- R. Two-way catalytic converter
- S. Vaporization
- T. Vent solenoid

_____ 1. A flexible valve made from spring steel that flexes to open or close, usually due to pressure changes.

_____ 2. A valve that controls the amount of crankcase ventilation flow that is allowed and varies with changes in manifold pressure.

_____ 3. The rapid burning of the air/fuel mixture that moves outward from the spark plug across the cylinder.

_____ 4. Pressure that leaks past the compression rings during compression and combustion.

_____ 5. The process of pulling stored fuel vapors from the charcoal canister and moving them into the engine to be burned.

_____ 6. A valve that is designed to release pressure if it gets above a calibrated level; used in a gas cap as a safety device.

_____ 7. A converter that changes only hydrocarbons and carbon monoxide into harmless elements.

_____ 8. A system that recirculates a portion of burned gases back into the combustion chamber to displace air and fuel and cool combustion temperatures.

_____ 9. A solenoid that allows fresh air to enter the evaporative system during a purge event. Also used for an evaporative system monitoring test.

_____ 10. A molecule made up of hydrogen and carbon. It is considered a harmful vehicle emission.

_____ 11. An engine that uses the heat of compression to ignite the air/fuel mixture; also known as a diesel engine.

_____ 12. Particles that are heavier than air that are released during combustion. It is the black smoke that is commonly emitted from a diesel engine.

_____ 13. A substance that will absorb large amounts of vapor, used in charcoal canisters to store and release fuel vapors.

_____ 14. A tapered valve that sits in a tapered seat to seal a passageway for air or fuel.

_____ 15. The ability of a liquid to evaporate.

_____ 16. A poisonous gas released during combustion of rich air/fuel mixtures.

_____ 17. The use of normal exhaust engine pulses to draw air into the exhaust stream.
_____ 18. A gas that is released to the atmosphere; usually refers to a harmful gas.
_____ 19. The movement of the piston in the engine from top dead center to bottom dead center, or vice versa.
_____ 20. The ratio of air to fuel at which all of the oxygen in the air and all of the fuel are completely burned.

Multiple Choice

Read each item carefully, and then select the best response.

_____ 1. The _____ standard is an automotive self-diagnostic system mandated by the EPA that requires a warning light to alert the driver of an emission system fault.
   A. clean air
   B. emission control
   C. OBDI
   D. OBDII

_____ 2. Oxides of nitrogen are categorized as _____ emission gases.
   A. nonharmful
   B. harmful
   C. debatable
   D. inert

_____ 3. _____ is a by-product of complete combustion.
   A. Water
   B. Nitrogen
   C. Carbon dioxide
   D. Both A and C

_____ 4. Since carbon monoxide is absorbed so easily by red blood cells, the _____ has set a permissible exposure limit in the workplace of 35 ppm for an 8-hour shift with a ceiling of 200 ppm for any length of time.
   A. Occupational Safety and Health Administration
   B. National Institute for Occupational Safety and Health
   C. American Conference of Governmental Industrial Hygienists
   D. Society of Automotive Engineers

_____ 5. Which corrosive compound, emitted into the atmosphere through exhaust, is a major environmental pollutant, coming back to earth in contaminated rainwater, called acid rain?
   A. Oxides of nitrogen
   B. Carbon monoxide
   C. Sulfur dioxide
   D. Hydrocarbons

_____ 6. What is produced in the atmosphere when unburned hydrocarbons and oxides of nitrogen react chemically with sunlight?
   A. Smog
   B. Acid rain
   C. Sulfur dioxide
   D. Carbon monoxide

_____ 7. The stoichiometric ratio for compressed natural gas is _____.
   A. 14.7:1
   B. 17.2:1
   C. 6.45:1
   D. 34.3:1

_____ 8. The operating condition in which a feedback system is controlling the outputs based on sensor feedback is called _____ operation.
   A. open-loop
   B. closed-loop
   C. variable-loop
   D. programmed-loop

_____ 9. The _____ is an example of a precombustion emission control system.
   A. exhaust gas recirculation system
   B. secondary air injection system
   C. heated air intake system
   D. both A and C

_____ 10. _____ catalytic converters contain both a reduction catalyst and an oxidizing catalyst.
   A. Two-way
   B. Three-way
   C. Four-way
   D. All of the above

_____ 11. The _____ is a type of positive crankcase ventilation system.
   A. fixed orifice
   B. variable orifice
   C. separator
   D. all of the above

_____ 12. The _____ is an example of an electrically operated exhaust gas recirculation valve.
   A. negative back pressure EGR valve
   B. positive back pressure EGR valve
   C. stepper motor EGR valve
   D. single-diaphragm EGR valve

_____ 13. The _____ system is designed to ensure hydrocarbons are not released into the atmosphere when fuel in the fuel tank begins to vaporize and build pressure.
   A. evaporative emission
   B. positive crankcase ventilation
   C. exhaust gas recirculation
   D. secondary air injection

_____ 14. The fuel cap may incorporate a(n) _____ to release a small amount of pressure to prevent the fuel tank from rupturing.
   A. expansion valve
   B. vacuum relief valve
   C. pressure relief valve
   D. positive back-pressure valve

_____ 15. Secondary air injection systems use a(n) _____ to control airflow in the system.
   A. air-switching valve
   B. air diverter valve
   C. check valve
   D. all of the above

True/False

If you believe the statement to be more true than false, write the letter "T" in the space provided. If you believe the statement to be more false than true, write the letter "F".

_____ 1. Humans breathe in oxygen and exhale carbon dioxide; trees and plants take in carbon dioxide and give back oxygen.

_____ 2. The EPA requires that when a vehicle's emissions deviate from the federal test procedure limit by 3.5 times, the malfunction indicator lamp must turn on and one or more specific diagnostic trouble codes be set in memory.

_____ 3. Particulate matter is graded in a size range from 10 nanometers to 100 micrometers in diameter; particulates of less than 10 micrometers are dangerous to humans.

_____ 4. Gasoline, diesel, LPG, and natural gas are all hydrocarbon compounds.

_____ 5. Exposure to carbon monoxide levels of 400 ppm may be fatal in as little as 3 hours, and 6400 ppm may be fatal in as little as 30 minutes.

_____ 6. Sulfur reduces catalyst efficiency in modern vehicles, and vehicles operating with higher sulfur gasoline have higher emissions than vehicles operating on lower sulfur gasoline.

_____ 7. Keeping the air/fuel ratio close to the stoichiometric point produces minimal hydrocarbon and carbon monoxide emissions.

_____ 8. The exhaust gas recirculation system is a postcombustion emission control system.

_____ 9. Any vehicle produced after 1996 monitors catalyst efficiency by using two oxygen sensors, one in front of the catalytic converter and one in the exhaust stream.

_____ 10. The positive crankcase ventilation system regulates the flow of blow-by gases between the crankcase and the intake manifold.

_____ 11. A PCV valve connects the exhaust port, or manifold, and the intake manifold.

_____ 12. The negative back-pressure exhaust gas recirculation valve has a single, spring-loaded diaphragm that seals off the vacuum chamber.

_____ 13. Older vehicles used vacuum controls solely to operate the exhaust gas recirculation system.

_____ 14. In modern vehicles the ECM controls the timing of when, how far, and how long the EGR valve opens, making it much more efficient than the thermo control (temperature-controlled valve) in vacuum-controlled EGR systems.

_____ 15. A vented filler cap allows air to enter the fuel tank to relieve the low pressure as fuel is drawn out of the tank as well as prevent the release of vaporous hydrocarbons.

_____ 16. In some earlier carbureted engine designs, the charcoal canister had a vapor line connected to the carburetor float bowl to absorb any fuel vapors from it.

_____ 17. Some manufacturers use dedicated sensors such as a purge switch, which is located in the purge line and senses the flow of gases or the pressure drop caused by the flow of gases that are being purged.

_____ 18. Aspirated air injection systems use an air pump to help facilitate the moving of air in the system.

_____ 19. Pinging can be referred to as the sound of marbles being shaken in a glass jar and is damaging to the engine if it is allowed to continue.

_____ 20. A five-gas analyzer is a diagnostic tool that uses sensors to measure the level of gases in the exhaust stream.

Fill in the Blank

Read each item carefully, and then complete the statement by filling in the missing word(s).

1. _____ _____ systems are designed to limit the pollution caused by the storing and burning of various fuels.

2. The _____ _____ is the maximum amount of emissions that a vehicle is permitted to emit.

3. The _____ _____ completes the combustion process by adding another oxygen molecule to the carbon monoxide to convert it to carbon dioxide.

4. Victims of _____ _____ poisoning can sometimes look healthy and pink-cheeked because concentrations of carbon monoxide in the bloodstream give the blood a brighter red color than normal.

5. _____ _____ _____ are claimed to be major contributors to photochemical smog, along with hydrocarbons and sunlight.

6. In spark-ignition engines, _____ are caused by incomplete combustion of rich air/fuel mixtures.

7. In a combustion chamber where surface temperatures are low, the combustion flame can go out or be _____.

8. Reducing the valve overlap reduces the _____ _____, which is when the exhaust gases moving out of the combustion chamber create a low-pressure area in the exhaust manifold, helping to pull air and fuel into the cylinder.

9. The _____ _____, also referred to as lambda, is the ratio of air to fuel at which all of the oxygen in the air and all of the fuel are completely burned.

10. The _____ system uses an oxygen sensor to detect rich and lean conditions in the exhaust stream.

11. A(n) _____ _____ _____ is a programmed chart that the manufacturer's engineers create to determine the approximate amount of fuel to deliver based on multiple sensors' input.

12. In older vehicles, crankcase vapors were vented directly to the atmosphere through a(n) _____ _____ or road-draft tube.

13. The _____ _____ _____ system was designed by automotive engineers in the 1970s to control the emission of oxides of nitrogen.
14. OBDII made monitoring of the EGR system mandatory; each manufacturer became responsible for ensuring its system runs a self-test called a(n) EGR _____ _____ test to verify that it is operating correctly.
15. Modern fuel tanks contain a(n) _____ _____, which is an air chamber that allows for the expansion of liquid fuel on hot days.
16. A(n) _____ _____ is connected to the vapor space in the fuel tank or the liquid vapor separator and carries fuel vapors from the fuel tank to a storage container called a charcoal canister.
17. Older vehicles use vacuum to control the opening and closing of the canister _____ _____, which is spring-loaded in the closed position.
18. Newer EVAP systems monitor purge operation by reading a _____ _____ pressure sensor.
19. The pulse air system uses the pulsations of the exhaust gas to open and close a(n) _____ _____, which is a flat metal valve that flexes when pressure is applied against it.
20. The _____ _____ intake system collects hot air from around the exhaust manifold and mixes it with outside air entering the air cleaner assembly.

Labeling

Label the following diagrams with the correct terms.

1. PCV system:

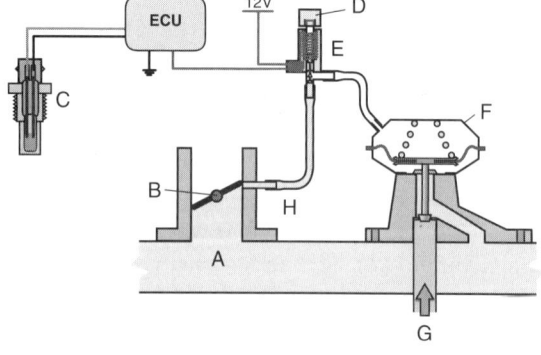

A. _____
B. _____
C. _____
D. _____

2. Computer-controlled EGR system:

A. _____
B. _____
C. _____
D. _____
E. _____
F. _____
G. _____
H. _____

3. Negative back-pressure EGR valve:

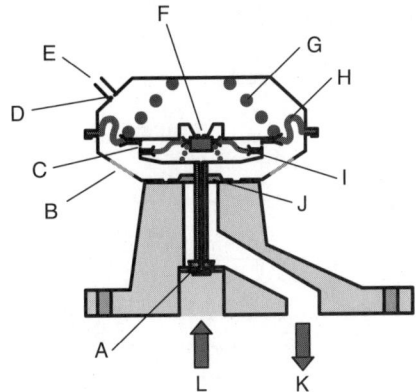

A. _____
B. _____
C. _____
D. _____
E. _____
F. _____
G. _____
H. _____
I. _____
J. _____
K. _____
L. _____

4. Pulse-width-modulated EGR valve and stepper motor EGR valve:

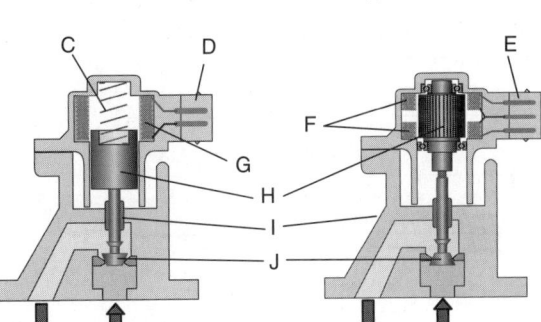

A. _____
B. _____
C. _____
D. _____
E. _____
F. _____
G. _____
H. _____
I. _____
J. _____
K. _____
L. _____

5. EGR vacuum control system:

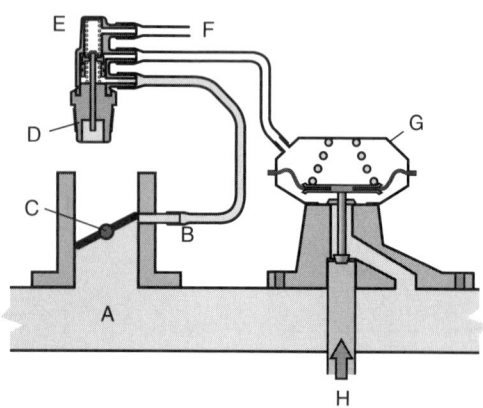

A. _____
B. _____
C. _____
D. _____
E. _____
F. _____
G. _____
H. _____

6. EVAP system:

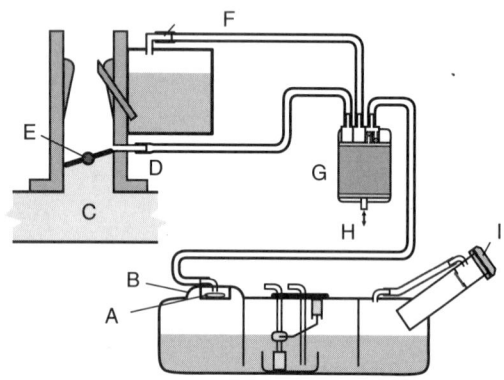

A. _____
B. _____
C. _____
D. _____
E. _____
F. _____
G. _____
H. _____
I. _____

7. Charcoal canister:

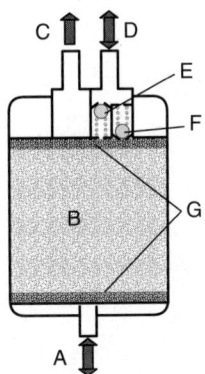

A. _____
B. _____
C. _____
D. _____
E. _____
F. _____
G. _____

8. Electrically operated vacuum purge system:

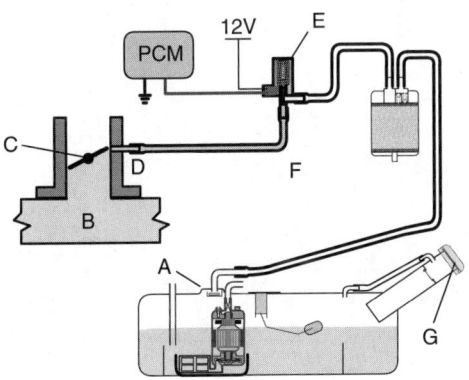

A. _____
B. _____
C. _____
D. _____
E. _____
F. _____
G. _____

9. Air diverter valve and check valve:

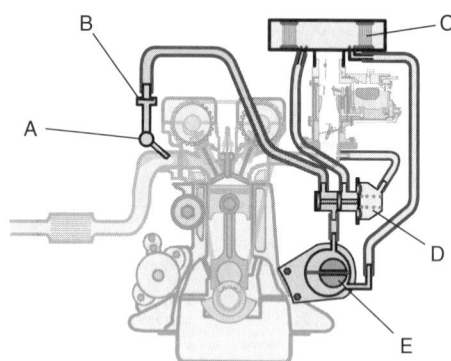

A. _____
B. _____
C. _____
D. _____
E. _____

10. Heated air intake system:

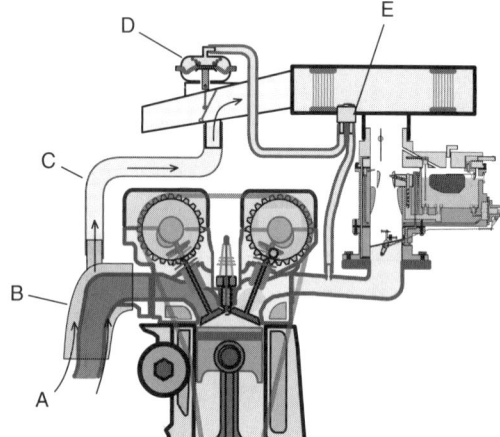

A. _____
B. _____
C. _____
D. _____
E. _____

Skill Drills

Test your knowledge of skill drills by filling in the correct words in the photo captions.

1. Using a Five-Gas Analyzer:

Step 1: Research the local _____ _____ for this vehicle in the applicable _____ or online sources. Warm up both the _____ and the exhaust gas _____. Leak test the analyzer _____ and _____.

Step 2: _____ the analyzer, if applicable. _____ the exhaust analyzer _____ in the vehicle's _____ _____ at least 18" (46 cm).

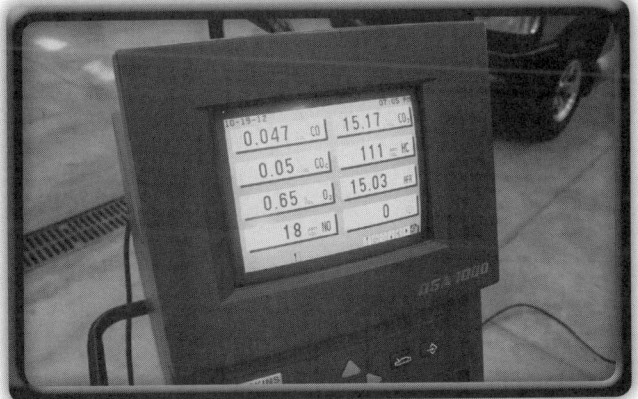

Step 3: Allow the vehicle to _____ and record the _____ readings. Compare your readings to the _____ and determine any necessary action(s).

2. Inspecting and Servicing the PCV System:

Step 1: Locate the _____ _____. With the engine idling, _____ the PCV _____ and check for the presence of a strong _____.

Step 2: Remove the _____ and check that it is still pliable and not _____ with _____ deposits.

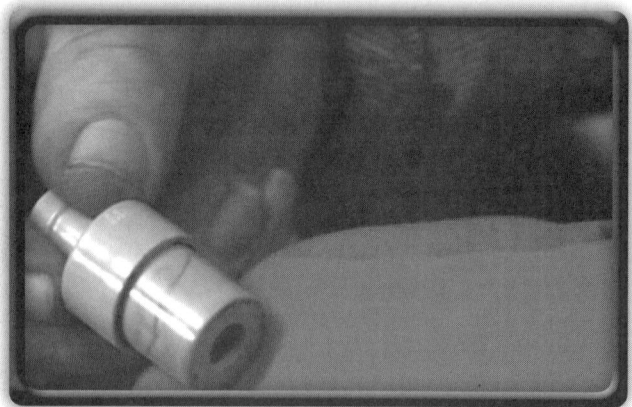

Step 3: Remove the PCV valve and _____ it for _____. If it is _____, is sludged up, or sticks, _____ it with a new one of the same type.

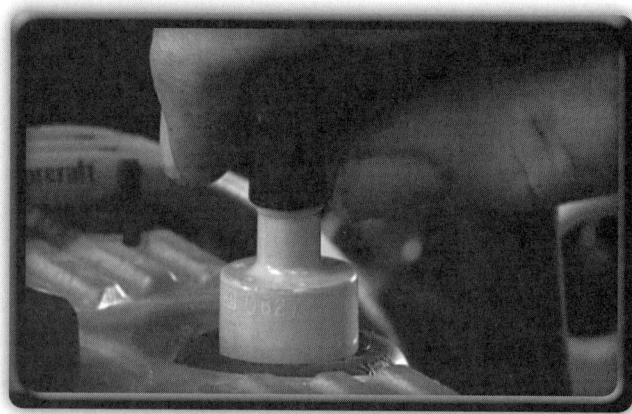

Step 4: Reinstall the PCV valve into the _____ _____, and remove the _____ hose from the _____ _____ assembly.

Step 5: With the engine _____, block off the breather hose and _____ for vacuum _____ up in the breather hose and _____, indicating that the PCV system can handle the amount of _____-_____ gases.

3. Inspecting and Servicing the Hoses of the EVAP System:

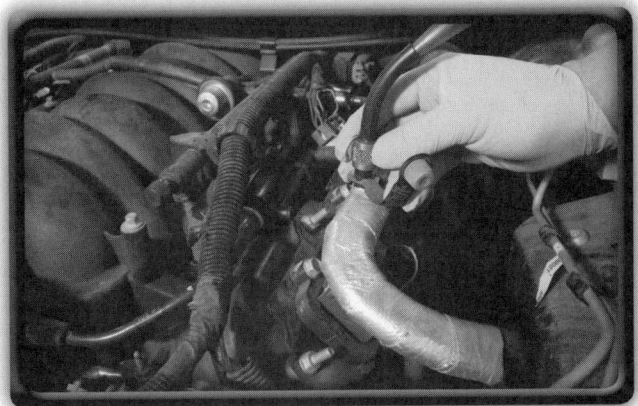

Step 1: Use a _____ _____ and check for _____ related to the EVAP system. If a code is found, follow the _____ procedures for that code. If the diagnostic code indicates a _____, connect an _____ tester to the EVAP test port.

Step 2: Use the EVAP system _____ _____ to determine if the _____ is in fact _____ and, if so, the _____ of the leak.

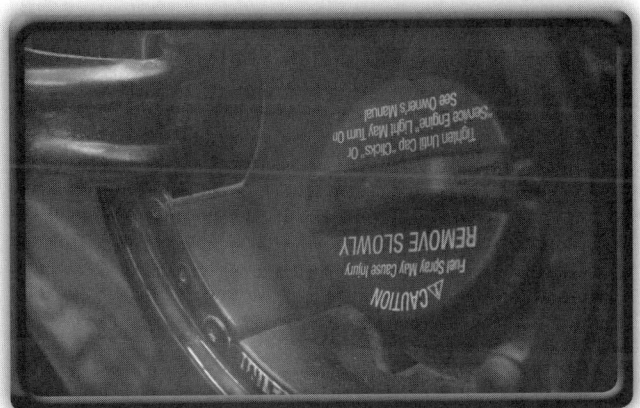

Step 3: Use a _____ _____ to help locate the leak. If smoke is detected, look in the area of the smoke to _____ the cause of the _____. If no smoke is detected, the system could be leaking _____ through a stuck-open _____ _____ or a _____ vent valve. Repair according to the manufacturer's procedure.

Crossword Puzzle

Use the clues to complete the puzzle.

Across

3. The operational status of the engine control system in which the oxygen sensor is providing feedback of rich or lean exhaust conditions and the control unit is varying fuel delivery based on that information.
4. A valve that changes secondary airflow from the exhaust stream to the air cleaner or vice versa.
7. A test in which a gas analyzer hose is placed into the exhaust stream before and after the catalytic convertor; it may require drilling an access hole in the exhaust pipe in front of the converter.
8. A valve used to control the flow of evaporative emissions from the charcoal canister to the intake manifold.
10. A mechanical part, usually controlled by electricity allowing a module to control something mechanical.
11. A device used to pump air to the exhaust stream to ensure the converter has enough oxygen to perform gas conversions.
13. A brown haze that hangs in the sky, typically seen over large cities. This is a major health issue to humans because it affects lung tissue.

Down

1. A tool that uses sensors to measure the level of gases in the exhaust stream.
2. A system that uses feedback to adjust what it is doing. Typically, a module gives a command and waits to see if the command was followed by use of a sensor.

3. The part of the carburetor that holds fuel to be burned in the engine; the bowl is at a constant level of fuel to ensure adequate fuel is present during driving.
5. The maximum amount of emissions that a vehicle is permitted to emit. Values are assigned to different classifications of vehicles.
6. A tube used in the PCV system to allow fresh air into the engine crankcase.
8. A device used to show the computer when purge is occurring. IT is used as a feedback device to allow the computer to determine whether flow is happening.
9. A valve that connects the exhaust port, or manifold, and the intake manifold.
12. A preignition or detonation concern that is damaging to the engine if it is allowed to continue. It is described as the sound of marbles being shaken in a glass jar.

ASE-Type Questions

Read each item carefully, and then select the best response.

_____ 1. Tech A says that the purge valve is part of the evaporative emission system. Tech B says that OBDII systems must illuminate the MIL when the emissions exceed 1.5 times the FTP. Who is correct?
   A. Tech A
   B. Tech B
   C. Both A and B
   D. Neither A nor B

_____ 2. Tech A says that carbon monoxide is partially burned fuel. Tech B says that oxides of nitrogen are unburned fuel. Who is correct?
   A. Tech A
   B. Tech B
   C. Both A and B
   D. Neither A nor B

_____ 3. Tech A says that hydrocarbons are a result of complete combustion. Tech B says that a catalytic converter creates a chemical reaction, changing carbon monoxide and hydrocarbons to water and carbon dioxide. Who is correct?
   A. Tech A
   B. Tech B
   C. Both A and B
   D. Neither A nor B

_____ 4. Tech A says that allowing hot exhaust gases into the engine through the EGR valve helps to warm up the engine during warm-up. Tech B says that rich fuel mixtures will create low amounts of carbon monoxide. Who is correct?
   A. Tech A
   B. Tech B
   C. Both A and B
   D. Neither A nor B

_____ 5. Tech A says that burning gasoline in an engine creates water as a by-product of combustion. Tech B says that variable valve timing can be used to reduce oxides of nitrogen, eliminating the need for an EGR valve on some engines. Who is correct?
   A. Tech A
   B. Tech B
   C. Both A and B
   D. Neither A nor B

_____ 6. Tech A says that the PCV system recirculates exhaust gases into the intake manifold. Tech B says that the PCV system recirculates blow-by gases into the intake manifold. Who is correct?
   A. Tech A
   B. Tech B
   C. Both A and B
   D. Neither A nor B

_____ 7. Tech A says that a closed loop system means the ECM is receiving feedback and adjusting the fuel injector on time as needed. Tech B says that lean means there is not enough air in the air/fuel mixture. Who is correct?
   A. Tech A
   B. Tech B
   C. Both A and B
   D. Neither A nor B

_____ 8. Tech A says that the EVAP system stores escaping hydrocarbons until the engine can burn them. Tech B says that catalytic converters were designed for regular leaded gas. Who is correct?
   A. Tech A
   B. Tech B
   C. Both A and B
   D. Neither A nor B

_____ 9. Tech A says that the EGR valve is open fully at idle so that the engine will not die. Tech B says that oxides of nitrogen are created in large amounts when the combustion temperature is above 2500°F (1400°C). Who is correct?
   A. Tech A
   B. Tech B
   C. Both A and B
   D. Neither A nor B

_____ 10. Tech A says that black exhaust indicates a very rich-running engine. Tech B says that white exhaust can indicate coolant in the exhaust. Who is correct?
   A. Tech A
   B. Tech B
   C. Both A and B
   D. Neither A nor B

Alternative Fuel Systems

CHAPTER 52

Tire Tread:
© AbleStock

Chapter Review

The following activities have been designed to help you refresh your knowledge of this chapter. Your instructor may require you to complete some or all of these activities as a regular part of your training program. You are encouraged to complete any activity that your instructor does not assign as a way to enhance your learning.

Matching

Match the following terms with the correct description or example.

A. Alternative fuel
B. Biodiesel
C. Burn rate
D. Catalyst
E. Continuously variable transmission
F. Emission standards
G. Ethanol
H. Hydrocarbon (HC)
I. Hygroscopic
J. Inverter
K. Mercaptan
L. Multiphase
M. Nitrogen oxides
N. Particulate matter (PM)
O. Pour point
P. Power-on-demand
Q. Reformer
R. Slow-fill
S. Stoichiometric ratio
T. Wood gas

_____ 1. Unburned fuel emitted from the tailpipe.
_____ 2. The various compounds of nitrogen that contribute to a smog-forming pollutant emitted from the tailpipe.
_____ 3. A feature that shuts down the engine when not needed to save fuel.
_____ 4. A transmission without individual gears or gear ratios.
_____ 5. A gas made by the carbonization or gasification of coal.
_____ 6. The temperature at which diesel fuel is not gelled.
_____ 7. The air/fuel ratio for complete burning and lowest pollution; the stoichiometric ratio for a gasoline engine is 14.7 to 1 by weight.
_____ 8. The rate of flame spread for liquid or gaseous fuel.
_____ 9. A device that converts direct current to alternating current.
_____ 10. The method of filling a compressed natural gas vehicle (such as overnight) so the cylinders are completely filled.
_____ 11. An electric motor that operates through more than one phase.
_____ 12. Renewable fuel made from organic feedstocks.
_____ 13. Cancer-causing unburned soot from the tailpipe.
_____ 14. A nonpetroleum-based motor fuel.
_____ 15. Alcohol-based motor fuel made from starches and sugars.
_____ 16. An on-board device that extracts hydrogen from another source material.
_____ 17. A device that changes chemical composition without self-destruction.
_____ 18. Cutoff points set by governmental agencies to limit tailpipe emissions.
_____ 19. An agent that smells like sulfur that is added to liquefied petroleum or natural gas to aid in detecting leaks.
_____ 20. The ability of a substance to attract and hold water molecules.

Multiple Choice

Read each item carefully, and then select the best response.

_____ 1. In 1992, Congress passed the _____, the intent of which was to advance, through mandates, the use of vehicles capable of using alternatives to gasoline and diesel.
   A. Clean Air Act
   B. Energy Policy Act
   C. Environmental Protection Act
   D. Alternative Fuel Act

_____ 2. What product of partially burned fuel displaces oxygen in the bloodstream and causes asphyxiation?
   A. Carbon dioxide
   B. Nitrogen oxides
   C. Hydrocarbons
   D. Carbon monoxide

_____ 3. In the United States, the _____ controls federally mandated emission standards.
   A. Emission Standards Agency
   B. Environmental Protection Agency
   C. Clean Air Agency
   D. Clean Energy Agency

_____ 4. Vehicles that are designed to operate on two different fuels blended together in various mixture amounts by percentage are called _____.
   A. alternative fuel vehicles
   B. biodiesel vehicles
   C. flexible-fuel vehicles
   D. hybrid vehicles

_____ 5. Which fossil fuel is primarily methane, but may also include ethane, propane, butane, and other gases such as carbon dioxide, nitrogen, and helium?
   A. Liquefied petroleum gas
   B. Natural gas
   C. Syngas
   D. Biodiesel

_____ 6. What type of fuel was used for lighting and internal combustion engine vehicles during the very early days before petroleum became the fuel of choice?
   A. Wood gas
   B. Natural gas
   C. Diesel
   D. Town gas

_____ 7. What type of gas consists mainly of varying combinations of hydrogen and carbon monoxide, and it can be produced in a number of different ways, including the Fischer-Tropsch method for the gasification of coal?
   A. Town gas
   B. Wood gas
   C. Compressed natural gas
   D. Syngas

_____ 8. What type of vehicle runs on two different fuels that are injected directly into the combustion chamber, almost at the same time?
   A. Dual-fuel
   B. Bi-fuel
   C. Hybrid
   D. Fuel cell

9. What kind of renewable fuel is made by chemically combining natural oils from soybeans, cottonseeds, canola, animal fats, algae, jatropha seeds, or even recycled cooking oil with an alcohol such as methanol or ethanol, and a catalyst like lye?
    A. Producer gas
    B. Biodiesel
    C. Syngas
    D. Bi-fuel

10. Some diesel engines can run on _____, but only if it has been properly treated to remove the glycerin, which could otherwise gunk up injectors and engine parts like piston rings.
    A. straight vegetable oil
    B. wood fuel
    C. waste vegetable oil
    D. either A or C

11. The _____ is the temperature below which diesel fuel will not flow at all.
    A. cloud point
    B. pour point
    C. freezing point
    D. viscosity threshold

12. Second-generation _____ is derived from a variety of cellulose-based plants such as switchgrass.
    A. biodiesel
    B. bio-ethanol
    C. methanol
    D. biomass

13. All of the following are basic elements of a fuel cell, *except*: _____.
    A. anode
    B. cathode
    C. electrode
    D. catalyst

14. A device called a(n) _____ is used to separate hydrogen from other fuels, such as gasoline.
    A. splitter
    B. inverter
    C. rectifier
    D. reformer

15. The "exhaust" produced by a fuel cell vehicle is _____.
    A. methane
    B. water
    C. carbon monoxide
    D. nitrogen

16. What type of electric motor does not use brushes or commutators?
    A. Permanent magnet
    B. Multiphase
    C. Brushless
    D. Synchronous

17. What type of vehicle uses a combination of electric power and an internal combustion engine?
    A. Fuel cell vehicle
    B. Hybrid electric vehicle
    C. Battery electric vehicle
    D. Flexible-fuel vehicle

_____ 18. In what type of hybrid drive configuration is the gasoline engine used only to drive a generator and charge a battery, which produces power to drive an electric motor?
   A. Series hybrid
   B. Parallel hybrid
   C. Series-parallel hybrid
   D. All of the above

_____ 19. In a hybrid vehicle the _____ feature temporarily shuts off the internal combustion engine when it is not needed, such as when idling or coasting.
   A. regenerative braking
   B. displacement-on-demand
   C. idle stop
   D. power-on-demand

_____ 20. What type of vehicle uses only one power source, the battery, to propel the vehicle for a certain distance before the batteries need charging and the internal combustion engine is needed?
   A. Plug-in hybrid electric vehicle
   B. Fuel cell vehicle
   C. Series-parallel hybrid vehicle
   D. Flexible-fuel vehicle

True/False

If you believe the statement to be more true than false, write the letter "T" in the space provided. If you believe the statement to be more false than true, write the letter "F".

_____ 1. Transitioning to cleaner alternative energy and fuel choices can help reduce the effects of pollution.

_____ 2. Although harmless in small amounts, too much carbon monoxide is a serious atmospheric concern. It is considered a greenhouse gas.

_____ 3. Emission standards set cutoff points for the amount of pollutants that vehicles may release into the environment.

_____ 4. Since methane has no odor, an odorant called mercaptan is added so humans may detect leaks.

_____ 5. Compressed natural gas is used in heavy-duty long-haul trucks, heavy-duty refuse and delivery vehicles, buses, and so forth.

_____ 6. Producer gas is a generic term that refers to a number of manufactured gases such as wood gas, town gas, and syngas.

_____ 7. Dual-fuel engines and vehicles are capable of running on either gasoline or natural gas, though some engine modifications are required for them to run efficiently.

_____ 8. In a dedicated natural gas conversion, the diesel injectors are replaced with spark plugs and the natural gas fuel is premixed with air before intake occurs.

_____ 9. There are two types of LPG storage container: portable universal cylinders and permanently mounted tanks.

_____ 10. Compressed natural gas valve installation and inspection specifications are contained in the National Fire Protection Association (NFPA) standards and published in their NFPA-52 document.

_____ 11. B100 fuel, the blend most commonly available, contains 10% biodiesel and 90% petrodiesel.

_____ 12. Methanol is highly poisonous and dangerous to the environment.

_____ 13. M85 vehicles are flex-fuel vehicles that can run on any mixture of methanol and gasoline/petrol, though the limit of methanol in the mixture is 85% in the United States.

_____ 14. Hydrogen is the lightest element, yet it has the highest energy content per unit weight/mass of all energy-based fuels—three times the energy of gasoline/petrol.

_____ 15. Each fuel cell consists of a membrane electrode assembly sandwiched between plates.

_____ 16. A parallel hybrid allows both the engine and the electric motor to drive the vehicle together, or either one by itself.

Chapter 52 Alternative Fuel Systems

_____ 17. Displacement-on-demand deactivates cylinders when operating under cruise or coasting conditions to save fuel.
_____ 18. Idle stop turns off the internal combustion engine when the vehicle is at a standstill.
_____ 19. The retarding effect (vehicle slowing) caused by regenerative braking provides the same deceleration as normal engine braking would.
_____ 20. Fuel cell vehicles are considered hybrid vehicles because they have two power sources.

Fill in the Blank

Read each item carefully, and then complete the statement by filling in the missing word(s).

1. The co-products of _____ production, such as dried distiller grains, are sold as feed for livestock.
2. Mainly produced by diesel engines, _____ _____ is literally fine soot that can become trapped in the lungs and cause health issues.
3. California's _____ _____ _____ standard includes six major emission categories, each with several targets depending on vehicle weight and cargo capacity.
4. _____ _____ _____ is heavier than air and is the least carbon-emitting hydrocarbon fuel available.
5. _____ _____ _____ vehicles use methane compressed for storage and use in light-duty passenger vehicles, delivery trucks, and buses.
6. A(n) _____-_____ fuel gauge is placed in the tank and raises or lowers in a tube for fuel level calculation.
7. _____ gas is a hydrogen and methane gas created by a process of thermal gasification of wood or some other biomass.
8. _____ gas is mostly hydrogen and carbon monoxide, and it is created by passing steam over red-hot coke.
9. A(n) _____ _____ _____ ensures that an LPG fuel storage tank cannot be filled past the safe fill limit of 80%.
10. The volume of space required for _____ _____ _____ storage tanks is almost twice that required for LNG tanks for delivering the equivalent mileage range.
11. Biodiesel burns with less particulate and with no _____ or _____, thus producing less harmful and irritating emissions.
12. Biodiesel, depending on the feedstock and process used, is a light yellow to dark gold liquid, with a(n) _____ very similar to petrodiesel fuel.
13. The cloud point is the temperature at which diesel fuel starts to appear cloudy, indicating that _____ crystals have begun to form.
14. _____ is a renewable energy source, plant, or animal material used as a source of fuel.
15. A fuel _____ sensor tells the engine computer what percentage of methanol is in the fuel, and it adjusts the injection quantity ignition timing accordingly.
16. A fuel cell is a(n) _____-_____ device that combines hydrogen and oxygen to produce water and in the process it produces electricity and heat.
17. When coupled with sustainable or renewable energy sources, the _____ _____ vehicle can truly serve as an emissions-free vehicle.
18. _____ braking occurs when the drive motor(s) act as generators to recharge the traction batteries during deceleration or when braking.
19. The internal combustion engine in a hybrid vehicle is designed to operate only within its most efficient operating range, typically somewhere between 2000 and 4500 rpm, where engine peak _____ efficiency and peak torque are developed.
20. The Chevy Volt is an example of a series _____-_____ electric vehicle.

Labeling

Label the following diagrams with the correct terms.

1. Liquefied natural gas fuel tank:

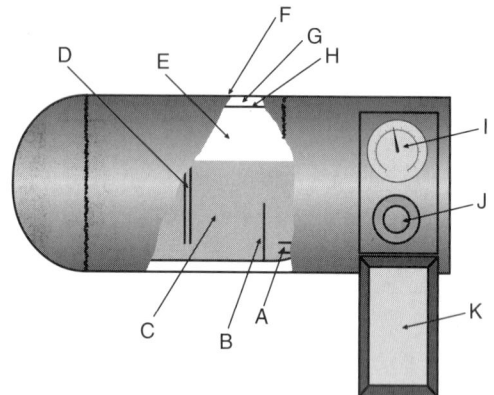

A. _____
B. _____
C. _____
D. _____
E. _____
F. _____
G. _____
H. _____
I. _____
J. _____
K. _____

2. Fuel cell creating electricity:

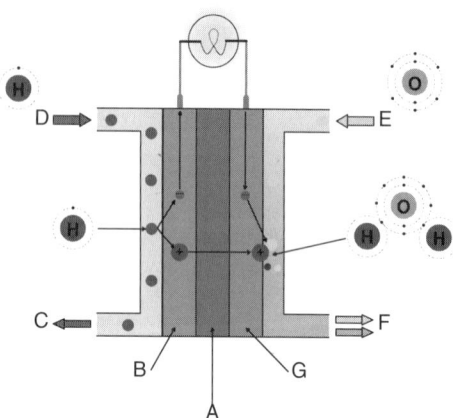

A. _____
B. _____
C. _____
D. _____
E. _____
F. _____
G. _____

3. Fuel cells in a stack:

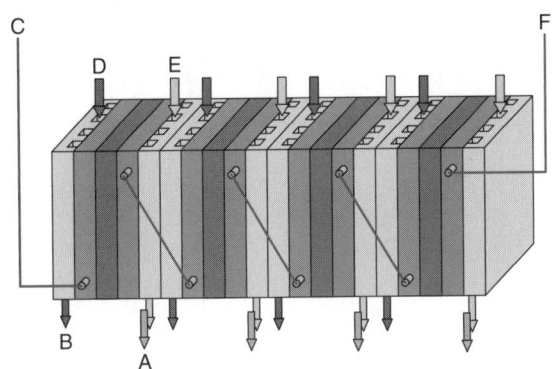

A. _____
B. _____
C. _____
D. _____
E. _____
F. _____

4. Power-splitting transmission:

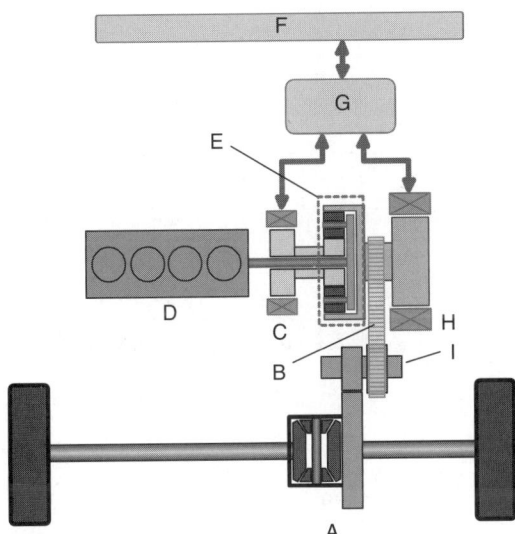

A. _____
B. _____
C. _____
D. _____
E. _____
F. _____
G. _____
H. _____
I. _____

Crossword Puzzle

Use the clues to complete the puzzle.

Across

5. The ability to lubricate moving parts.
6. A fuel that is 20% biodiesel blended with 80% petrodiesel.
7. A pollutant/irritant in diesel fuel.
8. A feature that turns off the internal combustion engine when the vehicle is at a standstill.
10. A toxic alcohol fuel.
12. Another name for a traction motor with regenerative capability.

Down

1. Two fuels used in a vehicle one at a time, such as compressed natural gas or gasoline.
2. Also known as an HEV, a vehicle that uses two power sources for propulsion, one of which is electricity.
3. A special refueling port and an overfill protection device fitted on LPG tanks, which ensure that the tank cannot be filled past the safe fill limit of 80%.
4. The temperature at which wax (paraffin) in diesel fuel starts to solidify.

6. Organic matter used for fuel.
9. A fuel that is 100% biodiesel ("neat").
11. An electro-chemical device that uses hydrogen and oxygen to create electricity.
12. A fuel with 85% ethanol and 15% gasoline blend.
13. A positively charged electrode or plate.

ASE-Type Questions

Read each item carefully, and then select the best response.

_____ 1. Tech A says that fuel cells only emit CO_2 to the atmosphere. Tech B says that fuel cells run on water. Who is correct?
   A. Tech A
   B. Tech B
   C. Both A and B
   D. Neither A nor B

_____ 2. Tech A says that ethanol-blended gas at 10% is a nationwide standard. Tech B says that 15% ethanol-blended gas has been approved for 2001 vehicles and newer. Who is correct?
   A. Tech A
   B. Tech B
   C. Both A and B
   D. Neither A nor B

_____ 3. Tech A says that high-voltage wires in hybrid and electric vehicles are usually a special color. Tech B says that special high-voltage shoes should be worn when working on high-voltage vehicles. Who is correct?
   A. Tech A
   B. Tech B
   C. Both A and B
   D. Neither A nor B

_____ 4. Tech A says that E85 vehicles are designed to use 85% ethanol efficiently. Tech B says that hydrogen cars are not zero-emission vehicles. Who is correct?
   A. Tech A
   B. Tech B
   C. Both A and B
   D. Neither A nor B

_____ 5. Tech A says that LPG-fueled engines are low on horsepower and torque, but are cost efficient. Tech B says that LPG-fueled vehicles are advantageous to a fleet as they can be fueled just about anywhere. Who is correct?
   A. Tech A
   B. Tech B
   C. Both A and B
   D. Neither A nor B

_____ 6. Tech A says that CNG fuel tanks hold 3600 psi (24,800 kPa). Tech B says that CNG engines emit white smoke, which is humidity. Who is correct?
   A. Tech A
   B. Tech B
   C. Both A and B
   D. Neither A nor B

_____ 7. Tech A says that LNG fuel is stored at approximately –263.2°F (–164°C). Tech B says that LNG fueling stations are the most common of all the alternative fuels. Who is correct?
   A. Tech A
   B. Tech B
   C. Both A and B
   D. Neither A nor B

_____ 8. Tech A says that dual-fuel–powered vehicles inject two fuels at the same time. Tech B says that bi-fuel engines inject two fuels at the same time. Who is correct?
   A. Tech A
   B. Tech B
   C. Both A and B
   D. Neither A nor B

_____ 9. Tech A says that when converting a CI diesel engine to an SI engine, you must lower the compression ratio in each cylinder. Tech B says that a diesel engine converted to use CNG and diesel fuel has a more complete and rapid combustion. Who is correct?
   A. Tech A
   B. Tech B
   C. Both A and B
   D. Neither A nor B

_____ 10. Tech A says that biodiesel mixed with diesel fuel will not reduce emissions. Tech B says that a 1% or more blend of biodiesel will extend the life of injectors. Who is correct?
   A. Tech A
   B. Tech B
   C. Both A and B
   D. Neither A nor B

Compression-Ignition Engines

CHAPTER 53

Tire Tread:
© AbleStock

Chapter Review

The following activities have been designed to help you refresh your knowledge of this chapter. Your instructor may require you to complete some or all of these activities as a regular part of your training program. You are encouraged to complete any activity that your instructor does not assign as a way to enhance your learning.

Matching

Match the following terms with the correct description or example.

- A. Cetane number
- B. Common rail
- C. Crankshaft
- D. Direct burning
- E. Direct fuel injection
- F. Exhaust valve
- G. Flame spread
- H. Ignition
- I. Ignition delay/lag period
- J. Indirect fuel injection
- K. Intake stroke
- L. Intake valve
- M. Main bearing
- N. Power chip
- O. Power stroke
- P. Prechamber
- Q. Scavenging
- R. Timing gear
- S. Valve
- T. Volumetric efficiency (VE)

_____ 1. The phase of combustion in which there is a sharp pressure rise in the combustion chamber because of the sudden combustion of the fuel.

_____ 2. An aftermarket performance programmable chip marketed to increase horsepower and mileage of electronically controlled diesel engines.

_____ 3. A valve used in an engine to provide a means for exhaust gases to exit the cylinder.

_____ 4. A gear that synchronizes the timed events of rotational engine components like camshafts and fuel injection pumps.

_____ 5. A type of injection system using a common fuel manifold from which individual injectors are fed high-pressure fuel.

_____ 6. A separate combustion area designed into a diesel cylinder head.

_____ 7. The burning of the fuel mixture ignited by the heat of compression.

_____ 8. The amount of air taken into the cylinder on the intake stroke, or intake port, opening in relation to the space in the cylinder.

_____ 9. A means for air to enter and exhaust gases to exit the cylinder.

_____ 10. A fuel injection system in which the fuel is injected directly into the combustion chamber.

_____ 11. A valve used in an engine to provide a means for air to enter the cylinder.

_____ 12. A rating of the ignition quality of diesel fuel.

_____ 13. The phase of combustion after ignition when the rest of the fuel is injected into the combustion chamber and there is a more gradual pressure change in the cylinder.

_____ 14. A fuel injection system in which the fuel is injected into a separate chamber in the cylinder head, often called a prechamber.

_____ 15. A shaft that converts reciprocating power strokes into rotating movement.

_____ 16. The removal of hot exhaust gases from the cylinder and the refilling of it with fresh air.

_____ 17. The time it takes for the fuel to ignite after the fuel is injected into the engine.

_____ 18. A bearing that supports the crankshaft in the block and provides a lubrication surface for the main journals.

_____ 19. The stroke in an engine cycle when the piston moves from top dead center (TDC) to bottom dead center (BDC).

_____ 20. The stroke in an engine cycle after ignition, when the piston again moves from top to bottom under the power.

Multiple Choice

Read each item carefully, and then select the best response.

_____ 1. For the piston in a four-stroke cycle to move up and down in four sequential strokes, the crankshaft must rotate _____ complete revolutions.
   A. one
   B. two
   C. four
   D. eight

_____ 2. The _____ stroke occurs when, after ignition, the piston again moves from top to bottom.
   A. intake
   B. compression
   C. power
   D. exhaust

_____ 3. The _____ is designed to control the ignition and flame front to develop maximum heat energy transfer and complete burning of the fuel with a minimum shock wave.
   A. fuel injector
   B. intake valve
   C. exhaust valve
   D. prechamber

_____ 4. All of the following are phases of the ignition and heat energy transfer process, *except*: _____.
   A. ignition delay/lag period
   B. compression
   C. flame spread
   D. direct burning

_____ 5. What condition is caused by ignition lag, which occurs between the start of injection and the start of ignition?
   A. Diesel clatter
   B. Engine knock
   C. Pre-ignition
   D. Flame spread

_____ 6. The holes left behind by the cooling chamber casting process are closed by using _____ plugs.
   A. freeze
   B. Welsh
   C. sand
   D. any of the above

_____ 7. The _____ may be formed into the piston crown in a variety of shapes and sizes.
   A. coolant passages
   B. combustion chamber
   C. prechamber
   D. fuel injectors

_____ 8. Small four-stroke engines usually have _____ valves per cylinder.
   A. one
   B. two
   C. three
   D. four

Chapter 53   Compression-Ignition Engines

_____ 9. Levers that transmit movement by the linkage system from the camshaft lobes to operate the valves are called _____.
   A. lifters
   B. pushrods
   C. rocker arms
   D. keepers

_____ 10. The _____ support the crankshaft in the engine block.
   A. main bearing journals
   B. flywheel
   C. throws
   D. thrust bearings

True/False

If you believe the statement to be more true than false, write the letter "T" in the space provided. If you believe the statement to be more false than true, write the letter "F".

_____ 1. Since 1988 diesel engine particulate matter emissions have been reduced by 83% and oxides of nitrogen by 63%.

_____ 2. Direct injection refers to the amount of fresh air taken into the cylinder on the intake stroke at a given revolutions per minute in relation to the space in the cylinder that exists at engine standstill.

_____ 3. Modern fuel systems use electronically controlled high-pressure fuel systems that make quiet and smoke-free direct injection possible, and desirable.

_____ 4. The two-stroke engine performs all of the events of a four-stroke engine, but they are accomplished with two strokes of the piston and one revolution of the crankshaft.

_____ 5. Diesel engine components are exposed to lower operating temperatures, pressures, and forces than gasoline engines of similar displacement.

_____ 6. The purpose of freeze plugs is to protect the cast engine block from freezing.

_____ 7. The exhaust valve is smaller because the exhaust is forced out by the movement of the piston.

_____ 8. The camshaft keeps the valves operating in the correct timing sequence.

_____ 9. Thrust bearings are usually located at the balance point of the crankshaft.

_____ 10. At 3000 rpm the pistons in high-speed diesel engines change direction 100 times per second.

Fill in the Blank

Read each item carefully, and then complete the statement by filling in the missing word(s).

1. In the diesel engine, a(n) _____ _____ is not needed for ignition, because the very hot air ignites the fuel when it is injected into the combustion chamber.

2. _____ occurs when fuel is injected, near the end of the compression stroke, into the very hot compressed air in the cylinder.

3. With indirect injection, fuel is sprayed into a robust and very hot antechamber in the cylinder head, often called a(n) _____.

4. The _____ process in a two-stroke diesel engine begins with an air pump or blower providing clean incoming air pressured slightly above that of outgoing exhaust gases.

5. The ignition delay period is directly affected by the _____ number of the fuel.

6. The _____ _____ provides the framework, strength, and structure needed for the compression ignition engine.

7. Valves are held in place by valve _____ located on the valve stem.

8. If a(n) _____ chain or belt is worn, valve timing can be thrown off, which affects engine power and emissions.

9. The connecting rod _____ support the connecting rods as the crankshaft turns because of forces transmitted through the connecting rods.

10. The use of a dual-mass _____ eliminates excessive transmission gear rattle, reduces gear change/shift effort, and increases fuel economy.

Fundamentals of Automotive Technology Student Workbook

Labeling

Label the following diagrams with the correct terms.

1. The four-stroke cycle:

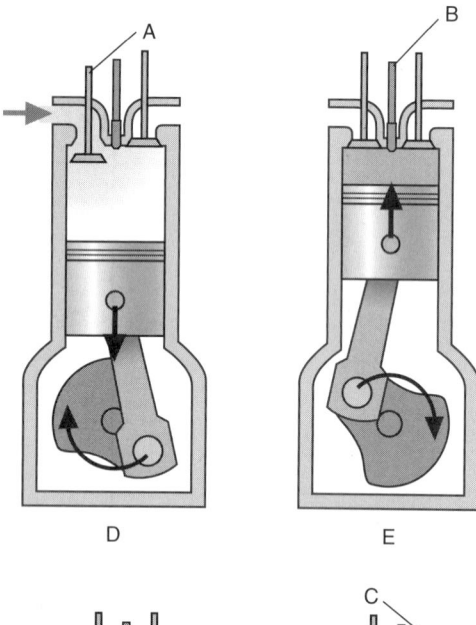
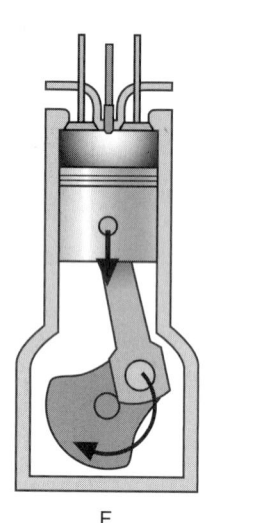
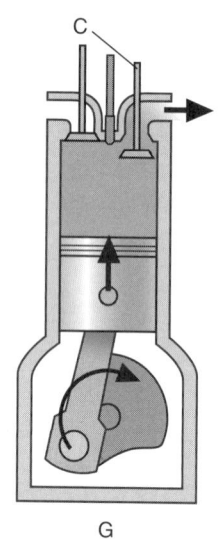

A. _____
B. _____
C. _____
D. _____
E. _____
F. _____
G. _____

2. Direct injection cylinder:

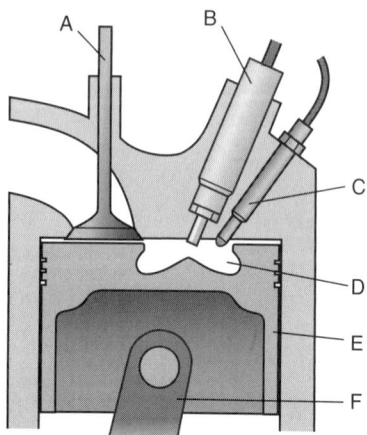

A. _____
B. _____
C. _____
D. _____
E. _____
F. _____

3. Indirect injection cylinder:

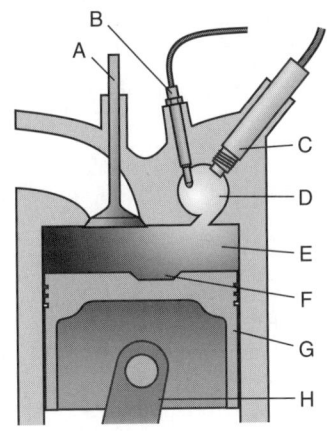

A. _____
B. _____
C. _____
D. _____
E. _____
F. _____
G. _____
H. _____

4. The two-stroke cycle:

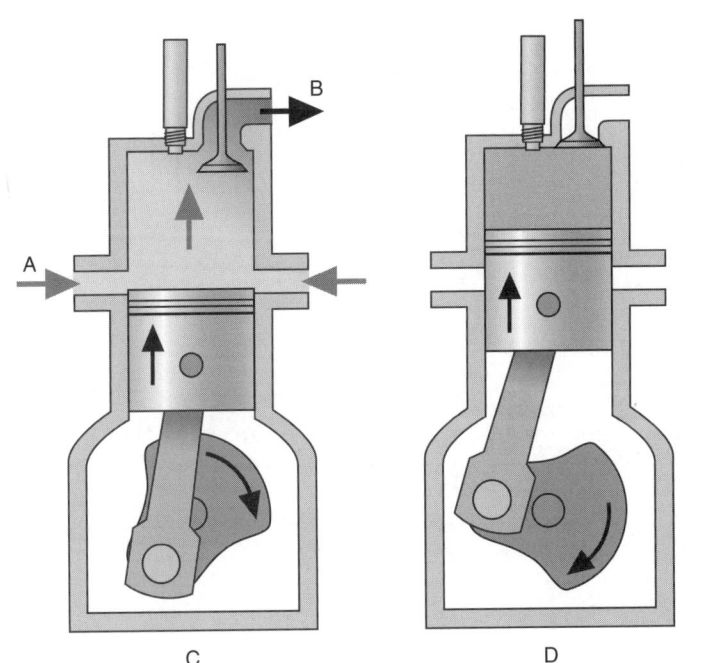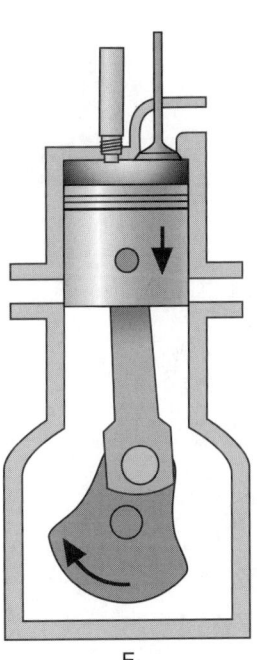

A. _____
B. _____
C. _____
D. _____
E. _____

5. Cooling passages:

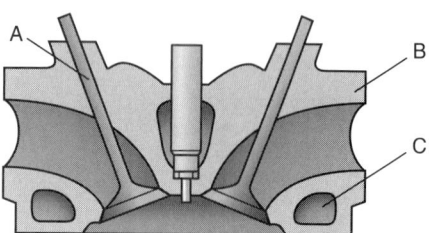

A. _____

B. _____

C. _____

6. Rod and main bearings:

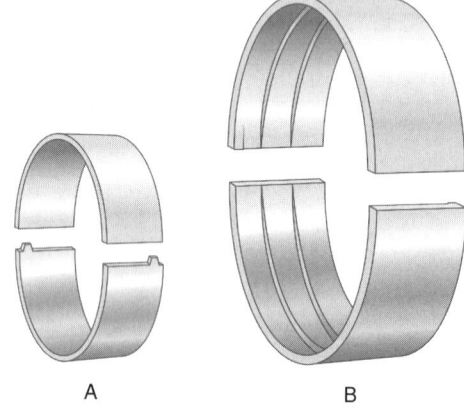

A. _____

B. _____

7. Connecting rod and bearings:

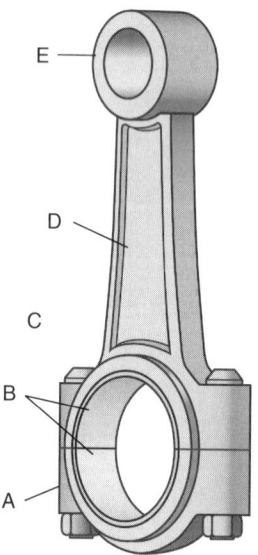

A. _____

B. _____

C. _____

D. _____

E. _____

8. Types of thrust bearings:

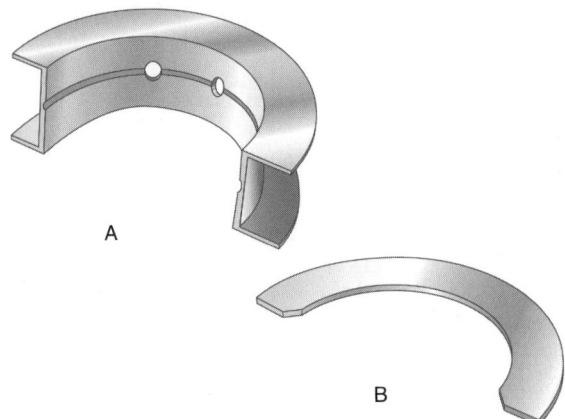

A. _____

B. _____

9. Piston with compression rings:

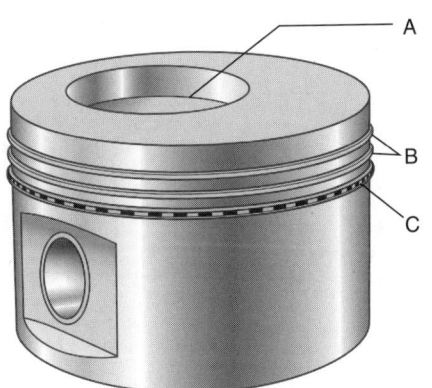

A. _____

B. _____

C. _____

Crossword Puzzle

Use the clues to complete the puzzle.

Across

4. A long, thin cylindrical shaft that provides a linkage between the lifters and the rocker arms to control valve opening and closing.
5. The stroke in an engine cycle when the piston moves up, with the exhaust valve(s) open, exhausting the burned gases out through the exhaust system.
6. A bearing surface that supports the crankshaft in the engine block.
9. A bearing that controls the lateral, or endwise, movement of a crankshaft.
10. A cylindrical component that rides on the camshaft lobe to transmit lobe movement to the valves through a linkage system.

Down

1. The stroke in an engine cycle when the piston moves up with the valves closed, compressing the air trapped in the cylinder.
2. A bearing surface that is offset from the centerline of the crankshaft.
3. A bearing insert located in the big end of the connecting rod.
7. The lever that transmits movement by the linkage system from the camshaft lobes to operate the valves.
8. A shaft with lobes that operates the valve mechanisms controlling valve opening time and duration. Also, it can be used to drive fuel injection pumps and mechanical fuel pumps.

ASE-Type Questions
Read each item carefully, and then select the best response.

_____ 1. Tech A says that the valves seal combustion pressure in the combustion chamber. Tech B says that the correct order of events to accomplish one cycle of operation is intake, compression, ignition, power, and exhaust. Who is correct?
   A. Tech A
   B. Tech B
   C. Both A and B
   D. Neither A nor B

_____ 2. Tech A says that it takes one revolution of the crankshaft to complete one four-stroke cycle. Tech B says that it takes two revolutions of the crankshaft to complete one two-stroke cycle. Who is correct?
   A. Tech A
   B. Tech B
   C. Both A and B
   D. Neither A nor B

_____ 3. How many revolutions does the camshaft make in one four-stroke cycle?
   A. ½
   B. 1
   C. 2
   D. 3

_____ 4. Tech A says that the three phases of combustion are ignition delay, flame spread, and direct burning. Tech B says that the cetane rating refers to the compression ratio of a diesel engine. Who is correct?
   A. Tech A
   B. Tech B
   C. Both A and B
   D. Neither A nor B

_____ 5. Tech A says that diesel direct injection sprays fuel directly into the combustion chamber. Tech B says that diesel indirect injection sprays fuel into the intake manifold. Who is correct?
   A. Tech A
   B. Tech B
   C. Both A and B
   D. Neither A nor B

_____ 6. Tech A says that a two-stroke engine must have intake valves to operate. Tech B says that a two-stroke engine could have four exhaust valves and no intake valves. Who is correct?
   A. Tech A
   B. Tech B
   C. Both A and B
   D. Neither A nor B

_____ 7. Tech A says that proper scavenging contributes to volumetric efficiency of the engine. Tech B says that a blower increases volumetric efficiency of the engine. Who is correct?
   A. Tech A
   B. Tech B
   C. Both A and B
   D. Neither A nor B

_____ 8. All of these describe the diesel process, *except:* _____.
   A. heat of compression
   B. spark plug
   C. glow plug
   D. direct or indirect ignition

_____ **9.** Tech A says that the crankshaft converts rotating motion into reciprocating motion. Tech B says that the crankshaft lateral movement is controlled by a thrust bearing. Who is correct?
   A. Tech A
   B. Tech B
   C. Both A and B
   D. Neither A nor B

_____ **10.** Tech A says that diesel engine pistons are normally made of aluminum alloys. Tech B says they are often made of machined stainless steel. Who is correct?
   A. Tech A
   B. Tech B
   C. Both A and B
   D. Neither A nor B

Diesel Fuel Systems

CHAPTER 54

Chapter Review
The following activities have been designed to help you refresh your knowledge of this chapter. Your instructor may require you to complete some or all of these activities as a regular part of your training program. You are encouraged to complete any activity that your instructor does not assign as a way to enhance your learning.

Matching
Match the following terms with the correct description or example.

A. American Petroleum Institute (API)
B. Carbon dioxide
C. Carbon monoxide
D. Common rail system (CRS)
E. Diesel exhaust fluid (DEF)
F. Diesel particulate filter (DPF)
G. Electronic diesel control (EDC)
H. Hydraulically actuated electronically controlled unit injector (HEUI)
I. Ignition delay period
J. Lift pump
K. Micron
L. Nitrogen oxide
M. Particulate matter
N. Selective catalytic reduction
O. Venturi effect

_____ 1. A urea chemical reactant specifically designed for use in selective catalytic reduction systems to reduce nitrogen oxides.
_____ 2. A unit of measurement equal to 0.000039" (0.001 mm).
_____ 3. The reduction in pressure that results when a fluid or air flows through a constricted section of pipe.
_____ 4. Particles that are heavier than air that are released during combustion; informally called soot.
_____ 5. The time it takes for the fuel to ignite after the fuel is injected into the engine.
_____ 6. A filter that converts particulate matter, or soot, into ash.
_____ 7. An active emission control system that injects a liquid reductant, or reducing agent, through a special catalyst into the exhaust stream of a diesel engine.
_____ 8. An exhaust emission that poses a serious health hazard to humans and can reduce oxygen delivery to organs.
_____ 9. The organization responsible for setting the standards for all American petroleum products including natural gas and biodiesel fuels.
_____ 10. An exhaust emission that results from high-temperature combustion; mixed with hydrocarbons and sunlight, they form ozone (smog) and acid rain.
_____ 11. An exhaust emission that causes changes to the atmosphere; also called greenhouse gas.
_____ 12. A computer or electronic control unit that controls diesel fuel injection systems.
_____ 13. A diesel fuel injection system that uses a pump to develop high-pressure fuel directed to an accumulator or rail.
_____ 14. A diesel fuel injection system that operates by drawing fuel from the tank using a tandem high- and low-pressure fuel pump.
_____ 15. A pump that transfers fuel from the fuel tank to the fuel injection system; also called a transfer pump or supply pump.

Multiple Choice

Read each item carefully, and then select the best response.

_____ 1. The injector pump, driven by the engine is a(n) _____ type pump.
   A. in-line
   B. axial piston
   C. rotary
   D. either A or C

_____ 2. Governors are operated _____.
   A. mechanically
   B. electronically
   C. hydraulically
   D. any of the above

_____ 3. When the temperature of diesel fuel drops to the fuel's _____, paraffin waxes that occur naturally in diesel fuel begin to crystallize and cling together.
   A. pour point
   B. cloud point
   C. viscosity
   D. freezing point

_____ 4. The result of complete combustion is water and _____.
   A. carbon monoxide
   B. hydrogen
   C. carbon dioxide
   D. nitrogen

_____ 5. The burning of diesel fuel creates exhaust emission pollutants containing _____.
   A. hydrocarbons
   B. nitrogen oxides
   C. carbon monoxide
   D. all of the above

_____ 6. Selective catalytic reduction is an active emission control system that injects a liquid _____ through a special catalyst into the exhaust stream of a diesel engine.
   A. reductant
   B. reactant
   C. particulate
   D. reagent

_____ 7. _____ increases volumetric efficiency and reduces nitrogen oxides.
   A. Catalytic reduction
   B. Turbocharging
   C. Biodiesel
   D. Electronic diesel control

_____ 8. The _____ is the temperature below which the fuel will not flow.
   A. cloud point
   B. viscosity point
   C. pour point
   D. freezing point

_____ 9. It is the job of the _____ to keep the injection pump full of fuel.
   A. lift pump
   B. fuel injector
   C. common rail system
   D. fuel tank

_____ 10. What type of lift pump is mounted on the in-line injection pump and is driven by a cam inside the in-line injection pump housing?
   A. Diaphragm lift pump
   B. Plunger lift pump
   C. Vane lift pump
   D. Priming pump

_____ 11. The primary job of the _____ is to maintain a set engine speed.
   A. lift pump
   B. governor
   C. rail system
   D. injection pump

_____ 12. What type of injector nozzles are commonly used in direct injection engines?
   A. Pintle-type
   B. Square-type
   C. Hole-type
   D. Elliptical

_____ 13. A _____ is a heating element that assists in raising the temperature in the combustion chamber, which makes the engine easier to start in cold weather.
   A. block warmer
   B. spark plug
   C. pintle
   D. glow plug

_____ 14. Another name that is used for the computer or electronic control unit is _____.
   A. power train control module
   B. body control module
   C. can-bus A
   D. diesel control unit

_____ 15. Which diesel fuel injection system operates by drawing fuel from the tank using a tandem high- and low-pressure fuel pump?
   A. Hydraulically actuated electronically controlled unit injector
   B. Detroit Diesel injection
   C. Cummins Pressure Time system
   D. Distributor-type injection system

True/False

If you believe the statement to be more true than false, write the letter "T" in the space provided. If you believe the statement to be more false than true, write the letter "F".

_____ 1. Just like most gasoline engines an air/fuel mixture is drawn into the cylinder on the intake stroke of diesel engines.

_____ 2. The basic diesel fuel injection system is divided into two sections: the low-pressure side, or delivery side, and the high-pressure side, or supply side.

_____ 3. Diesel engine parts are usually heavier and built more ruggedly than those of similar output gasoline engines.

_____ 4. The cetane number is a rating of a diesel fuel that expresses ignition quality and defines how easily the fuel will ignite when it is injected into the cylinder.

_____ 5. Diesel fuel should have a minimum octane number of 40 for direct injection diesel engines and 35 for indirect injection diesel engines.

_____ 6. Diesel fuel is vulnerable to contamination, particularly from water in the tank and from various types of sediment.

_____ 7. All highway diesel engines are required to have a diesel particulate filter; its job is to convert particulate matter, or soot, into ash.

_____ 8. Modern diesel engines need to make fuel adjustments in milliseconds, midrevolution of the crankshaft.

_____ 9. First-generation OBD systems are now standard on all diesel vehicles.

_____ 10. Biodiesel fuels made from yellow grease, or recycled cooking oil, perform worse than soybean-based biodiesel.

_____ 11. A diaphragm lift pump, also known as a transfer pump, is mounted on the input shaft and pumps fuel whenever the distributor pump is driven by the engine.

_____ 12. Some diesel engines use in-line fuel injection pumps to meter and pressurize the fuel.

_____ 13. A short effective stroke means a small amount of fuel is injected. A longer effective stroke lets more fuel be delivered.

642 Fundamentals of Automotive Technology Student Workbook

_____ 14. Mechanical governors in automotive use are called idling and maximum speed governors, because idling speed and maximum speed are all they control.

_____ 15. The injection pump delivers fuel to the injector.

_____ 16. The shape of the hole determines the shape of the spray and the atomization of the spray pattern of a hole-type injector nozzle.

_____ 17. In the Cummins Pressure Time system, each injector is like an individual plunger pump and fuel injector in one assembly.

_____ 18. Examples of input devices are throttle position, ambient temperature, engine coolant temperature, rpm, vehicle speed, and transmission gear.

_____ 19. Modern diesel vehicles often operate with several computers, or processing units.

_____ 20. The injectors used in a common rail diesel injection system are triggered externally by the electronic diesel control unit.

Fill in the Blank

Read each item carefully, and then complete the statement by filling in the missing word(s).

1. Leak-off pipes, also called fuel _____ _____, return excess fuel used for lubrication and cooling from the injection pump and injectors back to the tank.

2. All diesel engines use a(n) _____ to manage how much fuel is delivered from the injection pump, to the injectors, and into the engine.

3. To make cold starting easier, some diesel engines use preheaters, also called _____ _____.

4. Combustion _____ is the result of sudden and intense pressure changes within the cylinder.

5. Diesel fuel with a(n) _____ that is either too high or too low can cause serious damage to the engine's injection system.

6. In a modern quiet diesel engine, fuel injection timing and delivery are electronically controlled through _____-_____ _____, which is the time that the fuel injectors are turned off and on to inject a precise amount of fuel.

7. _____ _____ _____ has enabled cleaner, more complete combustion with lower emissions, greater economy and power, and lower noise levels not previously achievable with either direct or indirect combustion chambers.

8. The reductant, or reducing, agent is _____, which is a chemical reactant specifically designed for use in selective catalytic reduction systems to reduce nitrogen oxides.

9. The _____ injection system is a design that uses a separate chamber, or combustion area, to better control emissions and the noise level of the engine.

10. _____-_____ _____ II systems are now commonly used to monitor the entire power train and enable satellite connectivity.

11. _____ is a measure of a fuel's lubricating properties.

12. A(n) _____, or separator, removes any water and large particles in the fuel.

13. The most common type of filter material in light diesel vehicles is resin-impregnated _____, pleated to offer a large surface area for the fuel.

14. Without a(n) _____ pump, the engine would require excessive cranking, which could damage the starter motor and/or discharge the battery.

15. When the _____-_____ priming pump is not in use, the hand knob must be screwed closed.

16. The pneumatic governor has a manifold-mounted _____ unit, linked by tubing to a sealed diaphragm assembly on the in-line injection pump housing.

17. The _____-_____ injection pump uses a vane-type transfer pump to fill the single pumping element.

18. Glow plug resistor elements, also called _____, must be made of materials that are resistant both to heat and to oxidation.

Chapter 54 Diesel Fuel Systems 643

19. In the Cummins _____ _____ system, the fuel pressure in the fuel lines is very high, and the cam lobe has a low point that allows the injector needle to be lifted by the fuel pressure and enter the cylinder at a given time, which is determined by the cam lobe on the camshaft.

20. The _____ _____ system delivers a precisely controlled quantity of atomized fuel, which leads to better fuel economy, a reduction in exhaust emissions, and a significant decrease in engine noise during operation.

Labeling

Label the following diagrams with the correct terms.

1. Basic fuel injection system:

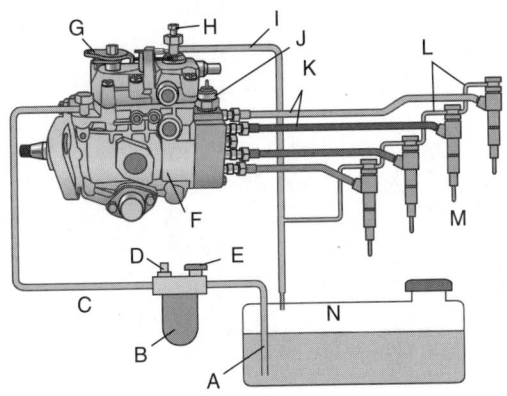

A. _____
B. _____
C. _____
D. _____
E. _____
F. _____
G. _____
H. _____
I. _____
J. _____
K. _____
L. _____
M. _____
N. _____

2. Diesel particulate filter:

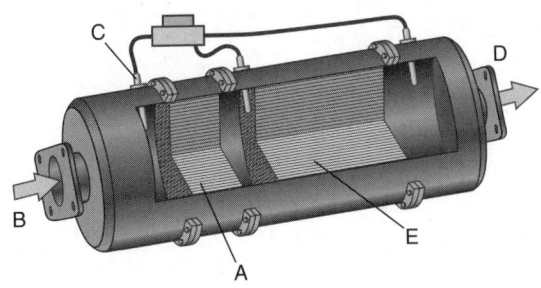

A. _____
B. _____
C. _____
D. _____
E. _____

3. Indirect fuel injection:

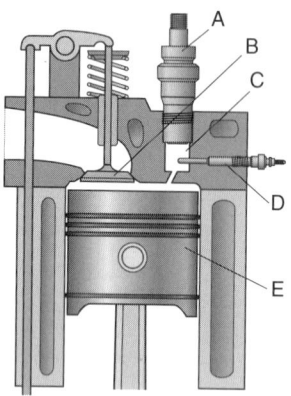

A. _____
B. _____
C. _____
D. _____
E. _____

4. Fuel filter combined with a water separator:

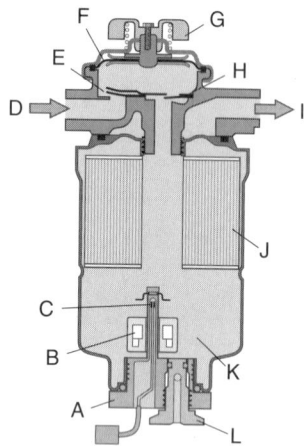

A. _____
B. _____
C. _____
D. _____
E. _____
F. _____
G. _____
H. _____
I. _____
J. _____
K. _____
L. _____

Chapter 54   Diesel Fuel Systems   645

5. Plunger-type lift pump:

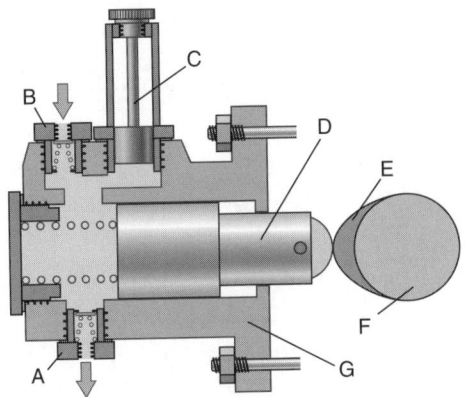

A. _____
B. _____
C. _____
D. _____
E. _____
F. _____
G. _____

6. Vane lift pump in a distributor-type injection pump:

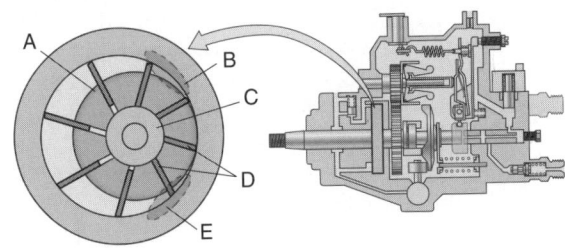

A. _____
B. _____
C. _____
D. _____
E. _____

7. In-line fuel injection pump:

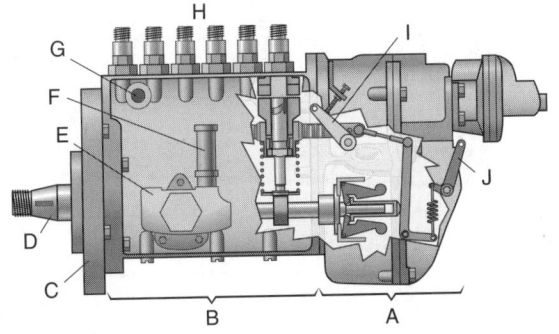

A. _____
B. _____
C. _____
D. _____
E. _____
F. _____
G. _____
H. _____
I. _____
J. _____

8. Fuel injection pumping element:

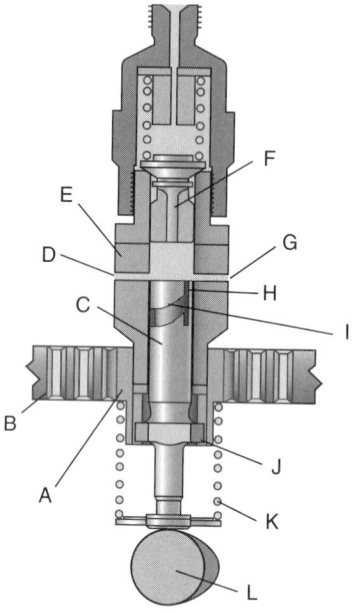

A. _____
B. _____
C. _____
D. _____
E. _____
F. _____
G. _____
H. _____
I. _____
J. _____
K. _____
L. _____

9. Typical mechanical governor:

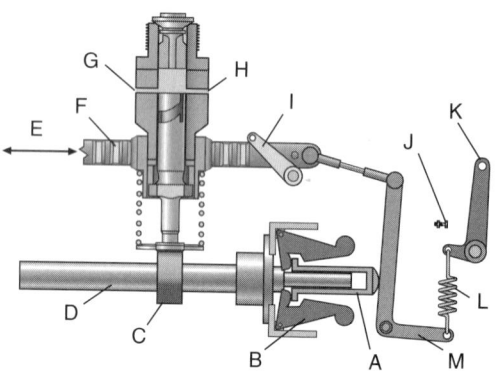

A. _____
B. _____
C. _____
D. _____
E. _____
F. _____
G. _____
H. _____
I. _____
J. _____
K. _____
L. _____
M. _____

10. Pneumatic governor:

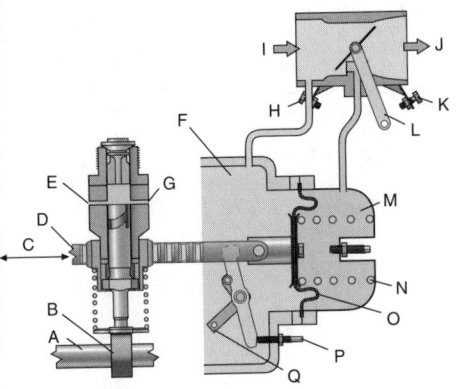

A. _____
B. _____
C. _____
D. _____
E. _____
F. _____
G. _____
H. _____
I. _____
J. _____
K. _____
L. _____
M. _____
N. _____
O. _____
P. _____
Q. _____

Crossword Puzzle

Use the clues to complete the puzzle.

Across

4. A characteristic at a fixed fuel setting where the amount of fuel delivered to the engine will increase as engine speed and pump speed increase.

6. A reducing agent.

9. An exhaust emission that mixes with nitrogen oxides and sunlight to form ozone (smog) and acid rain. Hydrocarbons result from incomplete combustion.

10. The temperature at which fuel starts to appear cloudy.

11. A plate installed vertically in the fuel tank to control sloshing of the fuel.

12. The time that the fuel injectors are turned off and on to inject a precise amount of fuel.

13. The duration of plunger stroke with the spill port covered, which determines the amount of fuel injected.

Down

1. Another term for pulse-width modulation.

2. Also known as PCM, a computer or electronic control unit that controls engine and related vehicle systems.

3. A measurement of resistance to diesel fuel flow.

5. A device that controls how much fuel is delivered to the injector.

7. A chemical reactant specifically designed for use in selective catalytic reduction systems to reduce nitrogen oxides. Also called diesel exhaust fluid.

8. The temperature below which the fuel will not flow.

ASE-Type Questions

Read each item carefully, and then select the best response.

_____ 1. Tech A says that a lift pump is another name for an in-line fuel injection pump. Tech B says that a lift pump injects fuel into the cylinder precisely at the right time with the exact amount of fuel needed. Who is correct?
   A. Tech A only
   B. Tech B only
   C. Both A and B
   D. Neither A nor B

_____ 2. Tech A says that high injection pressures are needed to overcome the compression and combustion pressures in the combustion chamber. Tech B says that high injection pressures are needed to heat the combustion chamber so that the fuel will ignite. Who is correct?
   A. Tech A only
   B. Tech B only
   C. Both A and B
   D. Neither A nor B

_____ 3. Tech A says that the cetane number expresses the moisture content of diesel fuel. Tech B says that the cetane rating expresses the ignition quality of diesel fuel. Who is correct?
   A. Tech A only
   B. Tech B only
   C. Both A and B
   D. Neither A nor B

_____ 4. Tech A says that "viscosity value" is a measure of a fuel's resistance to flow. Tech B says that "cloud point" is the temperature at which fuel turns cloudy. Who is correct?
   A. Tech A only
   B. Tech B only
   C. Both A and B
   D. Neither A nor B

_____ 5. Tech A says that the ignition delay period is the time it takes for the fuel to ignite after the fuel is injected into the engine. Tech B says that "lag time" is the time it takes for the fuel to ignite after the fuel is injected into the engine. Who is correct?
   A. Tech A only
   B. Tech B only
   C. Both A and B
   D. Neither A nor B

_____ 6. Tech A says that pulse-width modulation refers to the time that the fuel injectors are turned off and on. Tech B says that duty cycle is another term for pulse-width modulation. Who is correct?
   A. Tech A only
   B. Tech B only
   C. Both A and B
   D. Neither A nor B

_____ 7. Tech A says that nitrogen oxide emissions result from low-temperature combustion and are a serious health hazard to humans. Tech B says that carbon monoxide is the result of high-temperature combustion. Who is correct?
   A. Tech A only
   B. Tech B only
   C. Both A and B
   D. Neither A nor B

_____ 8. Tech A says that diesel exhaust fluid (DEF) is the liquid observed at the tailpipe during cold engine starts. Tech B says that a diesel particulate filter (DPF) is used to convert particulate matter, or soot, into ash. Who is correct?
   A. Tech A only
   B. Tech B only
   C. Both A and B
   D. Neither A nor B

_____ 9. Tech A says that an electronic communications network called the data bus allows exchange of data between computers. Tech B says that an electronic communications network called the data bus is necessary for efficient operation and fault diagnosis. Who is correct?
   A. Tech A only
   B. Tech B only
   C. Both A and B
   D. Neither A nor B

_____ 10. Tech A says that the final filtration of diesel fuel should remove particles down to between 50 and 75 microns. Tech B says that the final filtration of some modern diesel fuel injection systems can be 3 to 5 microns. Who is correct?
   A. Tech A only
   B. Tech B only
   C. Both A and B
   D. Neither A nor B